INTRODUCTION TO

ATOMIC and NUCLEAR PHYSICS

Introduction to
ATOMIC and NUCLEAR
PHYSICS

Harvey E. White Ph.D., Sc.D.

Professor of Physics and Director of the Lawrence Hall of Science, University of California

D. VAN NOSTRAND COMPANY, INC.

PRINCETON, NEW JERSEY

Toronto   New York   London

D. VAN NOSTRAND COMPANY, INC.
120 Alexander St., Princeton, New Jersey (*Principal office*)
24 West 40 Street, New York 18, New York

D. VAN NOSTRAND COMPANY, LTD.
358, Kensington High Street, London, W.14, England

D. VAN NOSTRAND COMPANY (Canada), LTD.
25 Hollinger Road, Toronto 16, Canada

PRINTED IN THE UNITED STATES OF AMERICA

To GLENN T. SEABORG

PREFACE

This book is written as the result of requests by many people for a really elementary text on atomic and nuclear physics. The subjects as they are presented here are the outgrowth of a course given at the University of California for students majoring in any one of a number of the physical or life sciences. The book is designed (1) to follow a one-semester or a one-year course in classical physics and (2) to confine the mathematics to algebra, geometry, and trigonometry.

The first three chapters on Gravitational Fields and Potential, Electrical Fields and Potential, and Magnetic Fields and Magnetic Moments, respectively, serve several important functions; they serve as a review of the meter-kilogram-second-ampere (mksa) system of units, commonly used in mechanics and electricity, and at the same time bring out the similarity between the basic treatments of gravitational and electrical phenomena. They also introduce, through elementary mechanics, the concepts of energy levels, a subject so fundamental to quantum theory and the structure of atoms and nuclei.

A chapter on Atoms with Two Valence Electrons is included for those who wish to go beyond the simplest atomic structures involving one valence electron.

A chapter on Moving Frames of Reference and one on Interferometers and Lasers provide the background material needed for a better understanding of the special theory of relativity on the one hand and the subject of lasers on the other.

The chapter on Photon Collisions and Atomic Waves brings out the apparent dual nature of matter, namely, particles vs. waves. A special chapter on Beta and Gamma Rays and one on Neutron and Gamma Ray Reactions present sufficient details to illustrate this duality and at the same time demonstrate the complexity of high-energy nuclear phenomena. Another chapter on Special Atomic and Nuclear Effects deals with the breadth of spectrum lines, gamma rays, resonance phenomena, and the Mossbauer effect.

The final chapter on Elementary Particles brings the subject of nuclear physics up-to-date with a qualitative presentation of a number of conservation laws, parity, antimatter, and the meaning of extremely short mean lives.

The author wishes to take this opportunity to thank the many friends, students, and teachers who have volunteered their valuable criticisms of the material as it is herein presented, and especially to his wife for the proofreading of the entire book.

HARVEY E. WHITE

Berkeley
March 1964

CONTENTS

APPENDICES

Gravitational Fields and Potential

As an introduction to atomic and nuclear physics it is appropriate that we begin with the basic principles of *gravitational, electric,* and *magnetic fields.* The introduction in this first chapter of dynamics and gravitational fields, will serve three functions: (1) it will provide a review of the meter-kilogram-second-ampere system of units (abbr. mksa); (2) it will introduce the concept of energy levels and potential wells; (3) it will provide the background for a clearer understanding of electric and magnetic fields and potentials so useful in the experimental study of atomic and nuclear physics.

1.1. The Force Equation

The basic principles of dynamics rightfully begin with Newton's second law of motion, which states that *when a body is acted on by an unbalanced force, its resultant acceleration is proportional to the force and inversely proportional to the mass.* As an equation,

$$F = ma \tag{1a}$$

where m is the mass in kilograms, a is the acceleration in meters/second², and F is the force in newtons. See Fig. 1A:

$$1 \text{ newton} = 1 \frac{\text{kilogram meter}}{\text{second}^2}$$

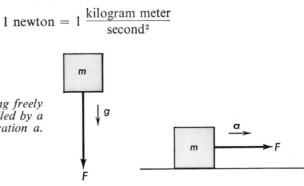

Fig. 1A *A mass m falling freely has an acceleration g; pulled by a force F it has an acceleration a.*

To abbreviate, we may write

$$1 \text{ newt} = 1 \frac{\text{Kg m}}{\text{sec}^2} \tag{1b}$$

If a mass is allowed to fall freely under the gravitational pull of the earth, we write

$$F = mg \tag{1c}$$

where F is the gravitational force on the mass m, and g is the downward acceleration due to gravity. As an average value over the surface of the earth the value

$$g = 9.80 \frac{\text{m}}{\text{sec}^2} \tag{1d}$$

is commonly used in calculations. The downward force F in Fig. 1A is also called the weight of the mass m.

> *Example 1.* Calculate the weight of a 5-Kg mass.
> *Solution.* By Eq. (1c) we obtain
>
> $$F = 5 \text{ Kg} \times 9.80 \frac{\text{m}}{\text{sec}^2} = 49.0 \frac{\text{Kg m}}{\text{sec}^2}$$
>
> $$F = 49.0 \text{ newt}$$

At a meeting held by the International Union for Pure and Applied Physics a few years ago, the following symbolism was adopted for general use:*

10^3	Kilo	K	10^{-3}	milli	m
10^6	Mega	M	10^{-6}	micro	μ
10^9	Giga	G	10^{-9}	nana	n
10^{12}	Terra	T	10^{-12}	pica	p

One of the reasons for adopting these notations was that B, for billion, stands for 10^9 in the United States, and for 10^{12} in Great Britain. Because the term "billion" is so thoroughly entrenched in the minds of scientists in the United States, only a few have adopted G and T, while the change from k to K for *one thousand* is becoming more common.

1.2. Gravitational Fields

A convenient and informative method of describing the gravitational attraction of one body for another at a distance is to describe what is called the *gravitational field*. To see how this concept arises and how it is used, consider the following development.

In Fig. 1B a mass M is shown exerting a gravitational force F on a small mass m. The magnitude of this force, as given by Newton's law of gravitation, is

$$F = -G \frac{Mm}{r^2} \tag{1e}$$

* See *Physics Today*, Vol. 9, p. 23, 1956; Vol. 10, p. 30, 1957.

Fig. 1B   *Gravitational forces are always those of attraction.*

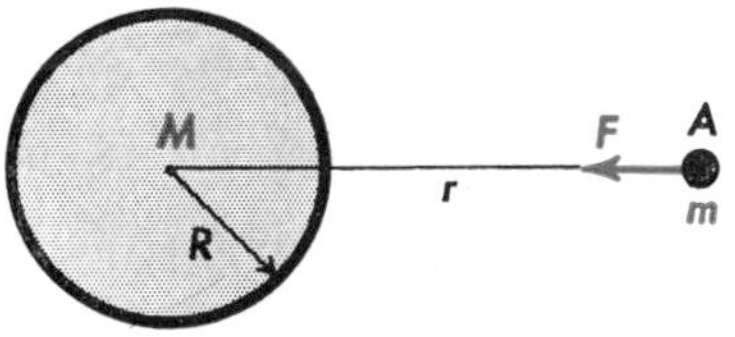

where M and m are in kilograms, r is in meters, F is in newtons, and the universal gravitational constant G is given by

$$G = 6.670 \times 10^{-11} \frac{m^3}{Kg\ sec^2} \tag{1f}$$

The minus sign in Eq. (1e) signifies the force as one of attraction.

The field intensity I at any point A in the space surrounding any mass M is defined as the force per unit mass acting on any object placed there.

$$I = \frac{F}{m} \tag{1g}$$

The small mass m is used in this definition only as a means of detecting and measuring the gravitational field at A. Whether m is large or small, the force per unit mass at the point is exactly the same. Double the mass m and F will be doubled, triple the mass m and F will be tripled, and so on.

To develop an equation for the field intensity I, we obtain a value of F/m from Eq. (1e). Transposing m to the left side, we find that

$$\frac{F}{m} = -\frac{GM}{r^2}$$

which gives

$$I = -\frac{GM}{r^2} \tag{1h}$$

A diagram of the gravitational field around a spherical mass M is shown in Fig. 1C. The arrows show the direction of the field as everywhere radially inward, and the spacing of the lines shows the field is strongest at the surface. For every point at the same distance from the center, the field intensity I has the same magnitude, but as the distance increases, the field intensity decreases.

As many radial lines as desired can be drawn to represent the field, but they must be equally spaced. The lines themselves are imaginary and do not actually exist; they are introduced as an aid to the understanding of gravitational forces and gravitational phenomena.

If m is transferred to the other side of Eq. (1g), we obtain the relation that

$$F = mI \tag{1i}$$

This equation says that any mass m placed in a gravitational field of intensity I experiences a force given by the product $m \times I$.

While Eqs. (1e) through (1i) apply to all masses m, we may use Eq. (1c) and

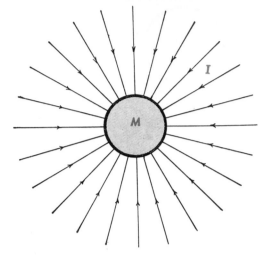

Fig. 1C   *The gravitational field around a spherical mass M is radially inward.*

obtain a special case for the earth. The downward force on any mass m at the earth's surface is given by

$$F = mg$$

Since by Eq. (1i), $F = mI$, we obtain the equality

$$I = g$$

Although the field intensity has the dimensions of an acceleration, it is more meaningful to express I, through Eq. (1g), as a force per unit mass:

$$1 \frac{\text{newton}}{\text{kilogram}} = 1 \frac{\text{meter}}{\text{second}^2}$$

Over a small volume of space, the gravitational field can be assumed to be constant. Throughout the space of one cubic mile at or near the earth's surface, for example, the direction of the field at all points is practically parallel, and the magnitude of I is practically constant. See Fig. 1D. Such a field is said to be *uniform*. The path taken by a mass m projected through a uniform gravitational

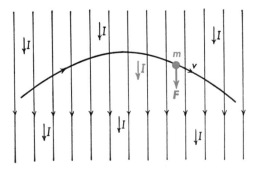

Fig. 1D   *The path of a projectile in a uniform gravitational field is a parabola.*

field is the result of a constant downward force F; and if air friction is negligibly small, the path is a parabola, as shown in the diagram.

1.3. Potential Energy

Potential energy is one of the two forms of mechanical energy and is defined as the energy a body has by virtue of its position or state:

$$\text{P.E.} = F \times s = E_p \qquad (1j)$$

A simple example is that of lifting a mass m against the pull of gravity as shown in Fig. 1E. An upward force F is applied to raise the mass m to a height

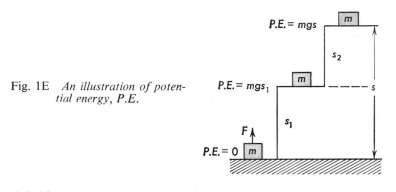

Fig. 1E   *An illustration of poten-*
tial energy, P.E.

s_1. By definition the *work done* is equal to the force times the distance moved, $W = F \times s_1$. Since the force is equal to mg, the work done is stored as potential energy E_p and we can write

$$\text{P.E.} - mgs_1 = E_1 \qquad (1k)$$

If the mass is lifted to the top level, an additional distance s_2, the work done $F \times s_2$, is stored as an additional amount of potential energy mgs_2. The total stored energy is then given by

$$\text{P.E.} = mgs = E_2 \qquad (1l)$$

where $s = s_1 + s_2$.

In the mksa system, m is in kilograms, g is in m/sec^2, s is in meters, and the potential energy E_p is in joules. The joule as a unit of potential energy has the following dimensions:

$$1 \text{ joule} = 1 \frac{\text{Kg m}^2}{\text{sec}^2}$$

The horizontal lines E_0, E_1, and E_2 in Fig. 1F constitute an energy-level diagram in which stored energy in joules is plotted vertically. The up arrows indicate the storing of energy by raising the mass m, and the down arrows the release of energy when it is allowed to come down.

Example 2. A 6-Kg mass is raised to a height of 5 m. Find the potential energy.
Solution. By direct substitution in Eq. (1j) we obtain

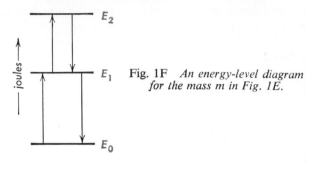

Fig. 1F   *An energy-level diagram for the mass m in Fig. 1E.*

$$E_p = 6 \text{ Kg} \times 9.8 \frac{\text{m}}{\text{sec}^2} \times 5 \text{ m}$$

$$E_p = 294 \frac{\text{Kg m}^2}{\text{sec}^2}$$

$$E_p = 294 \text{ joules}$$

abbreviated, $E_p = 294 \text{ j}$

If we select a base plane, such as sea level, as a level of zero potential energy, an object may have positive potential energy $+ mgs_2$, or negative potential energy mgs_1 as shown in Fig. 1G. Located at any point above the base plane, a

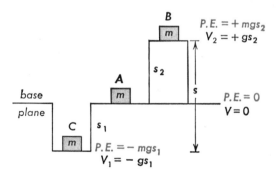

Fig. 1G   *Potential energy with respect to a base plane may be plus or minus.*

body has positive potential energy; at points below that level it has negative potential energy. In lifting the mass m from A to B, work is done. In returning from B to A, the mass releases energy. Similarly in going from A to C, the mass releases energy and ends up at C with less than it had at A. To raise it again to A, energy will have to be provided to do work.

From these considerations it is clear that we are free to choose the base plane of zero potential energy at any level and at the same time retain, unchanged, the energy differences between levels.

1.4. Kinetic Energy

The kinetic energy of a moving mass is defined as the energy a body possesses by virtue of its motion. See Fig. 1H. An airplane in straight and level flight has

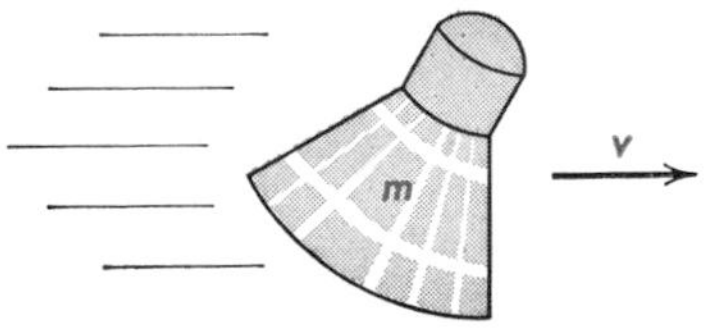

Fig. 1H   *A mass m moving with a velocity v has kinetic energy $\frac{1}{2}mv^2$.*

kinetic energy of translation, and a spinning wheel has kinetic energy of rotation. For a given mass m moving in a straight line with constant velocity v, the kinetic energy is given by

$$\text{K.E.} = \tfrac{1}{2}mv^2 = E_k \tag{1m}$$

Example 3. Calculate the kinetic energy of a 12-Kg mass, moving with a velocity of 3 m/sec.

Solution. By direct substitution in Eq. (1m), we obtain

$$E_k = \frac{1}{2}\, 12 \text{ Kg} \left(3\,\frac{\text{m}}{\text{sec}} \right)^2$$

$$E_k = 54\,\frac{\text{Kg m}^2}{\text{sec}^2}$$

$$E_k = 54\,\text{j}$$

Note that kinetic energy has the same fundamental units as potential energy, and in the mks system is in joules.

At times it is convenient to express the kinetic energy of a body in terms of *momentum*. In classical mechanics the momentum p of a moving body is defined as the mass m multiplied by the velocity v:

$$p = mv \tag{1n}$$

Multiplying numerator and denominator of Eq. (1m) by m and substituting p^2 for m^2v^2, we obtain for the kinetic energy

$$\text{K.E.} = \frac{p^2}{2m} \tag{1o}$$

1.5. Total Energy

When a body moves on an elevated plane, it possesses kinetic energy by virtue of its motion and potential energy by virtue of its position. The total energy of such a body is the sum of its separate energies:

$$\text{Total energy} = \text{K.E.} + \text{P.E.}$$

abbreviated

$$\boxed{E_t = E_k + E_p} \tag{1p}$$

As an illustration consider a car moving on different floor levels in a modern auto parking building. See Fig. 1I. Starting at rest at A, where the ground level is selected as the base plane, the velocity $v = 0$, the height $s = 0$, and both K.E. and P.E. are zero. The total mechanical energy is therefore zero:

$$E_0 = 0$$

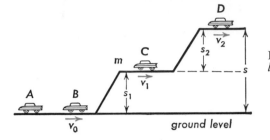

Fig. 1I   *A moving car at different levels has different amounts of energy.*

When the car is at B and moving with a velocity v_0 at the ground level, $s = 0$, and the total energy is all kinetic:

$$E_1 = \tfrac{1}{2}mv_0^2 + 0$$

When the car is at C, moving with a velocity v_1 at a height s_1 above the ground level, the total stored energy is

$$E_2 = \tfrac{1}{2}mv_1^2 + mgs_1$$

and when the car is moving with a velocity v_2 at D on the next floor up, the total energy is

$$E_3 = \tfrac{1}{2}mv_2^2 + mgs$$

If we assign specific values to m, v, and s in these equations, we can calculate each specific energy and construct an energy-level diagram of the kind shown in Fig. 1J. Energy in joules E_t is plotted vertically upward from zero and

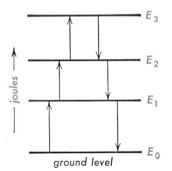

Fig. 1J   *Energy-level diagram for the total mechanical energy of the car shown in Fig. 1I.*

horizontal lines are drawn at the appropriate values. The up arrows show the energy stored as the car goes up from one level to the other, while the down arrows show the energy being liberated as the car descends. Note that although the floor levels might be equally spaced the energy levels need not be.

1.6. Gravitational Potential

The gravitational potential at any level is here defined as the work done per unit mass in carrying any object from the base plane to that level.

$$V_G = \frac{W}{m}$$

In terms of stored potential energy, we can also write

$$V_G = \frac{E_p}{m} \tag{1q}$$

At the lowest level in Fig. 1G, the gravitational potential $V_1 = -gs_1$ and at the highest level $V_2 = +gs_2$. The difference in potential between the highest and lowest levels is

$$V_G = +gs \tag{1r}$$

where $s = s_1 + s_2$.

If we transpose m to the other side of Eq. (1q), the resultant relation may be interpreted to mean that the energy expended in lifting any mass m through a difference of potential V_G, is given by the product $V_G \times m$:

$$E_p = V_G m \tag{1s}$$

Example 4. Find (a) the difference in gravitational potential between a point 6 m above sea level and a point 4 m below sea level, and (b) the energy expended in lifting a 6-Kg mass through this difference of potential.

Solution. (a) By Eq. (1r), we obtain

$$V_G = 9.8 \frac{m}{\sec^2} \times 10 \text{ m} = 98 \frac{m^2}{\sec^2}$$

(b) By Eq. (1s), we obtain

$$E_p = 98 \frac{m^2}{\sec^2} \times 6 \text{ Kg} = 588 \text{ j}$$

In developing the preceding potential energy equations it is assumed that the gravitational field is uniform, that is, that the field intensity I is constant over the distance s through which the force acts. If the distance is large, however, the field intensity I is not constant, but varies inversely as the square of the distance from the center of the earth. See Fig. 1C. Under these conditions it is more convenient to define gravitational potential in another way.

The gravitational potential of a point in space near a mass M may be defined as the work done per unit mass in bringing any mass from infinity up to that point. See Fig. 1K. In other words, we start with any small mass m at infinity, i.e.,

Fig. 1K   *The gravitational force acting on a mass m decreases with increasing distance from the surface of a body M and becomes zero at infinity.*

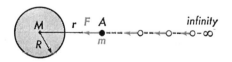

at a distance that is great compared with the size of M, and measure the work done in carrying it up to the point A.

It can be shown by use of the calculus that the energy expended is given by

$$E_p = F \times r \tag{1t}$$

where r is the distance from the center of mass M to the point A, and F has the magnitude of the gravitational force acting on m when it is at A. See Appendix IX.

This simple result makes it easy to employ Eq. (1e) and calculate E_p. By direct substitution of GMm/r^2 for F in Eq. (1t), we obtain

$$E_p = -G\frac{Mm}{r} \tag{1u}$$

This equation does two things. (1) It sets the base plane of zero potential energy at infinity, and (2) it ascribes a negative potential energy to all bodies m when they are at a finite distance from M. If m is close to the mass M, the distance r is small, and the magnitude of E_p is large but negative. As r gets larger and larger and approaches infinity, the potential energy E_p decreases in magnitude and approaches zero.

Returning to the definition of gravitational potential as given by Eq. (1q), we can divide both sides of Eq. (1u) by m, and obtain

$$V_G = -\frac{GM}{r} \tag{1v}$$

This simple relation assigns a specific value to the gravitational potential of every point in space at a distance r from any given mass M.

1.7. Mechanical Well Model

The preceding section shows that we on the surface of the earth are living at the bottom of a "gravitational well" thousands of miles deep. To reach the moon, the planets, or other worlds, we must climb out of this well onto a horizontal plane we call gravitational free space. Free space is the level of zero potential, and the level of zero potential energy.

To see why we speak of a gravitational well surrounding every mass M, we can plot a potential graph for the space around it. See Fig. 1L. Such a graph is obtained by using Eq. (1v). In this diagram V_G is plotted vertically, and the distance r is plotted horizontally.

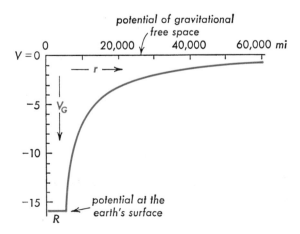

Fig. 1L *Potential-energy graph for the space around a mass M like the earth.*

The top line represents the zero level, $V_G = 0$, for the end-point energy where $r = \infty$. For smaller and smaller distances r, the magnitude of V_G increases but is negative. The color curve in the diagram, drawn for the earth, comes down to the earth's surface where $r = R = 3960$ miles, or 6.37×10^6 meters. Since $V_G \times r =$ a constant, this curve is an equilateral hyperbola, and by simply changing the V and r scales has the same shape for all bodies large and small.

A mechanical well model for demonstrating satellite orbits is obtained by rotating this potential-energy graph around the vertical axis. In so doing the curve describes a conelike surface as shown in Fig. 1M.

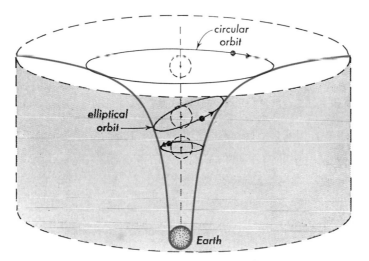

Fig. 1M   *Mechanical well model for demonstrating elliptical and circular orbits.*

If we construct a solid mass having the shape of the shaded section, a small steel or glass marble can be rolled around in the cone to describe satellite orbits. A great variety of orbits can be generated simply by varying the initial velocity of the marbles. When viewed from directly overhead, most orbits are elliptical or circular. Orbital distortion arising from rolling friction can be held to a minimum by using hard materials with smooth surfaces.

If the well model is made to demonstrate orbits around the earth, trajectories to and around the moon can be demonstrated by including a small potential crater far out on the periphery.

1.8. Orbiting and Escape Velocities

The orbiting velocity of a satellite moving in a circular orbit around any mass M is readily obtained from the basic principles of mechanics. The centripetal force acting on a body of mass m moving in a circle of radius r is given by

$$F = m \frac{v^2}{r} \qquad (1w)$$

Since this inward force is the force of gravitational attraction between two masses, by Newton's law, Eq. (1e), we may equate the right-hand terms of the two equations and obtain

$$m \frac{v^2}{r} = G \frac{Mm}{r^2}$$

Canceling like terms and solving for v, we find for the orbiting velocity,

$$v = \sqrt{\frac{GM}{r}} \tag{1x}$$

Orbiting velocity

Close to the earth's surface v is approximately 17,500 mi/hr.

If a mass m is to be projected from the surface of the earth with a velocity v great enough to enable it to escape into gravitational free space, its initial kinetic energy $\frac{1}{2}mv^2$ must be sufficient to lift it out of its potential well (see Fig. 1M). By the law of conservation of energy and Eq. (1u),

$$\frac{1}{2} mv^2 = G \frac{Mm}{r}$$

Canceling like terms and solving for v, we obtain for the escape velocity

$$v = \sqrt{2 \frac{GM}{r}} \tag{1y}$$

Escape velocity

From the earth's surface, v is approximately 25,000 mi/hr. Note that both Eq. (1x) and Eq. (1y) apply for all astronomical bodies M, and for all distances r.

PROBLEMS

1. Find (a) the difference in gravitational potential between a point 25 m above the ground and 10 m below ground, and (b) the energy required to lift a 50-Kg mass through this difference of potential.

2. A 1000-Kg elevator in a tall building goes from the basement 3 m below the ground floor to the 20th floor 65 m above the ground. Calculate (a) the gravitational potential difference between the lowest and highest points, and (b) the total energy expended in rising. *Ans.* (a) 666.4 m²/sec². (b) 666,400 j.

3. A racing car of 1500 Kg acquires a speed of 360 kilometers per hour on a straight track. Calculate its kinetic energy.

4. A truck of 2500 Kg coasts down a hill, acquiring at the bottom a speed of 72 Km/hr. Calculate its kinetic energy. *Ans.* 5×10^5 j.

5. An amphibious plane of 4000 Kg takes off from the surface of a lake 600 m above sea level with a speed of 50 m/sec. Flying to a height of 8000 m, it cruises at a speed of 80 m/sec. Calculate the gravitational potential at (a) the level of the lake and (b) the cruising altitude. Find the kinetic energy (c) at take-off and (d) while cruising. Determine the total energy (e) at take-off and (f) while cruising. Assume the base plane to be sea level.

6. A jet plane of 6000 Kg takes off from an airport runway in Denver, Colorado, 1600 m above sea level with a speed of 80 m/sec. Flying to an altitude of

10,000 m it cruises at a speed of 120 m/sec. Calculate the gravitational potential at (a) the runway and (b) the cruising altitude. Find the kinetic energy at (c) take-off and (d) while cruising. What is the total energy (e) at take-off and (f) while cruising? Assume the base plane to be at sea level. *Ans.* (a) 15,680 m²/sec². (b) 98,000 m²/sec². (c) 1.92×10^7 j. (d) 4.32×10^7 j. (e) 1.133×10^8 j. (f) 6.31×10^8 j.

7. A 1000-Kg car drives from the street level to the fourth floor of a car-parking building. The car maintains a speed of 3 m/sec on the ramps as well as on the floor levels. The latter are spaced 3 m apart, with the first floor at street level. (a) Calculate the total energy at each floor level. (b) Plot an energy level diagram for the car while moving on each of the floor levels.

8. An elevator car of 800 Kg goes from the basement level 4 m below street level to the fourth floor of an apartment house. The first floor is at the street level and all floors are 4 m apart. Starting from rest at the basement the elevator passes floors 1, 2, and 3 at the same speed of 4 m/sec, and stops at the fourth floor. Calculate (a) the potential energy at each floor level and (b) the kinetic energy at each floor level. (c) Plot an energy level diagram for the total energy at each floor if we select the base plane at the street level. *Ans.* (a) $-31,360$ j, 0, $+31,360$ j, $+62,720$ j, and $+94,080$ j, respectively. (b) 0, 6400 j, 6400 j, 6400 j, and 0, respectively. (c) $-31,360$ j, $+6400$ j, $+37,760$ j, $+69,120$ j, and $+94,080$ j, respectively.

9. Calculate (a) the gravitational potential and (b) the gravitational field intensity at a point 8000 mi above the earth's surface. Assume the earth's radius to be 4000 mi.

10. Calculate (a) the gravitational potential and (b) the gravitational field intensity on the surface of the moon. Assume the moon's radius to be 1.60×10^6 m, and its mass to be 7.32×10^{22} Kg. *Ans.* (a) 3.05×10^6 m²/sec². (b) 1.91 newt/Kg.

11. Using the values given in Problem 10, calculate (a) the acceleration due to gravity on the moon and (b) the relative weight of a man on the earth compared with a man on the moon.

12. Calculate the orbiting velocity of a satellite in a circular orbit 4000 mi above the earth's surface. Assume the earth's radius to be 3960 mi, or 6.371×10^6 m, and its mass to be 5.975×10^{24} Kg. *Ans.* 12,480 mi/hr, or 5570 m/sec.

13. Calculate the orbiting velocity of a satellite in a circular orbit 100 Km above the surface of the planet Mars. Assume the mass and radius of Mars to be 6.4×10^{23} Kg and 3332 Km, respectively.

14. Calculate the speed of a satellite orbiting the planet Jupiter at a distance of 200 Km above the surface. Assume the mass and radius of Jupiter to be 1.90×10^{27} Kg and 69,892 Km, respectively. *Ans.* 42,500 m/sec, or 95,200 mi/hr.

15. Calculate the escape velocity for a missile leaving the moon. See Problem 10.

16. Find the escape velocity for a missile leaving the surface of Mars. See Problem 13. *Ans.* 5050 m/sec.

2

Electric Fields and Potential

When an electric charge is at rest, it is spoken of as static electricity, but where it is in motion, it is called an electric current. In most cases, an electric current is described as a flow of electric charge along a metallic conductor.

Many years ago, before it was known which of the electric charges (+) or (−) moved through a wire, there seemed to be some evidence that it was the positive and not the negative. This notion became so entrenched in the minds of those interested in electrical phenomena that in later years, when it was discovered that the electrons move in solids, and not the positively charged nuclei, it became difficult to change.

The convention that electric current flows from plus to minus is still to be found in many books and is used by some electrical engineers in designing electrical machines and appliances. The rapid growth and the importance of electronics, however, is gradually bringing about a change in this practice, and we shall hereafter in this text speak of a current in a metallic conductor as one of electron flow from (−) to (+) and call it electron current. When we refer to liquids or gases in which positive and negative ions may be involved, then charges and motions will be specifically described in whatever action they manifest.

2.1. The Electric Field

In the space around every charged body is an invisible something we call an electric field. This field is just another way of describing the action of one charge upon another at any distance.

The basic law concerning the forces between charged bodies, known as Coulomb's law, states that

The force acting between two charges is directly proportional to the product of the two charges, and inversely proportional to the square of the distance between them. See Fig. 2A. As an equation,

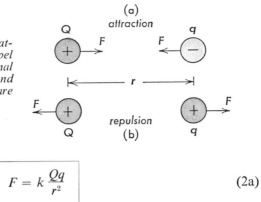
(a)
attraction
repulsion
(b)

Fig. 2A *Two unlike charges attract and two like charges repel
each other with a force proportional
to the product of the charges and
inversely proportional to the square
of the distance between them.*

$$F = k \frac{Qq}{r^2}$$
(2a)

In the mksa system of units, F is in newtons, Q and q are the charges in coulombs, r is the distance between them in meters, and k is a proportionality constant,

$$k = 8.9878 \times 10^9 \frac{\text{newton meter}^2}{\text{coulomb}^2}$$

For most calculations it is common practice to use the rounded value,

$$k = 9 \times 10^9 \frac{\text{newton meter}^2}{\text{coulomb}^2}$$
(2b)

If both charges Q and q are positive or both are negative, F is positive, and we have repulsion. If one charge is negative, F is negative, and we have attraction.

Experiments described in Chapter 4 show that electrons are all alike and that each carries a charge:

$$e = 1.6019 \times 10^{-19} \text{ coulomb}$$
(2c)

This means that if a body has one coulomb of negative charge, it has an excess of 6.24×10^{18} electrons, and a body charged positively with 1 coulomb has a deficiency of 6.24×10^{18} electrons:

$$1 \text{ coulomb} = 6.24 \times 10^{18} \times e$$
(2d)

The electric field around isolated charged bodies is shown diagrammatically in Fig. 2B.

The intensity of the electric field at any point in the neighborhood of a charged body is equal to the force per unit charge exerted on any charge placed at that point:

$$\varepsilon = \frac{F}{q}$$
(2e)

This concept is analogous to gravitational field intensity as given by Eq. (1g) in § 1.2. Since force is a vector quantity, an electric field has *magnitude* and

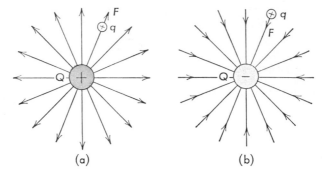

Fig. 2B   *Diagrams of the electric field around isolated charged bodies.*

direction. The field about a positive charge is described as radially outward as shown in diagram (a) of Fig. 2B. The direction is an arbitrary assignment and is based upon the custom of finding the direction in which a force would act upon a positive test charge $+q$.

By a similar reasoning the field about a negative charge is radially inward as shown in diagram (b). These lines are sometimes called *electric lines of force*. It is to be noted that as many lines as desired can be drawn and that no two lines ever cross. Furthermore, the lines are imaginary and do not actually exist. They were first introduced by Michael Faraday about 1820 as an aid to the understanding of various electrical phenomena.

Since $\mathcal{E}$, the electric field intensity, is defined as the force per unit charge for any charge placed there, Eq. (2e), Coulomb's law may be used to obtain a formula for the field intensity at any point near a body with a charge Q. Transposing q to the left side of Eq. (2a) gives

$$\frac{F}{q} = k\,\frac{Q}{r^2}$$

or

$$\mathcal{E} = k\,\frac{Q}{r^2} \tag{2f}$$

If the charge Q is in coulombs, and the distance r is in meters, the field intensity $\mathcal{E}$ is in newtons per coulomb. Since F represents the force on any charge q placed at that point,

$$\boxed{F = q\mathcal{E}} \tag{2g}$$

$$1 \text{ newton} = 1 \text{ coulomb} \times 1\,\frac{\text{newton}}{\text{coulomb}}$$

Example 1. Alpha rays are atomic particles each having a mass of 6.6×10^{-27} Kg and a positive charge of $+3.20 \times 10^{-19}$ coulomb. Calculate the force on an alpha particle in an electric field $\mathcal{E}$ having an intensity of 8000 newtons per coulomb.

Solution. By direct substitution in Eq. (2g)

$$F = 3.2 \times 10^{-19} \text{ coulomb} \times 8000 \frac{\text{newtons}}{\text{coulomb}}$$

$$F = 2.56 \times 10^{-15} \text{ newt}$$

2.2. Electric Potential

If the connection of a charged body to the ground by a metallic conductor would cause electrons to flow to that body from the ground, the body is at a positive potential. See Fig. 2C(a). Conversely, if the connection of a body to

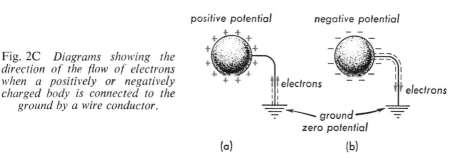

Fig. 2C *Diagrams showing the direction of the flow of electrons when a positively or negatively charged body is connected to the ground by a wire conductor.*

the ground would cause electrons to flow from that body into the ground, the body is at a negative potential. See diagram (b) of Fig. 2C.

In these definitions it is assumed that the earth is at zero potential. The bodies therefore have positive and negative potentials, respectively, before they are grounded, because after they are grounded, the flow of electrons to or from the ground will bring them to zero potential.

Electrical potential is analogous to gravitational potential in mechanics. See § 1.6, and Eq. (1q). *The electrical potential V of a body is equal to the amount of work done, i.e., energy expended, per unit positive charge in carrying any charge q from the ground up to the body.* See Fig. 2D:

$$V_E = \frac{E_p}{q} \tag{2h}$$

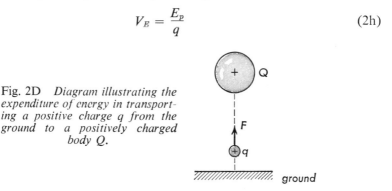

Fig. 2D *Diagram illustrating the expenditure of energy in transporting a positive charge q from the ground to a positively charged body Q.*

In the mksa system of units E_p is in joules, q is in coulombs, and V_E is in volts:

$$1 \text{ volt} = 1 \frac{\text{joule}}{\text{coulomb}} \tag{2i}$$

If Q is positive, energy is stored in carrying a positive charge q up to the body, and the potential V is positive. If Q is negative, energy is liberated in carrying a positive charge q to the body, and the potential is negative. It must be pointed out that potential as here defined applies to the body of charge Q before the small test charge q arrives at the surface; the instant that contact is made, the total charge Q has been altered by an amount q.

The difference of potential V between two bodies having potentials V_1 and V_2, respectively, is defined as the energy expended per unit positive charge in carrying any charge from one body to the other:

$$V = V_2 - V_1 \tag{2j}$$

For example, the difference of potential between the two terminals of a car storage battery is 12 volts. This means that the energy expended per unit positive charge in carrying a charge q from one terminal to the other is 12 joules per coulomb.

Transposing Eq. (2h), we obtain

$$E_p = Vq \tag{2k}$$

If we connect the negative terminal of a 12-volt battery to the ground, this brings that terminal to zero potential, and the positive terminal to $+12$ volts. If the positive terminal is grounded, that terminal will come to zero potential and the negative terminal to -12 volts. The difference of potential between the two terminals will be the same, 12 volts, in either case.

Consider an example in which a 12-volt battery is connected to two metal plates V_1 and V_2 as shown in Fig. 2E. The grounding of V_1 maintains that plate

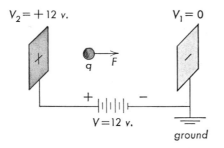

Fig. 2E *Energy is expended or liberated when a charge q is transported through a difference of potential V.*

at zero potential and the battery maintains V_2 at $+12$ volts. If we now carry a negative charge $-q$ from V_2 to V_1, an amount of energy $E_p = Vq$ is consumed, but in carrying it back again from V_1 to V_2 the same amount of energy is liberated.

In carrying a charge $+q$ from V_2 to V_1 an amount of energy Vq is liberated, but in carrying it from V_1 to V_2 the same amount is consumed. If V_1 and V_2 are connected by a metallic conductor, electrons will flow toward V_2 and the liberated energy will be converted into heat.

2.3. Potential at a Point

Just as every point in the space around a mass M has a gravitational potential V, so every point in the space around a charge Q has an electrical potential V. See Fig. 2F. *The electrical potential at any point in an electric field may be defined*

Fig. 2F *The electrical potential at a point A at a distance r from the center of charge Q is given by* kQ/r.

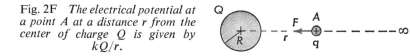

as the work done per unit positive charge in carrying any charge from infinity up to that point. By means of the calculus it can be shown that the stored energy under these circumstances is just $F \times r$. See Appendix IX:

$$E_p = F \times r$$

Making use of Coulomb's law, and substituting Eq. (2a) for F, we obtain

$$E_p = k \frac{Qq}{r} \qquad (2l)$$

Returning to the concept of electrical potential as given by Eq. (2h), we can divide both sides of Eq. (2l) by q, and obtain

$$V = k \frac{Q}{r} \qquad (2m)$$

If q is close to the charge Q, the distance r is small and the magnitude of V is large. For positive Q, V is positive; for negative Q, V is negative. Note that by definition of electrical potential the charge q is positive. As r gets larger and approaches infinity, the potential V approaches zero.

Mathematically then, the ground, referred to in the preceding section as having zero potential, is the same as the potential of a point infinitely far away.

If the charge Q is located on a spherical conductor of radius R, the potential at all points outside the sphere is the same as though the charge Q were concentrated at the geometrical center. At the surface where $r = R$ the potential will be

$$V = k \frac{Q}{R} \qquad (2n)$$

while inside the sphere it will remain the same as at the surface.

2.4. Uniform Electric Field

In many experimental studies of atomic structure, a great deal of knowledge can be obtained by observing the behavior of charged atomic particles traversing a uniform electric field. To obtain such a field, that is, a field constant in magnitude and direction over a specified volume of space, two flat metal plates are set up parallel to each other as shown in Fig. 2G.

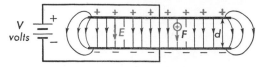

Fig. 2G   *The electric field between two parallel charged plates is uniform.*

When the terminals of a battery are connected to these plates as indicated in the diagram, a uniform electric field $\mathcal{E}$ is produced between the plates. Outside the plates and near the ends, the field is not uniform.

In mechanics, work done is given by force times distance, $E_p = F \times d$. The electrical equivalent of this equation can now be obtained by using Eq. (2k) for the energy E_p, and Eq. (2g) for the applied force F:

$$Vq = \mathcal{E}q \times d$$

or

$$\boxed{\mathcal{E} = \frac{V}{d}}$$

(2o)

Between parallel plates

If V is in volts and d is in meters, $\mathcal{E}$ is in volts/meter:

$$1 \frac{\text{volt}}{\text{meter}} = 1 \frac{\text{newton}}{\text{coulomb}}$$

(2p)

Example 2. Two flat metal plates 2 cm apart are connected to a 1000-volt battery. A proton with a positive charge of 1.6×10^{-19} coul is located between these plates. Find (a) the electric field intensity between the plates and (b) the force on the proton in newtons.

Solution. By direct substitution of given quantites in Eq. (2o) we obtain for (a)

$$\mathcal{E} = \frac{1000 \text{ volts}}{0.02 \text{ meter}} = 50,000 \frac{v}{m}$$

By direct substitution in Eq. (2g), we obtain for (b)

$$F = 50,000 \frac{\text{volts}}{\text{meter}} \times 1.6 \times 10^{-19} \text{ coulomb}$$

$$F = 8.0 \times 10^{-15} \text{ newt}$$

Just as the path of a mass m moving in a uniform gravitational field is a parabola, so the trajectory of a charge q moving in a uniform electric field is a parabola. This behavior will be considered in detail in Chapter 4.

2.5. Total Energy

Consider the illustration in Fig. 2H of a small charge $-q$ moving in a circular orbit of radius r around a fixed charge $+Q$. The motion of this charge $-q$, moving in the radial electric field of Q, is analogous to the motion of a satellite of mass m moving in the radial gravitational field of a planet.

The total energy of the moving charge $-q$ is composed of two parts: the

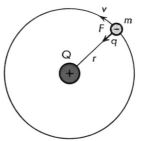

Fig. 2H   *A mass m of charge −q
is shown moving in a stable orbit in
the radial electric field of the fixed
body of charge +Q.*

kinetic energy E_k of the mass m, given by $\frac{1}{2} mv^2$, and the electrical potential

energy E_p given by $k \frac{Qq}{r}$. See Eq. (21). Adding these two energies

$$E_t = \frac{1}{2} mv^2 - k \frac{Qq}{r} \tag{2q}$$

If the orbit is mechanically and electrically stable, the centripetal force $m \frac{v^2}{r}$

keeping the mass m in its orbit is the same as the electrostatic force of attraction given by Coulomb's law. We can therefore write

$$m \frac{v^2}{r} = k \frac{Qq}{r^2} \tag{2r}$$

By canceling r from both sides, we obtain

$$mv^2 = k \frac{Qq}{r}$$

The term $k \frac{Qq}{r}$ can be substituted for mv^2 in Eq. (2q), and we obtain

$$E_t = \frac{1}{2} k \frac{Qq}{r} - k \frac{Qq}{r}$$

which simplifies to

$$E_t = -\frac{1}{2} k \frac{Qq}{r} \tag{2s}$$

PROBLEMS

1. A charge of $+12 \times 10^{-8}$ coulomb is carried from a distant point up to a
charged body. What is the potential of that body if the work done is 5×10^{-4}
joule?

2. A spherical conductor 8 cm in diameter has a charge of 5×10^{-9} coulomb. Calculate the potential of (a) the spherical conductor and (b) a point
12 cm from the center. *Ans.* (a) 1125 volts. (b) 375 volts.

3. A metal sphere with a charge of 4×10^{-8} coulomb is supported on the
end of a thin glass rod. What is (a) the electric field intensity, and (b) the potential, of a point 20 cm from the center of the sphere?

4. A hollow metal ball 6 cm in diameter is given a charge of -5×10^{-8} coulomb. What is (a) the electric field intensity, and (b) the electric potential at a

point 10 cm from the center of the ball? *Ans.* (a) 4.5×10^4 newt/coul. (b) 4500 v.

5. Calculate the energy expended in carrying a charge of 2×10^{-8} coulomb from the ground to a metal sphere of 5 cm radius and charge 5×10^{-7} coulomb.

6. A 2000-volt battery is applied to two parallel metal plates 4 cm apart. A small metal sphere with a charge of 2.5×10^{-9} coulomb is located midway between the plates. Find (a) the electric intensity between the plates in newtons per coulomb and (b) the force on the metal sphere. *Ans.* (a) 50,000 newt/coul. (b) 1.25×10^{-4} newt.

7. Two flat metal plates 0.5 cm apart are connected to a 400-volt battery. A proton with its charge of $+1.60 \times 10^{-19}$ coulomb passes between these plates. Find (a) the electric field intensity between the plates and (b) the force on the protons.

8. Two flat metal plates 3 cm apart are connected to a 120-volt battery. A small metallic sphere with a charge of 2×10^{-9} coulomb is located between the plates. Find (a) the electric field intensity and (b) the force on the charge q. *Ans.* (a) 4000 newt/coul. (b) 8×10^{-6} newt.

9. Two small metal spheres 20 cm apart each have a positive charge of 8×10^{-9} coulomb. Calculate (a) the field intensity and (b) the potential at a point 2 cm from the center of each sphere.

10. Two small metal spheres 30 cm apart have charges of $+5 \times 10^{-8}$ coulomb and -5×10^{-8} coulomb, respectively. Calculate (a) the field intensity and (b) the potential at a point 30 cm from each sphere. *Ans.* (a) 5×10^3 newt/coul, parallel to line joining the charges. (b) zero.

11. Two insulated spheres 20 cm apart each have a charge of -5×10^{-8} coulomb. Calculate (a) the field intensity and (b) the potential at a point on a straight line through the centers of the spheres, and 20 cm beyond the nearest sphere.

12. A metal sphere is suspended by a silk thread and charged positively. In carrying a positive charge of 5×10^{-8} coulomb from the ground to the metal sphere 2×10^{-5} joules of energy is expended. (a) What is the potential of the sphere? *Ans.* 400 v.

Magnetic Fields and Magnetic Moments

The early historical development of the idea of magnetic poles has in recent years been deemphasized because of the realization that magnetism is to be attributed to moving electrons. The idea of poles, however, is useful in setting up the basic concepts of magnetic fields, magnetic moments, and action at a distance, and for this reason will be used as an introduction to these subjects.

3.1. Permanent Magnets

Until recent years the best permanent magnets were made of tempered steel. Today, however, the strongest ones are made of alloys containing aluminum, cobalt, nickel, and iron.

While magnets are made in all sizes and shapes, several typical forms are shown in Fig. 3A, and labeled with N and S poles to indicate the regions that exhibit the strongest external magnetic properties.

Fig. 3A *Sample shapes of perma-nent magnets.*

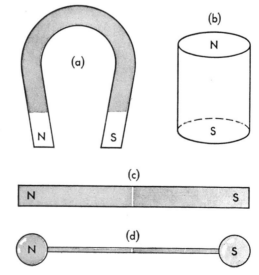

Everyone is familiar with the experimental fact that like poles of two different magnets repel each other and unlike poles attract. Since N and S poles cannot be separated by breaking magnets apart, poles can be isolated for study by adopting the form shown in Fig. 3A, diagram (d).

When two long, thin, dumbbell shaped magnets are arranged in the configuration shown in Fig. 3B, the two poles nearest each other will exert by far the

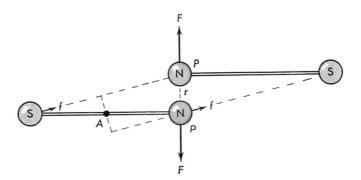

Fig. 3B   *Geometry of the forces and torques involved in measuring magnetic repulsion (Hibbert balance).*

greatest forces on each other. These forces, represented by F in the diagram, are an action and reaction pair.

To measure the force F exerted on the lower N pole, the lower magnet can be initially balanced on a pivot A. If the upper magnet is then brought into place, with two like poles adjacent to each other as shown, the repulsion will throw the lower magnet off balance. The weights that must then be added to restore balance are a direct measure of the force of repulsion. With such an experiment one can establish the following relationship.

The force acting between two magnetic poles is proportional to the product of the pole strengths and inversely proportional to the square of the distance between them. As an equation,

$$F = \kappa \frac{Pp}{r^2} \qquad\qquad (3a)$$

where, in the rationalized mksa system of units, F is in newtons, P and p are the pole strengths in *ampere meters*, r is the distance between poles in meters, and the constant κ is given by

$$\kappa = 1 \times 10^{-7} \frac{\text{newton}}{\text{ampere}^2} \qquad\qquad (3b)$$

If N poles are assigned a $(+)$ sign and S poles a $(-)$ sign, a positive F signifies repulsion and a negative F attraction. Eq. (3a) is called Coulomb's law.

3.2. Magnetic Fields

In the space surrounding every magnet there exists an invisible something we call a magnetic field. Although this field cannot be seen, we can describe its action on other magnetic fields and objects, as well as on charged particles.

If an isolated N pole of a magnet is placed at some point near the N pole of another magnet and then moved in the direction of the repelling force acting upon it, the isolated pole will trace out a smooth line called a *magnetic line of force*. Starting at various points, many such lines of force may be drawn. See Fig. 3C.

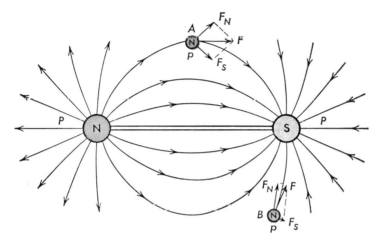

Fig. 3C   *Diagram of the magnetic field around a straight magnet with well-defined poles.*

Observe in the diagram that the magnetic lines are directed outward from the N pole and inward toward the S pole. To see why they are called magnetic lines of force consider the small test pole of strength p located at point A equidistant from the two magnet poles P and P. By Coulomb's law the N pole of the magnet repels p by a force F_N and the S pole attracts it with a force F_S of equal magnitude. The length of the resultant F represents the magnitude of the force at A, and the direction of F gives the direction of the line of force.

At other points, like B, the two forces F_N and F_S are of different magnitude, and F gives the resultant magnitude and direction of the line of force through that point.

Comparisons of this force field with the electric fields around charged bodies in Fig. 2B and the gravitational field around a mass M in Fig. 1C show striking similarities both in shape and concept.

An excellent demonstration of a magnetic field and its shape can be performed by laying a plate of glass or sheet of paper over a magnet and then sprinkling iron filings over the top. By gently tapping the glass or paper, the iron filings turn and line up with the field as shown in the photograph of Fig. 3D.

Each iron filing becomes magnetized by the field in which it is located, and

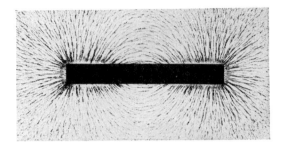

Fig. 3D *Photograph of the iron filings lined up by the magnetic field of a permanent straight-bar magnet.*

its *N* and *S* polar ends experience forces tending to turn it parallel to the field. This action is described in § 3.7.

3.3. Magnetic Induction B

It is common practice to refer to the strength or the intensity of a magnetic field as the *magnetic induction*. The magnetic induction *B* at any point in a magnetic field may be defined as the force per unit *N* pole acting on any pole placed at that point. Stated as an equation,

$$B = \frac{F}{p} \tag{3c}$$

Suppose, for example, that an *N* pole of 2 ampere meters, when placed at a given point in a magnetic field, experiences a force of 10 newtons. The magnetic induction *B* at that point is therefore 5 newtons/ampere meter.

If *p* is transposed to the left side of Eq. (3a), we obtain

$$\frac{F}{p} = \kappa \, \frac{P}{r^2}$$

This formula gives the magnetic induction at any point near a magnetic pole *P*:

$$B = \kappa \, \frac{P}{r^2} \tag{3d}$$

In graphically representing magnetic fields by lines of force it is convenient to represent the magnitude of the induction *B* by the number of lines per square meter of area. Where the lines come close together, *B* is large and where they are far apart, *B* is small. In the mksa system of units each line of force, or line of induction, is called a *weber*. Instead of expressing *B* in newtons/ampere meter, it is convenient to specify *B* in webers/meter². By definition, therefore, we are saying that

$$1 \, \frac{\text{newton}}{\text{ampere meter}} = 1 \, \frac{\text{weber}}{\text{meter}^2} \tag{3e}$$

A field, for example, for which the magnetic induction is 5 newtons/ampere meter, is written

$$B = 5 \, \frac{\text{webers}}{\text{meter}^2}$$

3.4. Electricity and Magnetism

The first discovery of any connection between electricity and magnetism was made by Oersted* in 1820. Oersted succeeded in showing that around every current-carrying wire there is a magnetic field.

The direction of this field at every point in space can be mapped by means of iron filings. See Fig. 3E. If a wire is mounted vertically through a hole in a

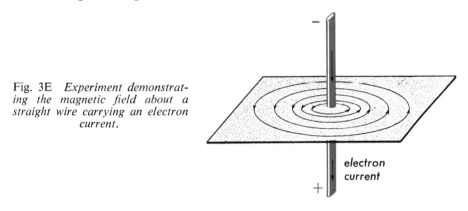

Fig. 3E *Experiment demonstrating the magnetic field about a straight wire carrying an electron current.*

electron current

plate of glass or other suitable nonconductor and iron filings are sprinkled on the plate, the e will be a lining-up of the filings parallel to the magnetic field.

The result shows that the magnetic lines of induction are concentric circles whose planes are at right angles to the current. This is illustrated by circles in Fig. 3F. A left-hand rule that can always be relied upon to give the direction of the magnetic field due to an electron current is the following: Grasp the wire with the left hand, the thumb pointing in the direction of the electron current, (−) to (+), and the fingers will point in the direction of the magnetic induction.

The magnitude of B at any point W, close to a long straight wire, is given by the relation

$$B = \kappa \, \frac{2I}{r} \tag{3f}$$

* Hans Christian Oersted (1777–1851) was a Danish scientist. Born the son of an apothecary, Oersted spent part of his boyhood teaching himself arithmetic. At the age of 12 he assisted his father in his shop and there became interested in chemistry. Passing the entrance examinations at the University of Copenhagen at the age of 17, he entered the medical school, and graduated six years later with his doctorate in medicine. At age 29 he returned to the university, this time as professor of physics. It was at one of his demonstration lectures on chemistry and metaphysics that he discovered the magnetic effect bearing his name. The discovery not only brought him many endowments and prizes, but also made him one of the most eminent personalities in his own country.

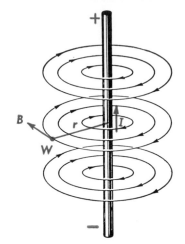

Fig. 3F   *The magnetic field arouna a long, straight conductor carrying an electron current I.*

where I is the electron current in amperes, r is the perpendicular distance to the point W in meters, B is the magnetic induction in webers/meter^2, and κ is given by Eq. (3b).

In formulas for the magnetic field produced by electric currents it is customary to replace the constant κ by another constant μ_0, defined as

$$\mu_0 = 4\pi\kappa \qquad\qquad\qquad (3g)$$

Substitution of the value of κ from Eq. (3b) gives

$$\mu_0 = 12.57 \times 10^{-7} \frac{\text{newton}}{\text{ampere}^2}$$

which, in terms of webers, becomes

$$\mu_0 = 12.57 \times 10^{-7} \frac{\text{weber}}{\text{ampere meter}}$$

Replacing κ in Eq. (3f) by its equivalent from Eq. (3g), we obtain

$$B = \mu_0 \frac{I}{2\pi r} \qquad\qquad\qquad (3h)$$

For a straight wire

For a wire conductor that is wound in the form of a flat circular coil of radius r, and carries a steady current I, the magnetic field in and around the coil has the form shown in Fig. 3G. The small compass needles, shown at various points in the diagram, line up with the field and indicate its direction.

Along the axis of the coil the magnetic induction B is perpendicular to the plane of the coil and its magnitude is given by

$$B_x = \mu_0 \frac{NIr^2}{2(r^2 + x^2)^{3/2}} \qquad\qquad\qquad (3i)$$

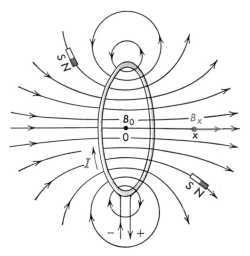

Fig. 3G *Diagram of the magnetic field through and around a flat circular coil carrying an electron current I.*

where N is the number of turns, r is their radius in meters, I is the current in amperes, and x is the distance from the coil center to the axial point where B_x is to be calculated.

At the coil center, where $x = 0$, Eq. (3i) simplifies to

$$B_0 = \mu_0 \frac{NI}{2r}$$

(3j)

Center of a flat coil

3.5. Uniform Magnetic Fields

While there are numerous ways of producing uniform magnetic fields, the methods commonly employed in the research laboratory are few in number. One of these, shown in Fig. 3H, consists of sending a current through a long

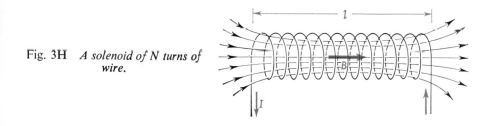

Fig. 3H *A solenoid of N turns of wire.*

straight solenoid. As indicated by the lines of induction drawn in the figure, the number of lines per square meter of cross section, i.e., webers/meter², is fairly constant over a considerable volume of space at the center.

The magnetic induction at the center of such a solenoid is given by

$$B = \mu_0 \frac{NI}{l} \qquad (3k)$$

Long solenoid

where N is the number of turns, I is the current in amperes, and l is the length of the coil in meters.

A second experimental arrangement frequently employed to produce uniform magnetic fields is shown in Fig. 3I. Two identical flat coils of radius r, placed a distance r apart, form what are called *Helmholtz coils*.

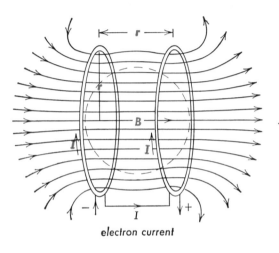

Fig. 3I *Diagram of the magnetic field around Helmholtz coils, showing the uniformity of field over the large central volume of space.*

electron current

The magnetic induction B at the geometrical center of this arrangement, and over a considerable volume of space around this center, is given by

$$B = 1.43\mu_0 \frac{NI}{2r} \qquad (3l)$$

Helmholtz coils

This equation is derived from Eq. (3i) by setting $x = r/2$ and doubling the result, since the two coils contribute equal amounts to the field at the center.

When uniform fields of great strength are required, electromagnets are commonly employed. See Fig. 3J. As in any electromagnet, the soft iron core increases the magnetic induction to several thousand times that produced by the coils alone.

Fields up to approximately 1.5 webers/meter^2 are produced by this means. Since the resultant field strengths thus produced depend a great deal upon the nature of the iron core itself, the magnitude of B between the poles is usually measured by experimental methods.

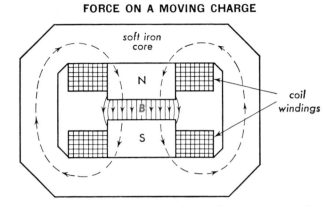

Fig. 3J   *Diagram of an electromagnetic for producing strong uniform magnetic fields*

In the cgs system of units, each line of induction is called a *maxwell*, and B is given in *maxwells/cm²*, or *gauss*. For convenience of comparison with some of the older books, note that

$$1 \text{ maxwell/centimeter}^2 = 1 \text{ gauss}$$

and that

$$1 \text{ weber/meter}^2 = 10,000 \text{ gauss} \qquad (3m)$$

3.6. Force on a Moving Charge

A charged body in an electric field experiences a force acting upon it whether it is at rest or moving. See Eq. (2g). A charged body in a magnetic field, however, experiences a force only when it is moving.

A charge q, moving with a velocity v, through a magnetic field at right angles to B, experiences a force F given by

$$\boxed{F = Bqv} \qquad (3n)$$

Force on moving charge

In the mksa system, F is in newtons, B is in webers/meter², v is in meters/second, and q is in coulombs. The vectors B, v, and F are all mutually perpendicular to each other. If the velocity vector makes an angle θ with B, as indicated in Fig. 3K, the magnitude of the force is proportional to the component of the velocity perpendicular to B.

$$F = Bqv \sin \theta \qquad (3o)$$

If the charge q is positive, the force F is opposite in direction to the one shown in the diagram of Fig. 3K. When a charged body moves parallel to the field—that is, along the magnetic lines of induction, $\sin \theta = 0$—there is no force.

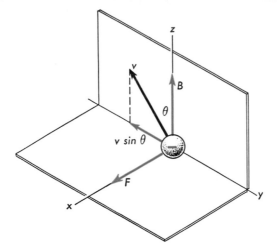

Fig. 3K *Force on a negative charge moving at an angle to a magnetic field.*

3.7. Magnetic Moments

When a bar magnet is located in a uniform magnetic field as shown in Fig. 3L, it is acted upon by a torque which tends to line it up with the field. The force acting on each pole is given by Eq. (3c):

$$F = pB \tag{3p}$$

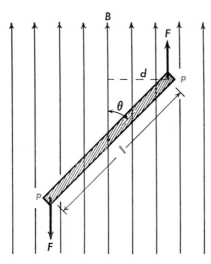

Fig. 3L *A bar magnet in a magnetic field is acted upon by a torque.*

where p is the pole strength in ampere meters. Assuming the distance between pole centers to be l, each of these oppositely directed forces has a lever arm $d = \frac{1}{2}l \sin \theta$. The total torque L is equal to $2(F \times d)$:

$$L = 2(pB \times \tfrac{1}{2}l \sin \theta)$$
$$L = plB \sin \theta$$

The product pl is a property of the magnet alone, and by analogy with mass moment and force moment in mechanics it is called the *magnetic moment:*

$$M = pl \tag{3q}$$

The torque may therefore be written simply as

$$L = MB \sin \theta \tag{3r}$$

When the magnet is at right angles to B,

$$L = MB \tag{3s}$$

If a current-carrying loop of wire is located in a magnetic field as shown in Fig. 3M, it too is acted upon by a torque, tending to line its axis up with the

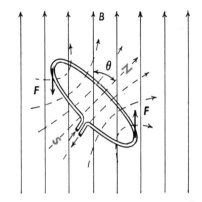

Fig. 3M   *A current-carrying loop of wire in a magnetic field is acted upon by a torque.*

field B. This torque is given by Eq. (3r), where the magnetic moment is given by the very simple relation

$$\boxed{M = IA} \tag{3t}$$

Here I is the current in amperes, and A is the loop area in meters². If the flat coil has a number of turns N, the equation is multiplied by N, so that $M = NIA$.

Because the magnetic field set up by a current loop is similar to the field around a straight bar magnet, the quantity M is frequently called the *magnetic dipole moment.*

The magnetic properties of all substances are due entirely to the motions of electrons within the atoms, and magnets of all kinds are the result of the lining-up of electron current loops, or spinning electrons. Each tiny elementary magnet has a magnetic moment M and can be thought of as a magnet, as given by Eq. (3q), or as a current loop, as given by Eq. (3t).

PROBLEMS

1. Three N poles of 8 ampere meters each are located at the corners of an equilateral triangle 4 cm on a side. Find (a) the force between each pair of poles and (b) the resultant force on each pole.

2. Three S poles of 5 ampere meters each are located at the corners of a right triangle whose sides are 6 cm, 8 cm, and 10 cm. Find the magnitude of the force on the pole at the 90° corner. *Ans.* 7.97 × 10⁻⁴ newt.

3. A straight bar magnet has two poles of 50 amp-m each, with centers 10 cm apart. Calculate the magnetic induction at a point in line with the two poles, (a) 5 cm beyond the *N* pole and (b) 10 cm beyond the *S* pole. (c) Calculate the magnetic moment.

4. A straight bar magnet has two poles, 20 cm apart, each with a strength of 10 amp-m. Calculate (a) the magnitude of the magnetic induction at a point 12 cm from the *S* pole and 16 cm from the *N* pole, and (b) the magnetic moment. *Ans.* (a) 7.97 × 10⁻⁵ web/m². (b) 2 amp-m².

5. A straight bar magnet has two poles, each of 20 amp-m, located 10 cm apart. (a) Find the magnitude of the magnetic induction at a point in line with the poles at a distance beyond the *N* pole of 1, 5, 10, 15, and 20 cm. (b) Plot a graph of *B* vs. *x* where *x* is the distance from the center of the magnet.

6. Two long, straight, parallel wires 4 cm apart each carry an electron current of 50 amp. Calculate the magnetic induction at a point between the wires, 1 cm from one and 3 cm from the other, when the currents are (a) in the same direction, and (b) in opposite directions. *Ans.* (a) 6.67 × 10⁻⁵ web/m². (b) 13.3 × 10⁻⁴ web/m².

7. Two long, straight, parallel wires 6 cm apart each carry an electron current of 40 amp. Calculate the magnetic induction at a point between the wires, 2 cm from one and 4 cm from the other, when the currents are (a) in the same direction, and (b) in opposite directions.

8. A wire 10 m long is wound into a flat coil 10 cm in diameter. If an electron current of 10 amp flows through the coil, (a) what is the number of turns? (b) What is the magnetic induction at the center? (c) What is the magnetic moment? *Ans.* (a) 31.8 turns. (b) 4.0 × 10⁻³ web/m². (c) 2.5 amp m².

9. A wire 60 m long is wound into a flat coil 12 cm in diameter. If an electron current of 5 amp flows through the coil, (a) what is the number of turns? (b) What is the magnetic induction at the center? (c) What is the magnetic moment?

10. If the magnetic induction at the center of a solenoid 50 cm long is to be 2 × 10⁻³ web/m² when an electron current of 5 amp is flowing through it, how many turns must it have? *Ans.* 159 turns.

11. A flat coil of 200 turns has a diameter of 4 cm and carries an electron current of 7 amp. Find (a) the magnetic induction at the center and (b) the magnetic moment.

12. If the magnetic induction at the center of a solenoid 60 cm long is to be 5 × 10⁻³ web/m² when an electron current of 6 amp is flowing through it, how many turns must it have? *Ans.* 398 turns.

13. A solenoid 40 cm long has 1000 turns of wire. What electron current is required to produce a magnetic induction of 4 × 10⁻³ web/m² at its center?

14. A solenoid 30 cm long has 70 turns of wire. What electron current is required to produce a magnetic induction of 6 × 10⁻⁴ web/m² at its center? *Ans.* 2.05 amp.

15. A flat coil of 20 turns and radius 10 cm carries a current of 16 amp. Find (a) the magnetic induction at its center in gauss and (b) the magnetic moment.

16. A solenoid 75 cm long carries an electron current of 8.5 amp. How many turns are required if the magnetic induction at the center is to be 100 gauss? *Ans.* 702 turns.

17. A flat coil of 20 turns and 20 cm radius carries a current of 10 amp. Find the magnetic induction at (a) the center, and (b) a point on the axis 20 cm from the center.

18. A flat coil of 50 turns has a diameter of 10 cm and carries a current of 2 amp. Calculate the magnetic induction at (a) the center, and (b) a point on the axis 5 cm from the center. *Ans.* (a) 1.26×10^{-3} web/m². (b) 0.45×10^{-3} web/m².

19. Helmholtz coils, each with 25 turns and radius of 30 cm, carry a current of 5 amp. Calculate the magnetic induction at the center.

20. Helmholtz coils, each with 40 turns and radius of 20 cm, carry a current of 2.5 amp. Find the magnetic induction at the center. *Ans.* 4.50×10^{-4} web/m².

21. An alpha particle with a positive charge 3.2×10^{-19} coulomb moves through a uniform magnetic field of 5×10^{-2} web/m² at right angles to the lines of induction. If the particle has a velocity of 3×10^6 m/sec, find the force on the particle in newtons.

22. An electron with a charge of -1.6×10^{-19} coulomb moves through a uniform magnetic field of 2.5×10^{-2} web/m², at right angles to the lines of induction. If the electron has a velocity of 5×10^6 m/sec, find the force on the electron in newtons. *Ans.* 2.0×10^{-14} newt.

23. An electron with a charge of -1.6×10^{-19} coulomb and mass 9.1×10^{-31} Kg, moves through a uniform magnetic field of 0.1 web/m² with a velocity of 5×10^6 m/sec at right angles to B. Find the radius of its circular path.

The Discovery of the Electron

Modern physics, dealing principally with atoms, molecules, and the structure of matter, has developed at such a tremendous rate within the past three-score years that it now occupies the center of attention of many leading scientists the world over. Recent discoveries in atomic physics have had and will continue to have a tremendous influence on the development of civilization. Because the subject of atomic physics is relatively new, it is logical to treat the subject matter associated with each major discovery in roughly chronological order.

4.1. Electrical Discharge Through a Gas

In 1853 an obscure French scientist by the name of Masson sent the first electric spark from a high-voltage induction coil through a partially evacuated glass vessel and discovered that, instead of the typical spark observed in air, the tube was filled with a bright glow. Several years later, Heinrich Geissler, a German glass blower in Tübingen, developed and began to manufacture gaseous discharge tubes. These tubes, made in diverse sizes, shapes, and colors of glass, and resembling the modern neon and argon signs used in advertising, attracted the attention of physicists in the leading scientific institutions and universities of the world. They purchased many of these "Geissler tubes" and used them for study and lecture demonstrations.

In 1869, W. Hittorf of Munster, with improved vacuum pumps, observed a dark region near one electrode of the electrical discharge that grew in size as the exhaustion was continued. This is but one of a number of phases of the study of electrical discharges through gases that were observed and studied a few years later by Sir William Crookes.*

* Sir William Crookes (1832–1919), English physicist and chemist. At the age of 22 he became an assistant at the Radcliffe Observatory in Oxford. He was knighted in 1897, received the Order of Merit in 1910, and was president of the Royal Society from 1913 to 1915. He invented and made the first focusing type of X-ray tube. His experiments with electrical discharges through rarefied gases led to his discovery of the dark space that now bears his name.

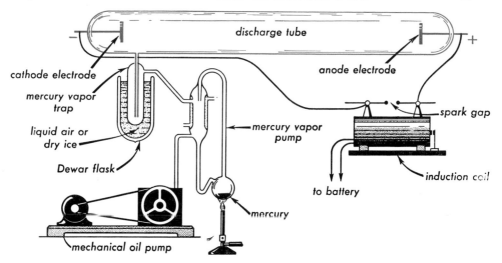

Fig. 4A   *Diagram of a gaseous discharge tube, showing the electrical connections as well as the vacuum pumps and accessories.*

In Fig. 4A, a long glass tube about 4 cm in diameter and 150 cm long is shown connected to a mercury diffusion pump and a mechanical vacuum pump. The purpose of the pumps is to enable one to observe continuously the changes in the electrical discharge as the air is slowly removed from the tube. The purpose of the *trap* is to freeze out any mercury vapor and to prevent it from reaching the discharge. High voltage from an induction coil is shown connected to the two electrodes, one at either end of the tube.

Although an induction coil does not deliver direct current, its characteristics are such that the potentials are higher on half of the alternations than they are on the other, and the two electrodes act nearly the same as if a high-voltage direct current were used. The negative electrode under these circumstances is called the *cathode*, and the positive electrode the *anode*.

As the long tube is slowly pumped out, an emf of 10,000 to 15,000 volts will produce the first discharge when the pressure has dropped to about $\frac{1}{100}$ of an atmosphere, i.e., at a barometric pressure of about 8 mm of mercury. This first discharge, as illustrated in diagram (a) of Fig. 4B, consists of long, thin streamers. As the gas pressure drops to about 5 mm of mercury, sometimes called a Geissler-tube vacuum, the discharge widens until it fills the whole tube as shown in diagram (b). At a still lower pressure of about 2 mm, a dark region called the *Faraday dark space* appears in the region of the cathode, which divides the bright discharge into two parts, a long pinkish section called the *positive column* and a short bluish section called the *negative glow*. As the pressure drops still further, the Faraday dark space grows in size and the negative glow moves away from the cathode, producing another dark space between it and the cathode. With the appearance of this second dark region,

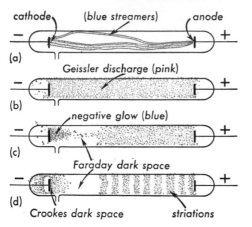

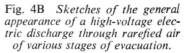

Fig. 4B *Sketches of the general appearance of a high-voltage electric discharge through rarefied air of various stages of evacuation.*

called the *Crookes dark space*, the positive column divides into a number of equally spaced layers, called "striations."

As the pumping proceeds, the striations and the negative glow grow fainter, and the Crookes dark space widens, until finally, at a pressure of about 0.01 mm, it fills the whole tube. At this point a new feature appears: the whole glass tube itself glows with a faint greenish light.

4.2. Cathode Rays

The green glow in the final stage of the gaseous discharge just described was soon found to be a *fluorescence of the glass produced by invisible rays emanating from the cathode itself.* These *cathode rays,* as they are called, believed by Sir William Crookes to be an "ultra gaseous state" and by Johann W. Hittorf to be a "fourth state" of matter, were discovered to be tiny corpuscles which we now call *electrons.* In the relatively free space of a highly evacuated tube, cathode particles, torn loose from the atoms of the cathode, stream down the length of the tube and seldom collide with a gas molecule until they hit the glass walls.

The first important discovery concerning the nature of cathode rays was that

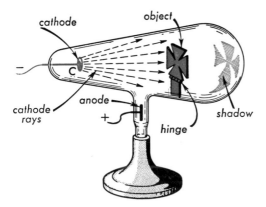

Fig. 4C *A Crookes discharge tube for demonstrating that cathode rays travel in straight lines.*

they travel in straight lines. This was first revealed by Hittorf, in 1869, by casting shadows of objects placed inside the discharge tube. Hittorf's discovery is usually demonstrated by a tube of special design, as shown in Fig. 4C.

Where the rays strike the walls of the tube, the glass fluoresces green, while in the shadow it remains dark. Under continuous bombardment of the walls by cathode rays, the fluorescence grows fainter because of a fatigue effect of the glass. This effect is demonstrated by tipping the object down on its hinge, thus permitting the rays to strike the fresh glass surface. Where the shadow appeared previously, a bright green image of the object is clearly visible.

That *cathode rays have momentum and energy* was first demonstrated in 1870, by Crookes, who used a tube of special design as illustrated in Fig. 4D. Leaving

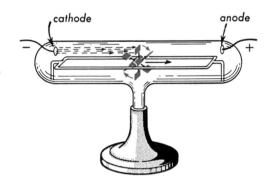

Fig. 4D   *Experiment showing that cathode rays have momentum and energy. Cathode rays striking the vanes of a small pinwheel cause it to roll from one end of the tube to the other.*

the cathode and acquiring a high speed on their way toward the anode, the rays strike the mica vanes of a small pinwheel and exert a force, causing it to turn and thus roll along a double track toward the anode. When it reaches the end of the track, a reversal of the potential, making the right-hand electrode the cathode, will send it rolling back toward the anode, now at the left. From this experiment Crookes concluded that cathode particles have *momentum,* and that they therefore have *mass, velocity,* and *kinetic energy* $\frac{1}{2}mv^2$.

That *cathode rays are negatively charged particles* was first discovered in Paris in 1895 by Jean Perrin. A discharge tube of special design usually used to demonstrate this property is illustrated in Fig. 4E. A beam of cathode rays is narrowed down to a thin pencil or ribbon of rays by a narrow slit near the

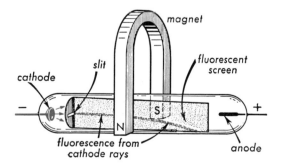

Fig. 4E   *The bending of a beam of cathode rays in the field of a magnet demonstrates that cathode rays are negatively charged particles.*

cathode. The path of the rays is made visible by allowing them to strike a long strip of metal painted with zinc sulfide, a fluorescent paint. By placing a horseshoe magnet over the outside of the tube, as illustrated, the path of the cathode rays is bent down. If the polarity of the magnet is reversed, the path is bent up. The bending shows that they are charged, and the direction of bending shows the kind of charge. Being charged, a stream of particles is like an electron current. From the direction of the magnetic field and the current, and by application of the left-hand rule (see § 3.6), the charge is found to be *negative*. (Remember that the left-hand rule applies to a current from $(-)$ to $(+)$.)

The penetrating power of cathode rays was first demonstrated by Heinrich Hertz and his assistant, P. Lenard, by passing cathode rays through thin aluminum foils. Out in the air, the rays were found to retain sufficient power to cause fluorescence and phosphorescence.

4.3. J. J. Thomson's Experiments

When, in 1895, it was discovered that cathode rays were negatively charged particles, the question immediately arose whether they were all alike. It was clear from the beginning that two things would have to be done: (1) measure the amount of charge on the particles and (2) measure the mass of the particles.

Although the first attempts to measure the electronic charge and mass were not entirely successful, J. J. Thomson* did succeed, in 1897, in determining the velocity of the rays and in measuring the ratio between their charge and mass.

The discharge tube designed for these experiments is shown in Fig. 4F.

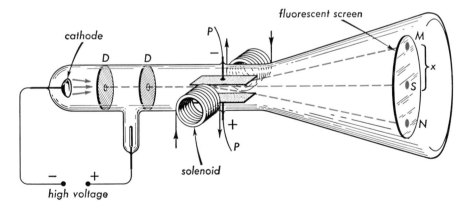

Fig. 4F   *Diagram of discharge tube used by J. J. Thomson to measure the velocity of cathode rays.*

* Sir Joseph John Thomson (1856–1940), English physicist, educated at Owens College, Manchester, and at Trinity College, Cambridge. He was appointed Cavendish professor at Cambridge in 1884, and professor of physics at the Royal Institution, London, in 1905. He was awarded the Nobel Prize in physics in 1906, was knighted in 1908, and elected to the presidency of the Royal Society in 1915. He became master of Trinity College in 1918 and helped to develop at Cambridge a great research laboratory attracting scientific workers from all over the world.

Cathode rays, originating at the left-hand electrode and limited to a thin pencil of rays by pinholes in diaphragms DD, are made to pass between two parallel metal plates and the magnetic field of two external solenoids to a fluorescent screen at the far end.

When the two metal plates P are connected to a source of high voltage, the particles experience a downward force, and their path curves to strike the screen at N. Without a charge on the plates, the beam passes straight through un-deviated and strikes the screen at S.

When the magnetic field alone is applied, so that the magnetic lines are perpendicular to the plane of the page, the path of the rays curves upward to strike the fluorescent screen at some point M. If both the electric field and the magnetic field are applied simultaneously, a proper adjustment of the strength of either field can be made, so that the deflection downward by the one is exactly counteracted by the deflection upward of the other. When this condition is attained, a measurement of the magnetic induction B and the electric intensity ε permits a calculation of the velocity of cathode rays.

4.4. Deflection in an Electric Field

In Chapter 2, on the theory of electricity, it is shown that if e is the charge on a body located in an electric field of strength ε, the force exerted on the body is given by

$$F_E = e\varepsilon \tag{4a}$$

See Eq. (2g).

As a charged particle like an electron enters the electric field between two charged plates (shown in Fig. 4G), this force acts straight downward, parallel

Fig. 4G *Electrons in a uniform electric field E follow a parabolic path.*

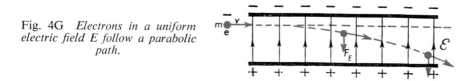

to the field lines at all points. The net result is that the particle traverses a parabolic path in much the same way that a projectile follows a parabolic path in the earth's gravitational field.

4.5. Deflection in a Magnetic Field

In Chapter 3, Eq. (3n), it is shown that if e is the charge on a body moving through a magnetic field B with a velocity v, the force acting upon it is given by

$$F_B = Bev \tag{4b}$$

Since this force is always at right angles to both the magnetic induction and the direction of motion, the particle will traverse a circular path. (See Fig. 4H.)

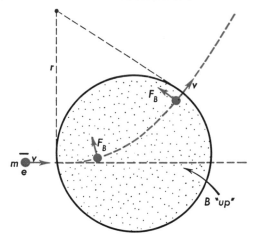

Fig. 4H   *Electrons in a uniform magnetic field B follow a circular path.*

By counterbalancing the two forces F_E and F_B, i.e., by making them equal in magnitude and opposite in direction, the two relations can be set equal to each other:

$$e\mathcal{E} = Bev \tag{4c}$$

By canceling the charge e on both sides of this equation, we obtain

$$\mathcal{E} = vB$$

from which

$$v = \frac{\mathcal{E}}{B} \tag{4d}$$

where $\mathcal{E}$ is in volts per meter, B is in webers per meter^2, and v is in meters per second. If we insert the known values of $\mathcal{E}$ and B, the *velocity v can be calculated*. The results show that cathode rays generally travel with a speed of several thousand miles per second, about $\frac{1}{5}$ *the velocity of light*. Furthermore, the velocity is not always the same but depends upon the voltage applied between the anode and cathode. By increasing this voltage, the velocity of the rays is increased.

Tubes used for scanning and observing moving pictures by modern television receivers are quite similar in shape and principle to J. J. Thomson's cathode-ray tube of Fig. 4F.

4.6. The Ratio of Charge to Mass, e/m

The next step taken by Thomson was to measure the deflection of the cathode beam produced by a magnetic field alone and, from this, to calculate the ratio between the charge e and the mass m of the electron. To do this, he reasoned that, if a charged particle moving through a uniform magnetic field has a force exerted on it at right angles to its direction of motion, causing it to move in the arc of a circle, the force is of the nature of a centripetal force. Calling F_B a centripetal force,

$$F_B = m \frac{v^2}{r}$$

and from Eq. (4b), we obtain

$$Bev = m \frac{v^2}{r} \qquad (4e)$$

Transposition of m to the left side, and B and v to the right side, in the second equation results in

$$\frac{e}{m} = \frac{v}{Br}$$

where r is the radius of the circular arc in meters through which the particles are deviated; v is the velocity of the particles in meters per second, as measured in the last section; m is the particle mass in kilograms; and B is the magnetic induction in webers per meter². With all of these known, the value of e/m can be calculated. It is found to be

$$e/m = 1.7589 \times 10^{11} \frac{\text{coul}}{\text{Kg}} \qquad (4f)$$

Such a large number means that the mass of a cathode ray particle in Kg is extremely small, compared with the charge it carries in coulombs. If now it were possible by some experiment to measure the charge e alone, the value could be substituted in Eq. (4f) and the mass m calculated.

4.7. Millikan's Oil-Drop Experiment

Millikan* began his experiments on the electronic charge e in 1906. His apparatus is illustrated by the simple diagram in Fig. 4I. Minute oil drops from an atomizer are sprayed into the region just over the top of one of two circular metal plates, V^+ and V^-. Shown in cross section, the upper plate is pierced with a tiny pinhole P through which an occasional oil drop from the cloud will fall. Once between the plates, such a drop, illuminated by an arc light from the side, is observed by means of a low-powered microscope.

With the switch S in the "up" position, the capacitor plates are grounded so that they are not charged. Under these conditions, the oil drop falling under the pull of gravity has a constant velocity. This *terminal velocity*, as it is called, is reached by the drop before it enters the field of view and is of such a value

---

* Robert Andrews Millikan (1868–1953), American physicist, educated at Oberlin College and Columbia University, for 25 years professor of physics at the University of Chicago and for 30 years president of the Norman Bridge Laboratory at the California Institute of Technology in Pasadena. He served during World War I in the research division of the Signal Corps with the rank of lieutenant colonel. His principal contributions to science were his measurement of the charge on the electron, his photoelectric determination of the energy in a light quantum, and his precision study of cosmic rays. He was the second American to be awarded the Nobel Prize in physics (1923). He was also awarded the Edison Medal, the Hughes Medal of the Royal Society, the Faraday Medal, and the Mattenci Medal.

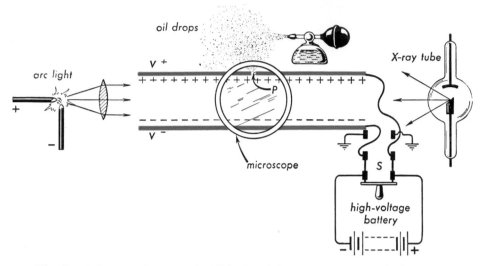

Fig. 4I *Schematic diagram of Millikan's oil-drop experiment. With this experiment the charge on the electron was determined.*

that the downward pull of gravity, F_G, in Fig. 4J(a), is exactly equalized by the upward resisting force of the air. By measuring this velocity of fall, the force F_G can be calculated and from it the mass of the oil drop determined. The velocity of the drop can be determined by using a stop watch to measure the time required for the drop to fall the distance between the two cross hairs illustrated in Fig. 4K.

As the drop nears the bottom plate, the switch S is thrown "down," charging the two parallel plates positive and negative. If now the drop has a negative charge, as illustrated in diagram (b), there will be an upward electrostatic force F_E, acting to propel the drop up across the field of view. The drop will move upward with a constant velocity if F_E is greater than the gravitational force F_G. Again using the stop watch, this time to measure the velocity of rise, we can

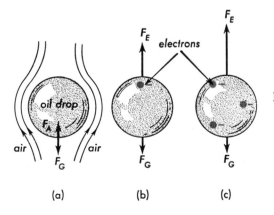

Fig. 4J *Diagrams of oil drop with extra electronic charges.*

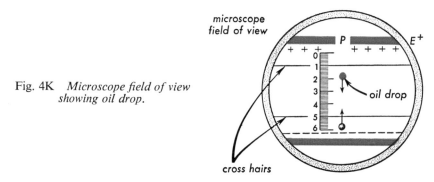

Fig. 4K *Microscope field of view showing oil drop.*

calculate the upward force F_E. Knowing the force, and the voltage on the capacitor plates, we can compute the charge on the drop.

As the drop nears the top plate, the switch S is thrown "up" and the plates are again grounded. Under these conditions the drop falls again, under the pull of gravity alone. Upon nearing the bottom plate, the switch is again thrown "down" and the drop rises once more. When this process is repeated, a single drop may be made to move up and down many times across the field of view. Each time it falls, the velocity is measured and the mass computed, while each time it rises the velocity is measured and the charge computed.

Millikan found that, if X rays were allowed to pass through the apparatus while an oil drop was being observed, the charge on the drop could be increased or decreased almost at will. One time, on rising, the velocity would be low due to a small charge (see diagram (b) in Fig. 4J), while the next time the velocity would be high due to a larger charge, as in diagram (c). Regardless of the amount of charge, the rate of fall for a given drop will always be the same, because the total mass of a number of electrons is so small compared with the mass of the oil drop that their added mass is not perceptible.

Millikan, and numerous other experimenters who have repeated these experiments, have found that the charge on a drop is never less than a certain minimum value, and is always some integral multiple of this value. In other words, any one electron is like every other electron, each carrying this minimum charge called e:

$$e = -1.6019 \times 10^{-19} \text{ coul} \tag{4g}$$

This is the most recent and probable value of the electronic charge.

4.8. The Mass of the Electron

From Millikan's determination of the charge on the electron and Thomson's measurement of e/m, the mass of the electron can be calculated by dividing one value by the other. Using the most accurately known values for both e and e/m, we obtain

$$m = \frac{e}{e/m} = \frac{1.6019 \times 10^{-19} \text{ coul}}{1.7589 \times 10^{11} \text{ coul/Kg}} = 9.1072 \times 10^{-31} \text{ Kg}$$

that is,

$$m = 9.1072 \times 10^{-31} \text{ Kg} \tag{4h}$$

This mass is unbelievably small; its value has been determined many times and by many experimenters, and yet it is always the same.*

PROBLEMS

1. Electrons with a velocity of $\frac{1}{20}$ the velocity of light enter a uniform magnetic field at right angles to the magnetic induction. What will be the radius of their circular path if $B = 4.0 \times 10^{-3}$ web/m²?

2. Electrons entering a uniform magnetic field where $B = 50$ gauss, in a direction at right angles to the lines of induction, have a velocity of 4.8×10^9 cm/sec. Calculate the radius of their circular path. *Ans.* 5.46 cm.

3. Two flat parallel metal plates 25 cm long and 5 cm apart (see Fig. 4G) are connected to a 45-volt battery. If electrons enter this field with a velocity of 2.5×10^9 cm/sec, how far will they be deviated from their original straight line path by the time they reach the other end?

4. In J. J. Thomson's experiment shown in Fig. 4F, a magnetic induction field of 2.4×10^{-3} web/m² is employed. If electrons entering this field have a velocity of 3.4×10^9 cm/sec, what potential difference applied to the parallel plates will keep their path straight? Assume the plates to be 0.8 cm apart. See Eq. (2o). *Ans.* 652.8 volts.

5. Electrons, moving in a uniform magnetic field where $B = 50$ gauss, follow a circular path of 10 cm radius. Calculate their velocity.

6. Electrons, moving in a uniform magnetic field $B = 6 \times 10^{-4}$ web/m², follow a circular path of 20 cm radius. Find the velocity. *Ans.* 2.11×10^7 m/sec.

7. If one-half gram of free electrons could be bound together on the moon and another one-half gram of electrons on the earth, what would be their force of repulsion? Earth-moon distance is 239,000 mi.

8. Make a diagram and briefly explain the experiment by which J. J. Thomson measured the velocity of electrons.

9. Make several diagrams showing the main features of an electrical discharge through a gas-filled tube as the pressure is lowered.

10. Diagram and briefly explain the three experiments given in this chapter to demonstrate that cathode rays (a) travel in straight lines, (b) have momentum and energy, and (c) are negatively charged particles.

11. Explain and give a diagram of Millikan's oil-drop experiment. What conclusions were reached by Millikan in this experiment?

12. Electrons are injected with a speed of 2×10^7 m/sec into a uniform magnetic field at right angles to the lines of force. If the magnetic induction is 2×10^{-3} web/m², find the diameter of the circular path. *Ans.* 11.36 cm.

13. If a beam of electrons, moving with a speed of 2×10^7 m/sec, enters a uniform magnetic field at right angles to the lines of force and describes a circular path with a 30 cm radius, calculate the magnetic induction.

14. A 300-volt battery is connected to two flat, parallel metal plates 8 cm long and 1.5 cm apart. If electrons enter this field from one end, moving with a constant velocity of 4×10^7 m/sec, how far will they be deviated from their original straight line path by the time they reach the other end? *Ans.* 0.702 cm.

15. Electrons are injected with a speed of 8×10^6 m/sec into a uniform

* For a more complete and elementary treatment of these early experiments, see *Electrons +* and −, by R. A. Millikan, University of Chicago Press.

magnetic field at right angles to the lines of induction. If the magnetic induction B is 4×10^{-3} web/m², find the diameter of their circular path.

16. If a beam of electrons with a speed of 3×10^7 m/sec enters a uniform magnetic field at right angles to B and describes a circular path of 10 cm radius, what is the value of B? *Ans.* 1.70×10^{-3} web/m².

Atoms and the Periodic Table

Although no one has ever seen individual atoms, there is no doubt that such particles really exist. To the physicists and chemists who have built up and established the present-day theories of the structure of matter, atoms are as real as any material objects large enough to be seen with the eyes or to be felt with the hands. Their reality is evidenced by hundreds of experiments that can be planned and executed in the research laboratory.

As the subject of atomic physics is developed in this and the following chapters, it will become more and more apparent that, although a physicist requires an extremely imaginative mind, the accumulated knowledge of atoms, their structure, and their behavior under a multitude of conditions is based upon exact results of experiments performed with the greatest of accuracy and precision.

5.1. The Discovery of Positive Rays

During the latter part of the nineteenth century, when many physicists were investigating the properties of cathode rays, Goldstein designed a special discharge tube; with it he discovered new rays called *canal rays*. The name "canal rays" is derived from the fact that the rays, traveling in straight lines through a vacuum tube in the opposite direction to cathode rays, pass through and emerge from a canal or hole in the cathode. A tube designed to illustrate this is shown in Fig. 5A.

Shortly after the measurement of the electronic charge by J. J. Thomson in 1896, W. Wien deflected a beam of canal rays in a magnetic field and came to the conclusion that the rays consisted of positively charged particles. Owing to this and other experiments, canal rays have become more commonly known as *positive rays*.

Since the time of Goldstein's discovery, positive rays have been found to be charged atoms of different weights. The origin of the charge carried by such atoms is explained briefly as follows. As the electrons from the cathode stream

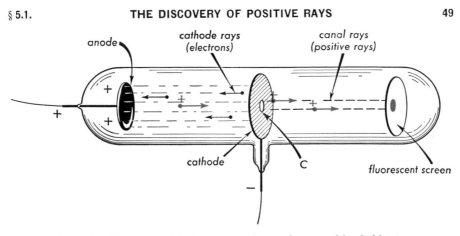

Fig. 5A *Experiment illustrating canal rays discovered by Goldstein.*

down the tube toward the anode, they occasionally collide with the atoms and molecules of the small quantity of remaining gas, knocking electrons from them. This process, called *ionization*, is illustrated by a schematic diagram of a single oxygen atom in Fig. 5B. Before the collision, the atom as a whole, with

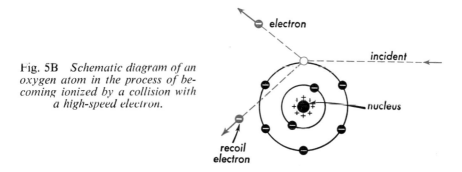

Fig. 5B *Schematic diagram of an oxygen atom in the process of becoming ionized by a collision with a high-speed electron.*

its eight electrons and eight equal positive charges on the nucleus, has no net charge. After one of the electrons is removed by collision, it has but seven electrons and therefore a net positive charge equivalent in amount to the charge of one electron.

Since the atom is now positively charged, the anode repels and the cathode attracts such atoms, accelerating them toward the cathode. There exist, therefore, between the anode and cathode, two streams of particles: electrons moving toward the anode, and positively charged atoms or molecules moving toward the cathode.

Of the many particles striking the cathode in Fig. 5A, the ones moving toward the small opening C, constituting the observed canal rays, pass straight through to the fluorescent screen. As each atom or molecule strikes the screen, a tiny flash of light is produced. These tiny flashes, which can be seen individually in the field of view of a microscope, are called *scintillations*.

Any process by which an electron is removed from an atom or molecule is called *ionization*, and the resulting charged particle is called a *positive ion*. The amount of charge carried by an electron is a unit called *the electronic charge*.

5.2. The Thomson Mass Spectrograph

Ever since the time canal rays were shown to be positively charged atoms or molecules of the gas contained within the discharged tube, physicists have tried to determine with ever-increasing accuracy the mass and charge of the individual ray particles. Although the charge and mass of every electron were known from Thomson's and Millikan's experiments to be the same as those for every other electron, i could be postulated that the mass of the positive rays should be different for the atoms of different chemical elements. The further postulation could be made that if each positive ion were produced by the removal of one electron from a neutral atom, all positive ions should have the same net charge. This, in part, is anticipating what is now known.

In 1911, J. J. Thomson developed a method of measuring the relative masses of different atoms and molecules by deflecting positive rays in a magnetic and an electric field. The apparatus he developed for doing this is shown schematically in Fig. 5C; it is called *Thomson's mass spectrograph*.

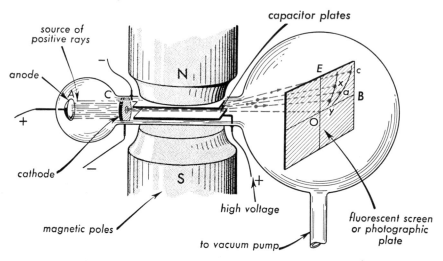

Fig. 5C   *Diagram of J. J. Thomson's mass spectrograph.*

The entire spectrograph, enclosed in an airtight glass chamber, is first thoroughly evacuated; then a small quantity of the gas, the masses of whose atoms are to be measured, is admitted to the bulb at the left. When a high voltage is applied to this chamber, electrons from the cathode ionize the atoms and molecules in the region between the anode A and the cathode C. Traveling to the right, many of these positively charged particles pass through the narrow hole in the cathode, thus forming a very narrow pencil of rays. Leaving the cathode with a constant velocity, they then pass between the poles of an electro-magnet and the parallel plates of a capacitor, and thence to a fluorescent screen at the far end of the chamber.

The two parallel plates, when charged, exert an upward force on the particles, deflecting them from the point O toward E. The magnetic field, on the other hand, with its magnetic lines vertically downward and in the plane of the page, exerts a force at right angles to this, deflecting the particles "into" the page from the point O toward B.

Suppose now that the apparatus contains a pure gas like helium, all of the atoms of which have exactly the same mass. Of these atoms, the ones that are ionized in a region near the cathode C cannot attain a very high velocity before reaching the cathode. Since these atoms remain longer in the deflecting fields, their paths are bent considerably up and back to a point such as c on the screen. Particles ionized near the anode A, on the other hand, attain a high velocity upon reaching the cathode and, being under the influence of the deflecting fields for a shorter time, have their paths bent only a little, to a point like a on the screen. Since the velocities of the particles vary considerably, a bright streak or line of fluorescence will appear on the screen. From a calculation of the forces exerted by both fields, it is found that the line on the screen should have the shape of a parabola.

If the gas in the apparatus is not pure but contains two kinds of atoms, the positive ions passing through the cathode will have two different masses. Although each ion will contain the same positive charge, and will therefore experience the same electric and magnetic forces when passing through the fields, the heavier particles will not be deflected as much as the lighter ones. The net result is that the heavier particles form one parabolic curve like xy, and the lighter particles another curve like ac.

By substituting a photographic plate for the fluorescent screen and exposing it to the rays for several minutes, photographs like those reproduced in Fig. 5D are obtained. The continual bombardment of the photographic plate by atoms and molecules has the same effect as does light, and images are produced upon development. The upper half of each picture is taken with the connections as shown in Fig. 5C, and the lower half by reversing the polarity of the electro-magnet and exposing for an equal length of time.

When photograph (a) was taken, the spectrograph contained *hydrogen*, *oxygen*, and *mercury*, and the magnetic field was relatively weak. From the known strengths of both the electric and magnetic fields, and the assumption that each atom carries a unit positive charge, the mass of the atoms producing each parabola can be calculated. The results of these calculations show that the two largest parabolas are due to ionized hydrogen atoms (H^+) of mass 1, and ionized hydrogen molecules (H_2^+) of mass 2. The next three are due to ionized atoms (O^+) of mass 16, ionized oxygen molecules of mass 32, and ionized mercury atoms (Hg^+) with a mass of approximately 200.

When photograph (b) was taken, the mass spectrograph contained carbon

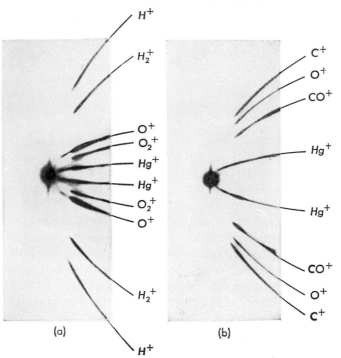

Fig. 5D *Reproductions of the photographs of parabolas made with Thomson's mass spectrograph.*

monoxide gas and mercury vapor, and the magnetic field was relatively strong. Upon calculating the masses of the particles producing the different parabolas, the four intense lines were identified as due to ionized carbon atoms (C^+) of mass 12, ionized oxygen atoms (O^+) of mass 16, ionized carbon monoxide molecules (CO^+) of mass 28, and ionized mercury atoms (Hg^+) of mass about 200. The three faint parabolas which show in the original photograph but probably not in the reproduction, are due to doubly ionized atoms of *carbon, oxygen,* and *mercury.*

A doubly ionized atom or molecule is one that has lost two electrons rather than only one and, having a net positive charge of two units, is designated by two ($^+$) signs as superscripts. Since the particles have double charges, the electric and magnetic forces exerted on them are double those for singly ionized atoms and they produce larger parabolas, because they undergo greater deflections.

The principal conclusion to be drawn from Thomson's experiments is: *Positive rays or canal rays are charged atoms or molecules of whatever gas is present in the apparatus.*

It is significant to point out that, while Thomson found many atoms could be doubly and some even triply ionized, hydrogen could never be found more than singly ionized or helium more than doubly ionized. The reason for this, as will be seen later, is that neutral hydrogen atoms have but one electron and neutral helium atoms but two. All other elements have more than two electrons.

5.3. The Periodic Table of Elements

From present-day knowledge of physics, chemistry, and astronomy, it is quite certain that the entire universe is made up of 80 to 90 stable elements. By an element is meant a substance composed of atoms having identical chemical properties. All but two or three of these elements have been found in the earth's crust, some of them in much greater abundance than others. Silicon and iron are examples of abundant elements, whereas platinum is an example of a rare element.

Long before the Thomson mass spectrograph had been devised and used to measure the relative masses of atoms, the chemist had arranged all of the elements in a table according to their atomic weights. The most common form of this arrangement is given in Appendix VII. Divided as they are into eight separate groups, all elements in the same column have similar chemical properties. In Group I, for example, the elements Li, Na, K, Rb, and Cs, known as the *alkali metals*, have one set of chemical properties, whereas the elements Be, Mg, Ca, Sr, and Ba in Group II, known as the *alkaline earths*, have another set of chemical properties. The largest group of elements having similar chemical properties are the fourteen rare earth elements listed by themselves at the bottom of the table.

The names of the elements are all indicated by one-letter and two-letter symbols. (The full names are given in the first column of Appendix I.) The number in the third column following each abbreviation is the order number of that element and is called *the atomic number*. The average weight of atoms of that element called *the atomic weight*, is given in the last column.

The atomic weights of all elements are based upon the weight of carbon 12. See § 5.7. This is purely an arbitrary selection of a unit of weight but one which has considerable significance when it is noted that the weights of the first 25 elements, with the exception of chlorine (Cl), atomic number 17, are very close to whole numbers. This suggests the possibility that the weights of all atoms are really whole number units of the unit of weight, the hydrogen atom, and that those weights of an element which differ considerably from whole numbers are incorrectly determined values. On the strength of this, Prout was the first to propose the hypothesis that all elements are made of hydrogen atoms as building stones. These suppositions, as will be seen later, are only partly true.

5.4. Thomson's Discovery of Isotopes

In 1912, Thomson, in comparing the mass of the neon atom with the known masses of other elements, discovered two parabolas for neon in place of one. Upon computing the masses of the particles involved, the stronger of two parabolas was found to be due to particles of mass 20 and the other, a fainter parabola, to particles of mass 22.

Since the atomic weight of neon was then known to be 20.2, Thomson expressed the belief that neon is composed of two kinds of atoms, 90% of which have a mass of 20 and the other 10% a mass of 22. Because these two kinds of atoms exist as a mixture and cannot be separated chemically, their atomic weight, when measured by chemical methods, is found to be their average value, 20.2.

The discovery of two kinds of neon atoms, identical chemically but differing in atomic weight, suggested the possibility that all other elements whose atomic weights were not whole numbers might also be mixtures of atoms that do have whole number weights. Not only has this been confirmed by experiment, but a large majority of the elements have been found to be mixtures of from two to ten different kinds of atoms.

To all atoms of different weight belonging to the same element, Soddy gave the name *isotopes*. The external structures of all isotopes of a given element are identical. The two atoms, Ne-20 and Ne-22, shown in Fig. 5E, are neon isotopes.

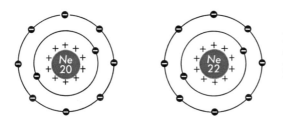

Fig. 5E *Schematic diagrams of two kinds of neon atoms, one of mass 20 and the other of mass 22. The external electron structures of two such isotopes are identical.*

Each of these neutral atoms, before it is ionized to become a positive ray, has ten external electrons and ten positive charges on the nucleus. They differ only in the weight of the nucleus.

Atoms having different weights but belonging to the same chemical element have the same atomic number and are called isotopes.

5.5. Aston's Mass Measurements

Immediately following World War I, in 1919, F. W. Aston* developed a new and improved type of mass spectrograph, employing both the electric and magnetic fields. The chief improvement of this device over Thomson's mass spectrograph was the "focusing" of the rays of different velocities to the same point on the screen or photographic plate. This had two important effects: (1) it made it possible to observe rare isotopes which might otherwise escape

* Francis William Aston (1877–1945), British scientist, born in Birmingham and educated at Malvern College and Cambridge University. He became assistant lecturer in physics at the Birmingham University in 1909, and received the Mackenzie Davidson Medal of the Röntgen Society in 1920. In 1922, he was awarded the Hughes Medal of the Royal Society and the coveted Nobel Prize in chemistry for his work on atomic mass measurements. He wrote an authoritative book entitled *Isotopes*, in which a full account of his work is given.

detection, and (2) it produced sharper images of the different masses on the photographic plate, so that their masses could be more accurately measured.

An Aston mass spectrogram is reproduced in Fig. 5F(a). In taking this particular photograph, Aston had introduced into his apparatus, among other things, a little *hydrochloric acid* (HCl), *carbon monoxide* (CO), and *sulfur dioxide* (SO_2). Being close together in the periodic table, these elements furnish an excellent demonstration of the linear shift of atoms and molecules, differing in mass by one unit. It is found from this, and other photographs, that sulfur has three isotopes with masses 32, 33, and 34, and that chlorine has two isotopes of mass 35 and 37.

Since the atomic weight of chlorine is 35.453, then for every atom of mass 37 in a given quantity of chlorine gas there are four of mass 35. Mixed together in these proportions, they give an average mass of 35.4.

The photographic lines corresponding to masses 28, 36, and 38 are due to diatomic molecules CO and HCl, each molecule having the combined weight of its constituent atoms. Since there are two relatively abundant chlorine isotopes, there are two kinds of HCl molecules. One type, H^1Cl^{35}, has a mass of 36; and the other type, H^1Cl^{37}, a mass of 38.

A CO molecule of the type producing the strong line at mass 28 in Fig. 5F(a) is shown schematically in Fig. 5G. Since the molecule is neutral, there are just as many electrons surrounding the two bound atoms as there are positive charges on the nuclei (six on the carbon nucleus and eight on the oxygen nucleus). When the molecule becomes ionized and is moving through the apparatus as a positive ray, it contains one less electron than the number shown.

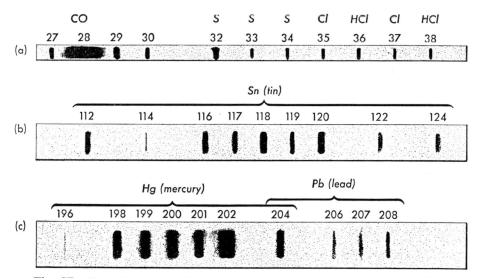

Fig. 5F *Reproductions of photographs taken with a mass spectrograph illustrating the linear shift of atoms differing by one unit of mass. (a) Carbon monoxide, sulfur, chlorine, and HCl lines. (b) Isotopes of tin. (c) Isotopes of mercury and lead.*

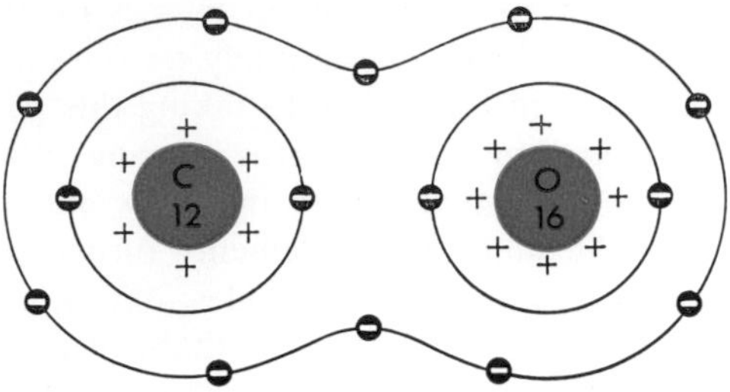

Fig. 5G *Schematic diagram of a diatomic molecule, carbon monoxide (CO).*

Since the mass of the electrons is negligibly small, the mass of the molecule is 12 + 16 or 28 mass units.

So successful was Aston with his mass measurements and his determination of isotopes of different elements that he attempted an investigation of the entire periodic table. All of the known elements are listed in Appendix III, with all of their observed stable isotopes. In each case, the most abundant isotope is indicated in heavy type, while the very rare isotopes, i.e., those present to less than 1%, are listed in parentheses. Where more than one isotope is set in heavy type, the isotopes occur with almost equal abundance. The masses printed in italics represent unstable atoms which are responsible for *radioactivity*, the subject of Chapter 23. Recent developments in mass spectroscopy have made it possible to detect exceptionally rare isotopes. In neon, for example, an isotope of mass number 21 has been found, making three in all, with relative abundances as follows:

Isotope.............	Ne-20	Ne-21	Ne-22
Abundance, %......	90.92	0.26	8.82

In a pure carbon monoxide gas, all of the molecules are diatomic and alike in every respect except for mass. Since there are two carbon isotopes, 12 and 13, and three oxygen isotopes, 16, 17, and 18 (see Appendix III), there are six different combinations of atoms to form molecules. These are $C^{12}O^{16}$, $C^{12}O^{17}$, $C^{12}O^{18}$, $C^{13}O^{16}$, $C^{13}O^{17}$, and $C^{13}O^{18}$. The relative abundances of all but the C^{12} and O^{16} isotopes are so small, however, that more than 90% of the molecules in a given quantity of gas are of the type $C^{12}O^{16}$, with a mass of 28.

5.6. Isobars

Another mass spectrograph of remarkably high precision was devised in 1933 by an American physicist, K. T. Bainbridge.

Two photographs taken with this instrument are shown in Fig. 5F. The middle picture (b) shows the many isotopes of tin, and the lower plate (c) the isotopes of mercury and lead. The rare lead isotope 204 falls on top of the strong mercury isotope 204. Such coincidences are called *isobars*.

Atoms having the same mass but belonging to different chemical elements are called isobars.

The first pair of isobars (see Appendix III) occurs in argon and calcium. The principal isotope of argon, atomic number 18, has a mass of 40, as does also the principal isotope of calcium, atomic number 20. Other examples are Cr^{54} and Fe^{54}, Ge^{76} and Se^{76}, Rb^{87} and Sr^{87}, Zn^{92} and Mo^{92}. The isobars Hg^{204} and Pb^{204} are illustrated in Fig. 5F(c).

5.7. Unit Atomic Mass and the Hydrogen Atom

Until 1927, all oxygen atoms were thought to have the same mass and were arbitrarily chosen to be the standard by which all atomic masses were measured. At this time Giauque and Johnson discovered the existence of two rare oxygen isotopes with masses 17 and 18. So rare are these heavier particles that in every ten thousand oxygen atoms twenty of them have a mass of 18, and only four a mass of 17.

Retaining natural oxygen with its mixture of three isotopes as a standard, the chemists continued to use $\frac{1}{16}$ the weight of this mixture as *unit atomic weight*. The physicists, on the other hand, found it most convenient for their purposes to adopt a mass scale based upon the mass of the most abundant oxygen atom as exactly 16, and $\frac{1}{16}$ of this as *unit atomic mass*.

As isotope mass determinations (measured largely by research physicists) became more and more accurate, and mass differences became more important, the slight differences between the two scales became more and more of a problem.

In 1960 the International Union of Pure and Applied Physics (IUPAP), met in Ottawa and adopted the atomic mass scale based upon the carbon-12 isotope as having a mass of exactly 12. The same scale was adopted by the International Union of Pure and Applied Chemistry (IUPAC), at their Montreal meeting in 1961. While this new atomic weight scale differs ever so slightly from the older $O = 16$ scale, the accurate masses of isotopes show greater differences, as can be seen in Appendix II.

Standard unit atomic mass is now generally accepted to be $\frac{1}{12}$ of the mass of the carbon-12 isotope. On this basis very accurate measurements give for the mass of the hydrogen atom 1.0081456 atomic mass units (amu), a value nearly 1% higher than unity.

For many practical purposes it is convenient to know the masses of atoms in kilograms. For easy calculations the following value may be used:

$$M = 1.660 \times 10^{-27} \text{ Kg} \qquad (5a)$$
$$\textbf{Unit atomic mass}$$

This number multiplied by the atomic mass of any atom will give its mass in kilograms.

Compared with the mass of the electron, namely,

$$m = 9.1072 \times 10^{-31} \text{ Kg} \qquad (5b)$$
$$\textbf{Electron mass}$$

an atom of unit mass would be 1824 times as heavy. The hydrogen atom is slightly heavier than one unit mass and is about 1836 times as heavy as the electron. This latter number is convenient to remember, for it is often quoted to illustrate the enormous difference between the mass of the nucleus of a hydrogen atom and the mass of its one and only electron.

Atomic number is defined as the number ascribed to an element specifying its position in the periodic table of elements. (See column 1, Appendix III.)

Mass number is defined as the whole number nearest the actual mass of an isotope measured in atomic mass units. (See column 4, Appendix III.)

Atomic weight is defined as the average weight of all the isotopes of an element, weighted according to relative abundance and expressed in atomic mass units. (See column 5, Appendix III.)

QUESTIONS

1. Who discovered isotopes? What are isotopes?

2. What is meant by (a) *atomic number*, (b) *mass number*, and (c) *atomic weight*?

3. How do isotopes of any given element differ from each other?

4. What is meant by *relative abundance*?

5. What are isobars? Give an example.

6. How much greater is the mass of a hydrogen atom than the mass of an electron?

7. What do the more or less equal spacings of the mass spectrograms shown in Fig. 5F suggest regarding the relative masses of atoms?

PROBLEMS

1. The atomic weight of cobalt is 58.9332. Find the mass in grams of one cobalt atom.

2. If the atomic weight of gold is 196.967, how many atoms are there in 1 gm of gold metal? *Ans.* 3.05×10^{21} atoms.

3. The atomic weight of iodine is 126.9044. How many atoms are there in 1 gm of iodine crystals?

4. If the atomic weights of silicon and oxygen are 28.086 and 15.9994, respectively, find the mass in grams of a silicon dioxide molecule, SiO_2. *Ans.* 9.96×10^{-23} gm.

5. The atomic weights of hydrogen, carbon, and oxygen, are 1.00797, 12.01115, and 15.9994, respectively. How many ethyl alcohol molecules are there in 1 gm of ethyl alcohol (C_2H_5OH)?

6. Define or briefly explain the meaning of the following: (a) *isotopes*, (b) *isobars*, (c) *positive rays*, (d) *atomic weight*, and (e) *unit atomic mass*.

7. Name five members of each of the following classifications: (a) alkali metals, and (b) alkaline earths. (See Appendix VII.)

8. What chemical element has the greatest number of isotopes? (See Appendix III.)

9. Make a list of elements having (a) atoms of one mass only, and (b) only two isotopes. (See Appendix III.)

10. Carbon has two isotopes, 12 and 13, while oxygen has three, 16, 17, and 18. Find to three figures only the mass in kilograms for each of the six possible CO molecules. *Ans.* $C^{12}O^{16} = 4.65$, $C^{12}O^{17} = 4.81$, $C^{12}O^{18} = 4.98$, $C^{13}O^{16} = 4.81$, $C^{13}O^{17} = 4.98$, $C^{13}O^{18} = 5.15 \times 10^{-26}$ Kg.

11. Gallium has two isotopes, 69 and 71. Calculate the mass in kilograms to three figures only for each of the three possible kinds of diatomic molecules.

12. The two stable isotopes of copper have masses of 63 and 65 atomic mass units. The normal mixture of these atoms has an atomic weight of 63.54. What percentage of any given amount of the normal metal is composed of Cu^{63} atoms? *Ans.* 73%.

13. The forty-seventh element in the periodic table, silver, has two stable isotopes with masses of 107 and 109 atomic mass units, respectively. The normal mixture of these atoms has an atomic weight of 107.870. Calculate the percentage of any given amount of the normal silver composed of Ag^{109} atoms.

Diffraction and Interference of Light

When light passes close to the edge of any object, it is bent in its path and travels on in a new direction. This bending of light around corners is called *diffraction*.

To observe the phenomenon of diffraction, the following simple experiment may be performed in a darkened room. A box containing a light bulb and a pinhole is placed on one side of the room and a ground-glass observing screen or photographic film is placed on the other. Objects of various kinds are then placed about halfway between the source and the screen as shown in Fig. 6A.

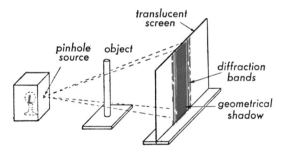

Fig. 6A *The shadow cast by the light from a small source is not sharp at the edges, but exhibits a banded structure.*

This is the arrangement used in obtaining the original photograph reproduced in Fig. 6B. Notice that the edges of the shadows are not sharp but are bounded at the edges by narrow bands or fringes. At the center of two figures the fringes form sets of concentric rings.

6.1. Huygens' Principle

The phenomenon of diffraction was explained in the time of Newton by assuming that light is composed of small particles or corpuscles obeying the ordinary laws of mechanical motion. And so it was that Newton and his fol-

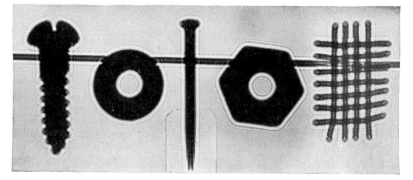

Fig. 6B   *Photographs of the shadows cast by small objects. The narrow bands are due to the diffraction of light.*

lowers held for many years to the idea that a source of light is a source of high-speed particles radiated in all directions.

Although such a viewpoint was accepted for many years, it was later abandoned in favor of a wave theory of light, according to which a beam of light is made up of many waves of extremely short wavelengths. By adopting the wave hypothesis, a complete and adequate account of the phenomena of reflection, refraction, diffraction, interference, and polarization was finally formulated on a mathematical basis at the beginning of the nineteenth century by Augustin Fresnel, a French physicist. The wave theory of light was first proposed by the English physicist Robert Hooke in 1665 and improved twenty years later by the Dutch scientist and mathematician, Christian Huygens.*

Everyone has at some time or another dropped a stone in a still pond of water and watched the waves spread slowly outward in ever-widening concentric circles. In the analogous case of a point source of light, the spreading waves form concentric spheres moving outward with the extremely high velocity of 186,300 mi/sec. This is represented diagrammatically in Fig. 6C. Each circle represents the crest of a wave so that the distance between consecutive circles is one wavelength.

According to Huygens' principle, every point on any wave front may be regarded as a new point source of waves. Regarding each of any number of points like *a*, *b*, *c*, etc., as point sources like *S*, secondary wavelets spread out simultaneously as shown. The envelope of these an instant later is the new wave front *A*, *B*, *C*, etc., and still later the wave front *L*, *M*, *N*, etc. Although Huygens'

* Christian Huygens (1629–1695), famous Dutch physicist and contemporary of Isaac Newton. Born at The Hague in 1629, young Christian got his first ideas about waves and their propagation by watching the ripples on the canals about his home. Although his chief title-deed to immortality is his development of the wave theory of light, he made many and valuable contributions to mathematics and astronomy. He improved upon the method of grinding telescope lenses and discovered the Orion nebula, part of which is now known by his name. He was elected to the Royal Society of London in 1663, and delivered before that august body the first clear statement of the laws governing the collision of elastic bodies. He died a confirmed bachelor at The Hague in 1695.

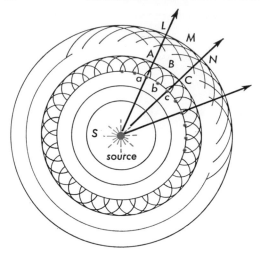

Fig. 6C *Diagram of waves spreading out from a point source. The secondary wavelets and new wave fronts illustrate Huygens' principle.*

principle at first glance might seem to be a useless play with circles it has quite general application to many optical phenomena.

A direct experimental demonstration of Huygens' principle is illustrated in Fig. 6D. Plane waves approaching a barrier *AB* from the left are reflected or

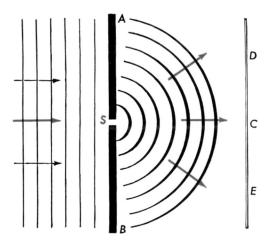

Fig. 6D *Diagram of the diffraction of waves at a small opening. Huygens' principle.*

absorbed at every point except at *S*, where they are allowed to pass on through. When the experiment is carried out with water, one can see the waves spreading out in all directions as if *S* were a point source.

If *AB* is an opaque screen and *S* is a pinhole small in comparison to the wavelength of light, the light waves will spread out in hemispheres with *S* at their center. If *S* is a long narrow slit (perpendicular to the page), the waves spread out with cylindrical wave fronts. Cross sections of all of these cases are represented by the semicircles, the light traveling in the direction of the arrows.

The action of a converging lens on light waves is illustrated in Fig. 6E. If a point source of light is placed at the focal point, as in diagram (a), the expanding waves pass through the lens and come out as plane waves, i.e., as parallel light.

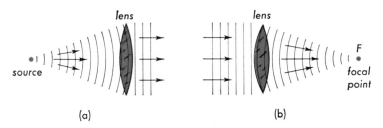

Fig. 6E   *The behavior of light waves as they pass through a converging lens.*

In diagram (b), incident plane waves are shown emerging from the lens as converging waves which come to a focus at *F*. The change brought about by the lens can be explained by the fact that light travels faster in air than it does in glass.

It is important to note that the time taken for light rays leaving the point source and arriving at the point image is the same for all paths.

6.2. Interference

When two stones are dropped simultaneously into a still pond of water, two sets of waves will spread outward as shown in Fig. 6F. As these waves cross

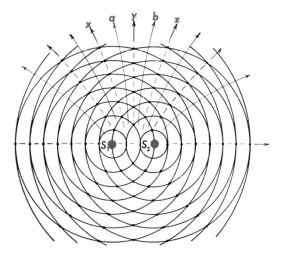

Fig. 6F   *Concentric waves traveling outward from a double source, producing what is called an interference pattern.*

each other, they act one upon the other, producing what is called an *interference pattern*. Where the crests of two waves come together at the dotted intersections, they are in step, or in phase, and the amplitude of the water surface is increased. Where the crest of one wave and the trough of another come together, they are out of step, or out of phase, and the amplitude of the water surface is reduced.

The in-phase regions of the waves move outward along the dotted lines, such as x, y, and z, and we have what is called *constructive interference*. The out-of-phase regions move outward along the solid lines, such as a and b, and we have what is called *destructive interference*.

An instantaneous photograph of such a wave pattern is shown in Fig. 6G.

Fig. 6G   *Ripple tank photograph of the interference of water waves from two sources. (Courtesy, Physical Sciences Study Committee Project.)*

Note how clearly the interference regions of the waves stand out. Photographs of this kind, as well as direct observations of such wave patterns, are readily made as follows. A glass tray for maintaining a shallow water layer can be made from a piece of window glass and a wooden frame. A thin metal strip, clamped at one end and set vibrating up and down over the water, is used as a source. A piece of wire, fastened to the vibrating end of this strip, should have one wire end dipping into the water for a single-wave source, and both ends dipping into the water for a double source. Intermittent viewing of the waves through a slotted disk or illumination by means of a stroboscopic light source makes the wave pattern appear to stand still or to progress in slow motion.

6.3. Coherence

Coherence is a condition that must exist between two or more waves if a steady state of interference is to be observed, and is a condition involving the relative phases of waves wherever they are brought together.

If two wave sources are set up to emit identical frequencies, and the sources are made to vibrate in step, out of step, or to maintain a constant phase difference between them, the waves they emit are said to be *coherent*.

In the case of the water waves as demonstrated in Fig. 6G, the two sources are made to vibrate in phase with each other by mounting the two prongs on one vibrating metal strip. When two separately mounted metal strips are used, each with a single prong dipping in the water, each may be set vibrating with any random phase. As long as they have the same vibration frequency the sources will maintain the same phase difference, the waves will be coherent, and a steady interference pattern will be produced. With different phase angles between the sources, however, the direction lines, x, a, y, b, and z, as shown in Fig. 6F, will be different.

In order to obtain interference patterns with light waves, two or more coherent sets of waves are derived from the same light source. We find by experiment that it is impossible to obtain interference from two separate sources, such as two lamp filaments set side by side. This failure is caused by the fact that the light from any one source is not an infinite train of waves. On the contrary emission is random and there are sudden changes in phase occurring in very short intervals of time (of the order of 10^{-8} sec).

Thus, although interference patterns may exist for a short time interval, they will shift in position and shape each time there is a phase change, and no interference fringes will be seen. To produce a steady prolonged interference pattern the difference in phase between any pair of points in the two sources must remain constant. It is characteristic of any interference experiment that the sources must have a point-to-point phase relation, and sources that have this relation are called coherent sources.*

6.4. Young's Double-Source Experiment

Young's double-source experiment, first performed with sunlight and pinholes in 1801, served as a crucial test deciding between Newton's corpuscular theory of light and Huygens' wave theory. This is represented schematically in Fig. 6H.

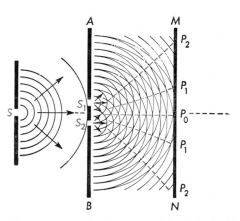

Fig. 6H   *Diagram of Young's double-slit experiment illustrating the interference of light waves.*

* For a detailed treatment of interference phenomena, see *Fundamentals of Optics* by Jenkins and White, published by the McGraw-Hill Book Company.

Sunlight from a pinhole S was allowed to fall on a distant screen containing two pinholes, S_1 and S_2. The two sets of spherical waves emerging from the two holes interfered with each other in such a way as to form a symmetrical pattern of bands on another screen MN.

For convenience it is now customary to perform Young's experiment with narrow slits in place of pinholes. If S, S_1, and S_2 in Fig. 6H represent the cross sections of three narrow slits, the light falling on the farther screen MN has the appearance of equidistant bands or fringes, as shown by the photograph in Fig. 6I. The bright fringes correspond to the points P_0, P_1, P_2, etc., and the dark fringes to the points halfway between.

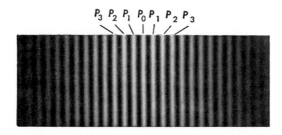

Fig. 6I *Interference fringes produced by a double slit as in Young's experiment.*

As the waves travel outward from each slit S_1 and S_2, they cross each other only at points that lie along the dotted lines shown in the diagram. These lines represent the areas where the crests of two waves come together and produce a maximum brightness. About halfway between these dotted lines lie other areas where the crest of one wave and the trough of another cancel each other and produce darkness. This is the same interference phenomenon as is illustrated by water waves in Fig. 6G. With light waves, where the bright fringes are formed there is constructive interference, and where the dark fringes appear there is destructive interference.

This experiment is frequently performed in the elementary physics laboratory, and from measurements of fringe and slit spacings, as well as double-slit to screen distance, the wavelengths of different colors of light are calculated.

A formula for the wavelength of light can be derived from the geometry of Fig. 6J. Let P be the position of any bright fringe on the screen, and d the distance between slit centers. P_0 is located on the perpendicular bisector of the double slit S_1 and S_2.

The two rays emerging from S_1 and S_2 parallel to CP are brought to a focus at P. Line S_2A, drawn perpendicular to the two rays, forms a right triangle S_1AS_2. By this construction the short side S_1A becomes the extra distance that light must travel from the upper slit to arrive at the screen MN.

To produce a bright fringe at P, the interval S_1A must be equal to one whole wavelength, two whole wavelengths, three whole wavelengths, etc., for only then will waves from S_1 and S_2 arrive at P in phase. From the right triangle S_1AS_2, we can write

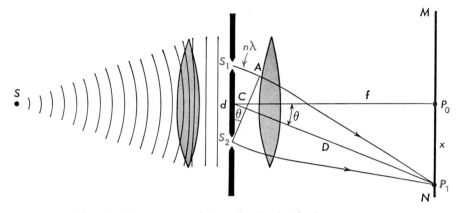

Fig. 6J　*Geometrical relations for the double-slit experiment.*

$$\frac{n\lambda}{d} = \sin \theta \qquad (6a)$$

where

$$n = 1, 2, 3, 4, \ldots$$

Since line CP_0 is perpendicular to MN and S_2A is perpendicular to CP, angle S_1S_2A equals angle P_0CP, and we can write

$$\frac{x}{D} = \sin \theta \qquad (6b)$$

Since the fringes are so close together, and the focal length f is long by comparison, D and f are so nearly equal we can write

$$\frac{x}{f} = \sin \theta \qquad (6c)$$

Combining left hand sides of Eqs. (6a) and (6c), we obtain

$$\frac{n\lambda}{d} = \frac{x}{f} \qquad (6d)$$

from which

$$x = \frac{n\lambda f}{d} \qquad (6e)$$

This formula shows that on increasing the slit spacing d, the fringe spacing decreases. On increasing the wavelength λ, the bands are farther apart.

By measuring the fringe spacing x, the distance between slit centers and f, the wavelength of different colors of light can be calculated. Repeated experiments, carefully performed, give the following results.

Violet, $\lambda = 0.000042$ cm　　　Yellow, $\lambda = 0.000058$ cm
Blue, 　$\lambda = 0.000046$ cm　　　Orange, $\lambda = 0.000061$ cm
Green, $\lambda = 0.000054$ cm　　　Red, 　$\lambda = 0.000066$ cm

As illustrated by the drawing of waves in Fig. 6K, red light has the longest waves.

These wavelengths, measured by the double-slit experiment, are therefore

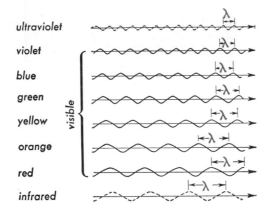

Fig. 6K *Diagram showing the relative wavelengths of light.*

average values, since each color corresponds to a range of different wavelengths.

6.5. Diffraction by a Single Slit

If a beam of parallel light is allowed to pass through a single narrow slit, the emerging light is spread out and forms a pattern of bands on a distant screen.

An experimental arrangement for observing such a diffraction pattern is shown in Fig. 6L. A single slit is a rectangular aperture, long in comparison to

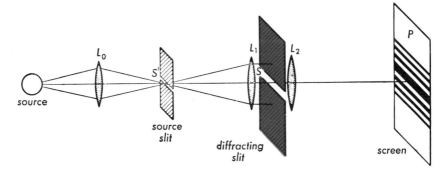

Fig. 6L *Experimental arrangement for obtaining the diffraction pattern of a single slit. Fraunhofer diffraction.*

its width. An adjustable slit S is set up with its long dimension horizontal, and illuminated by parallel light of one wavelength. This beam of monochromatic light is obtained by the use of a source of light with a filter, a very narrow slit S' and two lenses L_0 and L_1.

Actual photographs of the light falling on a screen such as the one at the right

are reproduced in Fig. 6M. The photograph (b), made with a shorter time
exposure, shows a band of light fading out at the edges. The longer exposure
(c) shows this central band somewhat widened, and narrow bands symmetrically

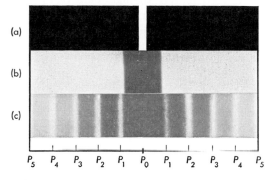

Fig. 6M *Photographs of the
single-slit diffraction pattern. (a)
The slit itself, (b) short exposure,
(c) long exposure.*

located on either side. For these photographs the distance $S'L_1$ was 25 cm, and
L_2P was 100 cm. The width of slit S' was 0.10 mm and S was 0.090 mm. When
S' was widened to more than 0.3 mm, the details of the pattern began to disappear. On the original photograph the total width of the central band was
9.68 mm. The light source was a small mercury arc and a violet glass filter
transmitting only the mercury violet light, $\lambda = 4358 \times 10^{-8}$ cm.

This diffraction pattern can be observed directly by ruling a groove on an
unused photographic plate with a penknife, holding it in front of the eye, and
then looking directly at the light coming through another such slit several feet
away.

The explanation of the single-slit pattern lies in the interference of many
waves and is similar in principle to the interference of two waves described in
§ 6.2. The many waves to be considered here can be thought of as sent out from
every point on a wave crest at the instant that it crosses the plane of the slit.

The cross-section diagram of Fig. 6N shows a slit AB, of width d, with
parallel wave crests approaching from the left. The wave crest at the slit is
shown divided into 12 imaginary segments of equal width. Each of these

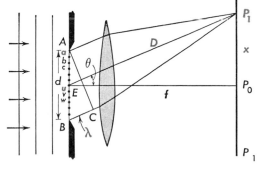

Fig. 6N *Geometry for the single
slit diffraction experiment.*

segments may be thought of as a new source for a secondary wavelet. Let us now choose a point P_1 on the distant screen where the light intensity is observed to be zero, corresponding to the point P_1 in Fig. 6M. This is a point in Fig. 6N where the length of the light path BCP_1 is one whole wavelength λ greater than the light path AP_1. In other words the path $BC = 1\lambda$.

To see why this is just the right condition for no light on the screen, consider the wave diagram in Fig. 6O. The points marked a, b, c, etc., correspond to the

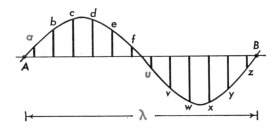

Fig. 6O *Graph of the single-slit wavelet contributions for the first dark band, point P_1 of Fig. 6N.*

relative phases of the wavelets arriving at P_1 at the same instant from all 12 of the slit elements. The heights of the vertical lines give the relative displacements of the instantaneous light contributions at P_1 due to the different slit elements. Note that the wave from a has a small upward displacement, while the wavelet from u has an equal downward displacement. One cancels the other to produce darkness. At this same instant the wavelet from b has a larger upward displacement while the wavelet from v has an equal downward displacement; thus they cancel each other. Similarly the elements c and w may be paired off with opposite displacements, and the process continued across the slit. Each pair of displacements is seen to cancel out and to produce destructive interference, or darkness, at P_1.

Suppose we now consider a point higher up on the screen corresponding to the center of the next dark band, P_2, of Fig. 6M. For this point the light path BC in Fig. 6N will be two whole wavelengths. If the slit is again divided into an equal number of small segments, the phases of the light wavelets arriving at P_2 will again be found to cancel in pairs.

When this treatment is applied to the point P_0 at the center of the screen, all the paths are the same, all wavelets arrive in phase, and we obtain the bright band center.

If the above treatment of pairing-off light contributions from small elements of a single aperture is carried out over all angles of θ, complete cancellation will be found to occur only when the path difference BC is exactly a whole-number multiple of one wavelength, i.e., $BC = 1\lambda$, 2λ, 3λ, 4λ, etc. For the dark bands, then, we can write $BC = n\lambda$, where $n = 0, 1, 2, 3$, etc. Since line AC is perpendicular to EP_1 and line AB is perpendicular to EP_0, the two triangles can be assumed to be similar, and the following proportions between corresponding sides can be written:

$$\frac{n\lambda}{d} = \frac{x}{D} \tag{6f}$$

But $x/D = \sin\theta$, and we may write

$$\frac{n\lambda}{d} = \sin\theta \tag{6g}$$

from which we can write

$$n\lambda = d\sin\theta \tag{6h}$$

Dark bands

This formula shows that by widening the slit the diffraction bands become narrower, and vice versa. By increasing the wavelength the bands become wider.

If the path difference BC in Fig. 6N is not equal to a whole number of wavelengths λ, the wavelet contributions to those points on the screen will not cancel out, and this accounts for the bright bands.

A more detailed theory than that given above shows that the light intensity on the screen should have a distribution like that graphed in Fig. 6P. If we call

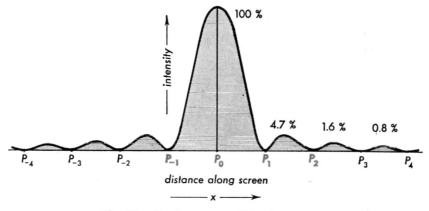

Fig. 6P   *Single-aperture diffraction pattern.*

the central intensity 100%, the maximum intensity of the side bands reaches the relatively low values of 4.7%, 1.6%, 0.83%, etc. Note carefully that the dark points P_0, P_1, P_2, etc., are equally spaced but that the maxima do not come exactly halfway between.

To find the distance from the center point P_0 to the first dark band, we place $n = 1$ in Eq. (6f), and transpose D to obtain

$$x_1 = \frac{\lambda D}{d} \tag{6i}$$

The value of x_1 gives directly the width of all side bands, as well as the half-width of the central bright band.

Example 1. A parallel beam of monochromatic light falls on a single slit 1 mm wide. When the diffraction pattern is observed on a screen 2 m away, the central band is found to have a width of 2.5 mm. Find the wavelength of the light.

Solution. The given quantities in this problem are $D = 200$ cm, $d = 0.10$ cm, $n = 1$, and $2x = 0.25$ cm. Upon substitution in Eq. (6f), we obtain

$$\frac{1\lambda}{0.10} = \frac{0.125}{200}$$

from which

$$\lambda = \frac{0.10 \times 0.125}{200} = 6.25 \times 10^{-5} \text{ cm}$$

The light, $\lambda = 6.25 \times 10^{-5}$ cm, is red in color.

6.6. Diffraction by a Circular Aperture

The diffraction pattern formed by light passing through a circular aperture is of considerable importance, as it applies to the resolving power of telescopes and other optical instruments. The resolving power refers to the ability of an instrument to reveal fine detail in the object being viewed.

Owing to the diffraction of light waves as explained in the preceding section, light through a circular aperture produces a diffraction pattern having the same general intensity variations as given by Fig. 6P. Being circular, however, the parallel bands of light from a slit aperture are replaced by concentric circles with a bright disk at the center. It is as if the graph of Fig. 6P were rotated around the center line.

A lens acts as a circular aperture for light passing through it, and the image it forms for every bright spot in any object is a tiny diffraction pattern. The photograph in Fig. 6Q shows that, with pinholes in a screen as objects, the

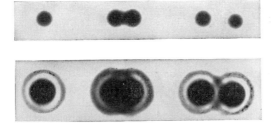

Fig. 6Q *Photographs of diffraction images of one point source taken with a circular aperture, two points close together, and two point sources farther apart. Top: Short exposure.* Bottom: *Longer exposure.*

images formed by a single lens are composed of tiny disks surrounded by faint concentric rings of light. The larger the lens aperture, the smaller the diffraction patterns. The distant stars act as point objects, and their images formed by a telescope objective are diffraction patterns of this kind.

We can see from this why the magnification by a telescope or microscope is limited, and cannot exceed certain values. If the eyepiece of an instrument has too high a magnifying power, each point in the object is observed as a disk, and the image appears blurred. The mathematical treatment of diffraction by a circular aperture requires Bessel functions and leads to the same formula as

Eq. (6i) but with the factor 1.22 introduced in place of $n = 1$ in Eq. (6f):

$$x = 1.22 \frac{\lambda D}{d} \qquad (6j)$$

where d is the diameter of the circular aperture and x is the radius of the disk of the diffraction pattern. Transposing D to the other side of the equation, and noting that D is usually hundreds of times greater than x, the quantity x/D can be replaced by θ_1, and we obtain

$$\theta_1 = \frac{1.22\lambda}{d} \qquad (6k)$$

where the angle θ_1 is in radians.

Sound waves from the circular aperture of a radio loud-speaker will form diffraction patterns of the same kind. Such behavior gives rise to marked changes in sound quality at different points around a room. The microwaves from a radar reflector radiate outward as a single-aperture diffraction pattern, with a central maximum radiated straight forward.

The *lobe* patterns shown for three different wavelengths in Fig. 6R, are polar graphs of the intensity contour shown in Fig. 6P. In such polar graphs the

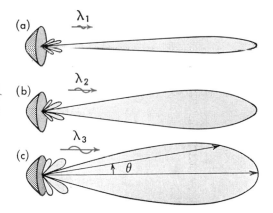

Fig. 6R *Polar diagrams of the diffraction patterns for waves of different wavelength from the same parabolic reflector.*

intensity in any direction making an angle θ with the central line is plotted from the center out. The length of any arrow drawn at any angle θ is therefore proportional to the intensity radiated in that direction.

The shorter the wavelength and the greater the diameter of the circular aperture emitting the waves, the narrower the lobe pattern. Light waves from a point source at the focus of a parabolic mirror will produce a very narrow beam as shown in diagram (a), while radar and microwaves, with their much longer wavelengths, will produce much wider beams as shown in diagrams (b) and (c).

6.7. The Diffraction Grating

The diffraction grating is an optical device widely used in place of a prism for studying the spectrum and measuring the wavelengths of light. Gratings are made by ruling fine grooves with a diamond point either on a glass plate to produce a transmission grating or on a polished metal mirror to produce a reflecting grating. As illustrated in Fig. 6S the rulings are all parallel and equally

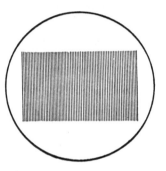

Fig. 6S *Schematic diagram of the grooves or rulings on a diffraction grating.*

spaced. The best gratings are several inches in width and contain from 5000 to 30,000 lines/in.

The transmission grating and its effect on light is idealized by the cross-section diagrams in Fig. 6T. The heavy black lines represent the lines which permit no

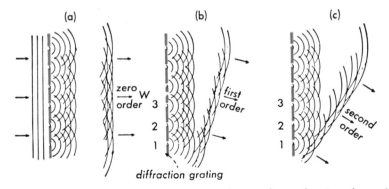

Fig. 6T *Diagrams showing the formation of wave fronts forming the various orders of interference observed with a diffraction grating.*

light to get through and the open intervals between them represent the un-disturbed parts of the glass which transmit the light and act like the parallel slits in Young's double-slit experiment. In diagram (a) parallel light is shown arriving at the grating surface as a succession of plane waves. The light then passing through the openings spreads out as Huygens' wavelets and forms new wave fronts parallel to the grating face. These wave fronts, parallel to the original waves, constitute a beam of light *W* traveling on in the same direction as the original beam.

These are not the only wave fronts, however, for other beams of parallel light are to be found traveling away from the grating in other directions. Two other such wave fronts are illustrated in diagrams (b) and (c). In (b), a dotted line is drawn tangent to the seventh wave from opening 1, the eighth wave from opening 2, the ninth wave from opening 3, etc., to form what is called a wave front of the first order of interference. In (c), a line is drawn tangent to the fourth wave from opening 1, the sixth wave from opening 2, the eighth wave from opening 3, etc., to form what is called a wave front of the second order of interference. Similarly, by taking every third wave or every fourth wave from consecutive slits, other parallel wave fronts corresponding to the third or fourth orders are found moving off at greater angles. By symmetry all of the orders found on one side of the zeroth order are also found at the same angle on the other side.

Experimentally there are two methods of observing the various orders of interference from a small diffraction grating; one is to place the grating directly in front of the eye and the other is to place it in the parallel beam of light between two lenses as shown in Fig. 6U. In the latter case the second lens is shown converging the various wave fronts of the different orders to a focus on a distant screen. If the source is a slit as shown at the left, and a colored glass filter is used to let through light of any one color, say violet, the light falling on the screen will appear as shown in the top photograph in Fig. 6V. Each vertical line is an image of the slit source and is violet in color.

If the three diagrams in Fig. 6T are redrawn for light of a longer wavelength, i.e., a greater distance between waves, the central beam of light W would travel on in the same direction as before, but the various *orders of interference* would be diffracted out at greater angles. Should green light of one wavelength be used in Fig. 6U, the slit images formed on the screen would be farther apart than for violet light, as illustrated by the images marked G in Fig. 6V. This lower photograph was taken with both violet and green light from a mercury

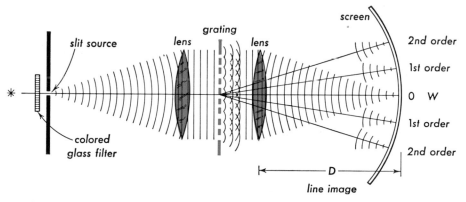

Fig. 6U   *How the wave fronts of the various orders of interference from a diffraction grating are brought to a focus by the same lens.*

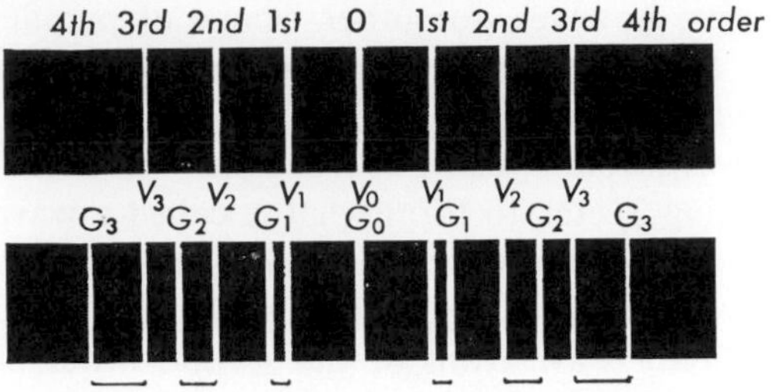

4th 3rd 2nd 1st 0 1st 2nd 3rd 4th order

V_3 V_2 V_1 V_0 V_1 V_2 V_3
G_3 G_2 G_1 G_0 G_1 G_2 G_3

Fig. 6V *Photographs of the different orders of interference of violet and green light obtained with a diffraction grating as shown in Fig. 6U.*

arc passing through the grating. These line images are called spectrum lines.

It will be noted that the separation of the spectrum lines V and G in the *third order* is three times as great as in the *first order*. In other words, any two spectrum lines are separated by an amount that is proportional to the order of interference.

If white light is sent through a grating, all of the different wavelengths, corresponding to the different colors, form their own characteristic wave fronts and produce a complete and continuous spectrum in each order of interference. This is illustrated by a diagram in Fig. 6W. Since the zeroth order for all colors

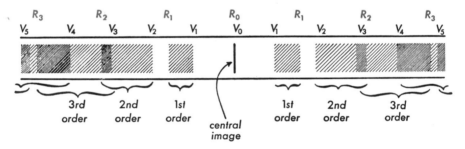

R_3 R_2 R_1 R_0 R_1 R_2 R_3
V_5 V_4 V_3 V_2 V_1 V_0 V_1 V_2 V_3 V_4 V_5

3rd   2nd   1st   1st   2nd   3rd
order   order   order   order   order   order
central
image

Fig. 6W *Diagram of the first several orders of the continuous spectrum as displayed by a diffraction grating.*

comes to the same point, the central image is white. Because the width of each spectrum is proportional to the order, the higher orders overlap one another more and more. The violet of the third order, V_3 for example, falls on the red of second order, R_2. For this reason only the first and second orders of the spectrum from any grating are generally used in practice.

The general appearance of a spectrum, produced by a diffraction-grating spectrograph, can be seen in the photographs reproduced in Chapter 9.

6.8. Mathematical Theory of the Diffraction Grating

The theory of the diffraction grating is similar to that of the double slit and is shown in its simplest form in Fig. 6X. These diagrams derive their construction from Fig. 6T. The wave fronts for the first order emerge at such an angle θ that the difference in path between the rays from any two consecutive openings, like 1 and 2, is just one wavelength. Since any tangent drawn to any circle is always

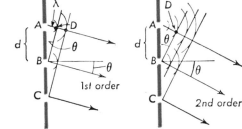

Fig. 6X *Geometry for the wave theory of the diffraction grating.*

perpendicular to the radius drawn through the point of contact, triangle ABD is a right triangle, and $\sin \theta = \lambda/d$. Transposing, we obtain

$$\lambda = d \sin \theta \qquad (6l)$$

where λ is the wavelength of the light, d is the grating spacing, and θ is the angle that the emergent light of the first order makes with the grating normal.

By similar reasoning, and by the use of diagrams like the one shown for the second order, it will be seen that spectra of the second, third, fourth, etc., order are formed at such angles θ that the difference in path between consecutive slits is 2λ, 3λ, 4λ, etc. In general, the side AD of the right triangle ABD must be equal to $n\lambda$, where $n = 1, 2, 3, 4$, etc., and $\sin \theta = n\lambda/d$. If we transpose as before, we obtain the general formula

$$\boxed{n\lambda = d \sin \theta} \qquad (6m)$$

In this general grating formula, n is the spectrum order.

Example 2. Red light of one particular wavelength falls normally on a grating having 4000 lines per cm. If the second-order spectrum makes an angle of 36° with the grating normal, what is the wavelength of the light?

Solution. Since the grating has 4000 lines per cm, the spacing between the lines is 1/4000, or $d = 0.00025$ cm. The other given quantities are $\theta = 36°$ and $n = 2$. Substituting in Eq. (6m), and solving for λ, we get

$$\lambda = \frac{0.00025 \times \sin 36°}{2} = \frac{0.00025 \times 0.588}{2} = 0.0000735 \text{ cm}$$

PROBLEMS

1. Green light of wavelength 5×10^{-5} cm is used in observing the interference fringes produced by a double slit. If the centers of the two slit openings are 0.3 mm apart, and the distance to the observing screen is 1.5 m, what is the fringe spacing?

2. Yellow light of wavelength 5.7×10^{-5} cm falls on a double slit; 2 m away, on a white screen, interference fringes are formed 4 mm apart. Calculate the double-slit separation. *Ans.* 0.285 mm.

3. Monochromatic light falls upon a double slit. The distance between the slit centers is 0.8 mm, and the distance between consecutive fringes on a screen 5 m away is 0.4 cm. What are the wavelength and the color of the light?

4. A beam of parallel light, $\lambda = 6 \times 10^{-5}$ cm, falls normally on a grating, and the third order is diffracted at an angle of 40° with the grating normal. How many lines per cm are on the grating? *Ans.* 3570.

5. Parallel yellow light of wavelength 5.8×10^{-5} cm falls normally on one side of a diffraction grating having 6000 lines per cm. Calculate the angle between the first-order spectrum on opposite sides of the grating normal.

6. A diffraction grating with 10,000 lines per cm is used with two lenses, each of 2-m focal length, as shown in Fig. 6U. Find the width of the first-order spectrum of white light as it is formed on the screen. Assume that $\lambda = 4 \times 10^{-5}$ cm and $\lambda = 7 \times 10^{-5}$ cm for the shortest and longest wavelengths and a curved screen of radius 200 cm. *Ans.* 72.5 cm.

7. A grating having 1000 lines per cm is set up with 1-m focal-length lenses as shown in Fig. 6U. (a) What fourth-order wavelength will be diffracted at the same angle as the third-order of $\lambda = 60 \times 10^{-5}$ cm? (b) What is the angle?

8. Red light of wavelength 6500×10^{-8} cm from a narrow slit falls on a double slit of separation 0.025 cm. If the interference pattern is formed on a screen 100 cm away, what will be the linear separation between fringes on the screen? *Ans.* 0.26 cm.

9. Green light of wavelength 5500×10^{-8} cm is incident on a double slit of separation 0.28 mm. If the interference pattern is formed on a screen 80 cm away, what will be the linear separation between fringes?

10. Yellow light of wavelength 5800×10^{-8} cm is incident on a double slit. If the over-all separation of 10 fringes on a screen 160 cm away is 1.2 cm, find the double-slit separation. *Ans.* 0.0773 cm.

11. Violet light of wavelength 4.3×10^{-5} cm falls as a parallel beam on a diffraction grating containing 8000 lines per cm. At what angle will the second-order spectrum be located?

The Polarization of Light

The experiments described in the preceding chapter illustrating the *diffraction* and *interference* of light are generally regarded as proof that *light is a wave motion*. Although such experiments enable the experimentalist to measure accurately the wavelengths of light, they give no information of the kinds of waves involved. The reason for this is that all types of waves, under the proper conditions, will exhibit diffraction and interference. The desired information in the case of light waves is found in another group of phenomena known as *polarized light*. Some of the phenomena, which will be described in this chapter, are considered to be a proof that *light is a transverse wave motion* in contrast with the longitudinal wave motion in sound.

In the case of longitudinal waves the vibrations are always parallel to the direction of propagation, so that in a plane at right angles to the direction of travel there is no motion and hence there is perfect symmetry. If light is a transverse wave motion, the vibrations of a beam of light are all at right angles to the direction of propagation and there may or may not be perfect symmetry around the direction of travel. If perfect symmetry does not exist for a beam of light the beam is said to be *polarized*.

The experimental methods by which light may be polarized are classified under one of the following heads: (1) *reflection*, (2) *double refraction*, (3) *selective absorption*, and (4) *scattering*.

7.1. Plane-Polarized Light

A better understanding of the experiments to be described in this chapter can best be attained by first presenting the graphical methods of representing transverse waves. We assume at the outset that each light wave is a transverse wave whose vibrations are along straight lines at right angles to the direction of propagation. Furthermore, we assume that a beam of ordinary light consists of millions of such waves, each with its own plane of vibration, and that there are waves vibrating in all planes with equal probability. Looking at such a beam end-on as in Fig. 7A, there should be just as many waves vibrating in one plane as there are vibrating in any other. This then can be referred to as perfect symmetry.

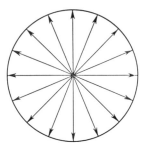

Fig. 7A *End-on view of a beam of unpolarized light illustrating schematically the equal probability of all planes of vibration.*

If, by some means or other, all the waves in a beam of light are made to vibrate in planes parallel to each other, the light is said to be plane-polarized. Diagrams illustrating such light are shown in Fig. 7B. The top diagram (a)

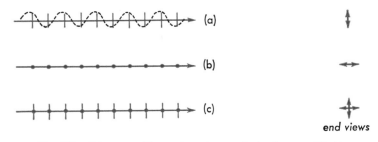

end views

Fig. 7B *Diagrams illustrating plane-polarized rays of light.*

represents plane-polarized light waves traveling to the right and vibrating in a vertical plane, while the second diagram (b) represents a ray of plane-polarized light vibrating in a horizontal plane. The dotted line indicating waves in diagram (a) is usually omitted.

It can be shown that a beam of ordinary unpolarized light, vibrating in all planes, may be regarded as being made up of two kinds of vibrations only, half of the waves vibrating in a vertical plane as in diagram (a) and the other half vibrating perpendicular to it as in diagram (b). The reason for this is that waves not vibrating in either of these two planes can be resolved into two components, one component vibrating in a vertical plane and the other vibrating in a horizontal plane. Although these two components may not be equal to each other, the similarly resolved components from all waves will average out to be equal. Diagram (c) is regarded therefore as being equivalent to ordinary unpolarized light.

7.2. Polarization by Reflection

When ordinary unpolarized light is incident at an angle of about 57° on the polished surface of a plate of glass, the reflected light is plane-polarized. This fact was first discovered by Etienne Malus, a French physicist, in 1808. The experiment usually performed to demonstrate his discovery is illustrated in Fig. 7C.

A beam of unpolarized light *AB* is incident at an angle of 57° on the first glass surface at *B*. This light is again reflected at the same angle by a second glass plate *C* placed parallel to the first, as in diagram (a). If now the lower plate is rotated about the line *BC* by slowly turning the pedestal on which it is mounted, the intensity of the reflected beam *CD* is found to decrease slowly and vanish completely at an angle of 90°. With further rotation the reflected beam *CD* appears again, reaching a maximum at an angle of 180° as shown in diagram (c). Continued rotation causes the intensity to decrease to zero again at 270°, and to reappear and reach a maximum at 360°, the starting point as in diagram (a). During this one complete rotation the angle of incidence on the lower plate, as well as the upper, has remained at 57°.

If the angle of incidence on either the upper or lower plate is not 57°, the beam *CD* will go through maxima and minima every 90° as before, but the minima will not go to zero. In other words, there will always be a reflected beam *CD*.

A complete mathematical theory of the polarization of light by reflection was first given by Fresnel in 1820. The remarkable confirmation of this theory, in

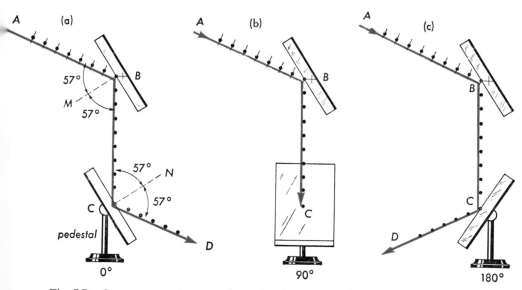

Fig. 7C   *Common experiment performed to demonstrate the polarization of light by reflection from a smooth glass surface.*

every detail, by experimental observations on the behavior of light and measurements, establishes Fresnel as the greatest contributor to the whole field of optics.

The explanation of the above experiment is made clearer by a detailed study of what happens to ordinary light when it is reflected at the polarizing angle of 57° from glass. As illustrated in Fig. 7D, 8% of the light is reflected as plane-

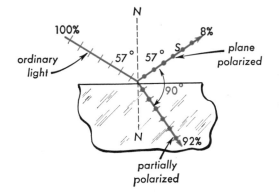

Fig. 7D *Light reflected from glass at an angle of 57° is plane-polarized, while the refracted light is only partially plane-polarized.*

polarized light vibrating in the plane at right angles to the plane of incidence, and the other 92% is refracted as partially plane-polarized light, 42% vibrating perpendicular to the plane of incidence and 50% vibrating parallel to the plane of incidence. The plane of incidence is defined as the plane passing through the incident ray and the ray normal *NN*. In nearly all diagrams the plane of the page is the plane of incidence.

If in Fig. 7D the angle of incidence is changed to some other value than 57°, the reflected beam will not be plane-polarized but will contain a certain amount of light vibrating parallel to the plane of incidence. In general, the light reflected from a transparent medium like glass or water is only partially plane-polarized; only at a certain angle, called the *polarizing angle*, is it plane-polarized. It was Sir David Brewster, a Scottish physicist, who first discovered that *at the polarizing angle the reflected and refracted rays are 90° apart*. This is now known as *Brewster's law*. The polarizing angle for water is 53°, for at this angle the reflected and refracted rays make an angle of 90° with each other.

Because these two rays make 90° with each other, the angle of incidence *i* and the angle of refraction *r* are complements of each other and sin *r* in *Snell's law* (sin *i*/sin *r* = μ) can be replaced by cos *i*, giving

$$\frac{\sin i}{\cos i} = \mu \quad \text{or} \quad \boxed{\tan i = \mu} \tag{7a}$$

This formula is useful in calculating the angle of polarization. For example, with water, $\mu = 1.33$, angle $i = 53°$; whereas for glass with $\mu = 1.52$, angle $i = 57°$.

Returning to the experiment demonstrated in Fig. 7C, we observe that the reflected light from the first mirror is plane-polarized as shown, and that the refracted light goes into the glass plate where it is absorbed by the black paint on the back face. The second mirror acts as a testing device or analyzer for polarized light. A certain fraction of the incident waves is reflected when the vibrations are perpendicular to the plane of incidence, and all are refracted (to be absorbed) when the vibrations are parallel to the plane of incidence.

7.3. Polarization by Double Refraction

The double refraction of light by Iceland spar (calcite) was first observed by a Swedish physician, Erasmus Bartholinus, in 1669, and later studied in detail by Huygens and Newton. Nearly all crystalline substances are now known to exhibit the phenomenon. The following are but a few samples of crystals that show this effect: *calcite, quartz, mica, sugar, topaz, selenite, aragonite,* and *ice.* Calcite and quartz are of particular importance because they are used extensively in the manufacture of special optical instruments.

Calcite, as found in nature, always has the characteristic shape shown in Fig. 7E(a), whereas quartz has many different forms, the most complicated of

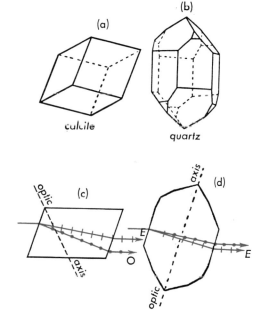

Fig. 7E   *Diagrams and cross sections of calcite and quartz crystals showing double refraction and polarization.*

which is illustrated in diagram (b). Each face of every calcite crystal is a parallelogram whose angles are 78° and 102°. Chemically, calcite is a hydrated calcium carbonate, $CaCO_3$; and quartz is silicon dioxide, SiO_2.

Not only is light doubly refracted by calcite and quartz, but both rays are found to be plane-polarized. One ray, called the *ordinary ray,* is polarized with its vibrations in one plane; and the other ray, called the *extraordinary ray,* is polarized with its vibrations in a plane at right angles to the first. This polarization is illustrated in diagrams (c) and (d) by *dots* and *lines* and can be proved by a glass plate rotated as plate *C* in Fig. 7C, or with some other analyzing device like a *Nicol prism* or a *polarizing film*. These devices will be described in the next two sections.

Since the two opposite faces of a calcite crystal are always parallel to each other, the two refracted rays always emerge parallel to the incident light and are therefore parallel to each other. If the incident light falls perpendicularly upon the surface of the crystal, as in Fig. 7F, the extraordinary ray will be refracted

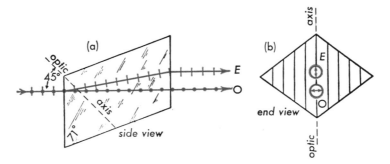

Fig. 7F   *Double refraction in calcite. At normal incidence the O ray travels straight through and the E ray is refracted to one side.*

away from the normal and will come out parallel to, but displaced from, the incident beam. The ordinary ray will pass straight through without deviation.

In general, the *O* ray obeys the ordinary laws of refraction, and in this way the crystal acts like glass or water, whereas the *E* ray obeys no such simple law and behaves quite abnormally.

In other words, the *O* ray travels with the same velocity regardless of its direction through the crystal, whereas the velocity of the *E* ray is different in different regions. This is the origin of the designations ordinary and extraordinary.

One important property of calcite and quartz is that there is one and only one direction through either crystal in which there is no double refraction. This particular direction, called the *optic axis*, is shown by the dashed lines in Fig. 7E. The optic axis, it should be noted, is not a single line through a crystal, but a direction.

A plane passing through the cyrstal parallel to the optic axis and perpendicular to one face of the crystal is called a *principal section*. The plane of the page in Fig. 7F(a) is but one of any number of principal sections which, from the end view in diagram (b), appears as a vertical line. A useful rule always to be remembered is that the vibrations of the *O* ray are always perpendicular to the optic axis.

7.4. The Nicol Prism

The Nicol prism is an optical device made from a calcite crystal and used in many optical instruments for producing and analyzing polarized light. Such a prism, as illustrated in Fig. 7G, is made by cutting a crystal along a diagonal and

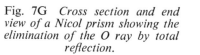
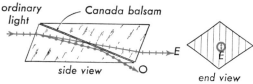

Fig. 7G *Cross section and end view of a Nicol prism showing the elimination of the O ray by total reflection.*

cementing it back together again with a special cement called *Canada balsam.* Canada balsam is used because it is a clear transparent substance whose reflective index is midway between that of the calcite for the *O* and *E rays.*

Optically the Canada balsam is more dense than calcite for the *E* ray and less dense for the *O* ray. There exists, therefore, a critical angle of refraction for the one *O* ray but not for the *E* ray. After both rays are refracted at the first crystal surface, the *O* ray is *totally reflected* by the first Canada balsam surface, as illustrated in the diagram, while the *E* ray passes on through to emerge parallel to the incident light. Starting with ordinary unpolarized light, a Nicol prism thus transmits plane polarized light only.

If two Nicols are lined up one behind the other as in Fig. 7H, they form an

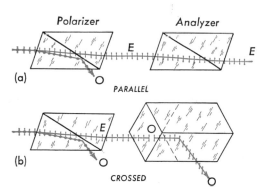

Fig. 7H *Two Nicol prisms mounted as polarizer and analyzer. (a) Parallel Nicols. (b) Crossed Nicols.*

optical system frequently used in specially constructed microscopes for studying the optical properties of other crystals. The first Nicol which is used to produce plane-polarized light is called the *polarizer,* and the second which is used to test the light is called the *analyzer.*

In the parallel position, diagram (a), the polarized light from the polarizer passes on through the analyzer. Upon rotating the analyzer through 90°, as in diagram (b), no light is transmitted. For the same reason that the *O* vibrations in the original beam were totally reflected in the polarizer, the E vibrations are totally reflected as O vibrations in the analyzer.

Rotated another 90°, the light again gets through the analyzer just as in the parallel position in diagram (a). Still another 90° finds the Nicols crossed again, with no light passing through.

7.5. Polarization by Selective Absorption

When ordinary light enters a crystal of tourmaline, double refraction takes place in much the same way that it does in calcite, but with this difference: one ray, the so-called O ray, is entirely absorbed by the crystal, while the other ray, the E ray, passes on through. This phenomenon is called "selective absorption" because the crystal absorbs light waves vibrating in one plane and not those vibrating in the other.

Tourmaline crystals are therefore like Nicol prisms, for they take in ordinary light, dispose of the O vibrations, and transmit plane-polarized light as illustrated in Fig. 7I(a). When two such crystals are lined up parallel, with one behind

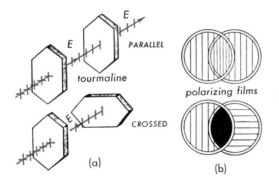

Fig. 7I *Diagrams illustrating the polarization of light by (a) tourmaline crystals, and (b) polarizing films.*

the other, the plane-polarized light from the first crystal passes through the second with little loss in intensity. If either crystal is turned at 90° to the other, i.e., in the *cross position*, the light is completely absorbed and none passes through.

The behavior of tourmaline and similar optical substances is due to the molecular structure of the crystal. To draw an analogy, the regularly spaced molecules of a single crystal are like the regularly spaced trees in an orchard or grove. If one tries to run between the rows of trees carrying a very long pole held at right angles to the direction of motion, the pole must be held in a vertical position. If it is held in the horizontal plane, the runner will be stopped.

The reason tourmaline is not used in optical instruments in place of Nicol prisms is that the crystals are yellow in color and do not transmit white light.

A more satisfactory substance for this purpose, which does transmit white light, is a relatively new manufactured material known as "Polaroid." This material is made in the form of very thin films, which have the general appearance of the more common substance "Cellophane," and is made from small needle-shaped crystals of an organic compound *iodosulphate of quinine*. Lined up parallel to each other and embedded in a *nitrocellulose mastic*, these crystals act like tourmaline by absorbing one component of polarization and transmitting the other. Two such films mounted separately in rings between thin glass plates are shown schematically in Fig. 7I(b). In the crossed position no light can pass through both films, whereas in the parallel position white light vibrating in the plane indicated by the parallel lines is transmitted. Many practical applications are being found for polarizing films of this kind, particularly wherever glaring light is not desired. The glaring light reflected at an angle from a table top, a book, a window pane, the water, or the road ahead when one is driving a car, is polarized and can be partly eliminated by polarizing films.

7.6. Dispersion by a Calcite Prism

In Fig. 7J a prism is shown cut from a calcite crystal with the optic axis parallel to the refracting edge A. The optic axis, being perpendicular to the page,

Fig. 7J *Refraction of white light by a prism cut from a calcite crystal.*

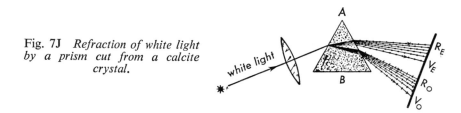

is represented by dots. (For the direction of the optic axis in calcite, see Fig. 7E.)

When white light is incident on one side of this prism, two completely separated spectra emerge from the other side. Not only is each spectrum complete in all its colors from red to violet but the light in each is plane-polarized. This can be demonstrated with an analyzing device like a Nicol prism, or polarizing film. By inserting the analyzer anywhere in the light beam and rotating it, one spectrum disappears first; then, 90° from it, the other fades and disappears while the first returns to full intensity.

The vibrations of all colors in the lower spectrum in Fig. 7J are perpendicular to the optic axis and are O vibrations. The upper spectrum with all vibrations parallel to the optic axis consists of E vibrations. If a prism is cut so that the refracted light as it travels through the crystal is parallel to the optic axis, only one spectrum is produced.

7.7. Scattering and the Blue Sky

The blue of the sky and the red of the sunset are due to a phenomenon called "scattering." When sunlight passes through the earth's atmosphere, much of the light is picked up by the air molecules and given out again in some other direction. The effect is quite similar to the action of water waves on floating objects. If, for example, the ripples from a stone dropped in a still pond of water encounter a small cork floating on the surface, the cork is set bobbing up and down with the frequency of the passing waves.

Light is pictured as acting in the same way on air molecules and fine dust particles. Once set into vibration by a light wave, a molecule or particle can send out the absorbed light again, sometimes in the same direction but generally in almost any other direction. This is illustrated schematically in Fig. 7K. Waves of light are shown being scattered at random in all directions.

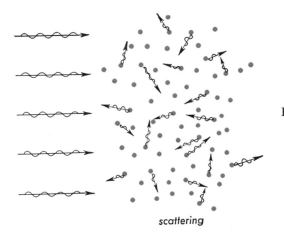

Fig. 7K *Light waves are scattered by air molecules.*

scattering

Experiments show, in agreement with the theory of scattering, that the shortest waves are scattered more readily than longer waves. To be more specific, the scattering is inversely proportional to the fourth power of the wavelength:

$$\text{scattering} \propto \frac{1}{\lambda^4} \tag{7b}$$

According to this law the short waves of violet light are scattered ten times as readily as the longer waves of red light. The other colors are scattered by intermediate amounts. Thus when sunlight enters the earth's atmosphere, *violet* and *blue light* are scattered the most, followed by *green, yellow, orange,* and *red,* in the order named. For every ten violet waves ($\lambda = 0.00004$ cm) scattered from a beam, there is only one red wave ($\lambda = 0.00007$ cm).

violet	blue	green	yellow	orange	red
10	7	5	3	2	1

At noon on a clear day when the sun is directly overhead, as illustrated by an observer at *A* in Fig. 7L, the whole sky appears *light blue.* This is the composite color of the mixture of colors scattered most effectively by the air molecules. It can be demonstrated that light blue is obtained by the added mixture of *violet, blue, green,* and *yellow.*

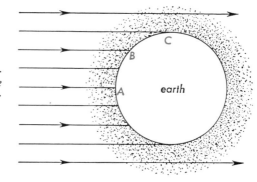

Fig. 7L *Schematic diagram show-ing the scattering of light by the air molecules of the earth's atmos-phere.*

7.8. The Red Sunset

The occasional observation of an orange-red sunset is attributed to the *scattering of light* by fine dust and smoke particles near the earth's surface. This is illustrated in Fig. 7M. To an observer at *A*, it is noonday and the direct

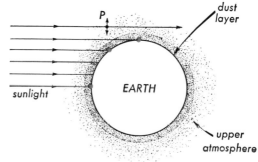

Fig. 7M *The scattering of light by a layer of dust near the earth's surface causes the sun to turn yellow, then orange, and finally red at sunset.*

sunlight from overhead, seen only by looking directly at the sun itself, travels through a relatively short dust path. As a result, very little violet and blue are scattered away and the sun appears white.

As sunset approaches, however, the direct sunlight has to travel through an ever-increasing dust path. The result is that an hour or so before sundown, when the observer is at *B*, practically all of the blue and violet have been scattered out and, owing to the remaining colors, red, orange, yellow, and a little green, the sun appears yellow. At sunset, when the observer is at *C*, the direct rays must travel through so many miles of dust particles that all but red are completely scattered out and the sun appears red. At this same time the sky overhead is still light blue. If the dust blanket is too dense, even the red will be scattered appreciably from the direct sunlight and the deepening red sun will become lost from view before it reaches the horizon.

An excellent demonstration of scattering by fine particles is illustrated in Fig. 7N. A parallel beam of white light from a carbon arc and lens L_1 is sent

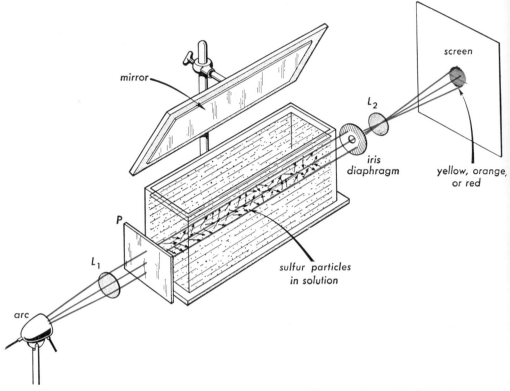

Fig. 7N   *The sunset experiment. Demonstration of the scattering and polarization of light by small particles.*

through a water trough with glass sides. After passing through an iris diaphragm at the other end, a second lens L_2 forms an image of the circular opening on the screen. To produce the fine particles for scattering, about 40 gm of photographic fixing powder (hyposulfite of soda) are first dissolved in about 2 gal of water. Next, about 1 to 2 cm³ of concentrated sulfuric acid is added and the contents thoroughly mixed in the trough.*

As the microscopic sulfur particles begin to form, scattered blue light will outline the parallel beam through the trough. A little later, when more particles have formed, the entire body of water will appear light blue, due principally to multiple scattering. Light scattered out of the central beam of light is scattered again and again before emerging from the trough. At first the transmitted light which falls on the screen appears white. Later, as more scattering takes out the shorter wavelengths, this image representing the sun turns yellow, then orange, and finally red.

---

* The correct amount of acid to produce the best results is determined by trial. The first visible precipitate should appear after 2 or 3 min.

7.9. Polarization by Scattering

If the blue of the sky is observed through a Nicol prism or a piece of polaroid, the light is found to be partially plane-polarized. This polarization can also be seen in the scattering experiment described above. Observed through a polaroid film, the beam in the tank appears bright at one orientation of the polaroid and disappears with a 90° rotation.*

PROBLEMS

1. Find the polarizing angle for violet light reflected from diamond, with its refractive index of 2.458.

2. If the polarizing angle of a transparent plastic is found to be 55°, what is the refractive index? *Ans.* 1.428.

3. The polarizing angle for yellow light incident on a clear transparent solid is 56°. Calculate the refractive index.

4. Find the polarizing angle for a clear plastic with a refractive index of 1.418 for green light. *Ans.* 54.8°.

5. What is the polarizing angle for a dense flint glass having a refractive index of 1.662?

6. What is the polarizing angle for water, with its refractive index of 1.33? *Ans.* 53.1°.

7. Calculate the polarizing angle for blue light incident on crown glass. Assume a refractive index of 1.531.

8. The polarizing angle for red light incident on a clear transparent material is 67° 28'. Calculate the refractive index. *Ans.* $\mu = 2.410$.

9. What is the refractive index of flint glass if the polarizing angle is 60°?

10. If the refractive index for a plastic is 1.45, what is the angle of refraction for a ray of light incident at the polarizing angle? *Ans.* 34.6°.

11. The polarizing angle of a piece of glass, for green light, is 61°. Find the angle of minimum deviation† for a 50° prism made of the same glass.

12. If the polarizing angle of a piece of clear plastic is 60°, find the angle of minimum deviation† for a 50° prism made of the same plastic. *Ans.* 44.2°.

13. If the critical angle of a clear crystal for green light is 26°, calculate the polarizing angle. For the critical angle r, $\mu = 1/\sin r$.

14. The critical angle for violet light incident on a piece of glass is 36.1°. (a) Find the refractive index and (b) calculate its corresponding polarizing angle. For the critical angle r, $\mu = 1/\sin r$. *Ans.* (a) 1.70, (b) 59.5°.

15. Find the polarizing angle for the boundary separating water of index 1.33 from glass of index 1.60. Assume the incident ray to be in water.

16. Find the amount of light, relative to yellow light, scattered by each of the following wavelengths of light: ultraviolet light 2.0×10^{-5} cm, violet light 4.0×10^{-5} cm, yellow light 5.8×10^{-5} cm, red light 7.0×10^{-5} cm, and infrared light 10×10^{-5} cm. *Ans.* 70, 4.4, 1.0, 0.47, and 0.11 times, respectively.

17. Zinc sulfide deposited on a glass surface is a clear transparent material having a refractive index of 2.50. Calculate the polarizing angle for this medium.

* For a more complete account of polarized light, see *Optics* by Jenkins and White, McGraw-Hill.

† For minimum deviation $\mu = \sin \frac{1}{2}(A + \delta)/\sin \frac{1}{2}A$, where A is the prism angle, and δ is the angle of minimum deviation.

18. A 60° calcite prism is cut with its faces parallel to the optic axis. Calculate the angle of minimum deviation for yellow light for each of the two polarized rays. The refractive index for calcite for the O ray is 1.658, and for the E ray, 1.486. *Ans.* 52° and 36°.

19. What is the ratio of the scattering of light waves between red light of wavelength 7×10^{-5} cm and blue light of wavelength 4.5×10^{-5} cm?

20. For every 200 waves of red light scattered by the air, how many waves of orange light will be scattered? Assume the wavelengths to be 7×10^{-5} cm and 6×10^{-5} cm, respectively. *Ans.* 371.

21. Calculate the ratio of the numbers of light rays scattered by the air for violet light ($\lambda = 4 \times 10^{-5}$ cm) and yellow light ($\lambda = 5.8 \times 10^{-5}$ cm).

22. Find the ratio of light waves scattered by fine particles between blue light ($\lambda = 4.3 \times 10^{-5}$ cm) and red light ($\lambda = 7.0 \times 10^{-5}$ cm). *Ans.* 7.0:1.

8

Light Sources and Spectrographs

When a block of metal like iron or copper is heated slowly to incandescence, the first noticeable change in its appearance occurs at a temperature of about 1000°K. At this temperature, the metal appears with a dull red glow. As the temperature continues to rise, the color changes slowly to orange, then to yellow, and finally to white.

If the metal, as it is slowly being heated, is observed through a prism, the first appearance of visible light will be found at the extreme red end of the spectrum. As the temperature rises, the light spreads slowly out across the spectrum until, at white heat, the entire band of visible colors from red to violet is seen. At the orange stage where the temperature is about 1500°K, the pure spectrum colors contain red, orange, and yellow; when the yellow stage is reached where the temperature is about 2000°K, the spectral green is included. When the white stage is reached at about 3000°K, and the spectrum is complete, a further rise in temperature continues to increase the intensity of each color without a noticeable change in color.

What the prism has done in such an experiment is to separate all of the light waves according to their wavelengths, the longest waves of red light at the one side, the shortest waves of violet light at the other, and the intermediate waves at their proper places in between. The fact that the color is continuous from red through violet is characteristic of the spectrum of all solids and liquids; this means that there is a continuous set of different wavelengths present.

8.1. The Spectrum

To demonstrate the existence of an ultraviolet and infrared spectrum, an experiment of the type illustrated in Fig. 8A may be performed. The visible light from a carbon arc lamp is made to pass through a quartz lens and prism to be focused on a nearby screen.

If at the violet end of the spectrum the screen is painted with luminous paint,

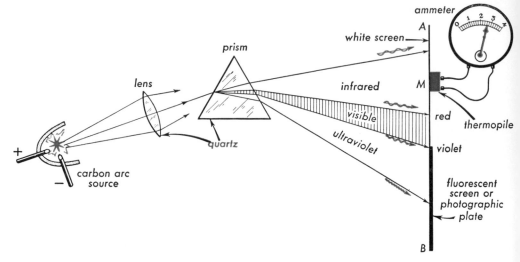

Fig. 8A　*Experiment demonstrating the existence of the ultraviolet and infrared rays beyond the visible spectrum.*

a bright fluorescence will be observed for a short distance beyond the visible violet. When the screen is replaced by a photographic plate, the exposed and developed picture will again show the extension of the spectrum into the ultraviolet.

To detect the presence of the infrared radiations, a thermopile is conveniently used, as shown at the top of the screen. Connected to an ammeter, a thermopile measures the amount of light energy falling upon its front face. If the thermopile is first placed to receive violet light, and then slowly moved across the visible spectrum out into the infrared region beyond, the ammeter will show a steady rise in current. The current will continue to rise until a maximum is reached at a point in the region of M, and then it will drop off slowly as the thermopile approaches the end of the screen at A. A graph of the energy from the carbon arc source, for the different parts along the screen, is shown by the 3000°K curve in Fig. 8B.

Each curve represents the amount of energy given out over the entire spectrum by a solid at different temperatures. Studying these curves, one will observe that at low temperatures very little light is emitted in the visible spectrum. At 1000°K, only the visible red is seen and even that is very faint. At 2000°K, not only does the brightness of the red increase, but the other colors, orange, yellow, and green, appear. At 3000°K, the temperature of a low-current carbon arc or tungsten filament light, all of the visible spectrum is emitted, but the maximum radiation is in the infrared. At 6000°K, the temperature of the surface of the sun, the maximum energy is radiated in the green of the visible spectrum, with an appreciable amount of ultraviolet light on the one side and the infrared on

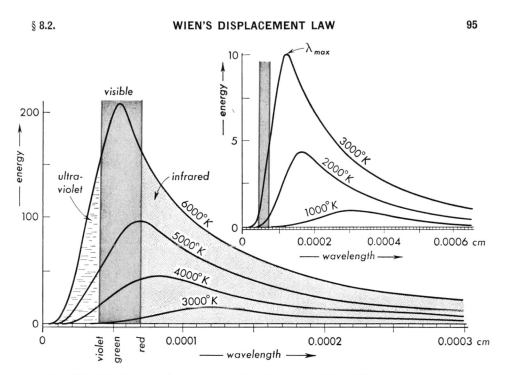

Fig. 8B　*Distribution of the energy emitted by a hot solid at different temperatures.*

the other. Thus the visible spectrum, as seen by the human eye, is but a small band out of all the waves emitted by a body as hot as the sun.

8.2. Wien's Displacement Law

It is an interesting fact that the maximum energy radiated by a hot body shifts to shorter and shorter waves as the temperature rises. See Fig. 8B. To be more exact, if the temperature of a body is doubled, the radiated energy maximum, λ_{max}, shifts to $\frac{1}{2}$ the wavelength. If the temperature is tripled, the energy maximum shifts to $\frac{1}{3}$ the wavelength, etc. This is known as Wien's* displacement law, and is written as an algebraic equation

$$\lambda_{max} T = C \qquad (8a)$$

where C is a constant, found by experiment to have a value of 2.897×10^{-3} m degrees, T is the absolute temperature, and λ_{max} the wavelength in m at which the maximum energy is radiated. By substituting the value of the constant C

---

* Wilhelm Wien (1864–1928), German physicist, chiefly known for important discoveries with cathode rays, canal rays, and the radiation of light. He was awarded the Nobel Prize in physics in 1911 for his discovery of the displacement law of heat radiation named in his honor.

in Eq. (8a), the wavelength maximum radiated by a hot body can be calculated for any temperature.

8.3. Emission and Absorption

The rate at which a body radiates or absorbs heat depends not only upon the absolute temperature, but upon the nature of the exposed surfaces as well. Objects that are good emitters of heat are also good absorbers of the same kind of radiation. This is known as *Kirchhoff's law of radiation*. A body whose surface is blackened is an excellent emitter as well as an excellent absorber. If the same body is chromium-plated it becomes a poor emitter and a poor absorber.

If the outside surface of a hot coffee cup were painted a dull black, the rate of cooling would be more rapid than if it were chromium-plated. A highly polished surface, as in the Dewar-flask, would help by reflection to keep radiant heat from crossing the boundary.

Black clothes should not be worn on a hot day since black is a good absorber of the sun's radiant heat. While black is also a good emitter, the external temperature is higher than the body temperature and the exchange rate is therefore such as to heat the body. White clothes are worn in hot climates because white is a good reflector and therefore a poor absorber.

8.4. Black Body Radiation

The relation between the total radiant heat energy E emitted by a body and its temperature was first made through the extensive laboratory experiments of Josef Stefan. The same law was later derived from theoretical considerations by Ludwig Boltzmann, and is now known as the *Stefan-Boltzmann law:*[*]

$$\boxed{E = kT^4} \tag{8b}$$

Here E represents the energy radiated per second by a body at an absolute temperature T, and k is a proportionality constant. The law applies only to so-called "black bodies."

A black body is defined as one that absorbs all of the radiant heat that falls upon it.

Such a perfect absorber would also be a perfect emitter.

If E represents the heat in calories radiated per second per square centimeter of a black body, then $k = 1.36 \times 10^{-12}$. If E is measured in ergs/cm² sec, then $k = 5.7 \times 10^{-5}$. If E is measured in joul/m² sec, then $k = 5 \times 10^{-8}$.

The best laboratory approach to a black body is a hole in a blackened box. Practically all heat entering such a hole would be absorbed inside. Black velvet

---

[*] Ludwig Boltzmann (1844–1906), Austrian theoretical physicst, was educated at Linz and Vienna. At 23 he was appointed assistant at the physical institute in Vienna. Later he became professor at Graz, then at Munich, and finally back at Vienna. His first publication was on the second law of thermodynamics, and was followed by numerous papers on molecular motion, on viscosity and diffusion of gases, on Maxwell's electromagnetic theory, on Hertz's electrical experiments, and on Stefan's law for black body radiation.

cloth or a surface painted dull with lampblack will absorb about 97% of the radiant heat falling on it, and may for many purposes be considered a black body. Polished metal surfaces, however, are far from black bodies; they absorb only about 6% of the incident energy and reflect the remainder. Most other substances have absorption ratios between these two extremes.

8.5. The Quantum Theory

The first successful attempt to explain the shape of the black body radiation curves as shown in Fig. 7B was made by Max Planck* in 1900. Planck's theoretical considerations led him to the conclusion that matter is composed of a large number of oscillating particles and that all conceivable vibration frequencies of these particles are possible. Although the frequency of any one particle according to classical theory could have any value, he assumed that the vibration energy must be given by

$$E = nh\nu \tag{8c}$$

where ν is the frequency of vibration, h is a constant, and n is a whole number,

$$n = 1, 2, 3, 4, 5, \ldots \tag{8d}$$

When an oscillator emits radiant energy, it does so in the form of electromagnetic waves and only in "chunks" given by Eq. (8c).

The whole number n is called a *quantum number*, h is called *Planck's constant*, and $h\nu$ is called a *quantum* of energy. All quanta of the same frequency have the same energy. Quanta of high frequency, such as gamma rays, have a large amount of energy, while those of low frequency, such as radio waves, have a small amount of energy.

From this theory Planck derived a radiant energy formula for which the rate of emission of a black body is a maximum at the value given by Wien's displacement law, Eq. (8a), and falls off at higher and lower wavelengths as shown in Fig. 8B. Planck's formula for the radiated energy E is

$$E = \frac{8\pi hc}{\lambda^5(e^{hc/\lambda kT} - 1)} \tag{8e}$$

where c is the velocity of light, λ is the wavelength, T is the absolute temperature, e is the base of the Naperian logarithms, k is the so-called *Boltzmann constant* determined from the general gas laws, and h is Planck's constant:

$$h = 6.6238 \times 10^{-34} \text{ joul sec}$$
$$k = 1.3805 \times 10^{-23} \text{ joul/°Abs.}$$
$$c = 3 \times 10^8 \text{ m/sec}$$
$$e = 2.7183$$

* Max Planck (1858–1947), German theoretical physicist, was born in Kiel, on April 23, 1858. He studied in Munich and Berlin and devoted much of his life to theoretical physics, particularly to thermodynamics. He published several books on this subject and is called the father of the quantum theory of radiation. For his contributions to the theory of black body radiation he was awarded the Nobel Prize in physics for the year 1918.

8.6. The Complete Spectrum

Visible, *ultraviolet*, and *infrared* light waves do not represent all of the known kinds of electromagnetic radiation. A complete chart of the known spectrum is shown in Fig. 8C. Beyond the visible and infrared toward longer wavelengths, we find the *heat waves* and the *wireless waves*, while beyond the ultraviolet toward shorter wavelengths we find the *X rays* and the *gamma rays*.

In spite of the tremendous expanse of wavelengths ranging all the way from the longest wireless waves several miles in length to γ-ray waves one-million-millionth of a centimeter in length, all electromagnetic waves travel with the same velocity in vacuum: 186,300 mi/sec, or 3×10^8 m/sec.

Although their velocities in a vacuum are all the same, the properties of the various waves differ considerably. One striking illustration of these differences is found in the response of the human eye. Of the entire spectrum, only one very narrow band of waves can be seen, all the rest being invisible. Another illustration is the passage of light waves through the atmosphere. With the exception of the band of waves known as the extreme ultraviolet, the air is fairly transparent to all electromagnetic waves. To waves of the extreme ultraviolet, the air is quite opaque. *Fog is opaque to all but the wireless waves.*

8.7. The Angstrom as a Unit of Length

Because the wavelengths of light are so very short, the physicist has adopted a smaller unit of length than the meter, centimeter, or millimeter. This unit is called the *angstrom* (abbr. A or Å), after the Swedish scientist by that name. In 1868 Ångstrom published a map of the visible spectrum of the sun, and on this map he labeled the wavelengths in ten-millionths of a millimeter. Since that time, light waves have been specified in these units.

In one meter there are 10,000,000,000 angstroms:

$$1 \text{ m} = 10^{10} \text{ A} \qquad (8f)$$

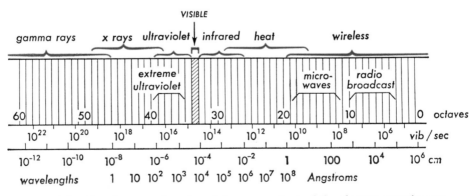

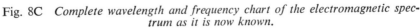

Fig. 8C   *Complete wavelength and frequency chart of the electromagnetic spectrum as it is now known.*

It is common practice among physicists to designate the wavelength of light waves by the Greek letter λ (lambda) and the frequency by the Greek letter ν (nu).

For all electromagnetic waves, visible or invisible, the velocity c in a vacuum is 3 × 10⁸ m/sec, and the following wave equation holds true:

$$c = \nu\lambda \qquad (8g)$$

where, for the mksa system, ν is the frequency in vib/sec and λ is the wavelength in meters. From this we see that the longer the wavelength, the lower the frequency; and the shorter the wavelength, the higher the frequency.

Here is a list of wavelengths of the approximate center of each color band.

Violet λ = 4200 A   Yellow λ = 5800 A
Blue   λ = 4600 A   Orange λ = 6100 A
Green λ = 5400 A   Red   λ = 6600 A

In other units green light, for example, has a wavelength of 5.4 × 10⁻⁵ cm, 5.4 × 10⁻⁷ m, or 0.54 μ.

8.8. Prism Spectrographs

One of the most common forms of prism spectrograph is shown in Fig. 8D. This *Littrow* type instrument, as it is called, employs a long-focus lens, a slit,

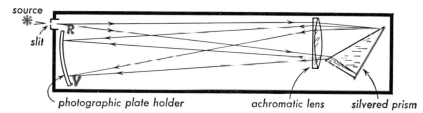

Fig. 8D *Littrow type of prism spectrograph showing dual use of a lens and prism.*

a 30° glass prism silvered on the back surface, and a curved photographic plateholder.

In this instrument a single lens serves a dual function; it provides a parallel beam of light incident upon the prism face, and it brings the emergent light of different wavelengths to a focus at different points on the photographic plateholder. Since the light crosses the glass-air boundary twice, the dispersion of the colors is the same as for an unsilvered 60° prism.

When a glass prism spectrograph is used to study a source of light, only a small part of the ultraviolet spectrum can be photographed. The reason for this is that glass absorbs ultraviolet light at wavelengths less than 3600 A. To photograph the shorter waves lenses and prisms made of quartz are frequently used. Quartz is transparent to ultraviolet light as far down in wavelength as 2000 A. Beyond this point the waves are absorbed by air.

Glass prism spectrographs are seldom used for the infrared spectrum since the dispersion for these wavelengths is so small.

8.9. Diffraction Grating Spectrographs

The diffraction grating is an optical device widely used in place of a prism for studying the spectrum and measuring the wavelengths of light.

The basic principles of how a grating spreads light out into a spectrum by diffraction have been explained in § 6.7. The principal advantage of the diffraction grating over the prism spectrograph is that it is capable of producing high dispersion for all regions of the spectrum and the light is distributed over a uniform wavelength scale. By comparison a prism instrument produces high dispersion at the violet end of the spectrum and crowds the longer red wavelengths into a shorter space.

Transmission gratings were first made by Fraunhofer, a German physicist, in 1819, and the first reflection gratings were made by H. A. Rowland,* an American physicist, in 1882. Although Rowland's first gratings were ruled on flat surfaces, his best ones were ruled upon the polished surfaces of concave mirrors.

One of the most useful of all spectrographs to be found in the research laboratory today is one whose design was originally devised by F. Paschen. A concave reflection grating with a radius of curvature of about 21 ft is mounted in one corner of a dark room. The slit and light source are located in another corner and a long plateholder directly opposite. See Fig. 8E.

Light from the source to be studied passes through the narrow slit and then

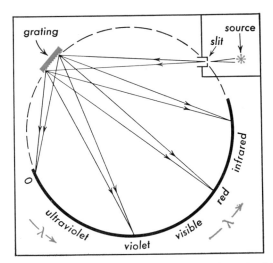

Fig. 8E *Diagram of a spectro-graph employing a concave diffraction grating. The Paschen mounting.*

---

* Henry A. Rowland (1848–1901), American physicist, is noted principally for his ruling of the first high-quality diffraction gratings and his publication of a large and detailed photograph of the sun's spectrum. He was the recipient of many honors, including the Rumford Medal and the Draper Medal.

falls on the grating to be diffracted. Note that the grating performs the double function of dispersing the light into a spectrum and of focusing it as well.

The slit, grating, and photographic plateholder are all located with high precision on the periphery of a circle whose diameter is equal to the radius of curvature of the grating. Such an arrangement brings the different wavelengths to focus all along the plateholder. This circle is called the *Rowland circle* in honor of H. A. Rowland.

When a photograph of any part of the spectrum is desired, a strip of plate or film is placed in the proper place in the plateholder and exposed to the spectrum. The original photographs in Figs. 13H and 13L were made with a 21-ft Paschen spectrograph.

Another useful type of mounting for a concave reflection grating is shown in Fig. 8F. Different wavelength regions of the spectrum are brought to focus on

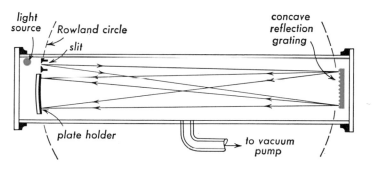

Fig. 8F   *Diagram of a spectrograph employing a concave reflection grating.*

the photographic plate by turning and moving the grating and plateholder by fine adjusting screws (not shown). Adjustments are made so that the slit, grating, and plateholder lie on the Rowland circle.

Since the light at no point, from source to photographic plate, traverses glass elements, such a spectrograph can be mounted in a suitable housing and highly evacuated for the study of the ultraviolet and extreme ultraviolet spectrum of any source.

Oxygen and nitrogen gases absorb broad wavelength regions of the ultraviolet, from 1900 A to approximately 50 A. Spectrum lines throughout this entire region are therefore to be photographed only by highly evacuated instruments of this kind. While ordinary clear dry air at and near sea level is transparent to ultraviolet light down to 1900 A, the ozone high in the stratosphere absorbs strongly in the region 1900 A to 3000 A, and sunlight in this band does not reach the earth's surface.

PROBLEMS

1. At what wavelength will the maximum energy be radiated by a solid piece of metal heated to a temperature of 2600°C?

2. Find the wavelength at which the maximum energy is radiated from a block of black carbon at a temperature of 227°C. What kind of light is it? *Ans.* 5.79 × 10⁴ A. Infrared.

3. The wavelength range of the light of the visible spectrum extends from 4000 A in the violet to 7500 A in the red. What are these same wavelength limits in (a) centimeters (b) meters, and (c) microns?

4. X rays from a certain X-ray source have a wavelength of 0.18 A. What is the wavelength in meters? *Ans.* 1.8 × 10⁻¹¹ m.

5. Find the frequency of blue light of wavelength 4600 A.

6. What is the frequency of X rays if the wavelength is 0.40 A? *Ans.* 7.5 × 10¹⁸ vib/sec.

7. Calculate the wavelength maximum for light radiated by a body at 10 million degrees absolute. To what kind of light is this equivalent?

8. A small copper ball 4 cm in diameter is coated with lampblack and heated to a temperature of 827°C. How much heat is radiated from this sphere in 1 min? *Ans.* 119 cal.

9. The rear wall of a brick fireplace has an effective open area of 5000 cm². Find the number of calories radiated per minute if this area is maintained at 127°C. Assume the surface to be a black body.

10. A silver ball 6 cm in diameter is coated with lampblack and its temperature maintained at 227°C. (a) How much heat will this body radiate each second of time? (b) If the walls of the room are maintained at 27°C, how much heat will this ball absorb each second of time? Assume black body surfaces. *Ans.* (a) 9.61 cal/sec, (b) 1.25 cal/sec.

11. Why must the air be removed from a spectograph that is used to observe spectrum lines in the region of 500 A to 1000 A?

12. Would a diffraction grating spectrograph have to be evacuated to photograph the spectra region 2500 A to 3000 A? Explain your answer.

Classification of Spectra

A spectrum may be defined as a smooth and orderly array of the wavelengths of light. Such an array is usually photographed with a prism or diffraction-grating spectrograph of the kind described in the preceding chapter.

Different sources of light produce different wavelength displays and hence display different spectra. All spectra may be grouped into five main classes:

(1) Continuous emission spectra
(2) Line emission spectra
(3) Continuous absorption spectra
(4) Line absorption spectra
(5) Band spectra

9.1. Continuous Emission Spectra

The continuous emission spectrum has already been treated in detail in § 8.1. There it was demonstrated that, when light from a hot *solid* like the tungsten filament of an electric light, or the positive carbon of an arc, is sent through a prism, a continuous band of color from red to violet is observed (see color plate, Fig. 9A). The intensity of such a spectrum depends upon the temperature and upon the hot body itself. All hot solids raised to the same temperature give very nearly the same continuous emission spectrum.

9.2. Line Emission Spectra

When the slit of a spectrograph is illuminated by the light from a mercury arc, a sodium lamp, a helium or a neon discharge tube, a number of bright lines appear on the photographic plate in place of a continuous spectrum. See Fig. 9B.

It is important to realize that line spectra derive their name from the fact that a slit is used whose image constitutes a line. If a small circular opening were used in place of a slit, a disk image would appear in the place of each line.

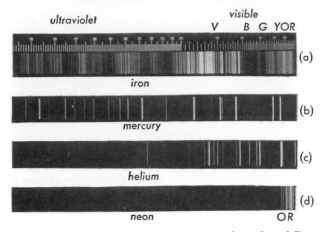

Fig. 9B *Photographs of the line emission spectrum from four different elements in the gaseous state.*

The ability to distinguish between two or more wavelengths, differing only slightly from one another, is therefore increased by the use of a slit.

The most intense spectrum lines are obtained from metallic arcs and sparks. The flame of a carbon arc may be used for demonstration purposes by previously soaking the *positive carbon rod* in various chemicals. (An experimental arrangement for projecting the spectrum on a large screen is shown in Fig. 9C.) Common salt water (sodium chloride in solution) gives a brilliant yellow line characteristic of sodium. Solutions of strontium or calcium chloride will show other strong spectrum lines in the red, green, and blue.

While a continuous emission spectrum arises from hot solids, a *line spectrum always arises from a gas at high temperatures*. It is the gas flame of the carbon arc that gives rise to the line emission spectrum in Fig. 9C.

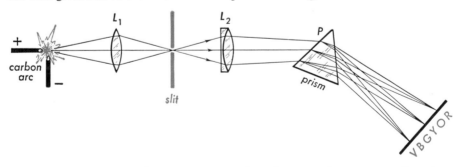

Fig. 9C *Experimental arrangement used in demonstrating spectrum lines in emission.*

9.3. Continuous Absorption Spectra

Continuous absorption spectra are usually produced by passing the light of a continuous emission spectrum through matter in the solid or liquid state. Good

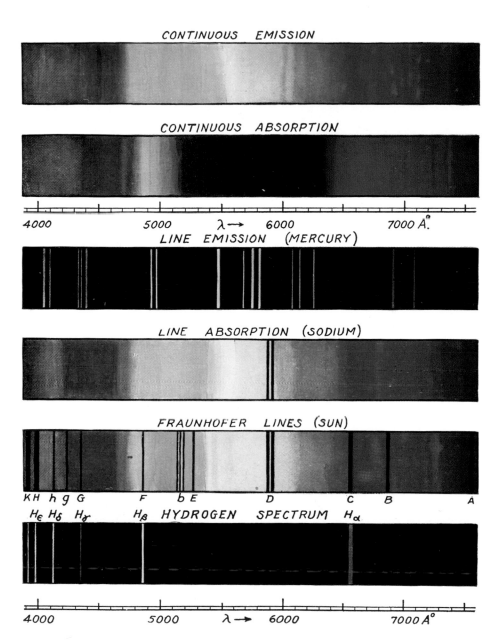

Fig. 9A   *Illustrations of continuous and line spectra.*

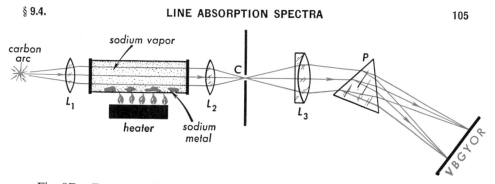

Fig. 9D   *Experimental arrangement for demonstrating the line absorption spectrum of sodium vapor.*

demonstrations can be performed by allowing white light to pass through colored glass. When the light is later dispersed by a prism, the missing colors will in general cover a wide band of wavelengths. A red piece of glass, for example, will absorb all visible light but the red. A magenta-colored piece of glass will absorb the whole central part of the visible spectrum.

9.4. Line Absorption Spectra

Line spectra in absorption are produced by sending continuous white light through a gas. Experimentally, the gas or vapor is inserted in the path of the light as shown in Fig. 9D. Light from a carbon arc, after passing as a parallel beam through a glass tube containing sodium vapor, is brought to a focus at the slit C. From there the light passes through a lens L_3 and a prism P to form a spectrum on the observing screen.

Sodium is chosen as an example for demonstration purposes because of its convenience. The vapor is produced by inserting a small amount of metallic sodium in a partially evacuated glass tube and heating it with a small gas burner. As the metal vaporizes, filling the tube with sodium vapor, a dark line will appear in the yellow region of the spectrum (see color plate, Fig. 9A).

If a photograph is taken of this absorption, and the photographic plate is long enough to extend into the ultraviolet, many absorption lines as shown in Fig. 9E are detected. A systematic array of absorption lines like this occurs with

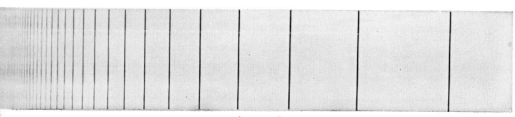

sodium series

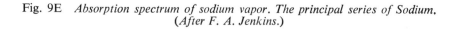

Fig. 9E   *Absorption spectrum of sodium vapor. The principal series of Sodium. (After F. A. Jenkins.)*

only a few elements, principally with the alkali metals, lithium, sodium, potassium, rubidium, and caesium. All elements in the gaseous state, however, give rise to a number of absorption lines, usually in the ultraviolet region of the spectrum. The absorption of yellow light by normal sodium atoms, for example, is a kind of resonance phenomenon. By virtue of their electronic structure atoms have definite and discrete natural frequencies to which they will vibrate in resonance. When light of one of these frequencies passes by, they respond to vibration and in so doing absorb the light energy.

9.5. The Sun's Spectrum

The solar spectrum, consisting of a bright colored continuous spectrum interspersed by thousands of dark lines, was first observed by Wollaston in 1802, and independently discovered and studied by Fraunhofer in 1817. Fraunhofer mapped out several hundred of these lines and labeled eight of the most prominent lines by the first letters of the alphabet. The strongest of these lines, now called Fraunhofer lines, are illustrated in Fig. 9F.

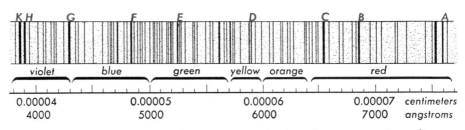

Fig. 9F   *Diagram of the solar spectrum indicating the most prominent lines labeled as they first were by Fraunhofer with the first letters of the alphabet.*

In 1882 the America physicist, H. A. Rowland photographed and published a 40-ft-long map of the sun's spectrum. Two small sections of Rowland's map are reproduced in Fig. 9G. These lines are explained as being due to the absorption of light by the solar atmosphere.

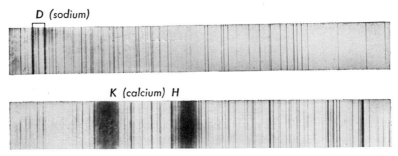

Fig. 9G   *Photographic reproductions of the solar spectrum taken from Rowland's original map.*

The surface of the sun at a temperature of 6000°K emits light of all wavelengths, i.e., a continuous emission spectrum. As this light passes out through the cooler gas layers of the solar atmosphere (see Fig. 9H), certain wavelengths

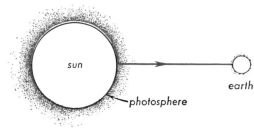

Fig. 9H *Light from the sun must pass through the solar atmosphere and the earth's atmosphere before reaching an observer on the earth's surface.*

are absorbed. Because the absorbing medium is in the gaseous state, the atoms and molecules there do not absorb all wavelengths equally, but rather they absorb principally those wavelengths they would emit if heated to a high temperature. Thus the atoms of one chemical element with their own characteristic frequencies absorb certain wavelengths, whereas the atoms of other elements absorb certain other wavelengths.

Before the sunlight reaches the earth's surface where it can be examined by an observer with a spectrograph, it must again pass through absorbing gases, this time the earth's atmosphere. Here, too, certain wavelengths are partially absorbed, producing other dark lines.

That the missing wavelengths correspond to definite chemical elements is illustrated by diagram in Fig. 9I. The center strip (b) represents a small section

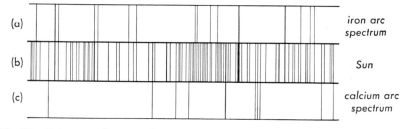

Fig. 9I *Schematic diagram illustrating the comparison of laboratory spectra from different elements with the many-line spectrum of the sun.*

of the visible spectrum as obtained with sunlight entering the slit of a spectrograph. The upper and lower strips, (a) and (c), represent the line spectrum observed when an iron arc and a calcium arc are successively placed in front of the same slit. Where each calcium line occurs in the laboratory source, an absorption line is found in the sun's spectrum. The same is true for each iron line. The remaining lines, not matched by an iron or calcium line, are due to other elements.

It has been possible by spectrum photographs of this kind to identify about two-thirds of the known chemical elements as existing on the sun. The reason why not all 90 or more elements are found is that some are too rare to produce absorption, whereas for other elements existing within the sun in large enough quantities the temperature is either too high or too low to bring out their lines.

Nine prominent Fraunhofer lines labeled in Fig. 9F have been identified as follows:

A, oxygen . . . . λ = 7594 A
B, oxygen . . . . λ = 6870 A
C, hydrogen . . . . λ = 6562 A
D, sodium . . . . λ = 5893 A
E, iron . . . . λ = 5270 A
F, hydrogen . . . . λ = 4861 A
G, iron . . . . λ = 4308 A
H, calcium . . . . λ = 3969 A
K, calcium . . . . λ = 3935 A

9.6. The Doppler Effect and the Sun's Rotation

It is not difficult to determine which of the Fraunhofer lines are due to absorption by the sun's atmosphere and which are due to the earth's atmosphere. The sun, like the earth, is rotating about an axis, and this motion produces for an observer on the earth a slight change of wavelength of light. This change in wavelength is due to the well-known Doppler effect so often observed with sound.

On one side of the sun, the east limb, the surface emitting light and the absorbing gases are approaching the earth at a high velocity, while on the other side, the west limb, the emitting surface and gases are receding from the earth.

On the Doppler principle the light waves emitted by the approaching side are crowded together to give a higher frequency, while those emitted by the receding side are lengthened out to give a lower frequency. Since the absorbing gases are also moving, the absorption lines due to these gases should be shifted in the solar spectrum.

The lines marked by arrows in Fig. 9J(a) show the Doppler shift. The lines marked by brackets, on the other hand, are produced by the earth's atmosphere which, because it is stationary with respect to an observer on the earth's surface, gives rise to unshifted lines. Remaining fixed, the unshifted lines serve as fiducial marks from which to observe and measure the shift in the solar lines.

The lower photograph (b) in Fig. 9J, shows that at the solar poles where there is relatively no motion toward or away from the earth, no Doppler shift is observed in the solar lines.

9.7. The Flash Spectrum

The Fraunhofer lines of the solar spectrum are not absolutely black, that is, devoid of all light; they are dark only in contrast with the far brighter colored

(a)

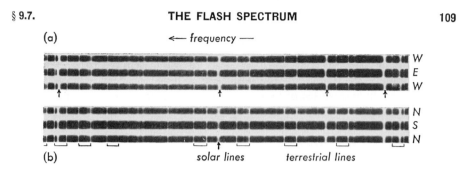

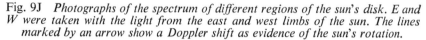

(b)   *solar lines*   *terrestrial lines*

Fig. 9J   *Photographs of the spectrum of different regions of the sun's disk. E and W were taken with the light from the east and west limbs of the sun. The lines marked by an arrow show a Doppler shift as evidence of the sun's rotation.*

background. Actually the sun's atmosphere is not cold but, as seen at the time of a total eclipse, is quite hot and emits light of its own. When this light is observed in a spectroscope, or photographed with a spectrograph, it consists of a bright-line spectrum called a *flash spectrum.*

The photograph of a flash spectrum reproduced in Fig. 9K was made in the

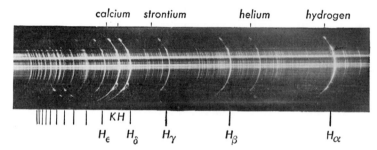

Fig. 9K   *Photograph of the flash spectrum of the sun taken at the time of a total eclipse. Note the prominences, called Halley's beads. (Courtesy of the Lick Observatory.)*

following way. During a solar eclipse, at just the instant before the moon's disk blanks out all of the sun's disk, leaving only a narrow bright crescent of the sun to be seen on one side, the light is allowed to enter a spectrograph.

With the customary slit and collimator lens removed from the instrument each spectral line becomes a crescent shaped image of the sun as cast by light of that wavelength. Note the high intensity of the **H** and **K** lines due to ionized calcium, and the solar prominences that appear at the same points on each image. Note also the high intensity of the lines marked H_α, H_β, H_γ, H_δ, and H_ϵ, as well as the succeeding lines, indicating a long series of spectral wavelengths. These lines are due to hydrogen and are known as the *Balmer series.*

The spectra of many of the distant stars are composed almost exclusively of the lines of the Balmer series, while others like our own sun exhibit more complex arrays.

9.8. The Balmer Series of Hydrogen

The first successful attempt to obtain a formula that represents the hydrogen series was made by Balmer in 1885. Since that time, these lines have become known as the Balmer series of hydrogen. Balmer's formula is written

$$\lambda = B\left[\frac{n_2{}^2 \times n_1{}^2}{n_2{}^2 - n_1{}^2}\right] \tag{9a}$$

where λ = wavelength in angstroms.

$$B = 911.27 \text{ A}$$
$$n_1 = 2$$
$$n_2 = 3, 4, 5, 6, 7, \ldots$$

If the number 3 is substituted for n_2 in the above formula, the wavelength λ of the first line of the series is calculated. Likewise, if the number 4 is substituted in its place, the wavelength of the second line can be calculated, etc. When these calculations are carried out, the following wavelengths are obtained for the first four lines:

	CALCULATED	MEASURED
$H_\alpha = B\frac{3.6}{5}$	= 6561.1 A	6562.1 A
$H_\beta = B\frac{6.4}{1.2}$	= 4860.8 A	4860.7 A
$H_\gamma = B\frac{1.0.0}{2.1}$	= 4339.4 A	4340.1 A
$H_\delta = B\frac{1.4.4}{3.2}$	= 4100.7 A	4101.3 A

These wavelengths, as well as those calculated for other lines of the series, agree remarkably well with the measured values. Balmer did not derive his formula from any theory, but simply formulated it from the measured wavelength for each series line. The meaning of those whole numbers n_1 and n_2 is given in § 10.4.

9.9. The Temperature Effect

The effect of raising the temperature of a source of light is to bring out new spectrum lines and cause others to disappear. This is illustrated by a small section of the spectrum of vanadium, element number 23, in Fig. 9L. The first three photographs were taken by heating small bits of vanadium metal in an electric furnace and photographing the light coming from the vapor just above the hot metal. The fourth photograph was made by placing a small piece of vanadium metal in the tip of the positive carbon of an arc, and the fifth by a high-voltage electric spark jumping between two vanadium metal rods.

Though the top spectrum taken at a temperature of 2000°K appears to be quite different from the lower one taken at a temperature at least twice as high, all of the lines arise from vanadium atoms and are characteristic of that element alone.

It is a well-founded and general rule that as the temperature of a gas is increased the maximum intensity shifts toward shorter wavelengths. At a low

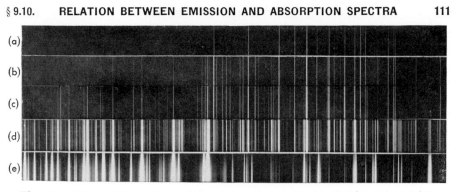

Fig. 9L   *Ultraviolet spectrum of vanadium, element 23. (a) 2000°K. (b) 2300°K. (c) 2600°K. (d) From a direct current arc between vanadium metal electrodes. (e) High-voltage electric spark between vanadium metal electrodes. (After A. S. King.)*

temperature, for example, the lines in the visible part of the spectrum of an element may be brightest, while at a higher temperature those in the ultraviolet become most intense.

9.10. Relation Between Emission and Absorption Spectra

When the line emission spectrum of an element is compared with the line absorption spectrum of the same element, the lines are found to coincide exactly. This is illustrated by a small section of the complex spectrum of iron in Fig. 9M.

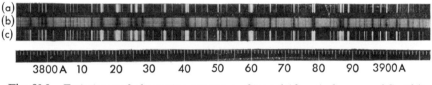

Fig. 9M   *Emission and absorption spectrum of iron. (After Anderson and Smith.)*

The strips (a) and (c), above and below, show the emission lines from the hot flame of an iron arc, while the middle strip (b) shows the absorption lines resulting from the passage of a continuous emission spectrum through cooler iron vapor.

To give a simple explanation of this we may picture the emission of light as due to the motions or vibrations of the electrons in the atoms of the source. These moving electric charges give rise to electromagnetic waves having the same frequencies as the electrons themselves. In any source of light only a small percent of the atoms have their electrons vibrating at any given time. These few, which we refer to as vibrating atoms, are called *excited atoms*, while those not vibrating are called *unexcited atoms*.

When yellow light from a sodium arc passes through sodium vapor, as in Fig. 9D, many individual atoms respond and become excited. In becoming excited the atoms have absorbed the incident light waves from the beam. This is like the resonance of two tuning forks that may be demonstrated for sound. Only when the sound waves from the first fork have exactly the same frequency as the second fork does the second fork pick up the motions and vibrate.

In the event that white light is sent through sodium vapor, only those light waves having frequencies equal to the natural frequencies of the atoms are absorbed. This is the explanation of the series of absorption lines in Fig. 9E. Again, when white light passes through iron vapor, only those light waves having frequencies equal to the natural frequencies of iron atoms are absorbed.

Although the atoms which are set into vibration by absorbing light from the beam emit the light again a small fraction of a second later, their chance of emitting it in some one direction is equally probable for all directions. Very little of this re-radiated light is therefore given out in the forward direction, parallel with the incident light.

9.11. Band Spectra

All of the line spectra described thus far are known to arise from single free atoms in a heated gas. Molecules of two or more atoms also give rise to spectrum lines grouped together into what are called bands. As shown by the reproductions in Fig. 9N these bands have the appearance of flutings.

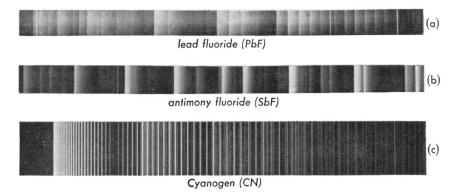

(a)

lead fluoride (PbF)

(b)

antimony fluoride (SbF)

(c)

Cyanogen (CN)

Fig. 9N *Band spectra taken with a large spectrograph. (After Jenkins.)*

Each fluting in the spectrum of a diatomic molecule is not a continuous band but a set of regularly spaced lines. The third photograph in Fig. 9N is a very greatly enlarged picture of a single band taken in the second order of a 21-ft grating spectrograph of the type shown in Fig. 8E.

The left-hand edge of the band is called the band head and the right-hand side the tail of the band. Note that one line near the band head is missing. The mystery of this missing line is now well understood, for it is the starting point of the band in place of the band head, and is called the band origin. The

theory of band spectra and the present-day knowledge of how diatomic molecules emit so many frequencies will be taken up in another chapter.

PROBLEMS

1. Calculate to five figures the wavelength of the fifth and sixth lines of the
Balmer series of hydrogen.

2. Calculate to five figures the wavelength of the tenth line of the Balmer
series of hydrogen. *Ans.* 3749.2 A.

3. Briefly describe an experimental arrangement by which a line emission
spectrum is produced.

4. Briefly describe an experimental arrangement by which a line absorption
spectrum is produced.

5. Find the wavelength of the ninth line of the Balmer series of hydrogen.

6. Calculate the wavelength of the eighth line of the Balmer series of hydrogen. *Ans.* 3796.9 A.

7. Determine the wavelength of the series limit of the Balmer series of hydrogen. (Note $n_1 = \infty$.)

8. From what kind of source does one obtain the most intense line emission
spectrum?

9. How can one produce a line absorption spectrum in the laboratory?

10. How is it possible to determine the elements existing in any given distant
star?

11. A given piece of metal is composed of several elements. How could one
determine spectroscopically whether it contains sodium?

10

The Photoelectric Effect

The photoelectric effect was discovered by Heinrich Hertz in 1887 when he observed that ultraviolet light, falling on the electrodes of a spark gap, caused a high-voltage discharge to jump greater distances than when it was left in the dark. One year later, Hallwachs made the important observation that ultraviolet light falling on a negatively charged body caused it to lose its charge, whereas a positively charged body was not affected. Ten years later J. J. Thomson and P. Lenard showed independently that the action of the light was to cause the emission of free negative charges from the metal surface. Although these negative charges are no different from all other electrons, it is customary to refer to them as "photoelectrons."

10.1. Photoelectrons

The photoelectric effect, in its simplest form, is demonstrated in Fig. 10A. Light from a carbon arc is focused by means of a quartz lens onto a freshly polished plate of zinc metal. When the plate is charged negatively and the light is turned on, the gold leaf of the attached electroscope slowly falls. It falls because the electrons, under the action of the light, leave the zinc plate at the illuminated spot *P*. When the plate is positively charged, the gold leaf does not fall, showing that the plate retains its charge. The same result of no discharge is observed if the zinc plate is negatively charged and a sheet of glass is inserted, as shown in the figure. When the glass is removed, the gold leaf again falls. Since common glass transmits visible and infrared light, but not ultraviolet, we conclude from the latter result that electrons are liberated only by ultraviolet light. This is also generally true for nearly all of the known metals.

A few elements—the alkali metals, *lithium, sodium, potassium, rubidium,* and *caesium*—are exceptions to this, for they will eject photoelectrons when visible light falls on them. For this reason the *alkali metals* are often used in the manufacture of photoelectric cells.

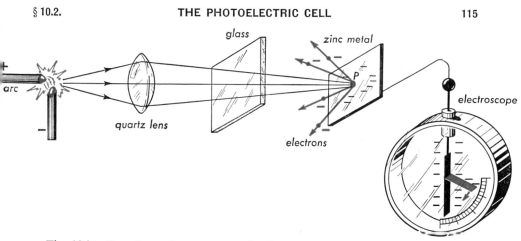

Fig. 10A *Experimental arrangement for demonstrating the photoelectric effect.*
When the glass plate is inserted, the effect stops.

10.2. The Photoelectric Cell

Photoelectric cells are usually made by depositing a thin layer of an alkali metal on the inner surface of a small vacuum tube (see Fig. 10B). If the cell is to operate in ultraviolet light, it is made of quartz, whereas if it is to be used in visible light it is made of common glass. The cell must be thoroughly evacuated, as the oxygen content of the air will combine chemically with the active

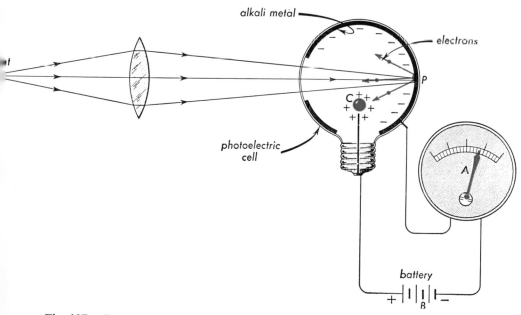

Fig. 10B *Diagram of a photoelectric cell, showing the light beam and electrical*
connections necessary for its operation.

metal layer, contaminating its surface and making it insensitive to visible light. A small section of the cell is always left clear to serve as a window for the incoming light. Photoelectrons, upon leaving the metal surface, are attracted and collected by the positively charged electrode C. The photosensitive surface, the cathode, and the collector, the anode, are maintained at a constant potential difference by the battery B.

A beam of light shining through the window of a photoelectric cell acts like a switch that completes an electric circuit. When the light strikes the metal P, there is a flow of electrons to the collector C, thus causing a current to flow around the circuit. This current can be measured by means of an ammeter at A. If the intensity of the light increases, the number of photoelectrons increases and the current therefore rises. When the light is shut off, the photo-electric action ceases and the current stops. If the metal film is positively charged, the cell becomes unreactive to light, since electrons released from the plate are held back by electrostatic attraction. All of these factors are readily demon-strated by a simple electrical circuit arranged as shown in Fig. 10B.

10.3. Practical Applications

Talking motion pictures, television, and burglar alarms are but three of the hundreds of practical applications of the photoelectric cell. The simplest of these is the burglar alarm, in which a beam of infrared light (invisible to the eye) is projected across the room into a photoelectric cell connected as shown in Fig. 10B. When an intruder walks through the beam, thus interrupting the beam for an instant, the photoelectric current ceases momentarily. An electric relay in place of the ammeter at A in the circuit moves, causing another electric circuit to be completed and thereby ringing an electric bell.

During the filming of talking motion pictures, a sound track is produced photographically on the side of the master motion picture film. Such sound tracks are shown in Fig. 10C. In strip (a), which is just a sample of one of the

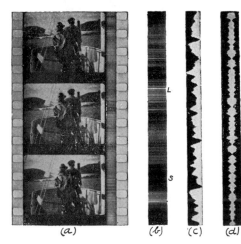

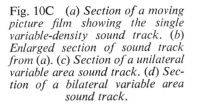

Fig. 10C   (a) *Section of a moving picture film showing the single variable-density sound track.* (b) *Enlarged section of sound track from* (a). (c) *Section of a unilateral variable area sound track.* (d) *Section of a bilateral variable area sound track.*

(a)   (b)   (c)   (d)

several kinds of sound tracks, the sound vibrations on the stage are converted into electrical vibrations by the stage microphone and then carried over wires to the camera taking the pictures. There the electrical impulses are made to move one of the jaws of a narrow slit through which a beam of light passes to the edge of the film. Loud sounds open the slit wide with each vibration, allowing a large amount of light through.

When the film is developed and positives are made for distribution, the loud sounds show up as periodic bands with considerable contrast as at *L* in strip (b). The latter strip is an enlarged section of the *single variable density* sound track seen on the right in photograph (a). Weaker sounds produce bands with less contrast as at *S*. Strips (c) and (d) are enlarged sections of two other types of sound track used in other patented recording systems.

When a sound film is projected on the screen in the theater (see Fig. 10D),

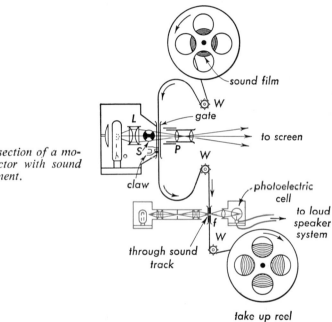

Fig. 10D   *Cross section of a motion picture projector with sound attachment.*

the film for the pictures themselves must of necessity move intermittently through the projection system *P* of the projection machine. As the film moves downward, each picture (frame) stops momentarily in front of the condensing lenses *L* and then moves on for the next frame. While the film is moving, the light is cut off by a rotating shutter *S*, and while it is at rest the light passes through to the screen. Thus the continuous motion seen on the screen is the result of a number of still pictures projected one after the other in rapid succession. To make this motion seem smooth and not jumpy, it is standard practice to project 24 frames each second.

To produce the sound, a small subsidiary beam of light, shown in detail in Fig. 10E, shines through the sound track at a point 25 frames farther along on the film where the motion is no longer intermittent, but smooth. As the sound

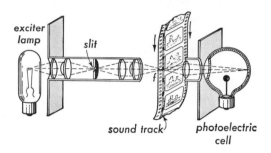

Fig. 10E   *Detail of sound pickup system of a moving-picture projector showing the exciter lamp, sound track, and photoelectric cell.*

track moves through the focus line f of the subsidiary light at constant speed, the transmitted light falling on the photoelectric cell fluctuates exactly as the sound track interrupts it. The photoelectric cell then changes the fluctuating light beam into a fluctuating electric current with the same variations. When transmitted to the radio amplifier and loud-speaker, the fluctuating current is changed into sound vibrations. Thus, sound vibrations have been carried over a light beam from the photographic film to the photoelectric cell and then by means of a loud-speaker system reproduced as sound.

10.4. Sound over a Light Beam

The sending of voice and musical sounds for several miles over a light beam is readily accomplished with a suitable light source as transmitter and a photoelectric cell as a receiver. A convenient laboratory demonstration can be made by using a small $\frac{1}{4}$-watt neon glow lamp as a source of light, as shown in Fig. 10F.

Sound waves entering the microphone M produce electric current fluctuations which, after being strengthened by a two-stage amplifier, cause the intensity of the neon glow lamp N to fluctuate accordingly. Made into a parallel beam by a lens L_1, the light travels across the room to a second lens L_2 and a photoelectric cell, where the light is changed back into a varying electric current. This faint signal is then amplified by a two-stage amplifier before it is delivered to the loud-speaker.

If the microphone is replaced by a phonograph pickup, records can be played

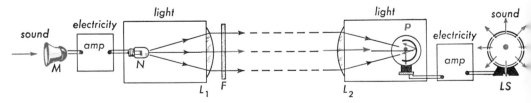

Fig. 10F   *Voice and musical sounds can be sent long distances over a beam of light.*

at the transmitter end, and excellent reproduction can be obtained from the loud-speaker. The light beam can be made completely invisible by placing an infrared filter in the light beam at *F*. Talking several miles over a beam of invisible light was developed to quite a high state of perfection during World War II. One system employs the infrared light from a glow discharge tube containing caesium, while several others, modulated by mechanically vibrating mirrors, employ the infrared from a tungsten filament lamp. Another system employs the invisible ultraviolet light from a glow discharge tube containing gallium.

10.5. Velocity of Photoelectrons

The first measurements of the velocity of photoelectrons led to the very startling discovery that the velocity does not increase as the intensity of the light increases. Increasing the intensity of the light increases the number of photoelectrons, but not their velocity. This discovery, as we shall see later, has had far-reaching implications in its result, for it has played an important role in the development of the quantum theory and our modern concepts of light and atomic structure.

Lenard's experiments, performed as far back as 1902, showed, that to increase the velocity of photoelectrons one must increase the frequency of the light, i.e., use shorter wavelengths. The shorter the wavelength of the light used, the higher the velocities of the electrons.

10.6. Einstein's Photoelectric Equation

Following an earlier idea of Planck's that light waves consist of tiny bundles of energy called *photons* or *quanta*, Einstein proposed an explanation of the photoelectric effect as early as 1905. His ideas were expressed in one simple relation, an algebraic equation destined to become famous in the annals of physics. Two Nobel Prizes, one to Einstein in 1921 and one to Millikan in 1923, have been granted on this, the photoelectric equation,

$$h\nu = W + \tfrac{1}{2}mv^2 \qquad (10a)$$

The first term, $h\nu$, represents the total energy content of a single quantum of light incident on a metal surface, as shown in Fig. 10G. The letter h is a constant, called *Planck's constant of action*, which has the same value for all light waves regardless of the frequency ν. At or beneath the surface of the metal, this *light quantum*, better known as a *photon*, is completely absorbed and, in disappearing, imparts its total energy to a single electron. Part of this energy W is consumed in getting the electron free from the atoms and away from the metal surface; the remainder is used in giving the electron a kinetic energy $\tfrac{1}{2}mv^2$, and therefore a velocity. For some metals like platinum, the energy required to pull an electron away from the surface is large, whereas for other metals like the alkalies it is quite small. W is called the *work function* of the metal.

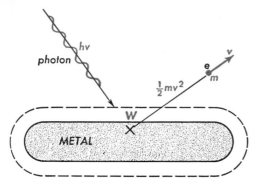

Fig. 10G *A light quantum (photon) of energy hv, incident on a metal surface, ejects an electron with a velocity v given by Einstein's equation.*

10.7. Millikan's Measurements of *h*

The letter *h* in Einstein's photoelectric equation is important because it is fundamental to the structure of all matter and is therefore *a universal constant.* Having first been introduced by Planck in 1901, the name *Planck's constant* has become firmly attached to this symbol *h*. The first experimental confirmation of Einstein's photoelectric equation came in 1912 when A. L. Hughes, and independently O. W. Richardson and K. T. Compton, observed that the energy of photoelectrons increased proportionately with the frequency. The constant of proportionality they found to be approximately equal to a constant, Planck's constant *h*.

Subsequently, Millikan carried out extensive experiments which established the photoelectric equation so accurately that his work is now regarded as giving one of the most trustworthy values for *h*.

To determine the value of *h*, it was necessary to measure the three factors, *v*, *W*, and $\frac{1}{2}mv^2$, and calculate *h* as the unknown quantity in Eq. (10a). A schematic diagram of part of Millikan's apparatus is shown in Fig. 10H. Light from a source *S* through a slit *T* is dispersed by means of a prism, and the spectrum is focused on a screen as shown. With a small aperture in this screen

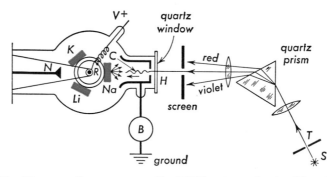

Fig. 10H *Diagram of apparatus used by Millikan in confirming Einstein's photoelectric equation.*

any desired frequency of light v could be admitted to the vacuum chamber through the window H.

Photoelectrons from any one of the three alkali metals, Na, K, or Li, could be obtained at will by turning the wheel R into the required position.

Previous experiments on the photoelectric effect had shown that good results could be obtained only when the metal surfaces were clean. By ingenious magnetic devices, operated from outside the vacuum chamber, Millikan was able to prepare uncontaminated metal surfaces just prior to each set of measurements.

By rotating R, one of the metal blocks could be brought opposite the knife N and a thin shaving of metal removed from the alkali metal. The fresh surface was then rotated 180° into a position directly in line with the light entering the window.

For each different frequency of light admitted to the chamber the velocity or energy of the photoelectrons had to be measured. This was accomplished by collecting the electrons in the cylinder C and measuring the accumulated charge by means of a sensitive electroscope or electrometer B.

By applying a positive potential to the metal block at V^+ the electrons would arrive at C with lower speeds owing to the retarding action of the charges. The positive charge on the Na block, for example, attracts the fast ejected electrons, slowing them down.

The velocity with which the electrons leave the metal can therefore be determined by measuring the potential difference V which is just great enough to prevent the electrons from reaching C. This stopping potential, applied between V^+ and the ground, can be equated to the photoelectrons ejected energy, by use of Eq. (2k). Replacing q by the electronic charge e, and the energy W by $\frac{1}{2}mv^2$, we obtain

$$Ve = \tfrac{1}{2}mv^2 \tag{10b}$$

In the mksa system, V is in volts, e is in coulombs, m is in kilograms, and v is in meters/second.

10.8. The Photoelectric Threshold

Having made the measurements described in the preceding section, Millikan calculated the photoelectron energies for different frequencies of light, and plotted the results on a graph as shown in Fig. 10I. The point at which the straight line intersects the bottom line determines the threshold frequency v_0. The photoelectric threshold v_0, is defined as the frequency of light which, falling on a surface, is just able to liberate electrons without giving them any additional kinetic energy. For such a frequency, the kinetic energy $\frac{1}{2}mv^2$ in Einstein's equation is zero and the energy of the photon, $hv_0 = W$. Eq. (10a) can therefore be written in the form

$$hv = hv_0 + \tfrac{1}{2}mv^2 \tag{10c}$$

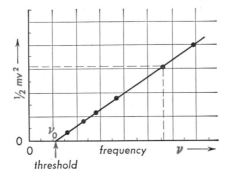

Fig. 10I *Graph showing the energy of photoelectrons ejected by light of different frequencies.*

The meaning of v_0 in this new equation is quite clear: For frequencies lower than v_0, electrons are not liberated, whereas for frequencies greater than v_0 they are ejected with a determined velocity.

The photoelectric threshold for most metals lies in the ultraviolet where the frequencies are relatively high. For the alkali metals the threshold lies in the visible and near infrared spectrum. In other words, it takes photons of less energy to free electrons from the alkali metals than it does to free them from most other metals.

Since every quantity in Eq. (10c), except h, is a measured quantity, the equation can be solved for h:

$$h = \frac{\frac{1}{2}mv^2}{v - v_0} \tag{10d}$$

Upon substitution of all measured quantities Millikan obtained the value $h = 6.56 \times 10^{-34}$ joule sec. The most recent value obtained for this universal constant is

$$h = 6.62 \times 10^{-34} \text{ joule sec}$$

Since the frequency of visible light is about 6×10^{14} vib/sec, the energy in a single photon or quantum of visible light is the product of these two numbers or 3.972×10^{-19} joule. In other words, it would take about 2.5×10^{18} photons to do one joule of work.

It was for his outstanding experimental work in the determination of the value of Planck's constant h, as well as the value of the electronic charge e, that Millikan was awarded the Nobel Prize in physics in 1923.

The photon, in ejecting an electron from a metal surface as in the photoelectric effect, disappears completely, i.e., it is annihilated. This is exactly the reverse of the process of the production of X rays, where a high-speed electron, upon hitting a metal target and being suddenly stopped, creates and emits a photon of high frequency.

10.9. Secondary Electrons

When electrons strike the surface of a metal plate, they knock additional electrons free from the surface. These are called *secondary electrons* and the process is called *secondary emission* (see Fig. 10J). As the speed of a primary or incident electron increases from zero to a few hundred volts, the number of secondaries increases toward a definite maximum. For most metal surfaces, this

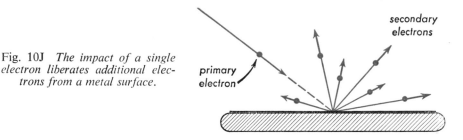

Fig. 10J *The impact of a single electron liberates additional electrons from a metal surface.*

maximum is in the neighborhood of two, while for certain alkali metal films it may be as great as eight or ten. In general, it is greatest for surfaces having a *low work function.*

10.10. Photo-Multiplier Tubes

The process of secondary electron emission is widely used in a special type of photoelectric cell used most effectively in detecting faint light. A cross-section diagram of such a photo-multiplier tube is given in Fig. 10K.

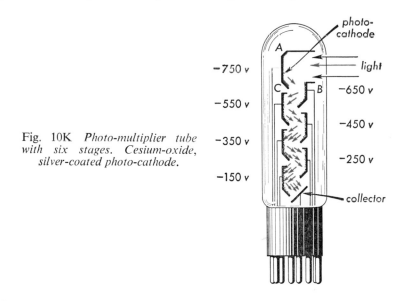

Fig. 10K *Photo-multiplier tube with six stages. Cesium-oxide, silver-coated photo-cathode.*

The number of photoelectrons from the photo-cathode *A* is proportional to the intensity of the incident light. These are attracted toward the next plate *B*, more positive by 100 volts, where upon impact additional electrons are liberated. Attracted to the next more positive plate *C*, still more electrons are liberated.

By the time the collector plate has been reached, a small avalanche of electrons has developed, and a correspondingly large charge and current are led off through that electrode to a suitable recording device.

If each electron on impact releases n secondaries, then, in a tube with k stages, the number arriving at the collector would be n^k. For example, if $n = 6$, and if $k = 5$, then $n^k = 7776$ electrons. This is an enormous gain over the signal obtained from a standard photo tube. Photo-multiplier tubes have been used most successfully with faint light, not only visible but infrared and ultraviolet as well.

PROBLEMS

1. Find the energy equivalent to a light wave of wavelength 6×10^{-7} m. Such light is in the green region of the visible spectrum.

2. Calculate the energy in joules of ultraviolet light photons of wavelength 3×10^{-7} m. *Ans.* 6.62×10^{-19} joules.

3. If X rays with a wavelength of 2.0×10^{-10} m fall on a metal plate, what would be the maximum velocity of the photoelectrons emitted? (Assume the work function to be negligibly small, $W = 0$.)

4. Assuming the work function of sodium to be negligibly small, what will be the velocity of photoelectrons emitted as the result of incident light of wavelength 3×10^{-8} m? This is ultraviolet light. *Ans.* 3.81×10^6 m/sec?

5. Find the energy equivalent to a γ ray whose wavelength is 0.5×10^{-3} A.

6. Calculate the energy in joules of a visible light photon of wavelength 7500 A. *Ans.* 2.65×10^{-19} joules.

7. If the photoelectric threshold of metallic silver is at $\lambda = 3500$ A, and ultraviolet light of wavelength 2500 A falls on it, find (a) the maximum kinetic energy of the photoelectrons ejected, (b) the maximum velocity of the photoelectrons, and (c) the value of the work function in joules.

8. If light of wavelength 6000 A falls on a metal surface and emits photoelectrons with a velocity of 4.0×10^5 cm/sec, what is the wavelength of the photoelectric threshold? *Ans.* 7690 A.

9. If X rays with a wavelength of 0.25 A fall on a metal plate, what would be the maximum velocity of the photoelectrons emitted? The work function can be assumed to be negligibly small.

10. Find the over-all gain of a ten-stage photo-multiplier tube if the average number of secondary electrons produced by each primary electron is six. *Ans.* 60.4×10^6.

11. An eight-stage, photo-multiplier tube has an over-all gain of 50,000. Find the average number of secondary electrons produced by each primary electron.

12. Make a diagram showing how a photoelectric cell could be used to count the number of cars passing a given point on a highway in a single day.

13. When X rays with a wavelength of 0.85 A fall on a copper plate, what will be the maximum velocity of the photoelectrons emitted? Assume the work function to be negligible.

The Structure of Atoms

Early in the twentieth century, while Einstein was working out his special theory of relativity, J. J. Thomson proposed a type of electron shell structure for all atoms. His model structures were worked out by mathematics from Coulomb's law for charged particles and soon became known as the *"plum-pudding atom."*

11.1. The Thomson Atom

Thomson visualized all of the positive charge of an atom as being spread out uniformly throughout a sphere about 10^{-8} cm in diameter, with the electrons as smaller particles distributed in shells somewhat as shown in Fig. 11A.

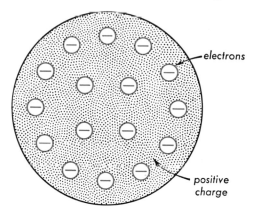

Fig. 11A   *Diagram of the Thomson atom model.*

While the net force exerted by the positively charged sphere on each electron is toward the center of the sphere, the electrons mutually repel each other and form shells.

An excellent demonstration of the tendency to form rings for a two-dimensional model is shown in Fig. 11B. A glass dish 15 to 20 cm in diameter is wound with about 30 turns of No. 14 insulated copper wire. The most common steel

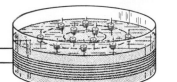

Fig. 11B *Floating needles in a magnetic field demonstrating the electron shell structure of the Thomson atom.*

sewing needles are then mounted in small corks (8 mm diameter and 8 mm long) as shown at the left, and magnetized by stroking from top to bottom with the N pole of a strong Alnico magnet.

With water in the dish, and a current of 1 to 2 amp through the coil, a single needle is placed upright in the water. Released, it will migrate to the center where the magnetic field is strongest. The addition of needles, one after another, near the edge of the dish, will result in the formation of geometrically symmetrical patterns and rings.

An increase or decrease in current will cause any given pattern to shrink or expand, corresponding to a greater or lesser positive charge. The stability of such ring patterns undoubtedly influenced the later extension by Bohr and Stoner of the quantized orbit model of the hydrogen atom to all atoms.

11.2. Bohr's Theory of the Hydrogen Atom

In 1913 Niels Bohr* proposed a theory of the hydrogen atom that marked the beginning of a new era in the history of physics. With his theory, Bohr gave not only a satisfactory explanation of the Balmer series of hydrogen but a model for the structure of all other atoms as well.

Starting with what should be the simplest of all atoms, Bohr assumed that a hydrogen atom, $Z = 1$, consists of a nucleus with one positive charge ^+e and a single electron of charge ^-e revolving around it in a circular orbit of radius r (see Fig. 11C). Because it is 1836 times heavier than the electron, the nucleus could be assumed at rest.

To keep the electron in its orbit and prevent it from spiraling in toward the nucleus, or away from it to escape, Bohr next assumed that the inward centripetal force is due to, and therefore is, the inward electrostatic force F. From dynamics we know that the centripetal force is mv^2/r, and from Coulomb's law, Eq. (2a), the electrostatic force is kee/r^2. Equating these two, we obtain

$$m\frac{v^2}{r} = k\frac{ee}{r^2} \tag{11a}$$

* Niels Bohr (1885–1962), Danish physicist, was born at Copenhagen, the son of Christian Bohr, professor of physiology at the University of Copenhagen. After taking his Ph.D. degree at Copenhagen in 1911, he studied for one year under J. J. Thomson at Cambridge, and one year under Ernest Rutherford at Manchester. Returning to Copenhagen in 1913, with the results of the Rutherford scattering experiments fresh in his mind, he worked out and published his now famous theory of the hydrogen atom. In 1920 Bohr was appointed head of the institute for theoretical physics at the University of Copenhagen. In 1921 he was awarded the Hughes Medal of the Royal Society, and in 1922 the Nobel Prize in physics.

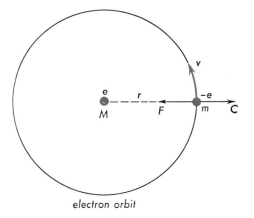

Fig. 11C *Orbital diagram of the hydrogen atom according to the Bohr theory.*

electron orbit

In the mksa system of units,

m is in Kg
v is in m/sec
e is in coulombs
r is in meters
$k = 9 \times 10^9$

At this point Bohr introduced his second assumption, *the quantum hypothesis.* The electron, he assumed, cannot move in any sized orbit, stable under the conditions of the equation above, but in just certain *definite and discrete orbits.* The sizes of these orbits are governed by Eq. (11a) and the rule that the *angular momentum of the electron in its orbit is equal to an integer n times a constant h divided by 2π:*

$$mvr = n\frac{h}{2\pi}$$ (11b)

$$n = 1, 2, 3, 4, 5, \ldots$$

In this equation n is called the principal *quantum number* and, because it can take whole number values only, it fixes the sizes of the allowed orbits. To find the radii of these "Bohr circular orbits," Eq. (11b) is solved for v, then squared and substituted in Eq. (11a) to give

$$r = \frac{n^2h^2}{4\pi^2me^2k}$$ (11c)

If we put into this equation the known values of the constants e, m, h, and k,

$e = -1.60 \times 10^{-19}$ coul
$m = 9.10 \times 10^{-31}$ Kg
$h = 6.62 \times 10^{-34}$ joule sec
$k = 9 \times 10^9$ newton m²/coul²

the orbits shown in Fig. 11D are calculated. The innermost orbit, with $n = 1$, has a radius $r = 0.528 \times 10^{-10}$ m, or 0.528 A, and a diameter of 1.056 A.

$$1 \text{ meter} = 10^{10} \text{ angstroms}$$

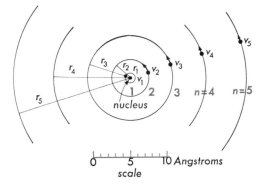

Fig. 11D *Scale diagram of the Bohr circular orbits of hydrogen.*

The second orbit is four times larger, and the third is nine times, etc. The constant h is *Planck's constant.*

The velocity of the electron, when it is in any one orbit, can be determined from Eqs. (11b) and (11c). By substituting the value of r from Eq. (11c) in Eq. (11b) and solving for the velocity v, we obtain

$$v = k\,\frac{2\pi e^2}{nh} \tag{11d}$$

In the innermost orbit, $n = 1$, the velocity v is $\frac{1}{137}$ the velocity of light. In the second orbit the speed is only $\frac{1}{2}$ as great, and in the third only $\frac{1}{3}$ as great, etc. With such small orbits and such high velocities, the number of revolutions per second becomes very high.

Since the circumference of any circular orbit is $2\pi r$, the frequency with which an electron goes around each orbit is given by

$$f = \frac{v}{2\pi r} \tag{11e}$$

In the second Bohr circular orbit, the frequency is calculated to be 10^{15} rps. This, by comparison with the frequency of vibration of visible light waves, is of the same order of magnitude.

It should be noted that the one and only electron in each hydrogen atom can occupy only one orbit at any one time. If the electron changes its orbit, it must move to one of the allowed orbits and never stop in between.

11.3. Electron Jumps

Bohr's third and final assumption regarding the hydrogen atom concerns the emission of light. Bohr postulated that light is not emitted by an electron when it is moving in one of its fixed orbits, but only when the electron jumps from one orbit to another, as illustrated in Fig. 11E. Bohr said that the frequency of this light is not determined by the frequency of revolution but by the difference in energy between the initial and final orbit,

$$E_2 - E_1 = hv \tag{11f}$$

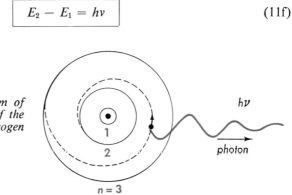

Fig. 11E *Schematic diagram of Bohr's quantum hypothesis of the radiation of light from a hydrogen atom.*

where E_2 is the energy of the *initial orbit*, E_1 the energy of the *final orbit*, h is Planck's constant, and v is the frequency of the light.

To illustrate this, let E_1, E_2, E_3, E_4, etc., represent the total energy of the electron when it is in the orbits $n = 1, 2, 3, 4$, etc., respectively. When, for example, the electron is in orbit $n = 3$ where its energy is E_3, and it jumps to orbit $n = 2$ where the energy is E_2 (see Fig. 11E), the energy difference $E_3 - E_2$ is ejected from the atom in the form of a light wave of energy hv called a *photon*. Here, then, is the origin of light waves from within the atom.

11.4. Bohr's Success

The success of Bohr's theory is not to be attributed so much to the mechanical picture or model of the atom just proposed, but rather to the development of an equation that agrees exactly with experimental observations.

By combining the equations presented in the preceding section, Bohr derived an equation for the frequency v of the light waves emitted by hydrogen atoms. This equation is

$$v = 3.28965 \times 10^{15} \left(\frac{1}{n_1{}^2} - \frac{1}{n_2{}^2} \right) \tag{11g}$$

where n_1 and n_2 represent the *principal quantum numbers* of two orbits.

If we introduce the wave equation, valid for all waves,

$$c = v\lambda \tag{11h}$$

and replace v by c/λ in Eq. (11g) it can be written

$$\lambda = 911.27 \text{ A} \left(\frac{n_2{}^2 \times n_1{}^2}{n_2{}^2 - n_1{}^2} \right) \tag{11i}$$

where λ is the wavelength of the light in angstroms, n_2 is the quantum number of any orbit of the hydrogen atom in which an electron is confined, and n_1 is the quantum number of the orbit to which the electron jumps to emit light of wavelength λ.

Bohr found that, if in Eq. (11i) he placed $n_1 = 2$ and $n_2 = 3$, the calculated wavelength, $\lambda = 6564.7$ A, is obtained, which is extremely close to the measured wavelength of the red spectrum line of hydrogen. If he placed $n_1 = 2$ and $n_2 = 4$, the calculated wavelength agreed exactly with the measured wavelength of the blue-green spectrum line of hydrogen.

In fact, the entire series of lines in the hydrogen spectrum are exactly represented by Eq. (11i), by setting $n_1 = 2$ and $n_2 = 3, 4, 5, 6$, etc. This series of lines, so prominently displayed by the sun and stars, as well as by any hydrogen discharge tube in the laboratory, is known as the *Balmer series*. (See § 9.8.)

These quantum number changes correspond, as shown in Fig. 11F, to an

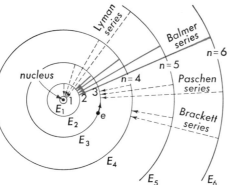

Fig. 11F   *Diagram of the Bohr circular orbits of hydrogen showing the various electron jumps that give rise to the emission of light waves of different frequency.*

electron jumping from any outer orbit n to the next to the smallest orbit $n = 2$. In any high-voltage electrical discharge in a glass tube containing hydrogen gas, many thousands of atoms may each have their one and only electron jumping from orbit 3 to 2, while in many other atoms the electron may be jumping from other orbits to $n = 2$. Hence, upon observing the light through a spectroscope, one may observe the entire Balmer series of lines.

11.5. Bohr's Predicted Series

Bohr's orbital model of the hydrogen atom not only accounts for the Balmer series of hydrogen, but also for many other observed lines as well.

By substituting $n_1 = 1$ and $n_2 = 2, 3, 4$, etc., in Eq. (11i), one obtains a series of spectrum lines in the ultraviolet region of the spectrum. These lines were first photographed by T. Lyman of Harvard University, and the wavelengths are found to check exactly with calculations. This series, now called the Lyman series, which can only be photographed in a vacuum spectrograph, is reproduced in Fig. 11G. On the orbital picture of Fig. 11F, the Lyman series of lines arises

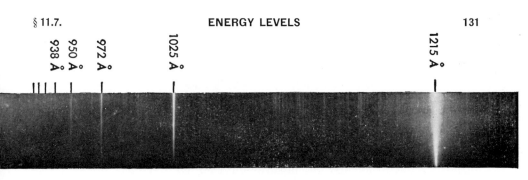

Lyman series of hydrogen

Fig. 11G   *Photograph of the extreme ultraviolet series of hydrogen, predicted by Bohr's theory and first observed by Lyman.*

from electron jumps from any outer orbit directly to the innermost orbit, the *normal state.*

If, in Eq. (11i), n_1 is set equal to 3 and n_2 to 4, 5, 6, etc., the calculated frequencies predict spectrum lines in the infrared spectrum. These lines were first looked for and observed, exactly as predicted, by F. Paschen; the series is now known by his name. Another series of lines arising from electron jumps, ending on orbit $n = 4$, was predicted and observed in the far infrared by Brackett.

11.6. Normal and Excited Atoms

When the single electron of a hydrogen atom is in the innermost orbit, $n = 1$, the atom is said to be in its normal state. As the name implies, this is the condition of most free hydrogen atoms in a gas under normal room temperature and pressure. If an electrical discharge is sent through a vessel containing hydrogen gas, cathode rays (electrons) moving at high speed make frequent collisions with electrons, knocking some of them out of the atom completely and some of them into one of the outer allowed orbits, $n = 2, 3, 4$, etc.

When the electron is completely removed from the atom, the atom is said to be *ionized;* whereas when it is forced into an outer orbit, the atom is said to be *excited.* Once in an excited state, an atom will not remain that way long, for the electron under the attraction by the nucleus will jump to an inner orbit. By jumping to an inner orbit, the electron loses all or part of the energy it had gained.

When an electron is in an excited state, it does not necessarily return to the innermost orbit by a single jump, but may return by several jumps, thereby emitting several different light waves, or quanta.

11.7. Energy Levels

By combining Bohr's equations, Eq. (11a) and Eq. (11b), the energy of an electron in a circular orbit of the hydrogen atom can be calculated. The total energy is just the sum of the kinetic energy $\frac{1}{2}mv^2$, and the potential energy:

$$E_t = E_k + E_p \tag{11j}$$

To find the potential energy of the electron in its orbit we use Eq. (2m). By this equation we see that the potential V, at any point at a distance r from the nuclear charge $+e$, is given by

$$V = k\frac{e}{r}$$

Since V is the work done per unit charge in carrying any charge from a distance r out to infinity, one must multiply by the electron's charge $-e$ to obtain as the stored potential energy, $-Ve$.

The two forms of stored energy are, therefore,

$$E_k = \tfrac{1}{2}\,mv^2$$

$$E_p = -k\frac{e^2}{r}$$

Substituting these values of E_k and E_p in Eq. (11j) and the value of r from Eq. (11c), we obtain, as Bohr did,

$$E_t = -\frac{2\pi^2 m e^4 k^2}{n^2 h^2} \tag{11k}$$

The minus sign signifies that one must do work on the electron to remove it from the atom.

With the exception of the principal quantum number n, all quantities in this equation are the same for all orbits. We can therefore write

$$E_t = -\frac{1}{n^2}\,R \tag{11l}$$

where R is constant and equal to

$$R = \frac{2\pi^2 m e^4 k^2}{h^2}$$

which, upon substitution of the known values of all the constants, gives

$$R = 2.1790 \times 10^{-18}\ \text{joules} \tag{11m}$$

Eq. (11l) is an important equation in atomic structure, for it gives the energy of the electron when it occupies any one of the different orbits of the hydrogen atom. Instead of drawing orbits to the scale of their radius as in Fig. 11D, it is customary to draw horizontal lines to an energy scale, as shown in Fig. 11H. This is called an *energy level diagram*. The various electron jumps between the allowed orbits of Fig. 11F now become vertical arrows between the energy levels.

The importance of this kind of diagram is to be attributed to Bohr's third relation, Eq. (11f), where the energy $h\nu$ of each radiated light wave is just equal to the difference between two energies. The energy of each radiated photon is, therefore, proportional to the length of its corresponding arrow.

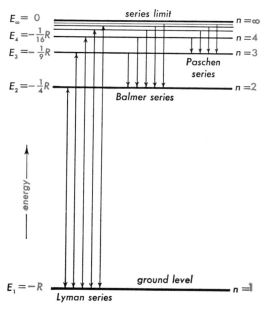

Fig. 11H *Energy-level diagram for the hydrogen atom. Vertical arrows represent electron jumps.*

The first line of the Balmer series $\lambda = 6561$ A, the red line in Figs. 9A and 9K, corresponds to the short arrow, $n = 3$ to $n = 2$. The second line of the same series is the blue-green line $\lambda = 4861$ A, and corresponds to the slightly longer arrow, $n - 4$ to $n - 2$, etc.

When an electrical discharge is sent through hydrogen gas, each atom, by collision with other atoms, has its only electron excited to an upper level, and then that electron jumps down again from one level to another, giving rise to the emission of light waves. If, as another experiment, a whole continuous spectrum of light waves is sent through a tube containing hydrogen gas, the hydrogen atoms will be in the ground level, $n = 1$, and by *resonance* may absorb frequencies corresponding to any one of the Lyman series. In absorbing one of these frequencies, the electron of that atom will jump to an upper energy level. The arrowheads at the top of these vertical lines, correspond, therefore, to resonance absorption. Those same excited electrons can then return by downward jumps, emitting light, and stopping finally on the ground level.

Resonance absorption is the explanation of the dark lines of the sodium spectrum shown in Fig. 9E and the solar spectrum in Fig. 9G.

11.8. Bohr-Stoner Scheme of the Building-Up of Atoms

Bohr and Stoner proposed an extension of the orbital model of hydrogen to include all of the chemical elements. As shown by the examples in Fig. 11I, each atom is composed of a positively charged nucleus with a number of electrons around it.

Although the nucleus is a relatively small particle less than 10^{-12} cm in diameter, it contains almost the entire mass of the atom, a mass equal in *atomic mass units* to the *atomic weight*.

The positive charge carried by the nucleus is equal numerically to the atomic number, and it determines the number of electrons located in orbits outside.

A helium atom, atomic number $Z = 2$, has two positive charges on the nucleus and two electrons outside. A lithium atom, atomic number $Z = 3$, contains three positive charges on the nucleus and three electrons outside. A mercury atom, atomic number 80, contains 80 positive charges on the nucleus and 80 electrons outside.

The orbits to which the electrons are confined are the Bohr orbits of hydrogen with $n = 1, 2, 3$, etc., and are called electron shells. Going from element to element in the atomic table, starting with hydrogen, electrons are added one after the other, filling one shell and then another. A shell is filled only when it contains a number of electrons given by $2n^2$. To illustrate this, the first shell $n = 1$ is filled when it has 2 electrons, the second shell $n = 2$ when it has 8 electrons, the third shell $n = 3$ when it has 18 electrons, etc. $2 \times 1^2 = 2$, $2 \times 2^2 = 8$, $2 \times 3^2 = 18$, etc.

quantum number}	$n = 1$	$n = 2$	$n = 3$	$n = 4$
number of electrons}	2	8	18	32

Among the heavier elements there are several departures from the order in which the shells are filled. Although these departures are not important from

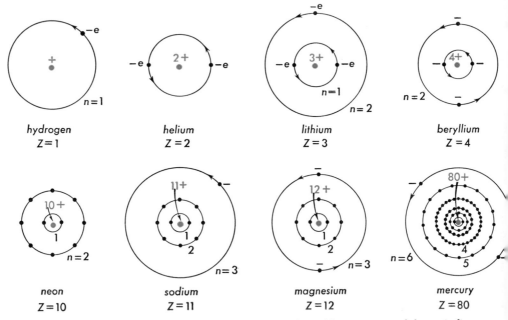

Fig. 11I　*Bohr-Stoner orbital models for the light and heavy atoms of the periodic table.*

the present standpoint, their nature is illustrated by the mercury atom, Fig. 11I. The four inner shells, $n = 1, 2, 3,$ and 4, are entirely filled with 2, 8, 18, and 32 electrons, respectively, while the fifth shell contains only 18 electrons and the sixth shell 2 electrons. The reasons for such departures are now well understood and are indicative of the chemical behavior of the heavy elements.

It is important to note that, as the nuclear charge increases and additional electrons are added in outer shells, the inner shells, under the stronger attraction of the nucleus, shrink in size. The net result of this shrinkage is that the heaviest elements in the periodic table are not much larger in diameter than the lighter elements. The schematic diagrams in Fig. 11I are drawn approximately to the same scale.

The experimental confirmation of these upper limits to the allowed number of electrons in each shell is now considered one of the most fundamental principles of nature. A sound theoretical explanation of this principle of atomic structure was first given by W. Pauli, in 1925, and is commonly referred to as the *Pauli exclusion principle.*

11.9. Elliptical Orbits

Within but a few months after Bohr (in Denmark) published a report telling of his phenomenal success in explaining the hydrogen spectrum with circular orbits, Sommerfeld (in Germany) extended the theory to include elliptical orbits as well. Because these orbits played such an important role in later developments in atomic structure, they deserve some attention here.

The net result of Sommerfeld's theory showed that the electron in any one of the allowed energy levels of a hydrogen atom may move in any one of a number of orbits. For each energy level $n = 1, n = 2, n = 3,$ etc., as shown in Fig. 11H, there are n possible orbits.

Diagrams of the allowed orbits for the first three energy levels are shown in Fig. 11J. For $n = 3,$ for example, there are three orbits, with designations $l = 2,$ $l = 1,$ and $l = 0.$ The diameter of the circular orbit is given by Bohr's theory, and this is just equal to the major axes of the two elliptical orbits. The minor axes are $\frac{2}{3}$ and $\frac{1}{3}$ of the major axis.

It is common practice to assign letters to the l-values as follows.

$l = 0$	$l = 1$	$l = 2$	$l = 3$	$l = 4$
s	p	d	f	g

According to this system, the circular orbit with $n = 3$ and $l = 2$ is designated $3d,$ while the elliptical orbit $n = 2$ and $l = 0$ is designated $2s,$ etc. n is the *principal quantum number* and l is the *orbital quantum number.* All orbits having the same value of n have the same total energy, the energy given by Bohr's equation for circular orbits, Eq. (11l).

Each of the allowed orbits of the Bohr-Sommerfeld model of the hydrogen atom becomes a subshell into which electrons are added to build up the elements

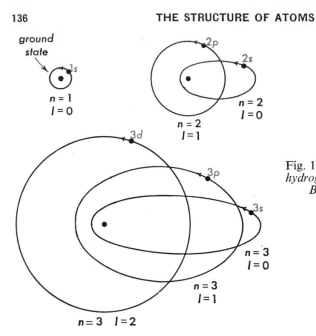

Fig. 11J *Electron orbitals for the hydrogen atom according to the Bohr-Sommerfeld theory.*

of the periodic table in the Bohr-Stoner scheme. These subshells are tabulated as follows:

		SUBSHELLS			
n	0	1	2	3	4
1	1s				
2	2s	2p			
3	3s	3p	3d		
4	4s	4p	4d	4f	
5	5s	5p	5d	5f	6g

SHELLS

The maximum number of electrons allowed in any one subshell is given by the relation

$$2(2l + 1)$$

This is called the *Pauli exclusion principle*, each subshell being filled when it contains the following number of electrons.

$l = 0$	1	2	3	4
Subshell s	p	d	f	g
Number of electrons 2	6	10	14	18

An orbital diagram of an argon atom is given in Fig. 11K. The electron configuration is given below, the exponents specifying the total number of electrons in that subshell. Because such elliptical orbits are not easily drawn, it is customary to group all subshells of the same n-value together and show them in rings as in Fig. 11I.

A complete table of subshell build-up of all the known elements is given in

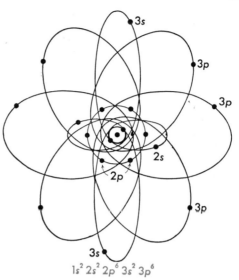

Fig. 11K *Electron configuration for an argon atom:* $Z = 18$.

$1s^2 2s^2 2p^6 3s^2 3p^6$

Appendix VI. The rules for the filling of subshells, and ones that hold throughout the periodic table are the following:

1. *Subshells are grouped under like values of $n + l$.*
2. *Groups are filled in the order of increasing $n + l$.*
3. *Within each $n + l$ group, subshells are filled in the order of decreasing l-values.*

PROBLEMS

1. Find the diameters of the 5th, 10th, and 20th circular orbits of hydrogen according to the Bohr theory.

2. What would be the approximate quantum number n for a circular orbit of hydrogen 0.0001 mm in diameter? This would be just big enough to see under a microscope. *Ans. $n = 31$.*

3. Calculate the wavelength of the fifth line of the Balmer series of hydrogen.

4. Find the wavelengths of the fourth and fifth lines of the Balmer series of hydrogen. *Ans. 4100.7 A; 3969.1 A.*

5. Compute the wavelengths of the 4th, 5th, and 6th lines of the Lyman series of hydrogen.

6. Find the wavelengths of the first three lines of the Paschen series of hydrogen. *Ans. 18,746 A; 12,815 A; 10,935 A.*

7. Make a diagram of a nickel atom (atomic number 28) according to the Bohr-Stoner scheme.

8. Make a diagram of a krypton atom (atomic number 36) according to the Bohr-Stoner scheme. *Ans. 2, 8, 18, 8.*

9. Calculate the wavelengths of the 10th and 20th spectrum lines of the Paschen series of hydrogen. See Eq. (11i).

10. Calculate the wavelength of the second spectrum line of the Brackett series of hydrogen. See Eq. (11i). *Ans. 26,245 A.*

11. Find the radius of the 100th orbit of the electron in a hydrogen atom.

12. Make a diagram of a zinc atom according to the Bohr-Stoner scheme.

13. Make a diagram of a xenon atom, showing the number of electrons in each shell according to the Bohr-Stoner scheme.

14. Determine the wavelength of the fourth line of the Brackett series of hydrogen. *Ans.* 19,440 A.

15. What would be the approximate quantum number n for a circular orbit of hydrogen 2 mm in diameter?

16. (a) Solve Eq. (11a) for r as the only unknown. (b) Solve Eq. (11b) for r as the only unknown.

17. Set the right-hand sides of the equations obtained in Problem 16 equal to each other, and solve for v as the unknown quantity.

18. Using Eq. (11d) for v and the known values of e, m, h, and k, calculate the electrons' velocity in the first Bohr circular orbit. *Ans.* 2.19×10^6 m/sec.

Spinning Electrons

With the development of the Bohr-Sommerfeld theory of the hydrogen atom in 1913, and its extension to the building-up of the electron structure of all atoms of the periodic table within but a few years, three new but important discoveries followed: (1) the discovery of the spinning electron, (2) the quantization of orbital electrons in a magnetic field, and (3) the direct experimental evidence for the existence of electron shells.

12.1. Orbital Mechanical Moment

The foundations of our present-day concepts of atomic structure were introduced into modern science when Bohr first proposed his theory of the hydrogen atom. (See Chapter 11.) According to Bohr's theory, the hydrogen atom is composed of a proton as a nucleus with a single electron revolving around it in a circular orbit. The quantum theory was introduced when Bohr assumed, through Eq. (11l), that the total energy of a hydrogen atom is given by

$$E_t = -\frac{1}{n^2} R \qquad n = 1, 2, 3, 4, \ldots$$

where n is the *principal quantum number*

$$R = \frac{2\pi^2 m e^4 k^2}{h^2}$$

and
$$h = 6.6238 \times 10^{-34} \text{ joule sec}$$

With the introduction of elliptical orbits as allowed states for the electron, a new quantum number l was introduced. This new quantum number, too, has integral values only, and is assigned letters as follows:

$$l = 0 \quad 1 \quad 2 \quad 3 \quad 4 \quad 5 \quad 6 \quad 7 \quad \ldots$$
$$ s \quad p \quad d \quad f \quad g \quad h \quad i \quad j \quad \ldots$$

In this newer notation, the orbital angular momentum is given by

$$p_l = l\frac{h}{2\pi} \tag{12a}$$

The angular momentum p_l is frequently called the *mechanical moment*, and unit angular momentum $h/2\pi$ is frequently abbreviated $\hbar$:

$$\hbar = \frac{h}{2\pi} \tag{12b}$$

where $\hbar = 1.054 \times 10^{-34}$ joule sec.

The mechanical moment of an *s*-electron orbit is 0; of a *p*-electron orbit, $1\hbar$; of a *d*-electron orbit, $2\hbar$, of an *f*-electron orbit, $3\hbar$, etc.

12.2. Orbital Magnetic Moment

Since an electron has a negative charge, its orbital motion, like that of an electron current in a loop of wire, sets up a magnetic field. The direction of this field, as shown in Fig. 12A, is given by the *left-hand rule*. Since angular

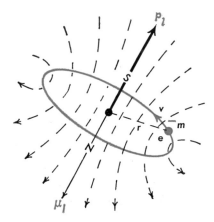

Fig. 12A   *An orbital electron produces a magnetic field.*

momentum is a vector quantity p_l is shown pointing up along its axis of rotation. The direction of this mechanical moment is given by the *right-hand rule*.

The magnetic field produced by the orbital electron is quite similar to the field around a bar magnet, and is therefore specified as a magnetic dipole moment. This magnetic moment μ_l is given by

$$\frac{\mu_l}{p_l} = \frac{e}{2m} \tag{12c}$$

where e and m are the charge and mass of the electron. The left-hand term μ_l/p_l is called the *gyromagnetic ratio*. It is the ratio between the magnetic moment and the mechanical moment. By substituting p_l from Eq. (12a), we obtain

$$\mu_l = l\hbar \frac{e}{2m}$$ (12d)

All symbols on the right, except l, are fixed atomic constants, and together they form a unit of magnetic moment called the *Bohr magneton*. The Bohr magneton is given by

$$\mu_B = \hbar \frac{e}{2m}$$ (12e)

which evaluated is equal to

$$\mu_B = 9.273 \times 10^{-24} \text{ ampere meters}^2$$ (12f)

The magnetic moment for a p-electron orbit is one Bohr magneton, for a d-electron orbit is two Bohr magnetons, etc. An s-electron orbit with $l = 0$, has no mechanical moment and no magnetic moment.

12.3. Spectral Series

When the spectra of most of the elements in the periodic table are examined, they are found to be complex arrays of hundreds of lines spaced in what appear to be random patterns. (See Fig. 12B(a).) A few of the elements, however, give

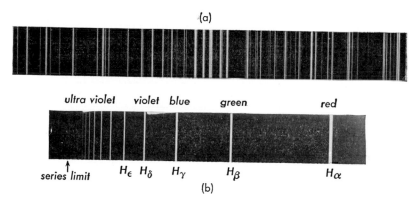

(a)

(b)

Fig. 12B   (a) *Small section of the spectrum of titanium, Z = 22.* (b) *Photograph of the Balmer series of hydrogen, Z = 1.*

rise to a much simpler looking spectrum, the lines arranging themselves into series like those in Figs. 12B(b), 9E(f), and 9K. In addition to hydrogen, the elements revealing such simple spectral series are the elements in the first column of the periodic table (Appendix VII).

	Li	Na	K	Rb	Cs	Fa
$Z =$	3	11	19	37	55	87

The spectrum arising from each of these, the alkali metals, is composed of several spectral series in different wavelength regions, but having the same general appearance as shown in Fig. 9E. If the wavelengths of any one of these series are measured, and the frequencies are calculated and plotted, one obtains a graph of the kind shown in Fig. 12C.

The first successful attempt to fit all such lines of a series into one general formula was made by Rydberg* in 1896. His formula is

$$v_n = v_\infty - \frac{\Re}{(n + \mu)^2} \qquad (12g)$$

where v_n is the frequency of the different lines n, and v_∞ is the frequency of the series limit. For example, for the *principal series of lithium*, shown in Fig. 12C,

$$\Re = 109{,}722 \text{ cm}^{-1}$$
$$v_\infty = 43{,}488 \text{ cm}^{-1}$$
$$\mu = 0.9596$$
$$n = 1, 2, 3, 4, \ldots$$

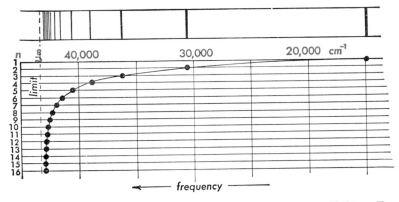

Fig. 12C   *Frequency plot of the principal series in the spectrum of lithium, $Z = 3$.*

By substituting $n = 1$ in Eq. (12g), one obtains the frequency of the first line of the lithium series, $v_n = 14{,}915$ cm^{-1}, corresponding to $\lambda = 6705$ A. By substituting $n = 2$, one obtains the frequency of the second line of the series, etc.

The success of the Rydberg formula is remarkable because by changing the constants v_∞ and μ it fits all series in all spectra. $\Re$ remains the same for all series in all elements. Because the actual frequencies of visible light waves are so extremely high, it is customary to divide them all by the speed of light. ($c = 3 \times 10^{10}$ cm/sec.) Such frequencies are then called *wave numbers*. The frequency of a spectrum line in wave numbers is therefore just equal to the number of wavelengths in a distance of 1 cm, instead of the number in 3×10^{10} cm. Frequencies, in wave numbers, have the units of $1/\text{cm}$, (cm^{-1}).

* The Rydberg constant $\Re$ used here can be obtained by dividing the value of R given in Eq.(11m) by Planck's constant h and the speed of light c.

12.4. The Spinning Electron

If the spectrum lines of hydrogen, lithium, sodium, potassium, etc., are observed under high magnification, each series member is found to be a double line. See Fig. 12D. These closely spaced doublets arise from the fact that all

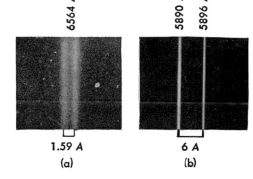

Fig. 12D *Highly magnified photographs, showing the doublet structure of (a) the first member of the Balmer series of hydrogen (the red line), and (b) the first member of the principal series of sodium (the yellow line).*

electrons are spinning. The single electron in the hydrogen atom, which is responsible for the observed spectrum, is spinning around its own axis as it moves in an orbit around the nucleus. Similarly, the single-valence electron in all sodium atoms is spinning as shown in Fig. 12E.

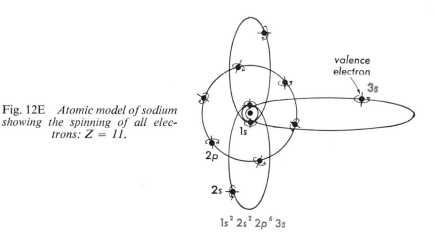

Fig. 12E *Atomic model of sodium showing the spinning of all electrons: Z = 11.*

Electrons in any completed (closed) subshell pair off with axes parallel but with opposite spin and orbit directions. Hence, in sodium: $Z = 11$, the one outermost electron $3s$ is unpaired, and its spin constitutes the resultant spin of the entire electronic system. Furthermore, it is the jumping of this valence electron from one orbit to another that is responsible for the observed spectrum.

The outermost electrons in any atom are largely responsible for its ability to combine chemically with other atoms to form molecules and are called *valence* electrons.

The spinning of electrons in atoms was first proposed by Goudsmit and Uhlenbeck, in 1925, to explain the double-line structure in the spectra of the alkali metals. Each electron, whether it is bound to an atom or a crystal, or is alone in free space, has an angular momentum. The spin angular momentum is given by

$$p_s = s\hbar \tag{12h}$$

where s is the spin quantum number and has the value $\frac{1}{2}$:

$$s = \tfrac{1}{2} \text{ only} \tag{12i}$$

Since an electron always has a negative charge, its spinning around its own mechanical axis generates a magnetic field like that shown in Fig. 12F. This field

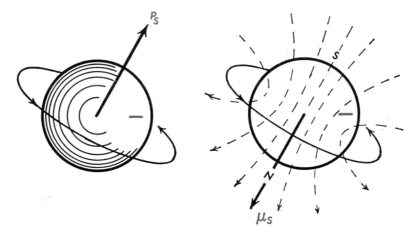

Fig. 12F   *A spinning electron has a mechanical moment p_s and a magnetic moment μ_s.*

is similar to the field around a bar magnet and may be specified by its equivalent *magnetic moment*. The magnetic moment μ_s of a spinning electron is given by

$$\frac{\mu_s}{p_s} = 2\,\frac{e}{2m} \tag{12j}$$

and is oppositely directed to its mechanical moment. Note that the *gyromagnetic ratio* is twice that for an electron orbit. Eq. (12c).

By substituting the value of p_s from Eq. (12h) in this formula, we obtain

$$\mu_s = \hbar\,\frac{e}{2m} \tag{12k}$$

Since this is the same as Eq. (12e), we see that while *a spinning electron has a mechanical moment of $\frac{1}{2}\hbar$, it has a magnetic moment of one Bohr magneton.*

12.5. Electron Spin-Orbit Interaction

The fact that the spectrum lines in the Balmer series of hydrogen, and those in the series of the alkali metals, are doublets, is interpreted to mean that each of the energy levels of these atoms is double. (See Fig. 12G for the energy level diagram of sodium.)

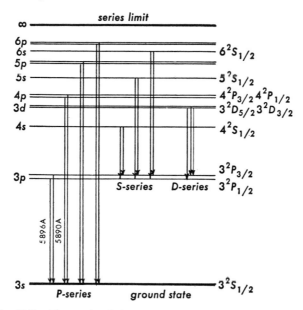

Fig. 12G *Energy-level diagram for the sodium atom, $Z = 11$.*

The doubling of energy levels in atoms having one valence electron is due to the interaction between the magnetic field of the electron orbit and the magnetic field of the electron spin. We have seen in § 3.7 that a bar magnetic located in a magnetic field has a torque exerted on it which tends to line it up parallel to the field. A spinning electron in a magnetic field behaves in exactly the same way; there is a torque acting upon it, trying to turn its axis parallel to the field. Owing to the mechanical properties of a revolving mass, the electron precesses around B in much the same way that a mechanical top precesses in a gravitational field.

A schematic diagram of the precession of a spinning electron is given in Fig. 12H. Such a motion is called a *Larmor precession*, and its frequency f is given by

$$f = \frac{e}{4\pi m} B \tag{12l}$$

A good demonstration of this precession can be made by a gyroscope of the kind shown in Fig. 12I. A nonconducting sphere, mounted free to turn in ball-

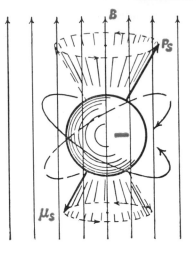

Fig. 12H *A spinning electron in a magnetic field precesses around an axis, parallel to the magnetic induction B.*

Fig. 12I *A spinning ball gyroscope will precess in a magnetic field. (Note: A small permanent bar magnet is mounted inside and on the axis of the ball.)*

bearings, is mounted in double gimbel rings and placed directly over the center of an electromagnet as shown. When the ball is set spinning in the position shown, and the magnetic field is turned on, the ball will retain its inclination angle as it precesses around the vertical axis. By reversing the magnetic field the precession will reverse direction.

Owing to orbital motion, as well as the positive charge on the nucleus, every electron in an atom is subjected to a magnetic field. To see how this field comes about, consider the simple case of a hydrogen atom with its one electron in an orbit around a positive charge. If we imagine ourselves riding around with the electron, looking out at the nucleus, we see this positive charge as though it were moving in an orbit around us. This moving charge gives rise to a magnetic field at the electron of the form shown in Fig. 12J. In this field the electron carries out a Larmor precession.

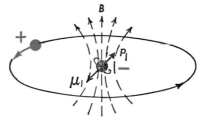

Fig. 12J   *The magnetic field at an electron in an atom is due to the positively charged nucleus that appears to be going around it in an orbit.*

Since an electron has a spin angular momentum $s\hbar$, as well as an orbital angular momentum $l\hbar$, the total angular momentum of the atom will be the vector sum of the two, and its magnitude will depend upon their relative orientations. If we represent $l\hbar$ and $s\hbar$ by vectors as shown in Fig. 12K, their

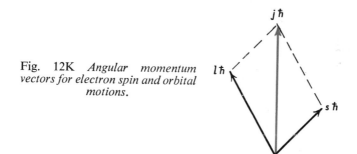

Fig.   12K   *Angular   momentum vectors for electron spin and orbital motions.*

vector sum $j\hbar$ will represent the total angular momentum. j is called the *total quantum number.*

The quantum theory requires that all possible vector sums for these two quantities differ from each other by $\hbar$. For each value of the orbital quantum number l, there are two possibilities,

either
$$j\hbar = l\hbar + s\hbar$$
or
$$j\hbar = l\hbar - s\hbar \tag{12m}$$

Since all angular momenta have the common factor $\hbar$, the quantum numbers l, s, and j, may be used as vectors, and we can write

$$j = l + s \quad \text{and} \quad j = l - s \tag{12n}$$

For a d-electron, for example, $l = 2$ and $s = \frac{1}{2}$,

$$j = 2 + \tfrac{1}{2} = \tfrac{5}{2} \quad \text{and} \quad j = 2 - \tfrac{1}{2} = \tfrac{3}{2}$$

All values of l, except $l = 0$, will give two j-values (see Fig. 12L), and these will always be half-integral. Since the total energy of the atom with l and s

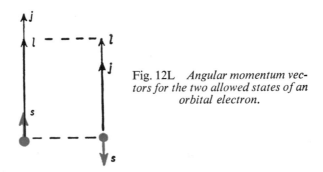

Fig. 12L *Angular momentum vectors for the two allowed states of an orbital electron.*

parallel will be different from when they are oppositely directed, all levels except s levels will be double.

12.6. Selection Rules

In doublet spectra arising from atomic systems containing but one valence electron, i.e., one electron in an incompleted subshell, the small letters s, p, d, f, g, etc., for the different electron orbits are replaced by the corresponding capitals ${}^2S, {}^2P, {}^2D, {}^2F, {}^2G$, etc., for the energy levels. The small superscript 2 in front of each capital letter indicates that the level in question, including S levels, has doublet properties and belongs to a doublet system.

Although all S levels are single, their doublet nature will later be seen to reveal itself when the atom is placed in a magnetic field. In order to distinguish between two fine-structure levels having the same n and l values, the cumbersome but theoretically important half-integral subscripts are used.

Observation shows that, for the transition of an electron from one energy state to another, definite selection rules are in operation. This is illustrated in Fig. 12M by six different sets of transitions. From these diagrams, which are based upon experimental observations, selection rules for doublets may be summarized as follows: In an electron transition

$$l \text{ changes by } +1 \text{ or } -1 \text{ only}$$

and is written

$$\Delta l = \pm 1$$

$$j \text{ changes by } 0, +1 \text{ or } -1 \text{ only}$$

and is written

$$\Delta j = 0, \pm 1 \qquad \text{with } 0 \to 0 \text{ forbidden.}$$

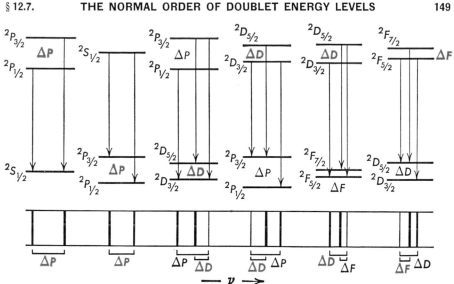

Fig. 12M   *Diagrams of selection rules for transitions between doublet energy levels, and the appearance of the corresponding spectrum lines.*

The total quantum number n has no restrictions and may change by any integral amount. Note that transitions involving a change of l by 2, e.g., $^2D \rightarrow {}^2S$, are forbidden.

Note how the energy level differences ΔP, ΔD, and ΔF show up as frequency differences between pairs of spectrum lines. This follows from the fact that energies of all emitted photons are given by $h\nu$.

The relative intensities of the observed spectrum lines are illustrated by the widths of the lines directly below each transition arrow at the bottom of the figure. Combinations between ^2P and ^2S always give rise to a doublet, whereas all other combinations give rise to two strong lines and one fainter line.

In designating any spectrum line like one of the yellow doublet in sodium, $\lambda = 5890$ A, the lower state is written first followed by the upper state thus, $3^2S_{1/2} - 3^2P_{3/2}$. The other line of the sodium doublet, $\lambda = 5896$ A, is written $3^2S_{1/2} - 3^2P_{1/2}$. Spectrum lines in absorption are written the same way, the lowest level first.

12.7. The Normal Order of Doublet Energy Levels

In the doublet energy levels of atomic systems containing one valence electron it is generally, but not always, observed that the energy level with $j = l - \frac{1}{2}$ lies deeper than the corresponding level $j = l + \frac{1}{2}$. For example, in the case of p and d electrons, $^2P_{1/2}$ lies deeper than $^2P_{3/2}$ and $^2D_{3/2}$ lies deeper than $^2D_{5/2}$.

As explained in § 12.5 and shown in Fig. 12J, the orbital electron is subjected to a magnetic field arising from the positively charged nucleus. In this field the most stable state of a given doublet will be the one in which the spinning electron, thought of as a small magnet of moment μ_s, is parallel to the magnetic field. This is the state in which $j = l - \frac{1}{2}$. In the state $j = l + \frac{1}{2}$, the spin magnetic moment μ_s is opposite in direction to the field. See Fig. 12N.

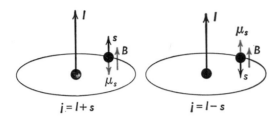

Fig. 12N   *Illustrating the mechanical and magnetic moments of the spinning electron for doublet energy levels* $i = l + s$ *and* $i = l - s$.

Of the two possible orientations of the electron the one $j = l - \frac{1}{2}$ is classically the more stable and lies deeper on an energy level diagram. To turn any magnet from a position where it lines up parallel to the field in which it is located to any position where it makes an angle with the field requires work to be done.

12.8. Space Quantization and the Zeeman Effect

When hydrogen atoms are located in a magnetic field B, the quantum theory requires that their electrons take on certain specified directions. These directions are determined as follows: The projection of the total angular momentum $j\hbar$ on the field direction B (see Fig. 12O) must take on half-integral values of $\hbar$, from

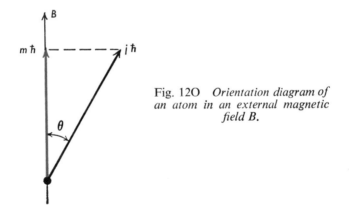

Fig. 12O   *Orientation diagram of an atom in an external magnetic field B.*

the largest to the smallest possible, and all half-integral values in between. As an equation

$$j\hbar \cos \theta = m\hbar \qquad (12o)$$

where m is the *magnetic quantum number*, and is given by

$$m = \pm\tfrac{1}{2}, \pm\tfrac{3}{2}, \pm\tfrac{5}{2}, \ldots \pm j \qquad (12p)$$

If, for example, the total quantum number $j = \frac{5}{2}$, the allowed orientations of j are six in number, and are specified by

$$m = +\tfrac{5}{2}, +\tfrac{3}{2}, +\tfrac{1}{2}, -\tfrac{1}{2}, -\tfrac{3}{2}, -\tfrac{5}{2} \qquad (12q)$$

These six orientations are shown in Fig. 12P, and the process is referred to as

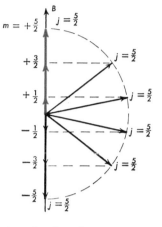

Fig. 12P   *Vector diagram repre-*
senting space quantization of an
atom in a magnetic field B.

space quantization. The minus values indicate only that the component of $j\hbar$ is opposite in direction to the field *B*.

A diagram of the precession of an atom in a magnetic field, when the electron is in a state $l = 2$, $s = \tfrac{1}{2}$, $j = \tfrac{5}{2}$, and $m = \tfrac{3}{2}$, is shown in Fig. 12Q. Because the

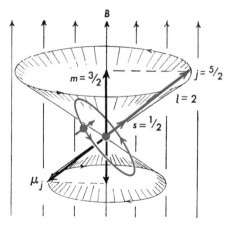

Fig. 12Q   *An electron spin and*
orbit precess together as a unit
around the magnetic field direc-
tion B.

energy of the atom differs slightly between the different orientations of the electron, energy levels as well as spectrum lines will be split into a number of equally spaced components. To observe this phenomenon, the light source must be placed in a strong and uniform magnetic field, and the light observed with a spectrograph. The phenomenon, called the Zeeman effect, is observed in the spectra of all elements. Some line patterns contain but a few lines, while others contain many. Typical Zeeman patterns, as seen under high magnification, are reproduced in Fig. 12R.

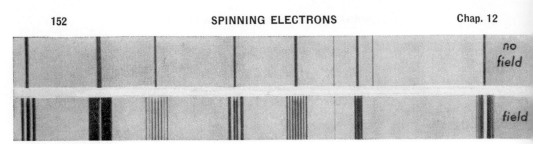

Fig. 12R *Small section of the spectrum of rhodium, Z = 45.* (Upper) *Ordinary lines with no magnetic field.* (Lower) *Zeeman patterns when light source is in strong magnetic field of 7.0 web/meter². (After Harrison and Bitter.)*

12.9. Pauli Exclusion Principle

We have now seen that four quantum numbers are required to specify the state of an electron in an atom. These are

n principal quantum number
l orbital quantum number
j total quantum number
m magnetic quantum number

(12r)

According to the Pauli exclusion principle, no two electrons in the same atom can have all four quantum numbers alike. They may have three alike but at least one must be different.

Consider, for example, the number of electrons that can have $n = 3$ and $l = 2$. Such electrons are designated $3d$. The two j-values possible are $j = l + s$ and $j = l - s$, i.e., $j = \frac{5}{2}$ and $j = \frac{3}{2}$. (See Eq. (12n).) For $j = \frac{5}{2}$ there are six possible values of m, and for $j = \frac{3}{2}$ there are four possible values of m. Together, these are

$$j = \tfrac{5}{2}, m = +\tfrac{5}{2}, +\tfrac{3}{2}, +\tfrac{1}{2}, -\tfrac{1}{2}, -\tfrac{3}{2}, -\tfrac{5}{2}$$
$$j = \tfrac{3}{2}, m = +\tfrac{3}{2}, +\tfrac{1}{2}, -\tfrac{1}{2}, -\tfrac{3}{2}$$

(12s)

or ten possibilities in all. Note that this is just the number of electrons that fills an $l = 2$ subshell in the building up of elements in the periodic table. In a similar way, an s-subshell can have two electrons, a p-subshell 6, and an f-subshell 14.

12.10. Elastic and Inelastic Impacts

When an electron collides with a neutral atom in a rarefied gas, the collision is either *elastic* or *inelastic*. An elastic collision is one in which the laws of conservation of momentum and conservation of mechanical energy are both upheld. In other words, the atomic particles behave as though they were perfectly elastic spheres; the total energy and total momentum before impact is equal to the total energy and total momentum after impact. (See Fig. 12S.)

An inelastic collision is one in which the impinging electron, in striking a neutral atom, hits one of the electrons and either knocks it into one of the outer

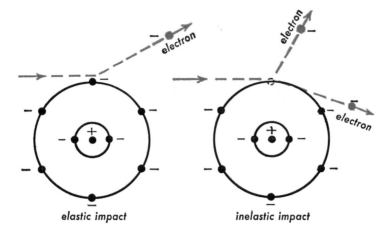

elastic impact   inelastic impact

Fig. 12S   *Schematic diagrams showing elastic and inelastic impacts.*

orbits (energy levels), or knocks it completely out of the atom. In the first instance, we say the atom has been *excited*, and in the second case it has been *ionized*. In either case it is usually a valence electron that is involved.

To raise an electron from its normal state to an excited state, or to remove it from the atom, requires the expenditure of energy; this is supplied by the impinging electron. As a consequence, some of the total energy before collision is used for excitation or ionization, and what is left is divided between the two particles. Because the masses of atoms are thousands of times that of the electron, nearly all kinetic energy before impact, and after, is confined to the electron. The recoil velocity and kinetic energy of an atom that has been hit by a moving electron is relatively small.

12.11. Franck-Hertz Experiments

One of the most direct proofs of the existence of energy levels or electron shells within an atom is to be found in the Franck-Hertz experiments. These experiments make it possible to measure the energy necessary to raise an electron from the ground state in an atom to an outer orbit, or state, or to remove it from the atom entirely.

In the Franck and Hertz experiments, a vapor like sodium or mercury is bombarded with electrons of known velocity. A diagram of the apparatus in its simplest form is given in Fig. 12T. Electrons from a hot filament F are accelerated toward a grid G by applying a voltage V between them. An opposing voltage V', much smaller than V, is applied between the grid G and the plate P. If sodium atoms are to be used, the gas pressure is reduced until the mean distance between atomic collisions (the mean free path) is considerably smaller than the filament-to-grid distance, and somewhat greater than the grid-to-plate distance.

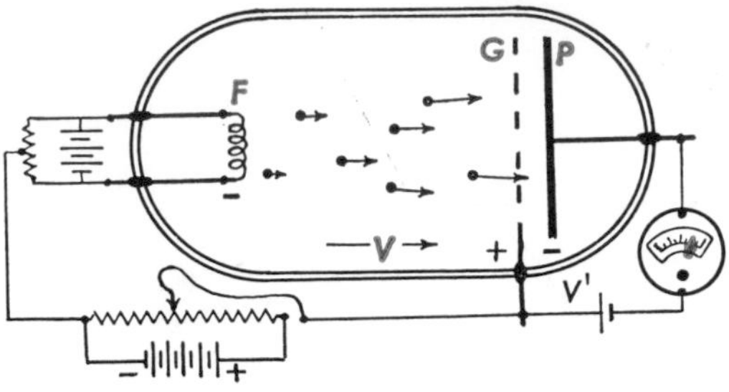

Fig. 12T *Diagram of the Franck and Hertz experiment.*

If an electron starts from rest at the filament and reaches the grid G without hitting a sodium atom, its velocity v is given by the equation

$$Ve = \tfrac{1}{2}mv^2 \tag{12t}$$

where V is in volts, e is in coulombs, m is in Kg, and v is in m/sec. See Eq. (10b). In order for an electron to collide inelastically with an atom, it must have sufficient kinetic energy $\tfrac{1}{2}mv^2$ to raise the valence electron from its ground level to the first excited level. To find how much energy this is, consider the example of sodium, $Z = 11$, in which the first energy state above the ground state is responsible for the yellow light we observe in any sodium arc light or lamp. This light arises from the jump of the valence electron from a $3p$ orbit, into the ground state, a $3s$ orbit. See Fig. 12G. This light of wavelength $\lambda = 5893$ A has a frequency $v = 5.091 \times 10^{14}$ vib/sec, and each photon an energy $hv = 3.359 \times 10^{-19}$ joules.

To find the voltage equivalent of this energy, we write

$$Ve = hv \tag{12u}$$

and substituting the electronic charge $e = 1.601 \times 10^{-19}$ coulombs and $h = 6.6238 \times 10^{-34}$ joule seconds, we obtain

$$V = 2.1 \text{ volts}$$

This means that electrons accelerated by a potential difference of 2.1 volts or more, upon collision with a normal sodium atom, can knock the single valence electron from its $3s$ orbit into the $3p$ orbit. In so doing the electron gives up kinetic energy equivalent to 2.1 electron volts and slows down accordingly.

The curve reproduced in Fig. 12U, showing the variation in plate current with the accelerating voltage V, is characteristic of the Franck and Hertz experiment with sodium. As V starts from zero and is slowly increased, the speed and number of electrons reaching the grid G increase and the plate current rises. When the velocity has increased sufficiently to excite sodium atoms, however, inelastic collisions occur.

With a further increase in V, more electrons reach the critical velocity, are stopped by inelastic collision, and, not being able to reach the plate P, cause a

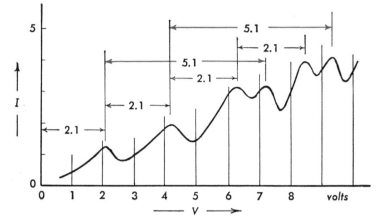

Fig. 12U   *Critical potential curve for sodium atoms showing the excitation potential of 2.1 volts and the ionization potential of 5.1 volts.*

drop in the current. This drop in current continues until the critical speed is attained far enough in front of the grid to collide inelastically again and reach the plate. The current, therefore, rises again and continues to rise until the electrons, after one inelastic collision, attain the critical speed and make a second inelastic collision.

The double peaks in Fig. 12U show that not only have collisions occurred in which the valence electron has been excited, but also collisions in which the electron is completely removed from the atom. Complete ejection of the electron from sodium requires an electron velocity equivalent to $V - 5.1$ volts. This then is the ionization potential of sodium, which in Fig. 12G is equivalent to raising the valence electron from its *ground state* to the *series limit*.

12.12. Ionization Potentials

The ionization potential of an element is defined as the energy in electron volts required to remove the most loosely bound electron from the normal atom. The *electron volt* is defined as the energy equivalent to the kinetic energy of an electron accelerated through a potential difference of 1 volt. In other words an energy of 1 ev is 1.60×10^{-19} joule.

A graph of the ionization potentials of the elements of the periodic table is given in Fig. 12V. It is clearly seen that the alkali metals, Li, Na, K, Rb, Cs, and Fa, have low values, about 5 volts, while the inert gases, He, Ne, A, Kr, Xe, and Rn, have the highest values. The inert gases represent those atoms in which all the electron subshells are complete. The adding of one more proton to the nucleus, and the addition of one more electron to the outer structure to make the next atom, an alkali metal, requires that electron to go into a new subshell, an outer orbit. Such an electron, being on the average farther away from the nucleus, requires less energy to remove it from the atom. As protons are added

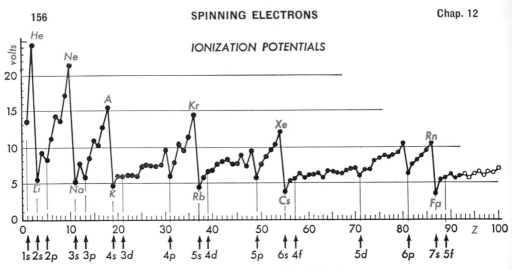

Fig. 12V   *Ionization potentials of the elements.*

to the nucleus, and electrons are added in this same subshell, the binding of the electrons grows stronger and stronger.

A direct correlation between the low ionization potentials of the alkali metals and the beginning of new subshells, as shown by the arrows at the bottom of Fig. 12V, leads to the building-up of the periodic table as given in Appendix VI.*

QUESTIONS AND PROBLEMS

1. What is the atomic unit of angular momentum? What is the unit of magnetic moment?

2. What is the Rydberg formula? What is Larmor precession?

3. What is a total quantum number? What is the Pauli exclusion principle? What are the four quantum numbers involved in the Pauli exclusion principle?

4. What is the Zeeman effect? What is space quantization?

5. What was the Franck-Hertz experiment?

6. What is an excitation potential? What is an ionization potential?

7. The Sharp Series of spectrum lines of lithium has a series limit at 28573 cm^{-1}. If the series constant $\mu = 0.5884$, what is (a) the frequency, and (b) the wavelength, of the second, third, and fourth lines of the series?

8. Using the Rydberg formula, calculate (a) the frequency, and (b) the wavelength of the second member of the principal series of lithium. *Ans.* (a) 30952 cm^{-1}, (b) 3231 A.

9. Make a space quantization diagram for the electronic state, $l = 4$, $s = \frac{1}{2}, j = \frac{7}{2}$.

10. Calculate the excitation potential for the first line of the principal series of lithium.* (*Note:* The energy level diagram for lithium is quite similar to that for sodium, Fig. 12G. See Fig. 12C.) *Ans.* 1.84 volts.

11. Calculate the ionization potential for lithium.† (*Note:* The limit of the principal series represents the valence electron at infinity.) See Fig. 12C.

---

* For a more complete account of atomic structure, see *Introduction to Atomic Spectra* by H. E. White, McGraw-Hill Book Company, New York.

† Frequency in wave numbers, cm^{-1}, when multiplied by the speed of light, $c = 3 \times 10^{10}$ cm/sec, gives the true frequency ν.

12. (a) Plot an energy-level diagram for potassium, $Z = 19$, using the following values for the term values in wave numbers.

CONFIG.	SYM.	J	TERM VALUES
4s	^2S	$\frac{1}{2}$	35005.9
4p	^2P	$\frac{1}{2}$	22020.8
		$\frac{3}{2}$	21963.1
5s	^2S	$\frac{1}{2}$	13980.3
3d	^2D	$\frac{3}{2}$	13470.3
		$\frac{5}{2}$	13467.5
5p	^2P	$\frac{1}{2}$	10304.4
		$\frac{3}{2}$	10285.7
4d	^2D	$\frac{3}{2}, \frac{5}{2}$	7608.3
6s	^2S	$\frac{1}{2}$	7555.7
4f	^2F	$\frac{5}{2}, \frac{7}{2}$	6878.5
6p	^2P	$\frac{1}{2}, \frac{3}{2}$	6009.3
5d	^2D	$\frac{3}{2}, \frac{5}{2}$	4821.9

(b) Show all allowed transitions based upon the selection rules. (c) Calculate the frequencies in wave numbers for the first members of the sharp, principal, diffuse, and fundamental series of spectrum lines. (d) Calculate the corresponding wavelengths to four significant figures. Make a table of your answers.

13. Calculate the excitation potential for the first line of the principal series of potassium. Use the energy level values given in Problem 12.

14. Calculate the ionization potential for potassium. See Problem 12, for energy levels. *Ans.* 4.76 v.

15. Calculate the excitation potential for the second line of the principal series of potassium. Use Problem 12 for energy levels.

Atoms with Two Valence Electrons

In the preceding chapter the detailed structure of the simplest of the known atomic systems has been presented. We have seen that a few elements, those classified as one-electron systems, always give rise to doublet energy levels, and that transitions of the single valence electrons between these levels give rise to observed doublet series of spectrum lines.

It is now well known that atoms like beryllium, magnesium, zinc, cadmium, mercury, calcium, strontium, barium, and radium, in Group II of the periodic table, contain two valence electrons and give rise to series of *singlet* and *triplet* spectrum lines and energy levels. See Appendix VII.

Like the doublet spectra of the alkali metals the singlet and triplet series may be grouped into four chief series: *sharp, principal, diffuse,* and *fundamental.* In the spectra of each element four chief series of singlets as well as four chief series of triplets have been identified.

If any series is plotted on a wavelength or frequency scale, as shown for lithium in Fig. 12C, a series limit is easily recognized, and when the Rydberg formula, Eq. (12g), is applied to the accurately determined frequencies, the entire series is found to conform. From the calculated constants in the Rydberg formula one can show that all series of any given element can be represented by transitions between a simpler set of energy levels.

13.1. Triplet Fine Structure

In examining the fine structure of triplets from any one element it is observed that (1) all members of the sharp series are composed of three spectrum lines with exactly the same frequency separation, (2) all members of the principal series are composed of three lines with decreasing frequency separations, and (3) all members of the diffuse series and fundamental series contain six lines, three strong lines and three fainter lines. Upon examining Fig. 13A some of these relationships can be observed.

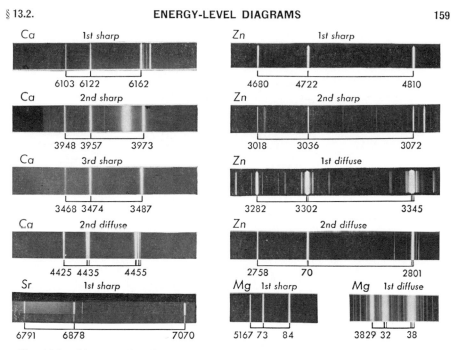

Fig. 13A   *Photographs of triplets taken from original spectrograms by Dr. A. S. King of the Mount Wilson Observatory.*

As a rule the over-all width of triplets for different elements increases rapidly with atomic number. Note in the figure the relatively narrow width of the sharp and diffuse triplets of magnesium, $Z = 12$, the wider spacing of the sharp and diffuse triplets of calcium, $Z = 20$, and the still wider spacing of the sharp and diffuse triplets of zinc, $Z = 30$.

13.2. Energy-Level Diagrams

An energy-level diagram of calcium is given in Fig. 13B. To avoid confusion the levels are not drawn across the page as they are for sodium in Fig. 12G, but are shortened and grouped in five columns. The ^1S and ^3S levels are on the left, followed by ^1P and ^3P, then ^1D and ^3D, and ^1F and ^3F. The fifth column contains a special group and will be described in § 13.8.

This typical diagram for two valence electron atoms reveals several general features of importance. With one exception, namely, the *normal state*, every singlet level has associated with it a triplet level of the same designation. Furthermore all ^3S levels are single. Triplet intervals as well as singlet-triplet intervals decrease as higher and higher levels are reached.

Since the energy of radiated light waves is given by the Bohr relation $E_2 - E_1 = h\nu'$, it is common practice to divide energies by h and obtain

$$\nu' = \frac{E_2}{h} - \frac{E_1}{h} \tag{13a}$$

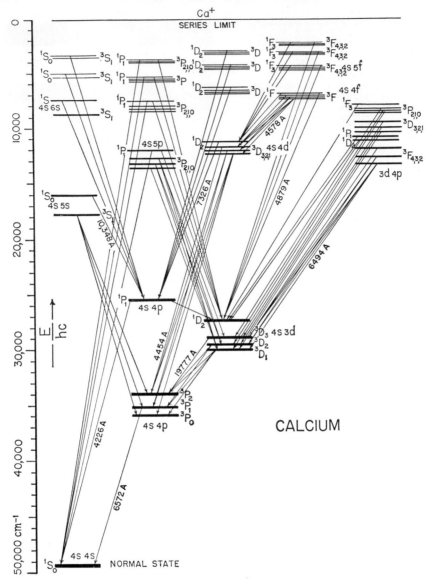

Fig. 13B *Energy level diagram for neutral calcium atoms, showing electron configurations and some of the more prominent spectral transitions.*

Since observed values of v', the actual radiated frequencies of light, are of the order of 10^{14} vib/sec, it is customary to divide the frequency by c, and obtain

$$v = \frac{E_2}{hc} - \frac{E_1}{hc} \tag{13b}$$

where v'/c is now written v and represents the frequency in *wave numbers*. Since the frequency of light is the number of waves in 3×10^{10} cm, dividing by c gives the number of waves per centimeter. For example, for green light of wavelength of 5000 A, the true frequency is 6×10^{14} vib/sec, and dividing by 3×10^{10} cm/sec gives 2×10^4 vib/cm, or 20,000 cm^{-1}.

By plotting the energy levels to a wave number scale, i.e., by plotting E/hc, the difference between any two levels gives directly the frequency of the corresponding spectrum line in wave numbers.

The principal series of singlet spectrum lines begins with a strong violet line at 4226 A and arises from transitions starting with the series of ^1P levels and ending on the normal state, ^1S. The sharp series of singlet spectrum lines begins with the infrared line at 10,348 A and arises from transitions starting on ^1S levels and ending on the lowest ^1P level.

The diffuse series of singlet lines, beginning with the deep red spectrum line at 7326 A, arises from transitions starting on the series of ^1D levels and ending on the lowest ^1P level, while the fundamental series of lines, beginning with the blue line at 4879 A, arises from transitions starting on the series of ^1F levels and ending on the lowest ^1D level.

The sharp, principal, diffuse, and fundamental series of triplets in a similar way always arise from transitions between triplet designated levels. In general, transitions between lower levels on an energy-level diagram give rise to the strongest observed lines, while transitions between higher levels, as well as higher series numbers, involve weaker lines.

Only occasionally does one observe lines arising from transitions between singlet and triplet levels. Such spectrum lines as 6572 A are called *intercombination lines* and are usually very faint. Many transitions between higher levels in Fig. 13B take place but are not shown, and series of lines ending on a common higher level are called *combination series*.

13.3. Selection Rules

To distinguish between the fine-structure levels of triplets, quantum numbers J are used as subscripts in a manner similar to that for doublets in § 12.6. These numbers are based upon the spin and orbital quantum numbers of both valence electrons, and will be shown in § 13.6 to result in the following assignments:

$$
\begin{array}{llll}
^1S_0 & ^3S_1 & & \\
^1P_1 & ^3P_2 & ^3P_1 & ^3P_0 \\
^1D_2 & ^3D_3 & ^3D_2 & ^3D_1 \\
^1D_3 & ^3F_4 & ^3F_3 & ^3F_2
\end{array}
\qquad (13c)
$$

The superscript gives the multiplicity to which the level belongs, and the subscript the J-value.

Typical diagrams of triplet energy levels, their appropriate fine-structure designations, allowed transitions, and typical observed spectrum line patterns are shown in Fig. 13C. In all energy-level diagrams, regardless of the number

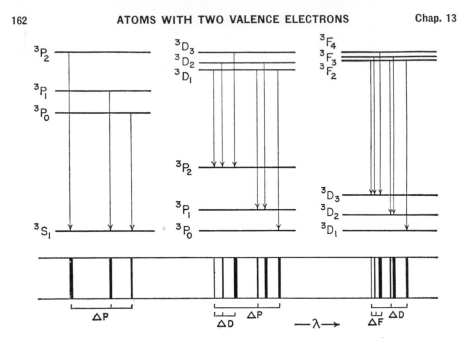

Fig. 13C   *Typical diagrams of triplet energy levels, their appropriate fine-structure designations, allowed transitions, and observed spectrum patterns.*

of valence electrons, the level designations, frequently called *term designations*, correspond to the orbital quantum numbers L as follows:

$$S \quad P \quad D \quad F \quad G \quad H \quad I \quad J \quad K \quad \cdots$$
$$L = 0 \quad 1 \quad 2 \quad 3 \quad 4 \quad 5 \quad 6 \quad 7 \quad 8 \quad \cdots \qquad (13d)$$

Whether the levels are singlets or triplets, transitions between them are observed to obey the following rule:

$$L \text{ changes by } +1 \text{ or } -1 \text{ only}$$
$$\Delta L = \pm 1 \qquad (13e)$$

According to this selection rule transitions between S and P, P and D, D and F, F and G levels can occur, but not between S and D, P and F, D and G levels.

Transitions between fine-structure levels obey the J-value selection rule, which states that

$$J \text{ changes by } +1, 0, \text{ or } -1 \text{ only}$$
$$\Delta J = 0, \pm 1 \qquad \text{with } 0 \to 0 \text{ forbidden.} \qquad (13f)$$

Applying this rule to the central diagram in Fig. 13C it will be seen that of nine possible transitions, the six shown are allowed, but that three, not shown, are forbidden. The forbidden transitions are, ${}^3P_0$–${}^3D_2$, ${}^3P_1$–${}^3D_3$, and ${}^3P_0$–${}^3D_3$. The right-hand diagram shows another line pattern arising from six allowed transitions, with three forbidden transitions not shown. Since these triplet patterns actually contain more than three lines, they are frequently called *multiplets*.

In atoms containing but one valence electron the superscript 2 signifies *doublets* and is a numerical representation of the quantity $(2s + 1)$, where s is the electron spin $\frac{1}{2}$. In § 13.6, we will see that for two electron systems, the two electron spins of $\frac{1}{2}$ may be added together to give a spin resultant of $S = 1$, or subtracted to give a spin resultant of $S = 0$. In the case $S = 1$, the quantity $(2S + 1) = 3$, and we have the superscript for the *triplet levels*, while for $S = 0$, the quantity $(2S + 1) = 1$, and we have the superscript for *singlet levels*. See § 13.4.

The selection rules for multiplicities may be written

$$\Delta S = 0$$

which means that triplets combine with triplets and singlets with singlets only. While this rule is observed to be violated in some cases, the observed spectrum lines are relatively very weak.

13.4. Intensity Rules

The relative intensities of the lines of any spectral series are such that the first member is the strongest, and succeeding members decrease smoothly and regularly as they approach the series limit.

Within the fine structure of any given triplet, as shown in Fig. 13C, the relative intensities are given by a set of formulas based upon what are called *the sum rules*. The sum rules are as follows: (1) The sum of the intensities of all lines of a multiplet, which start from a common initial level, is proportional to the quantum weight $(2J + 1)$ of the initial level. (2) The sum of the intensities of the lines of a multiplet which end on a common level is proportional to the quantum weight $(2J + 1)$ of the final level. The quantum weight $(2J + 1)$ yields the number of magnetic levels into which a given fine structure level J will split when the atom is located in a magnetic field.

Numbers satisfying the sum rules for a ^3P–^3D multiplet are presented in Table 13A. The strongest line $^3P_2$–$^3D_3$ should, for example, be 84 times more

Table 13A. Relative Intensities for
a ^3P–^3D Multiplet

	$^3D_3$	$^3D_2$	$^3D_1$		
$^3P_2$	84	15	1	100	5
					..
$^3P_1$		45	15	60	3
					..
$^3P_0$			20	20	1
	84	60	36		
	7 :	5 :	3		

intense than the line $^3P_2$–$^3D_1$. If these numbers are added *down* and *across*, as they are below and at the right of the array, the sum rules are seen to be obeyed.

Note that the three strongest lines of the multiplet of six lines are on the diagonal of the array and occur where both L and J change in the same way, both by $+1$ or both by -1. Jumping from a $^3D_3$ to a $^3P_2$ level, L changes by -1 and J changes by -1.

Formulas for relative intensities have been derived from quantum theory and transitions giving rise to all the lines of any one multiplet are given by the following:

For transitions $L - 1$ to L

$$J - 1 \text{ to } J: \quad I = +(L + J + S + 1)(L + J + S)(L + J - S)$$
$$(L + J - S - 1)/2J$$

$$J \text{ to } J: \quad I = -(L + J + S + 1)(L + J - S)(L - J + S)$$
$$(L - J - S - 1)(2J + 1)/2J(J + 1)$$

$$J + 1 \text{ to } J: \quad I = +(L - J + S)(L - J + S - 1)(L - J - S - 1)$$
$$(L - J - S - 2)/2(J + 1)$$

As an example, for the transition $^3P_2 \rightarrow {}^3D_3$, we have $L = 2$, $J = 3$, and $S = 1$, and the first equation is used to find $I = 84$. For the transition $^3P_1 \rightarrow {}^3D_1$, we have $L = 2$, $J = 1$, $S = 1$, and the second equation is used to find $I = 15$.

13.5. Two Valence Electrons

In the preceding chapter we have seen that, for hydrogen and the alkali metals, the interaction between the electron spin s and the orbit l of the single-valence electron splits each energy level P, D, F, etc., into two fine-structure levels.

The nine elements in Group II of the periodic table each contain two valence electrons and these are in s orbits. The complete electron configurations for the first four of these elements are listed in Table 13B. The exponents specify the total number of electrons in the subshell.

Table 13B. Electron Configurations for Group II Elements

4	Be	$1s^2\ 2s^2$
12	Mg	$1s^2\ 2s^2\ 2p^6\ 3s^2$
20	Ca	$1s^2\ 2s^2\ 2p^6\ 3s^2\ 3p^6\ 4s^2$
30	Zn	$1s^2\ 2s^2\ 2p^6\ 3s^2\ 3p^6\ 3d^{10}\ 4s^2$

It is the last two electrons in each of these configurations (set in heavy type) that are responsible for the chemical valence of two, and are chiefly responsible for the general characteristics of their spectra. An orbital diagram of calcium showing each of its twenty electrons is given in Fig. 13D.

Just as in the alkali metals, we may think of all but the last subshell of electrons as constituting a *core* of completed subshells, around and outside of which the valence electrons move in various types of orbits, some orbits penetrating the core and coming close to the nucleus and others remaining well outside the core.

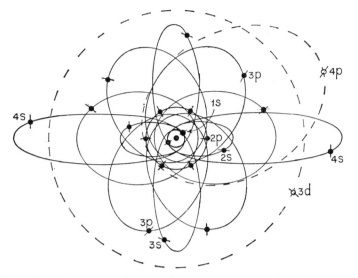

Fig. 13D *Orbital diagram of a neutral calcium atom, Z = 20, showing its twenty electrons and two of the many allowed virtual orbits in which the valence electrons may be excited.*

In calcium there are, in the normal state, two electrons in 4s orbits. If the calcium atom is excited by some means or other, either or both of these electrons may be excited to higher unoccupied orbits. Two such orbits, 3d and 4p, shown dotted in the diagram, represent but two of the many such virtual orbits in which the electrons might move. It turns out that all eight of the chief series of spectrum lines and energy levels result from the excitation of only one of the valence electrons.

Noting what subshells of electrons are already filled in calcium, one can predict with reasonable certainty that the first excited states of the atom should be those in which one valence electron is excited to the 3d or 4p orbit. See Appendix VI. Experiment confirms this as shown by the low-lying levels ^3P, ^1P, ^3D, and ^1D levels in 13B.

The lowest ^3P and the lowest ^1P are attributed to one of the valence electron's being in the 4s orbit and the other in a 4p orbit. The electron configuration for these states is $1s^2\ 2s^2\ 2p^6\ 3s^2\ 3p^6\ 4s\ 4p$, and is abbreviated 4s 4p. Similarly the lowest ^3D and ^1D levels are attributed to the abbreviated configuration 4s 3d.

By the same process the next higher ^3P and ^1P levels arise from the configuration 4s 5p. By continuing these assignments of the observed energy levels, successive members of the ^3P and ^1P series in calcium have the electron configurations 4s 4p, 4s 5p, 4s 6p, 4s 7p, etc., or in short, 4s np, where $n = 4, 5, 6, 7$, etc.

Regardless of the series of orbits through which we may imagine the one valence electron to be carried, we are finally left with one valence electron in a

4s orbit, and the other completely removed from the atom. In other words, the top level in Fig. 13B, which is the limit of all the chief series of levels, represents the lowest level of the ionized atom.

13.6. The Vector Model

Attempts to calculate the fine-structure separations of various energy levels arising from any given atom have been made by many investigators. Prior to the development of quantum mechanics, the vector model qualitatively accounted for the fine structure of all spectra.

We have already seen in the preceding chapter that in addition to the orbital angular momentum $lh/2\pi$ (abbrev. $l\hbar$)

$$l = 0 \quad 1 \quad 2 \quad 3 \quad 4 \quad 5 \quad \ldots$$
$$\text{for} \quad s \quad p \quad d \quad f \quad g \quad h \quad \text{electrons,}$$

each electron has a spin angular momentum $s\hbar$, and $s = \frac{1}{2}$ only. The angular momentum of any atom containing n electrons is therefore the sum of $2n$ angular momenta (n spins and n orbits). Since the spins and orbits of all closed subshells cancel out, the resultant angular momentum of the entire electron structure is determined by the two valence electrons when one or both are excited to unoccupied orbits.

Consider first the orbital motions of two electrons and let l_1 and l_2 represent their respective quantum numbers. For simplicity we may also let these numbers represent the respective orbital angular momenta $l_1\hbar$ and $l_2\hbar$. Following what is called the *Russell-Saunders*, or *LS-coupling scheme*, l_1 and l_2 are quantized in such a way that they form a resultant L, where different values of L are designated by the capital letters in Eq. (13d) and are the vector sums of l_1 and l_2.

Suppose for example that one electron is in a p orbit and the other is in a d orbit. Here the two vectors, one and two units long respectively, may orient themselves in any one of three positions as shown in Fig. 13E. The

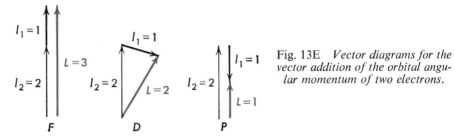

Fig. 13E   *Vector diagrams for the vector addition of the orbital angular momentum of two electrons.*

three orbital resultants $L = 1$, $L = 2$, and $L = 3$ correspond to P, D, and F levels, respectively.

In the case where one electron is in an s orbit, with $l_1 = 0$, and the other in a p, d, or f orbit, with $l_2 = 1$, 2, or 3, respectively, there is only one vector resultant possible, and $L = l_2$. If $l_1 = 0$ and $l_2 = 0$, the resultant $L = 0$ corresponds to an S level. With $l_1 = 0$, and $l_2 = 1$, the resultant $L = 1$ gives a P level, and with $l_1 = 0$, and $l_2 = 2$, the resultant $L = 2$ gives a D level, etc. For these cases, corresponding to the chief series in calcium, no vector diagram need be drawn. From the above results we obtain Table 13C.

**Table 13C. Energy Levels Arising from
Two Valence Electrons**

$ss \rightarrow S$	$pp \rightarrow S\ P\ D$	$dd \rightarrow S\ P\ D\ F\ G$
$sp \rightarrow P$	$pd \rightarrow P\ D\ F$	$df \rightarrow P\ D\ F\ G\ H$
$sd \rightarrow D$	$pf \rightarrow D\ F\ G$	$dg \rightarrow D\ F\ G\ H\ I$
$sf \rightarrow F$	$pg \rightarrow F\ G\ H$	$dh \rightarrow F\ G\ H\ I\ J$

With two electrons, each having a spin angular momentum of $s\hbar$, where $s = \frac{1}{2}$, there are two ways in which a resultant spin angular momentum $S\hbar$ may be formed. Let s_1 and s_2 represent the two respective spin vectors of the two electrons. Quantizing these as shown in Fig. 13F we obtain the two resultants

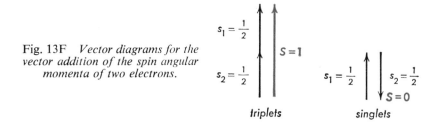

Fig. 13F *Vector diagrams for the vector addition of the spin angular momenta of two electrons.*

$s_1 = \frac{1}{2}$

$s_2 = \frac{1}{2}$

$S = 1$

triplets

$s_1 = \frac{1}{2}$ $s_2 = \frac{1}{2}$

$S = 0$

singlets

$S = 1$ and $S = 0$. The value $S = 1$ will now be shown to give rise to triplet levels, and $S = 0$ to singlet levels. Including doublet levels of one electron systems, where $s = \frac{1}{2}$, we can now write

$S = 0$	$S = \frac{1}{2}$	$S = 1$
Singlets	**Doublets**	**Triplets**

With the orbital motions of two electrons coupled together to give a resultant L, and the spins of the same two electrons coupled together to form a resultant S, both L and S will in turn be coupled together to form J, the vector sum of L and S. This resultant represents the total angular momentum of the atom. The quantum conditions imposed upon these vector additions are that J take on integral values only.

Consider the case of a ^3F state in which the spin resultant $S = 1$ and the orbital resultant $L = 3$. Three resultants are possible, $J = 4, 3$, and 2, as shown in Fig. 13G. These correspond to the three fine-structure levels $^3F_4$, $^3F_3$, and $^3F_2$.

For a ^3D level in which $S = 1$, and $L = 2$, three J values are possible, $J = 3$, $J = 2$, and $J = 1$. These are the values for the three levels $^3D_3$, $^3D_2$, and $^3D_1$, respectively.

If the spin resultant $S = 1$ and the orbital resultant $L = 0$ for an S level,

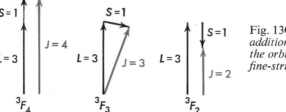

Fig. 13G   *Vector diagrams for the addition of the spin resultant S and the orbital resultant L for the three fine-structure levels of a ^3F energy state.*

the result J can have one value only, namely, $J = 1$. This corresponds to a $^3S_1$ level.

If the spin resultant $S = 0$ and the orbital resultant $L = 1, 2$, and 3, vector diagrams need not be drawn, since $J = 1, 2$, and 3, respectively, and each in turn corresponds to $^1P_1$, $^1D_2$, and $^1F_3$ levels.

If both the spin resultant $S = 0$ and the orbital resultant $L = 0$, the J value can only be zero, and the atom is in a $^1S_0$ state. Such is the case in the normal state of all atoms in Group II of the periodic table.

Table 13D. Energy Level Designations for Singlets and Triplets

$^1S_0$	$^3S_1$		
$^1P_1$	$^3P_2$	$^3P_1$	$^3P_0$
$^1D_2$	$^3D_3$	$^3D_2$	$^3D_1$
$^1F_3$	$^3F_4$	$^3F_3$	$^3F_2$
$^1G_4$	$^3G_5$	$^3G_4$	$^3G_3$

It should be noted that the level designations give all three quantum numbers, S, L, and J. The L value is given by the letter S, P, D, F, etc., the J value by the subscript, and the S value by the superscript. The value of the superscript is always given by $(2S + 1)$.

13.7. The Pauli Exclusion Principle

If two electrons have the same total quantum number n and the same orbital quantum number l, they are called *equivalent electrons*. Observations show that when two electrons are equivalent, certain levels are forbidden. For example, the normal state of each alkaline-earth element is given by two equivalent electrons in s orbits.

Of the two possible terms arising from the vector model of the preceding section—$S = 0$, $L = 0$, giving $J = 0$ for a $^1S_0$ term, or $S = 1$, $L = 0$, giving $J = 1$ for a $^3S_1$ term—only the $^1S_0$ level is observed. As soon as one of the electrons is excited to an s orbit of different n, the $^3S_1$ level is also observed. See Fig. 13B.

In order to apply the Pauli exclusion principle to electrons one must take into account the magnetic quantum numbers introduced in § 12.8. When this is done, it is found that the following levels are permitted by theory, and that exactly these and no others are found by experiment. See Table 13E.

Table 13E. Energy Levels for Two Equivalent and Two Nonequivalent Electrons

$s^2 \to {}^1S$	—							
$s \cdot s \to {}^1S$	${}^3S$							
$p^2 \to {}^1S$	—	—	${}^3P$	${}^1D$	—			
$p \cdot p \to {}^1S$	${}^3S$	${}^1P$	${}^3P$	${}^1D$	${}^3D$			
$d^2 \to {}^1S$	—	—	${}^3P$	${}^1D$	—	—	${}^3F$	${}^1G$
$d \cdot d \to {}^1S$	${}^3S$	${}^1P$	${}^3P$	${}^1D$	${}^3D$	${}^1F$	${}^3F$	${}^1G$	${}^3G$

The dot between two electrons of the same l-value indicates that the electrons have different principal quantum numbers n.

13.8. Double Electron Jumps

Not all of the prominent lines in the spectra of atoms with two valence electrons can be attributed to the chief series of singlets and triplets: *sharp, principal, diffuse,* and *fundamental.* Through the identification and classification of these additional lines into what are called *multiplets,* Russell and Saunders in 1925 were led to the key to complex spectra.

There are in the spectra of calcium, strontium, and barium, three prominent groups of very strong lines, recognized as forming a triad of *triplet multiplets,* and a triad of *singlets.* The triad of multiplets, one in the red and the other two in the green region of the spectrum of calcium, are reproduced in Fig. 13H.

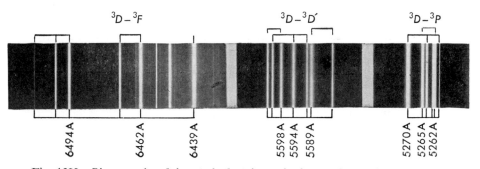

Fig. 13H   *Photographs of the triad of triplet multiplets in the visible spectrum of calcium.*

The significance of these multiplet groups is that they arise from the simultaneous excitation of both valence electrons in atoms when an electrical discharge is sent through a gas containing calcium.

Referring now to the energy level diagram of Fig. 13B, it will be seen that a group of twleve energy levels, high on the right of the diagram, ${}^1P$ ${}^1D$ ${}^1F$ ${}^3P$ ${}^3D$ and ${}^3F$, arise from the two valence electrons in the two allowed orbits $3d$ and $4p$.

From Table 13C, we see that p and d electrons give rise to P, D, and F levels, and with spins opposite, $S = 0$, we obtain the three singlet levels

$$^1P_1 \quad ^1D_2 \quad ^1F_3$$

and with spins parallel, $S = 1$, we obtain the three triplet levels

$$^3P_2 \ ^3P_1 \ ^3P_0 \qquad ^3D_3 \ ^3D_2 \ ^3D_1 \qquad ^3F_4 \ ^3F_3 \ ^3F_2$$

In jumping from these levels to the lower levels $^1D_2$ and $^3D_3$, $^3D_2$ and $^3D_1$, arising from $4s\ 3d$, either one or the other of two transitions takes place. Either the $4p$ electron jumps to a $4s$ orbit, or both electrons jump, the $4p$ electron to a $3d$ orbit and the $3d$ electron to a $4s$ orbit. The selection rules applied to orbital quantum numbers where two electrons are involved are

$$\text{One electron transition} \quad \Delta l = +1, -1$$
$$\text{Two electron transitions} \quad \begin{cases} \Delta l_1 = +1, -1 \\ \Delta l_2 = +2, 0, -2 \end{cases} \tag{13g}$$

Note that for double electron transitions such as those described above, lines arising from level transitions from ^3D above to ^3D below, constitute electron jumps in agreement with Eq. (13g), while the value of L remains unchanged. For double electron jumps the selection rules for all levels are

$$\Delta S = 0$$
$$\Delta L = +1, 0, -1 \tag{13h}$$
$$\Delta J = +1, 0, -1 \qquad \text{with } 0 \rightarrow 0 \text{ forbidden.}$$

The relative intensities, measured wavelengths, and calculated wave numbers for two of the calcium multiplets shown in Fig. 13H, are given in Table 13F.

13.9. Landé Interval Rule

Common regularities in the relative fine structure separations within triplet energy levels were first discovered by Landé before the advent of the quantum theory. According to Landé the intervals are proportional to the two larger of the three J-values. For a ^3P with J-values of 2, 1, 0, the intervals should be 2 to 1. For a ^3D with J-values of 3, 2, 1, the intervals should be 3 to 2, etc. These ratios are well borne out by a great number of observed frequency differences, particularly for the low lying levels and for the lighter elements of the periodic table. See Table 13G.

According to quantum mechanics electron spin and orbital angular momenta are not given by $s\hbar$, $l\hbar$, $j\hbar$, $S\hbar$, $L\hbar$, and $J\hbar$, but by the slightly different values, $s^*\hbar$, $l^*\hbar$, $j^*\hbar$, $S^*\hbar$, $L^*\hbar$, and $J^*\hbar$, respectively. These starred values are given by

$$s^* = \sqrt{s(s+1)}, \quad l^* = \sqrt{l(l+1)}, \quad j^* = \sqrt{j(j+1)} \tag{13i}$$
$$S^* = \sqrt{S(S+1)}, \quad L^* = \sqrt{L(L+1)}, \quad J^* = \sqrt{J(J+1)} \tag{13j}$$

Vector diagrams for the three levels of a ^3D state are given in Fig. 13I. Here with $S = 1$, $L = 2$, and $J = 1, 2$, and 3, we obtain by substitution in Eq. (13j), $S^* = 1.41$, $L^* = 2.45$, and $J^* = 3.46, 2.45, 1.41$.

**Table 13F. Relative Intensities, Measured Wavelengths, and Wave Numbers
of Two Calcium Multiplets Arising from the Double
Electron Jump 4p 3d to 3d 4s**

		3d 4s				
		$^3D_3$	21.75	$^3D_2$	13.88	$^3D_1$
	$^3F_4$	**150** 6439.09 A 15525.87 cm^{-1}				
	78.15					
3d 4p	$^3F_3$	**40** 6471.66 15447.72		**125** 6462.58 15469.44		
	88.28					
	$^3F_2$	**1** 6508.84 15359.40		**30** 6499.65 15381.20		**80** 6493.79 15395.08
	$^3D_3$	**80** 5588.74 17888.15		**25** 5581.97 17909.85		
	40.01					
3d 4p	$^3D_2$	**30** 5601.28 17848.11		**60** 5594.46 17869.87		**20** 5590.11 17883.79
	26.73					
	$^3D_1$			**20** 5602.83 17843.18		**50** 5598.48 17857.03

If we think of the resultant electron spin S^* as though it were a small magnet, in a magnetic field created by and parallel to the orbital motion L^*, the three different orientations of S^* with respect to L^*, should have different energies. See §§ 3.7 and 10.5, and Fig. 13K. The energy required to turn S^* with respect to L^*, from a position where their magnetic moments are parallel to each other to where they are oppositely directed, is given by classical theory as

$$\Delta E = k \cos \theta \tag{13k}$$

where k is a constant, and θ is the angle between S^* and L^*.

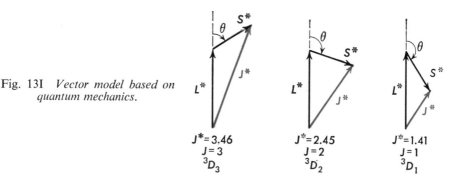

Fig. 13I *Vector model based on quantum mechanics.*

$J^*=3.46$
$J=3$
$^3D_3$

$J^*=2.45$
$J=2$
$^3D_2$

$J^*=1.41$
$J=1$
$^3D_1$

Table 13G. Triplet Level Separations in Wave Numbers Illustrating the Landé Interval Rule

Element, and Configuration	$^3P_0$–$^3P_1$ cm^{-1}	$^3P_1$–$^3P_2$ cm^{-1}	Obs. Ratio
Mg, $3s\,3p$	20.0	40.9	1:2
Ca, $4s\,4p$	52.3	105.9	1:2
Ca, $3d\,3d$	13.5	26.9	1:2
Zn, $4s\,4p$	190.0	389.0	1:2
Sr, $5s\,5p$	187.0	394.6	1:2

Element, and Configuration	$^3D_1$–$^3D_2$ cm^{-1}	$^3D_2$–$^3D_3$ cm^{-1}	Obs. Ratio
Ca, $4s\,3d$	13.6	21.7	2:3
Ca, $4p\,3d$	26.7	40.0	2:3
Ca, $4s\,4d$	3.8	5.6	2:3
Zn, $4s\,4d$	3.4	4.6	2:3
Cd, $5s\,5d$	11.7	18.2	2:3

If we apply the law of cosines to any triangle in Fig. 13I, we obtain $J^{*2} = L^{*2} + S^{*2} + 2L^*S^* \cos \theta$. Solving this equation for $\cos \theta$, we obtain

$$\cos \theta = \frac{1}{2} \frac{1}{L^*S^*} (J^{*2} - L^{*2} - S^{*2}) \tag{13l}$$

which upon substitution in Eq. (13k), gives

$$\Delta E = \frac{1}{2} \frac{k}{L^*S^*} (J^{*2} - L^{*2} - S^{*2})$$

For any given triplet the values of L^* and S^* are the same and we can replace k/L^*S^* by a constant A, and write

$$\Delta E = \tfrac{1}{2}A(J^{*2} - L^{*2} - S^{*2}) \tag{13m}$$

By substituting $S = 1$, $L = 1$, and $J = 0$, 1, or 2 in Eq. (13m), we find for $^3P_0$, $^3P_1$ and $^3P_2$, the values $\Delta E = -2A$, $-A$, and $+A$. Similarly, by substituting $S = 1$, $L = 2$, and $J = 1$, 2, or 3, we find for $^3D_1$, $^3D_2$, and $^3D_3$, the values $\Delta E = -3A$, $-A$, and $+2A$. When these values are plotted as in Fig. 13J,

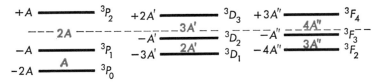

Fig. 13J   *Fine-structure level separations based on quantum mechanics and the vector model: The Landé interval rule.*

the level separations are seen to agree with observations. See Tables 13F and 13G.

13.10. Ionized Atoms

The nine elements in Group II of the periodic table of elements contain two valence electrons in *s* orbits. See Table 13B for the first four elements. If by some excitation process each of these atoms loses one of its valence electrons, the electron configurations, so far as the remaining electrons are concerned, become identical with those of the elements just preceding them in the periodic table.

While the process of excitation and emission of radiation from ionized atoms is similar to that for neutral atoms, the energies involved are considerably greater. If the excitation is brought about by a collision between a fast-moving electron and a neutral calcium atom, then the interchange of energy must be sufficient to completely remove one valence electron from the atom and to raise the other to an excited state.

An energy-level diagram for ionized calcium is presented in Fig. 13K. The normal state of the atom ^2S corresponds to the valence electron in a 4s orbit. Unlike the alkali metals, Li, Na, K, Rb, Cs, and Fa, the first excited state is a *D* level rather than a *P* level.

Owing to the fact that the electron in the 3d, ^2D level cannot return to the normal state 4s, ^2S, without violating the selection rule ($\Delta l = \pm 1$ only), this 3d, ^2D level is called a *metastable state*. Once the electron is in this metastable state, it will stay there for some time and may return to the normal state only by collision with another particle with the liberation of its energy without radiation. The energy released may excite the other atom to an upper energy level or the two atoms may recoil with added kinetic energy.

While the strong lines of the alkali metals, Group I, are to be found in or near the visible spectrum, the corresponding lines in the ionized alkaline earths, Group II, are found to be largely in the ultraviolet region. In potassium $Z = 19$, for example, the first member of the principal series is in the far red between 7000 and 8000 A, whereas the first line of the principal series of ionized calcium is in the ultraviolet just below 4000 A. This latter is a doublet and constitutes the two most prominent lines in the sun's spectrum, the *H* and *K* lines of Fig. 9G.

It should be noted that the normal state of ionized calcium is the series limit of the neutral calcium atom shown at the top of Fig. 13B. Furthermore, all levels of the ionized atom arise from one unbalanced electron as in the alkali metals, and all of its characteristic fine-structure is doublets.

13.11. Complex Spectra

Although the spectra of practically all neutral and ionized atoms in the periodic table have been thoroughly analyzed, the examples in this chapter will serve as an introduction to complex spectra. For elements in Group III of the periodic table, three valence electrons are responsible for their observed spectra, and their known energy-level diagrams reveal *doublet* and *quartet* fine structures.

This increase in multiplicity continues in each period until the valence electron subshell is half-filled and then decreases again to singlet and triplet levels with

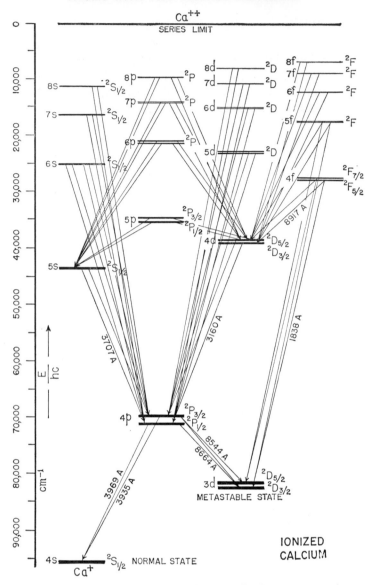

Fig. 13K *Energy-level diagram for singly ionized calcium atoms, showing most of the more prominent spectral transitions.*

the elements marking the completion of each subshell. Each new subshell then starts over again with doublets.

Photographs of two quartet multiplets in the spectrum of scandium, $Z = 21$, are reproduced in Fig. 13L. The first multiplet arises from transitions from four $^4F'$ levels and ending on four ^4F levels, while the second multiplet arises from transitions from four ^4G levels and ending on the same lower ^4F levels.

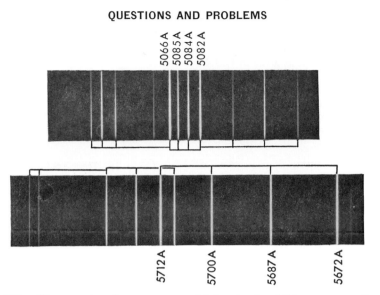

5066 A
5085 A
5084 A
5082 A

5712 A   5700 A   5687 A   5672 A

Fig. 13L *Photographs of two quartet multiplets in the visible spectrum of scandium, Z = 21.*

The two initial sets of quartet levels arise from the electron configuration $3d^2\,4p$ and the common lower ^4F levels arise from the configuration $3d^2\,4s$. All transitions follow the selection rules given in the preceding sections of this chapter.*

QUESTIONS AND PROBLEMS

1. Using the relative intensity formulas in § 13.4, calculate the relative intensities for a ^3D–^3F multiplet, and arrange the results as in Table 13A.

2. Using the relative intensity formulas in § 13.4, calculate the relative intensities for a ^3F–^3G multiplet and arrange the results as in Table 13A.

3. Write down the complete electron configuration for strontium, cadmium, and barium.

4. (a) Write down the complete electron configuration for mercury. (b) Write down the abbreviated configuration (the valence electrons only). (c) What abbreviated configurations would you expect the first and second set of excited states to arise from?

5. Make vector diagrams for the orbital quantum numbers for the electron configuration $3d\,4f$.

6. Make vector diagrams for the fine-structure levels of a ^3F state.

7. Make a vector diagram, based upon quantum mechanics, for the fine-structure levels of a ^3G state.

8. Using Table 13E, write down the energy levels arising from (a) two equivalent f electrons and (b) two nonequivalent electrons.

9. What energy levels can you expect to arise from the electron configuration $4f\,5g$?

10. Calculate the J values for the fine-structure levels of a quartet state, ^4F where $S = \tfrac{3}{2}$ and $L = 3$, and ^4G where $S = \tfrac{3}{2}$ and $L = 4$.

* For further studies of complex spectra and atomic structure see *Introduction to Atomic Spectra* by H. E. White, McGraw-Hill Book Company.

11. Make an array diagram like Table 13A for transitions between the two quartet levels given in Problem 10, and calculate the relative intensities of the allowed transitions as given in § 13.4.

12. (a) Plot an energy-level diagram for strontium, $Z = 38$, using the following values for the term values in wave numbers.

(b) Show transitions for the first members of the sharp, principal, diffuse, and fundamental series. (c) Calculate the frequencies for the first member of the sharp series of singlets and triplets. (c) Calculate the wavelengths for these same four lines.

CONFIG.	SYM.	J	TERM VALUE
$5s^2$	^1S	0	45925.6
$5s\ 5p$	^3P	0	31608.0
		1	31421.2
		2	31027.0
$5s\ 4d$	^3D	1	27766.0
		2	27706.4
		3	27606.0
$5s\ 4d$	^1D	2	25776.3
$5s\ 5p$	^1P	1	24227.1
$5s\ 6s$	^3S	1	16886.8
$5s\ 6s$	^1S	0	15334.5
$4d\ 5p$	^3F	2	12658.5
		3	12335.7
		4	12006.0
$5s\ 6p$	^3P	0	12098.8
		1	12057.4
		2	11952.8
$5s\ 6p$	^1P	1	11827.5
$5s\ 5d$	^1D	2	11110.0
$5s\ 5d$	^3D	1	10918.3
		2	10903.3
		3	10880.5
$4d\ 5p$	^3D	1	9661.2
		2	9543.4
		3	9365.9
$4d\ 5p$	^3P	0	8633.0
		1	8622.2
		2	8588.5
$5s\ 7s$	^3S	1	8500.9
$5s\ 7s$	^1S	0	7481.6
$5s\ 7p$	^1P	1	7019.0
$5s\ 7p$	^3P	0	6514.2
		1	6499.2
		2	6468.1

13. (a) Make a table like Table 13F, for the triplet multiplet, $5s\,4d$, ^3D—$4d\,5p$, ^3F in strontium. Use the table in Problem 12. (b) Calculate the relative intensities using equations in § 13.4. (c) Plot these six lines to a wavelength scale as shown at the bottom of Fig. 13C.

14. (a) Make a table like Table 13F, for the triplet multiplet $5s\,5p$, ^3P—$5s\,4d$, ^3D in strontium. Use the table in Problem 12. (b) Calculate the relative intensities using equations in § 13.4. (c) Plot these six lines to a wavelength scale as shown at the bottom of Fig. 13C.

15. The following is a list of the relative intensities, wavelengths, and wave numbers, of the ^3D–^3P multiplet shown at the right in Fig. 13H. Arrange these lines in a tabular array similar to that shown in Table 13F. Compute the average separations for the $^3P_{0,1,2}$ and $^3D_{1,2,3}$ levels.

INTENSITY	WAVELENGTH	WAVE NUMBER
2	5260.38	19004.77
20	5261.70	18999.99
25	5262.24	18998.05
20	5264.24	18990.89
40	5265.56	18986.07
60	5270.27	18969.08

14

X Rays

One of the most interesting episodes in the history of modern science began with the accidental discovery of X rays by Wilhelm Röntgen* in 1895. While studying the green fluorescent stage of an electrical discharge in a Crookes tube, Röntgen observed the bright fluorescence of some nearby crystals of barium platino-cyanide. Even though the discharge tube was in a darkened room, and entirely surrounded with black paper to prevent the escape of visible light, a distant screen covered with crystals would fluoresce brightly when the discharge was turned on. Röntgen reasoned, therefore, that some kind of invisible, yet penetrating, rays of an unknown kind were being given out by the discharge tube. These rays he called *X rays*, the letter *X* meaning, as it so often does in algebra, an unknown.

In the short series of experiments that followed his discovery, Röntgen found that the unknown rays were coming from the glass walls of the tube itself and, in particular, from the region where the most intense part of the cathode ray beam was striking the glass. So great was the importance of this discovery that, within but a few weeks of Röntgen's accouncement, X rays were being used as an aid in surgical operations in Vienna. This, along with other practical applications and uses that can be made of this single scientific discovery, is a good example of the role played by modern science in the rapid advancement of civilization.

* Wilhelm Konrad von Röntgen (1845–1923). Born at Lennep on March 27, 1845, Röntgen received his education in Holland and Switzerland. His scientific career began at the age of 25 when he became an assistant in the physics laboratory at Würtzburg, Germany. After a teaching period extending over a period of 25 years, which carried him to the University of Strasbourg, then to Hohenheim, back to Strasbourg, then to Giessen, and finally to Würzburg again, he discovered X rays in his laboratory at Würzburg in 1895. For this discovery, he received the Rumford Medal of the Royal Society in 1896 and the first Nobel Prize in physics in 1901. Röntgen also conducted researches in light, heat, and elasticity, but none of these works compares in importance with his discovery of X rays.

14.1. X-ray Tubes

The Crookes tube with which Röntgen made his discovery bears very little resemblance to the modern X-ray tube. In form it had somewhat the appearance of the tube shown in Fig. 4C. Within a short period of time after Röntgen's discovery, quite a number of noteworthy improvements upon tube design were made. The first important contribution in this direction came immediately after the discovery that it is the sudden stopping of electrons that gives rise to X rays.

In X-ray tubes of early design, the electrons from the cathode were not allowed to strike the glass walls, but were directed toward the anode as a target, as shown in Fig. 14A. By curving the cathode like a concave mirror, it was found that it

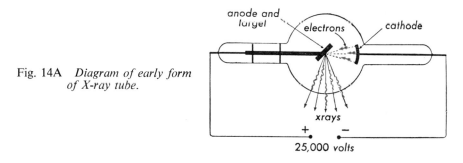

Fig. 14A   *Diagram of early form of X-ray tube.*

was possible to focus the electrons on one spot on the target, thus making of that spot a localized source of X rays. Radiating outward in all possible directions, these "Röntgen rays," as they are sometimes called, have no difficulty in passing through the glass walls of the tube.

The biggest improvement in X-ray tube design was made by Coolidge, an American physicist, in 1913. In the Coolidge tube, now a commercial product (see Fig. 14B), a tungsten wire filament is placed at the center of the cathode

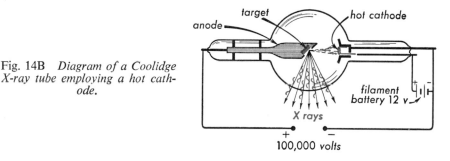

Fig. 14B   *Diagram of a Coolidge X-ray tube employing a hot cathode.*

and heated to incandescence by a storage battery or low-voltage transformer. This filament, being a copious source of electrons, gives rise at the target to a far more intense source of X rays than was previously possible with a cold cathode. Under the terrific bombardment of the target by so many electrons, most metals will melt. To overcome this difficulty, a metal with a high melting point, like tungsten or molybdenum, is imbedded in the face of a solid copper anode to become the target. Copper, being a good heat conductor, helps to dissipate the heat.

The early sources of high voltage applied to the anode and cathode of X-ray tubes were induction coils of various descriptions. Although some of these sources are still in use, they have been almost entirely supplanted by a more efficient high-voltage transformer. The emf generated by these transformers varies between 50,000 and 2,000,000 volts. The normal emf used for surgical work is about 100,000 volts, whereas for the treatment of diseases the higher emf's are employed. The high-voltage alternating emf supplied by a transformer is not applied directly to the X-ray tube, but is first changed into direct current by means of rectifier tubes.

14.2. Penetration of X Rays

Four useful and important properties of X rays are their ability (1) to penetrate solid matter, (2) to cause certain chemical compounds to fluoresce, (3) to ionize atoms, and (4) to affect a photographic plate. The penetration of X rays depends upon two things: first, the voltage applied between the anode and cathode of the X-ray tube; and second, the density of the substance through which the rays must travel. The higher the voltage applied to the tube, the greater is the penetration. *X rays of great penetrating power are called hard X rays, whereas those having little penetrating power are called soft X rays.*

The relation between density and penetration may be illustrated in several ways. When X rays are sent through a block of wood containing nails, or a closed leather purse containing coins, a clear and well-defined image of the nails, or coins, can be formed and observed on a fluorescent screen. The experimental arrangement is the same as that shown in Fig. 14C. When X rays are sent

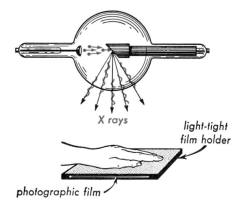

X rays

light-tight
film holder

photographic film

Fig. 14C   *Arrangement for taking X-ray photographs of the bones of the hand.*

through the hand or any part of the body to obtain photographs of the bones, it is the difference in penetration between the flesh and the bones that permits a picture to be made. Materials like paper, wood, flesh, etc., composed principally of light chemical elements like those at the beginning of the periodic table, are readily penetrated by X rays. In other words, they are poor absorbers of X rays. For materials like brass, steel, bone, gold, etc., composed partly of heavy elements, like those farther along and near the end of the periodic table, the penetration of X rays is very poor. Hence, heavy elements, or dense substances, are good absorbers.

The bones of the body, which contain large amounts of calcium, are relatively good absorbers of X rays, whereas the soft tissue, composed principally of much lighter elements—hydrogen, oxygen, carbon, and nitrogen—are poor absorbers. This explains the general appearance of X-ray photographs. X-ray pictures like

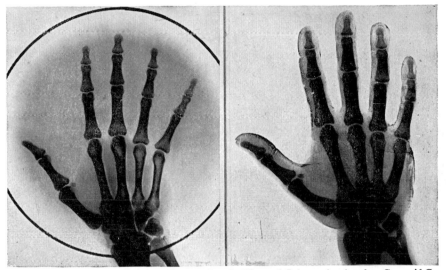

Courtesy, Stamford Research Laboratories, American Cyanamid Co.

Fig. 14D   *X-ray photographs of the wrist bones of the hand.* (a) *With hand in water.* (b) *With lead oxide ointment spread on hand.*

the ones in Fig. 14D are similar to shadows cast by the objects being photographed. The focus point on the X-ray target, being bombarded by high-speed electrons, acts as a point source of rays. These rays spread out in straight lines as shown in Fig. 14C. On passing through the hand to the photographic film, more X rays are absorbed by the bones than by the flesh. The shadow cast by the bones is therefore lacking in X rays, and the photographic film for these areas becomes transparent upon development.

Where only flesh is traversed, the X rays penetrate through to the photographic film, causing it to develop out black. The bones therefore appear white against a darker background. If this "negative film," as it is called, is printed on paper as in Fig. 14D, it becomes a "positive," the bones appearing black.

If the photographic film is placed farther away from the hand than shown in the diagram, the shadow picture will be larger and less distinct. The best pictures are obtained by placing the film as close in contact with the object to be photographed as is physically possible. Whenever a film is being exposed for an X-ray picture, it is mounted in a black paper envelope or thin aluminum box. This prevents visible light from reaching the film but allows the X rays to pass through.

14.3. Ionizing Power

As X rays pass through matter in the solid, liquid, or gaseous state, they are found to *ionize* atoms and molecules. This can be shown by charging a gold-leaf electroscope positively or negatively and placing it some 10 to 15 ft away from an X-ray tube. When the X-ray tube is turned on (see Fig. 14E), the gold leaf falls, showing discharge.

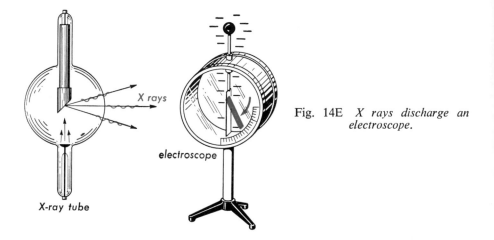

Fig. 14E   *X rays discharge an electroscope.*

The explanation of this experiment is as follows. X rays pass through the electroscope and ionize the air by removing electrons from many of the oxygen and nitrogen molecules. Leaving these particular molecules with a net positive charge, the freed electrons move about until they are picked up by other neutral molecules, which thus take on a net negative charge. The result is that the passage of X rays through matter produces both *positively charged* and *negatively charged ions*. If the electroscope is negatively charged, it attracts the positively charged ions to the gold leaf, neutralizing the charge and repelling the negatively charged ions to the "grounded" walls where they, too, become neutralized. If the electroscope is positively charged, it attracts the negative ions to it, again neutralizing the charge. The positive ions in this case are repelled to the walls. In either case, whether the electroscope is positively or negatively charged, the gold leaf falls, showing discharge.

It is the ionization of atoms and molecules in a substance that limits the penetrating power of X rays. Heavy elements contain more electrons than light elements, thus placing more electrons in the path of the X rays to stop them.

The stopping power of a thin sheet of lead, for example, is equivalent to the stopping power of a sheet of aluminum several times thicker. Lead atoms each contain 82 electrons, whereas aluminum atoms each contain only 13.

14.4. Practical Applications

During the first few weeks following Röntgen's discovery of X rays, reports from all over the world were received by the editors of scientific journals telling of how the new rays could be put to practical use. A few examples of the first applications were (1) the location of a bullet in a patient's leg, (2) the observation and photography of the healing of a broken bone, (3) the detection of contraband in baggage, (4) the distinction between artificial and real gems, (5) the detection of pearls in oysters, and (6) the examination of the contents in parcel post. In 1897 Dr. Morton exhibited in New York an X-ray picture of the entire skeleton of a living and fully clothed adult.

The biological effects became important when it was found that X rays killed off some forms of animal tissue more rapidly than others. This made them a possible means of cure for certain skin diseases. In particular, the application to the treatment of well-known forms of cancerous growths in animals and human beings has yielded amazing results, and oftentimes a cure. When an internal cancer is treated by sending a beam of X rays directly through the body, the cancerous tissue as well as the normal tissue is slowly killed off. It is principally because the normal tissue grows in again more rapidly than the cancerous tissue that it is possible to bring about a cure. Periodic radiation allows the normal tissue to build up in the intervals.

Although only certain diseases can be successfully treated by X rays, a great deal of research work is still being carried on with extremely high voltage X rays in the hope of discovering new and more effective medical aids. It is generally believed that the killing-off of cell tissue by X rays is due in part to ionization and in part to the formation of free radicals of the molecules within the individual cells.

The importance of X rays in some phases of the field of engineering cannot be overestimated. This can be appreciated when it is realized that metal castings or welded joints sometimes contain internal flaws or blowholes that otherwise escape detection. Because of the disastrous results that might occur by the insertion of defective castings or welded joints into a bridge or building, many such metal parts are examined by X rays before they are used.

14.5. X Rays Are Waves

Not long after Röntgen's discovery of X rays, there arose in scientific circles two schools of thought concerning the nature of these penetrating rays. One school held to the belief that X rays are high-speed particles like cathode rays, but more penetrating; and the other school held to the idea that they are electromagnetic waves of extremely high frequency. Although many experiments were performed to test these two hypotheses, several years passed before the wave theory was proven to be correct.

The crucial experiment came in 1912 when Von Laue* suggested to his associates, W. Friedrich and P. Knipping, that they try diffracting X rays by sending them through a thin crystal. Believing that the ultramicroscopic structure of a crystal is a three-dimensional array of regularly spaced atoms, Von Laue reasoned that the equally spaced layers of the atoms would act like a diffraction grating.

The experiment, as it was performed, is shown diagrammatically in Fig. 14F.

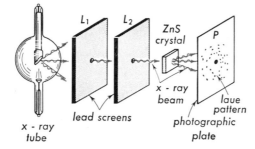

Fig. 14F *Experimental demonstration of the wave property of X rays. Diffracted by the atoms in a crystal, a Laue pattern is photographed.*

X rays from a cold cathode X-ray tube, and limited to a narrow pencil of rays by a pinhole in each of the two lead screens L_1 and L_2, are shown passing through a thin crystal to a photographic film or plate at P. In addition to the central beam, the major part of which goes straight through to produce a blackened spot at the center of the film, there are many other weaker beams emerging in different directions to produce other spots on the same film. The pattern of spots obtained in this way is always quite symmetrical, and is referred to as a *Laue pattern.*

Photographs of two Laue patterns obtained with single crystals are reproduced in Fig. 14G. The small number of spots in (a) is indicative of a relatively simple

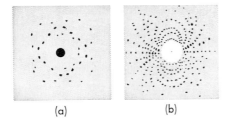

(a)   (b)

Fig. 14G *X-ray diffraction patterns from crystals: (a) Zinc sulfide crystal (face-centered cubic crystal). (b) Sugar crystal (a complex crystal structure).*

* Max von Laue (1879–1960). Born near Coblenz, Germany, in 1879, young Max was educated in the German universities of Strasbourg, Göttingen, and Munich. Following this, his teaching and research work carried him to the university at Munich, Zurich, Frankfurt on the Main, and finally Berlin. Since he was interested in theoretical physics, his early attentions were confined to various phases of Einstein's theory of relativity and to Bohr's quantum theory of atomic structure. His chief contribution to physics, however, was the instigation and supervision of experiments leading to the diffraction of X rays by crystals. For this work, which proved the wave nature of X rays, he was granted the Nobel Prize in 1914.

crystal structure for zinc sulfide, ZnS, and the large number of spots signifies a relatively complex crystal structure for sugar, $C_{12}H_{22}O_{11}$. While the picture for sugar was being taken, the central beam was masked off by a small lead disk placed just in front of the film to prevent excessive blackening. Simple Laue patterns in general arise from simple crystal structures. Common salt is an example of a simple crystal, containing sodium ions $(Na)^+$ and chlorine ions $(Cl)^-$ in equal numbers arranged in a three-dimensional cubic lattice. Fig. 14H

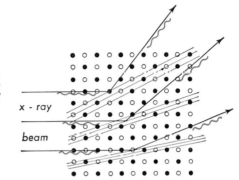

Fig. 14H *Illustration of the reflection of X rays from the various atomic planes in a cubic crystal lattice.*

is a cross section through such a crystal, showing the alternation of ions in two of the three directions. Here, in this two-dimensional array, the origin of the different spots on a Laue pattern is illustrated.

Each spot arises from the reflection of some of the incident X rays from one of the various sets of parallel crystal planes, three of which are shown by the sets of parallel lines. Always, the rays obey the law of reflection that the angle of incidence equals the angle of reflection. While the reflection planes shown in the diagram are all perpendicular to the plane of the page, there are many other planes in a three-dimensional lattice to reflect the rays off in other directions.

The success of the Laue experiment proves the correctness of two postulates: (1) that X rays are light rays of very short wavelength, and (2) that the ions of a crystal are arranged in a regular three-dimensional lattice. These are the results for which Von Laue was granted the Nobel Prize in physics in 1914. As a direct result of the Laue experiment, two new and important fields of experimental physics were opened up: (1) the study and measurement of X-ray wavelengths, and (2) the study of crystal structures by their action on X rays.

14.6. The X-Ray Spectrograph

No sooner had Von Laue, Friedrich, and Knipping announced the results of their experiments than many investigators began a study of the various phases of *X-ray diffraction* by crystals. The most outstanding of these experiments are those of W. H. Bragg* and his son, W. L. Bragg; they also developed the X-ray spectrometer and spectrograph.

A diagram of an X-ray spectrograph is shown in Fig. 14I. Instead of having pinhole screens, as in Fig. 14F, and sending a narrow pencil of rays through a crystal, the early spectrographs used screens with narrow slits and reflected the rays from one face of a crystal. The crystal is not fixed tightly in place but can be turned back and forth about a pivot C at the center of the front face. As this rocking motion takes place, the crystal acts somewhat like a mirror and causes the reflected X-ray beam to sweep back and forth along the photographic film from one end to the other. After the photographic film has been exposed to the rays for some time, and then developed, it is found to have the general appearance of the reproduction in Fig. 14J. This unusually clear photograph was originally taken by De Broglie who used an X-ray tube containing a tungsten-metal anode target. Instead of a general blackening from end to end, the film shows *bands* and *lines*, indicating that at certain orientation angles of the crystal the reflected rays were unusually intense, while at others there were apparently none. The lines, which are particularly noticeable at points marked $K\alpha$, $L\alpha$, $L\beta$, and $L\gamma$, are called *X-ray spectrum lines*.

The origin and interpretation of these spectrum lines are illustrated by a detailed diagram in Fig. 14K. To reflect X rays of one given wavelength from a crystal, a certain relation must exist between the direction of the incident rays and the distance d between surface layers of the crystal. This relation, known as

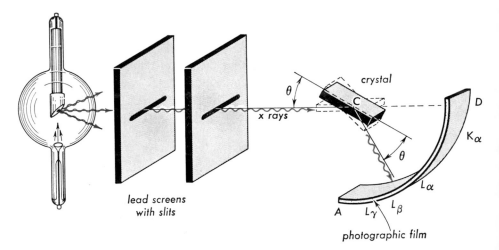

Fig. 14I *Schematic diagram of a Bragg X-ray crystal spectrograph.*

* Sir William Henry Bragg (1862–1942), British physicist and professor at the University of London. Bragg's researches on radioactive phenomena brought him early recognition from scientific societies at home and abroad. Joint work with his son, William Lawrence Bragg (1890–   ), on the arrangement of ions in crystals, and the development of the X-ray spectograph are his greatest scientific contributions. In 1915, father and son were jointly granted the Nobel Prize in physics, as well as the Barnard Gold Medal from Columbia University.

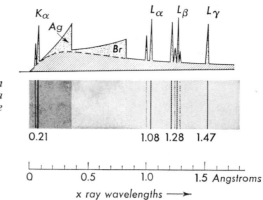

Fig. 14J *X-ray spectrogram taken with an X-ray tube containing a tungsten metal target. (After De Broglie.)*

the Bragg rule, requires the waves to be incident on the crystal face, at such an angle θ that the crests of the waves reflected from adjacent atomic layers move off together. This occurs when the additional distance traveled by ray (2), AMB in the diagram, is exactly one whole wavelength greater than that traveled by the ray (1) next above it. When the angle is adjusted so that this is true, other rays like (3), belonging to the same wave train as (1) and (2), will be reflected from the third crystal layer to be "in step" with the others.

Suppose now that the X-ray tube in Fig. 14I emits X rays of only one wavelength; then, as the crystal rocks back and forth, there will be no reflection except at one particular angle θ, and this will occur where the conditions of Bragg's rule are satisfied. At this particular position on the photographic plate, a single dark line will appear. If now the distance d between crystal layers is known, and the angle θ for the X-ray line measured, the wavelength of the X rays can be calculated. One wavelength, it will be noted in Fig. 14K, is equal to twice the length of the side AM of the right triangle AMC. Thus, with one side and two angles of a triangle known, either of the other sides can be calculated. Bragg's rule therefore becomes

$$2d \sin \theta = \lambda \qquad\qquad (14a)$$

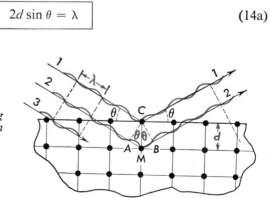

Fig. 14K *Illustration of the Bragg rule of reflection for X rays from the surface layers of a crystal.*

In the case of a sodium chloride crystal, NaCl, the atomic spacing is 2.81 A, or 2.81×10^{-10} m.

Since several spectrum lines appear in the photograph in Fig. 14J, there are several different wavelengths emitted by the same X-ray tube. The two fluted appearing bands between the K and L X-ray lines are not of interest here because they appear on all X-ray spectrograms; they are due to the strong absorption of X rays of many other wavelengths by the silver and bromine atoms in the photographic plate itself. Had the original photographic film been exposed for a much longer time, the spectrogram would have shown a general blackening over the whole plate. This blackening, illustrated by the shaded area in the curve above, is due to X rays of all different wavelengths being emitted by the X-ray tube; it is these which, although not very intense, strongly affect the photographic plate at the two bands, *Ag* and *Br*.

14.7. The Origin of X Rays

X rays, like visible light, originate from the jumping of an electron from one orbit to another. When high-speed electrons from the cathode of an X-ray tube strike the target, they ionize many of the atoms composing the surface layers of the metal.

Owing to their very high speeds (about $\frac{1}{10}$ the velocity of light), the electrons penetrate the atoms and remove an electron from the inner shells by collision. This is illustrated in Fig. 14L where an electron is knocked out of the K shell.

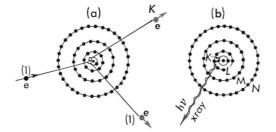

Fig. 14L *Schematic diagram illustrating (a) the ionization of an atom by a high-speed electron, and (b) the subsequent jumping of an inner electron with the simultaneous emission of an X ray.*

The designations K, L, M, N, O, P, etc., for the various electron shells originated with the X-ray spectroscopist and are identical with the quantum numbers $n = 1, 2, 3, 4, 5, 6$, etc. When an electron is missing in the innermost K shell, a nearby electron from the next shell beyond jumps into the vacant space, simultaneously emitting a photon of energy $h\nu$. Such X rays, arising from millions of atoms, produce the K lines shown in Fig. 14J.

Since the L shell now has one less electron, an M electron can jump into the L shell vacancy, with the consequent emission of another but different X-ray frequency. These are the L lines in Fig. 14J. The jumping process continues until the outermost shell is reached, where an electron jumping in gives rise to visible light. Thus we see how it is possible for a single atom to emit X rays of different wavelengths.

The continuous X-ray spectrum, illustrated by the shaded area under the curve in Fig. 14J, is due to another phenomenon often referred to as "Bremsstrahlung." These radiations are due to the slowing down of high-speed electrons as they pass close to the nuclei of the atoms within the target of the X-ray tube. The process is illustrated in Fig. 14M. As the electron passes through the atom, it is attracted by the positive charge of the nucleus and deflected in its path.

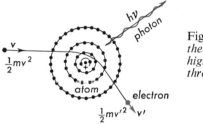

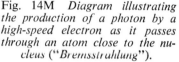

Fig. 14M *Diagram illustrating the production of a photon by a high-speed electron as it passes through an atom close to the nucleus ("Bremsstrahlung").*

During the deflection of the electron in the strong electric field of the nucleus, a light wave of energy is hv is emitted. Since the law of conservation of momentum must hold for such a collision, the electron is deflected off to one side of the atom, and the photon off to the other. Since the law of conservation of energy must hold, some of the energy of the incoming electron $\frac{1}{2}mv^2$ is given up to the newly created photon hv, and the remainder $\frac{1}{2}mv'^2$ is retained by the electron. Thus the electron is slowed down to a velocity v' by the encounter. The closer the electron comes to the nucleus, the greater is its loss in velocity and energy, and the greater is the frequency and energy of the radiated photon. By the conservation of energy,

$$\tfrac{1}{2}mv^2 - \tfrac{1}{2}mv'^2 = hv \qquad (14b)$$

The highest frequency that is possible is one in which the electron is completely stopped by the atom. In this special case,

$$\tfrac{1}{2}mv^2 = hv_{\max} \qquad (14c)$$

Since the kinetic energy of the electrons in the beam striking the target is given by the voltage V applied to the tube, we may use Eq. (10b),

$$Ve = \tfrac{1}{2}mv^2 \qquad (14d)$$

and obtain

$$Ve = hv_{\max} \qquad (14e)$$

While Eq. (14d) must be replaced by the relativistic equation, Eq. (20b), when voltages are greater than 20,000 volts, Eq. (14e) holds true for all voltages.

For a more complete treatment of X rays, see *X-Rays* by A. H. Compton and S. K. Allison, D. Van Nostrand, Princeton.

QUESTIONS AND PROBLEMS

1. What contribution did Coolidge make to the design and construction of X-ray tubes?

2. What are hard X rays? What are soft X rays? How is hardness or softness related to X-ray wavelengths?

3. How is X-ray absorption related to the periodic table of elements?

4. Which of the following materials is the best absorber of X rays: (a) beryllium, (b) magnesium, (c) calcium, (d) copper, (e) gold, (f) lead, or (g) uranium?

5. What is the process of ionization by X rays? What happens to the liberated free electrons?

6. Why does the skin show so clearly in the X-ray photograph in Fig. 14D(b)?

7. If a small child swallowed a safety pin, why would an X-ray photograph clearly show the location of the pin?

8. How is the penetrating power of X rays related to the voltage applied to the tube?

9. Briefly explain why an X-ray photograph of the hand shows the bones more clearly than the flesh surrounding them.

10. An X-ray photograph of a closed leather purse will readily show silver coins or other metal articles inside. Explain.

11. How and by whom were X rays discovered?

12. Diagram a modern X-ray tube of the type developed by Coolidge.

13. X rays sent through a Bragg crystal spectrometer using a rock salt crystal are reflected at an angle of $18°$. What is the wavelength of the X rays?

14. X rays sent through a Bragg crystal spectrometer show three spectrum lines at $4.28°$, $5.92°$, and $7.56°$, respectively. If the crystal used is rock salt, what are the wavelengths of the X rays? *Ans.* 0.420 A, 0.579 A, 0.740 A.

15. X rays of wavelength 1.60×10^{-8} cm are diffracted by a Bragg crystal spectrograph at an angle of $14.2°$. Find the effective spacing of the atomic layers in the crystal.

16. X rays having a wavelength of 0.36×10^{-8} cm are diffracted at an angle of $4.8°$ in a Bragg crystal spectograph. Find the effective spacing of the atomic layers in the crystal. *Ans.* 2.15 A.

17. When a nickel target is used in an X-ray tube, the two shortest wavelengths emitted are found with a Bragg crystal spectrograph to be diffracted at angles of $15.1°$ and $17.1°$, respectively. Find their wavelengths. Assume a crystal spacing of 2.81×10^{-8} cm.

18. When a tungsten target is used in an X-ray tube, the two shortest wavelengths emitted are found with a Bragg crystal spectograph to be diffracted at angles of $2°47'$ and $3°16'$, respectively. Find their wavelengths. Assume a crystal spacing of 2.81×10^{-8} cm. *Ans.* 0.273 A and 0.320 A.

15

Electromagnetic Waves and Vacuum Tubes

There is little doubt that *radio, radar,* and *television* are among the greatest miracles of modern science. Traveling with the speed of light, code signals, the human voice, and music can be heard around the world within the very second they are produced in the broadcasting studio. Through television, world events can be observed in full color at the same moment they occur hundreds and even thousands of miles away.

The more we learn of the fundamental principles of radio and its operation, the more amazing does their reality become. It is the purpose of this chapter to introduce some of the earlier fundamental principles of wireless telegraphy in the approximate chronological order in which they were discovered and developed, and in so doing to gain some familiarity with *electromagnetic waves.*

15.1. The Leyden Jar

A cross section of a "Leyden jar" of the type invented by the Dutch scientist Musschenbroek in 1746 is shown in Fig. 15A. Two metallic conductors forming the plates of a capacitor are separated by a glass bottle as a dielectric insulator. When such a capacitor is connected to a source of high potential, one plate will become positively charged and the other will be negative. If the source voltage is high enough, an electric spark will jump between the terminals indicating a sudden discharge of the capacitor, and an electron current will surge first one way and then the other around the circuit.

This oscillatory current was first postulated by Joseph Henry, then derived from theory by Lord Kelvin, and later proven experimentally by Fedderson. Fedderson, looking at a capacitor discharge with a rotating mirror, observed that each initial breakdown spark was followed by a succession of fainter sparks. The initial spark ionizes the air, making of it a good conductor and, of the entire system *ABCDEFGA,* a complete electrical circuit.

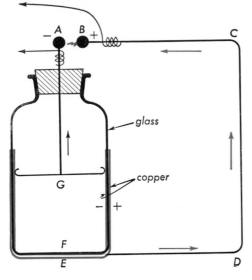

Fig. 15A *The discharge of a Leyden jar is oscillatory.*

15.2. The Oscillatory Circuit

The Leyden jar circuit in Fig. 15A contains, in addition to a *capacitance*, an *inductance* as well. The single loop *FGABCD* and *E* forms practically one turn of a coil. An inductance and capacitance, connected as shown in simplest, schematic form in Fig. 15B, form the necessary elements of all oscillating circuits.

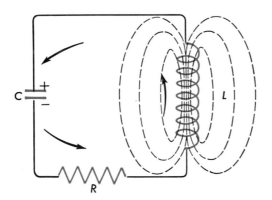

Fig. 15B *Schematic diagram of an oscillating circuit.*

If, initially, the capacitance is charged as indicated, the surplus electrons on the plate below cause a surge of negative charge counterclockwise around the circuit to neutralize the positives and, in so doing, set up a magnetic field in and around the inductance. When the positives become neutralized and the electron current tends to cease, the magnetic flux linking the circuit decreases and keeps the current flowing in the same direction. Once this field has vanished and the current has ceased, the capacitance is found to be in a charged condition, the upper plate negative and the lower plate positive.

Having reversed the charge on the capacitance, the above process will repeat itself, this time the electron current surging clockwise around the circuit. Thus the current rushes first in one direction, then the other, oscillating back and forth in an electrical way just as any spring pulled to one side and released vibrates in a mechanical way (see Fig. 15C).

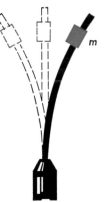

Fig. 15C *A vibrating spring is like an electrical oscillating circuit.*

When a straight spring is pulled to one side and released, the kinetic energy it gains upon straightening keeps the spring moving, and it bends to the other side. Just as the vibration amplitude of the spring slowly decreases because of *friction,* so also does the current in the electrical circuit decrease because of *electrical resistance.* A graph showing how current slowly dies out in an electric circuit is given in Fig. 15D. These are called *damped vibrations,* or *damped*

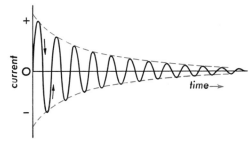

Fig. 15D *Graph of the damped oscillations of an electrical circuit.*

oscillations. If the resistance of the circuit is high, the damping is high and the current quickly dies out after but few oscillations. If the resistance is low, however, the damping is small, the amplitude decreases slowly, and there are many oscillations.

To calculate the frequency of an oscillating circuit, either of the following formulas may be used:

$$T = 2\pi\sqrt{LC} \qquad \boxed{f = \frac{1}{2\pi\sqrt{LC}}} \qquad (15a)$$

where L is the inductance in henries, C is the capacitance in farads, T is the time for one complete oscillation in seconds, and f is the number of oscillations per second. T is the period and f the frequency. The formula above is to be compared with the analogous formula for the period of a vibrating spring:

$$T = 2\pi\sqrt{m/k} \qquad (15b)$$

The mass m for the spring is analogous to the inductance L for the circuit, and the stiffness $1/k$ is analogous to the capacitance C. An increase of the inductance L, or capacitance C, or both, increases the period and decreases the frequency of the oscillating circuit.

> *Example.* A Leyden jar with a small capacitance of 0.01 μf is connected to a single turn of wire (about 6 in. in diameter) having an inductance of 1 microhenry. Calculate the natural frequency of the circuit.
>
> *Solution.* Since 1 henry $= 10^6$ microhenries and 1 farad $= 10^6$ microfarads, direct substitution for L and C in Eq. (15a) gives
>
> $$f = \frac{1}{2\pi\sqrt{LC}} = \frac{1}{2\pi\sqrt{1 \times 10^{-6}\,\text{h} \times 1 \times 10^{-8}\,\text{f}}}$$
> $$= 1,590,000 \text{ cyc/sec}$$
> or
> $$1.59 \text{ megacycles/sec}$$

15.3. Maxwell's Electromagnetic Wave Theory

In 1856 James Clerk Maxwell wrote his now famous theoretical paper on electromagnetic waves. In this scientific publication, he proposed the possible existence of electromagnetic waves and at the same time postulated that if such waves could ever be produced, they would travel through free space with the speed of light.

Light itself, said Maxwell, is propagated as an electromagnetic wave, and electrically produced waves should differ from light only in their wavelength and frequency. Because Maxwell gave no clues as to how such waves might be generated or detected, their real existence was not discovered until 32 years later when Heinrich Hertz made his important discovery.

15.4. Hertzian Waves

In 1888, a young German scientist, Heinrich Hertz,* began a series of experiments in which he not only produced and detected electromagnetic waves, but

---

* Heinrich Rudolf Hertz (1857–1894), German physicist born at Hamburg, February 22, 1857. He studied physics under Helmholtz in Berlin, at whose suggestion he first became interested in Maxwell's electromagnetic theory. His researches with electromagnetic waves which made his name famous were carried out at Karlsruhe Polytechnic between 1885 and 1889. As professor of physics at the University of Bonn, after 1889, he experimented with electrical discharges through gases and narrowly missed the discovery of X rays described by Röntgen a few years later. By his premature death, science lost one of its most promising disciples.

also demonstrated their properties of reflection, refraction, and interference. One of his experimental arrangements is diagramed in Fig. 15E.

The transmitter consists of two spheres QQ' located near the ends of two

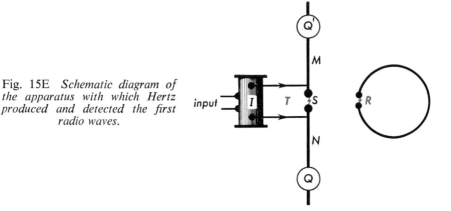

Fig. 15E *Schematic diagram of the apparatus with which Hertz produced and detected the first radio waves.*

straight rods MN separated by a spark gap S. With the two rods connected to an induction coil I, sparks jump across the gap S, giving rise to oscillating currents in MN. That such a generator is an oscillating circuit can be seen from the fact that the spheres QQ' form the plates of a capacitor and the rods form the inductance.

The receiver, or detector, consists of a single loop of wire with a tiny spark gap at R. This circuit, too, is an oscillating circuit with the spark gap as a capacitance and the loop as an inductance. Tuning the transmitter frequency to that of the receiver is accomplished by sliding the spheres QQ' along the rods MN, resonance being indicated by the appearance of sparks at R.

With apparatus of this general type, Hertz was able to transmit signals a distance of several hundred feet. He found that large metal plates would reflect the radiation, and that at normal incidence the reflected waves would interfere with those coming up to set up standing waves with nodes and loops. As the receiver was moved slowly away from the reflector, nodes and loops were located by the appearance of sparks only at equally spaced intervals.

With a large prism of paraffin he demonstrated refraction, and with a lens made of pitch he focused the waves as a glass lens focuses visible light.

15.5. Electromagnetic Waves

To visualize the production of waves by a Hertzian oscillator, consider the schematic diagram in Fig. 15F. Let the rods MN and spheres Q_1 and Q_2 be charged initially as indicated, and consider the electrostatic action of the charges on a small charge C located some distance away. The negative charge Q_1 attracts C with a force a and the positive charge Q_2 repels it with a force b. Since by symmetry these two forces are of equal magnitude, their resultant CE

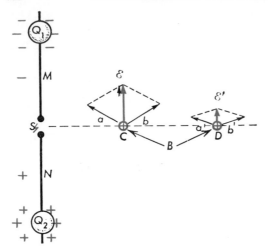

Fig. 15F *A Hertzian dipole.*

is parallel to *MN*. If the isolated charge is farther away as at *D*, the resultant force is also parallel to *MN*, but weaker. In other words, the electric field ε at points *C* and *D* is up and parallel to *MN*, decreasing in intensity as the distance from the transmitter increases.

Suppose that a spark jumps the gap *S* and oscillation sets in. One-half cycle after the condition shown in Fig. 15F, electrons have surged across the gap, charging Q_1 positively and Q_2 negatively. With reversed charges the resultant force on *C* and *D* will be down instead of up. Thus it is seen how oscillations in the transmitter, which constitute a surging of electrons back and forth between *M* and *N*, give rise to a periodically reversing electric field at distant points.

In addition to an electric field at *C* and *D*, the surging electrons in *MN* give rise to a magnetic field as well. When the electron current is down (using the conventional left-hand rule), the magnetic induction *B* at *C* or *D* is perpendicular to and into the plane of the page, and when the electron current is up the magnetic induction is out from the page. The surging of the charges therefore gives rise to a periodically reversing magnetic induction, the direction of which is at right angles to the electric intensity at the same points.

According to Maxwell's theory, the ε and *B* fields do not appear instantly at distant points; time is required for their propagation. The speed of propagation, according to Maxwell (and this has been confirmed by numerous experiments) is the same as the speed of light. The changing ε and *B* fields at *C* therefore lag behind the oscillating charges in *MN*, and those at *D* lag behind still farther.

Figure 15G is a graph of the instantaneous values of the electric and magnetic fields as they vary with distance from the transmitter. At certain points the fields are a maximum and at other points they are zero. As time goes on, these electric and magnetic waves move away from the transmitter with a speed of 186,300 mi/sec.

The mathematical theory of electromagnetic radiations shows that, close to

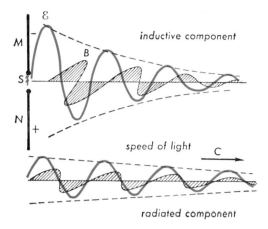

Fig. 15G *Graph of the electro-magnetic waves emitted by a Hertzian dipole.*

the transmitter, the $\mathcal{E}$ and B fields, called the *inductive components*, are 90° out of phase and that their magnitudes fall off very rapidly with distance. Farther out, however, the two get in step with each other and their amplitudes fall off more slowly, as shown in the diagram. The latter are called the *radiated components* and are the ones detected at great distances.

Suppose now that a series of sparks is made to occur in the gap S of a transmitter as in Fig. 15H. Each spark will give rise to a damped oscillation in MN

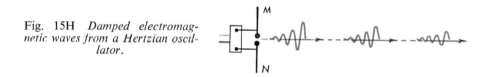

Fig. 15H *Damped electromagnetic waves from a Hertzian oscillator.*

which in turn sends out a damped electromagnetic wave. The succession of sparks sends out a train of such waves which, as they leave the antenna, decrease rapidly in magnitude at first, then more slowly as they get farther away. Only the electric component of such waves is shown in the diagram.

If an electrical conductor is located at some distant point, the free electrons within it will, as such waves go by, experience up and down forces tending to set them in oscillation. If the conductor is an oscillating circuit whose natural frequency is that of the passing waves, resonance will occur and large currents will be set up.

15.6. Air-Core Inductances and Transformers

When a high-frequency alternating current is sent through a solenoid with an iron core or the primary of an iron-core transformer, the back emf is so large that the current as well as the magnetic induction cannot build up to any appreciable value before it reverses in direction. The result is that in one ten-thousandth of a second or less the current hardly gets started in one direction before it stops and reverses.

In an iron-core transformer the back emf in the primary winding so retards the building up of strong fields that little or no induced currents can be "drawn" from the secondary. To overcome this difficulty, the iron core is done away with, so that we have what is called an *air-core transformer*. In the absence of any iron, the current in the coil may rise to an appreciable value each time it changes in direction. The rapidly increasing and diminishing flux that links both circuits induces a current in the secondary of exactly the same frequency.

Air-core transformers, consisting of nothing more than two coils of a few turns each, a primary winding and a secondary winding, are used extensively in radio, microwave, and television transmitters and receivers. In these instances the alternating currents with frequencies of thousands and even millions of cycles per second are usually referred to as *radiofrequencies*, and the transformers are referred to as *radiofrequency transformers*.

15.7. The Vacuum Tube Rectifier

While the great American inventor Thomas A. Edison was striving by a process of trial and error to produce a satisfactory electric light bulb, he made an accidental discovery, the importance of which was first recognized and used successfully by Sir John Fleming. Now called a vacuum tube rectifier or diode, the Fleming valve is used in nearly every radio and television transmitter and receiver to change alternating current into direct current.

The Fleming valve, as shown in Fig. 15I, consists of a highly evacuated glass bulb containing a wire filament that is heated electrically to incandescence.

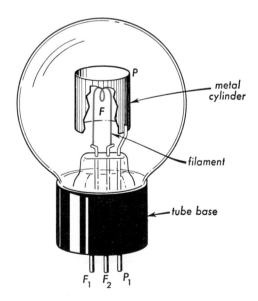

Fig. 15I *Diagram of a Fleming valve, or rectifier tube. Such tubes are now called diodes.*

Surrounding the filament and connected to the outside through the tube base and a prong P_1 is a cylindrical metal plate P. When the filament is heated to incandescence, it gives off large quantities of electrons in much the same way that water, when heated to the boiling point, gives off steam. The emission of electrons by a hot body is called *thermionic emission* and is due to the high temperature and not to the electric current. Heating a metal by any other means will produce the same effect. Electrons emitted from hot metal surfaces are called *thermoelectrons*.

The principal action of the *filament F* and *plate P* is explained by means of a typical electric circuit shown schematically in Fig. 15J. The circuit consists of a

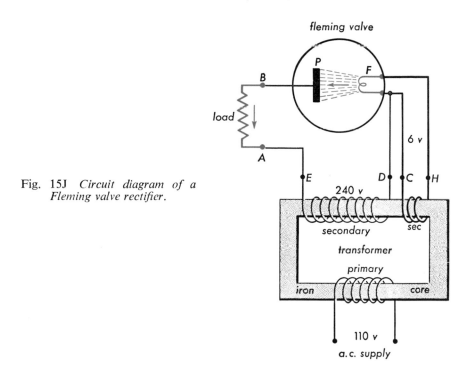

Fig. 15J   *Circuit diagram of a Fleming valve rectifier.*

transformer having *two secondary windings*, a *Fleming valve*, and a *load*. The latter, shown as a resistance, represents any electrical device requiring unidirectional current for its operation. With an alternating current of 110 volts supplied to the primary, a high voltage, 240 volts for example, is delivered by one secondary to the terminals ED and a low voltage of 6 volts alternating current is delivered by the other secondary to the terminals CH. The latter, called the *filament winding*, is for the purpose of heating the filament.

When for a fraction of a second the plate P of the tube is positively charged and the filament F is negatively charged, the electrons from F are attracted to the plate P and constitute a current flowing across the vacuum space PF and through the load from B to A. One-half cycle later, when the potential is reversed and P becomes negatively charged and F positively charged, the electrons from F are repelled by P and very little current flows.

The emf's in each part of the rectifier circuit are shown by graphs in Fig. 15K.

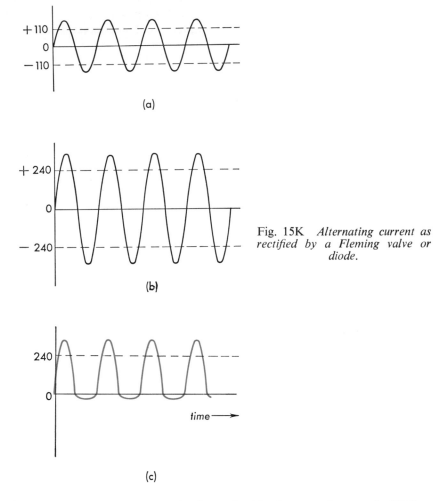

Fig. 15K   *Alternating current as rectified by a Fleming valve or diode.*

The primary emf of 110 volts is shown in (a), the secondary emf of 240 volts in (b), and the *rectified* or *pulsating emf* through the load AB in (c).

15.8. Full-Wave Rectifier

A full-wave rectifier tube, sometimes called a *duo-diode*, is essentially a double Fleming valve with two plates and two filaments. (See Fig. 15L.) The two prongs F_1 and F_2 in the base are connected to both filaments in series, while the prongs P_1 and P_2 are connected one to each plate.

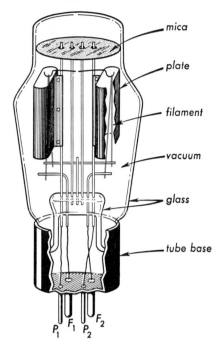

mica

plate

filament

vacuum

glass

tube base

Fig. 15L   *Drawing of a full-wave rectifier tube.*

A schematic diagram of a rectifier circuit employing such a tube is shown in Fig. 15M. Here an iron-core transformer with one primary and two secondary windings is used, differing from the single phase rectifier in Fig. 15J in that the center of each secondary winding is now connected to the load. *CHJ* is the filament winding and supplies current to both filaments (shown as one bent wire) while *GED* is the high-voltage winding. The latter supplies an alternating potential to the plates so that, when P_1 is $+$ and P_2 is $-$, electrons from the filament are attracted to P_1, and, when a moment later P_1 is $-$ and P_2 is $+$, electrons from the filament are attracted to P_2.

In the first instance a current flows around the circuit $F_1P_1GEABHJF_1$, and in the second it flows around the circuit $F_2P_2DEABHCF_2$. In each case the current has gone through the load AB in the same direction and has pulsating characteristics as shown in Fig. 15N.

If such a pulsating current were used to supply the direct current needed in every radio receiver, a loud objectionable hum with a frequency of 120 cycles would be heard. To make this current a steady smooth direct current, as illustrated by the straight line in the same graph, and thus eliminate the hum, a *filter circuit* as shown in Fig. 15O is used. The terminals A and B are connected to, and replace, the load A and B in Fig. 15M. K is an iron-core inductance and C_1 and C are capacitors of large capacity.

As the current through $AabB$ starts to flow, the capacitors become charged, as shown, and a magnetic field is created around K. This has a retarding action

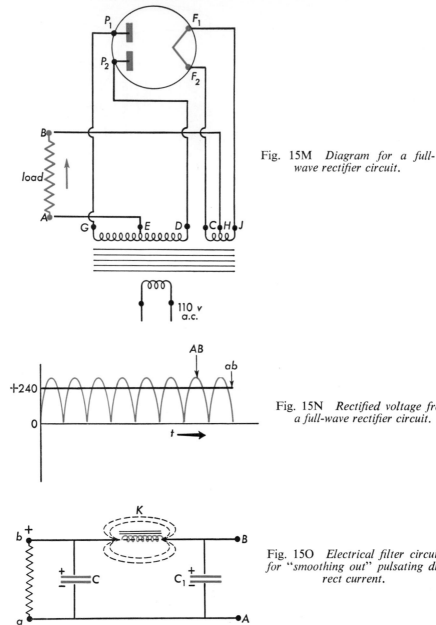

Fig. 15M *Diagram for a full-wave rectifier circuit.*

Fig. 15N *Rectified voltage from a full-wave rectifier circuit.*

Fig. 15O *Electrical filter circuit for "smoothing out" pulsating direct current.*

which prevents the current from reaching its otherwise peak value. When a moment later the filament-to-plate current drops to near zero, the capacitors discharge and the field around K collapses, thus sending a current through AB. This process is repeated with each pulse of electrons from either plate of the tube, and the current through *ab* remains steady. Large capacities and a large self-inductance deliver more constant voltage.

15.9. De Forest's Audion

Although the Fleming valve was originally developed for the purpose of detecting wireless waves, its operation as such did not prove to be very satisfactory until, in 1906, De Forest* invented the *audion*. By inserting a grid wire between the *plate* and *filament* of a Fleming valve, he created a device capable not only of detecting wireless waves, but of amplifying the signals as well. The purpose of the grid (see Fig. 15P) is to control the flow of electrons from the hot filament *F* to the plate *P*.

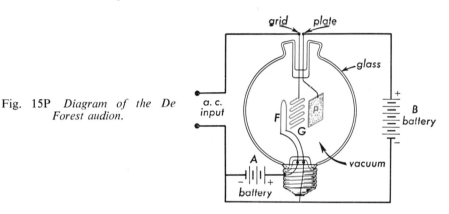

Fig. 15P   *Diagram of the De Forest audion.*

A circuit diagram showing how the audion may be used as a one-tube receiver of radio waves is given in Fig. 15Q. The filament *F* is heated by a 6-volt battery *A*, and the plate *P* is maintained at a positive potential of 45 volts or more by

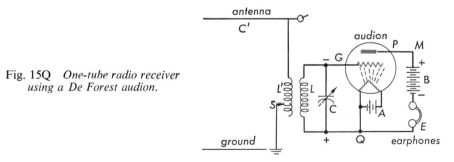

Fig. 15Q   *One-tube radio receiver using a De Forest audion.*

* Lee De Forest (1873–1961), American scientist; Ph.D. from Yale University, 1899. His most famous invention was the audion, considered by many to be the most important invention ever made in radio. He designed and installed the first five high-power radio stations for the U. S. Navy. After 1921 he devoted his time to the development of talking motion picture film. He was awarded gold medals at the St. Louis exposition in 1904, the Panama Pacific Exposition in San Francisco in 1915, and the Institute of France in 1923. He received the Cresson Medal of the Franklin Institute in 1921 for his important contributions to wireless.

the *B-battery.* The capacitor *C* and inductance *L* form an *oscillation circuit,* the natural frequency of which may be varied by changing the capacity of *C.* The arrow indicates a capacitor of variable capacitance.

Coaxial with *L* is another inductance *L′* (shown beside each other in the diagram), one end of which is connected to a wire in the air called the *antenna,* and a sliding contact *S* near the other end is connected to a metal conductor embedded in the ground. The antenna and ground form the two plates of a capacitor *C′* with the air as a dielectric. This capacity *C′* with the variable inductance *L′* forms a second oscillation circuit whose natural frequency is varied by the sliding contact *S.*

By varying the frequency of the *L′C′* circuit until it matches the frequency of any passing wave (shown as damped wireless waves in Fig. 15H), an oscillating current and magnetic field occur in *L′.* By tuning the *LC* circuit to this same frequency, resonance is set up and the oscillating current imposes alternate $(+)$ and $(-)$ charges on the grid *G.* During the time the grid *G* is negative, electrons from the filament are repelled and are unable to reach the plate *P.* When the grid is positive, however, the electrons from the filament are accelerated toward the plate and constitute a flow of current clockwise around the circuit *PMEQFP.*

Not only does the grid act as a rectifier valve and let the electron current flow in one direction only, from filament to plate, but it acts as an amplifier, allowing large currents from the high voltage *B*-battery to flow through when it is slightly positive, and practically no current when it is slightly negative.

Voltage graphs for the two parts of the receiver circuit are given in Fig. 15R.

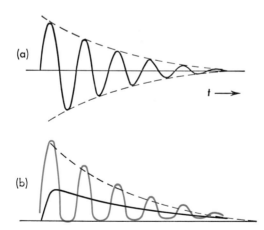

(a)

$t \longrightarrow$

(b)

Fig. 15R   (*a*) *Input and* (*b*) *output voltage curves for the De Forest audion used in the wireless receiver, Fig. 15Q.*

Diagram (a) represents the oscillating potentials in the *LC* circuit connected to the grid, and diagram (b) the current through the plate and earphones. Since the frequency of this rectified current is far too rapid for the electromagnets in the earphones, their metal diaphragms over the pole tips move up and down with the heavy solid line in (b), which is heard as a single audible click. A succession of damped waves is then heard as a noise, with each damped wave producing a single vibration of the receiver diaphragms.

15.10. Modern Vacuum Tubes

Every radio enthusiast today knows that there are hundreds of different kinds of radio tubes. Some contain two filaments and two plates, while others contain as many as three or four separate grids. Although a treatment of such complex tubes is out of place here, the fundamental principles of all of them are little different from De Forest's audion. One important difference, however, is illustrated in Fig. 15S, and that is the employment in some tubes of a cathode in place of a filament as a source of thermal electrons.

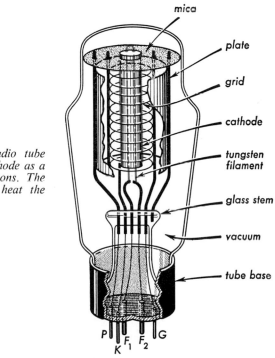

Fig. 15S   *A modern radio tube with a cesium-coated cathode as a source of thermal electrons. The filament serves only to heat the cathode.*

A fine tungsten wire filament is threaded through two small holes running lengthwise through a porcelainlike insulating rod. Fitting snugly around this rod is the cathode, a metal cylinder coated on the outside with a thin layer of thorium, strontium, or caesium oxide. These particular oxides are copious emitters of electrons when heated to a dull red heat. Insulated from the cathode, the filament as a source of heat can be, and generally is, connected directly to an appropriate transformer winding.

A schematic diagram of the cathode type of tube containing one plate and

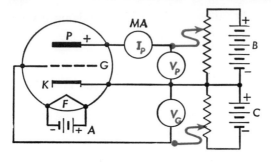

Fig. 15T   *Circuit diagram for determining the characteristics of a triode vacuum tube.*

one grid is given at the left in Fig. 15T. This circuit diagram shows the electrical connections and instruments needed to measure the *grid-voltage* vs. the *plate-current* curve for vacuum tubes in general. A variable voltage is applied to the plate P by a B-battery through a potential divider, and a variable voltage is applied to the grid G by a C-battery which is connected to another potential divider. The various applied grid potentials are read from the voltmeter V_G and the corresponding plate current is read from a milliammeter MA.

Four characteristic curves taken with the above circuit connections are reproduced in Fig. 15U, one for each of four plate voltages, 50, 100, 150, and 200 volts, respectively.

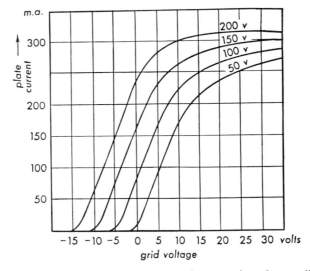

Fig. 15U   *Grid voltage–plate current curves for normal grid-controlled vacuum tube.*

The flattening-out of all the curves at the top indicates that, when the grid is highly positive, practically all of the electrons leaving the filament get to the plate and no further increase in plate current can occur. Should the cathode be heated to a higher temperature, however, more electrons will be emitted and the saturation current will occur at higher plate currents than those shown. The foot of each curve, near -2, -6, -11, and -16 volts, indicates what negative voltage on the grid will stop all electrons and prevent them from reaching the plate. The higher the $+$ potential on the plate, the more negative must be the grid to stop the electrons.

15.11. Vacuum Tube Oscillator

To broadcast the human voice by radio, a generator of alternating current of extremely high frequency and constant amplitude is required. In commercial broadcasting stations and amateur transmitters, this function is performed by a vacuum tube and circuit of relatively simple design.

One type of oscillator circuit is shown in Fig. 15V. When the switch S is

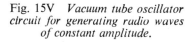

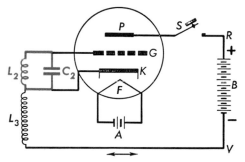

Fig. 15V   *Vacuum tube oscillator circuit for generating radio waves of constant amplitude.*

closed, connecting the B-battery to the plate of the tube, an electron current from the cathode K to the plate P starts a current in the circuit $PRVL_3K$. This growing current in L_3 creates an expanding magnetic field, which, cutting across L_2, induces a current in the grid circuit in such a direction that the grid becomes negative. A negative charge on the grid, as shown by the characteristic curves in Fig. 15U, causes the plate current to decrease. This decreasing current causes the field about L_3 to collapse, thus inducting a reversed current in the grid circuit and therefore a positive charge on the grid. Such a charge increases the plate current, and the above process is repeated.

If the two circuits, L_2C_2 and $PRVL_3$, are properly tuned by adjusting C_2, resonance will occur and energy from the B-battery will be continuously supplied to keep the oscillations going with constant amplitude. The graph of the continuous oscillations shown in Fig. 15W represents the voltage across L_3 as it varies in time. The L_2C_2 circuit controls the frequency by controlling the grid potential while the large voltage and current fluctuations take place in the L_3 circuit.

15.12. Radio Transmitter

To use an oscillating tube circuit, of the kind described above, as part of a radio transmitter, the high-frequency oscillations in the L_2C_2 circuit must be

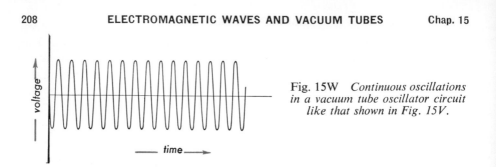

Fig. 15W   *Continuous oscillations in a vacuum tube oscillator circuit like that shown in Fig. 15V.*

modified by sound waves and then applied to an antenna and ground system, for broadcasting as electromagnetic waves. A simplified circuit diagram showing one of the many ways of doing this is given in Fig. 15X. There are three parts to

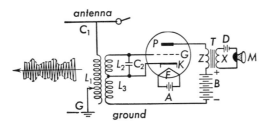

Fig. 15X   *Radio transmitter employing a microphone and only one tube as an oscillator.*

this particular hookup: (1) *the microphone circuit* containing a battery D and a transformer T, (2) *the oscillator circuit* in the middle, and (3) *the antenna circuit* $C_1 L_1 G$ at the left.

By talking or singing into the microphone, the diaphragm inside moves back and forth with the sound vibrations, thus altering the steady current previously flowing around the circuit DMX. An illustration of the pulsating current is shown in Fig. 15Y(a). Current pulsations in X, the transformer primary, cause similar pulsations in Z, the secondary circuit carrying the plate current. The effect of the relatively low frequency audio currents (a) on the high-frequency

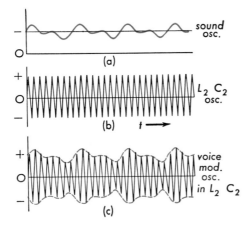

Fig. 15Y   *Graphs of (a) sound waves, (b) continuous oscillations in $L_2 C_2$ and (c) voice-modulated oscillations in $L_2 C_2$.*

oscillations (b) already there is to alter their amplitude as shown in diagram (c).

Through the *coupling* of L_3 and L_1, the modulated oscillations are induced in the antenna circuit by resonance, and are radiated as electromagnetic waves of the same frequency and form. The continuous wave produced by the radio-frequency oscillations alone is called the *carrier wave*, and the alteration of its amplitude by *audiofrequencies* is called *modulation*. Although radio transmitters with one vacuum tube have been used by radio amateurs, it is customary to find transmitters with half a dozen or more tubes. The principal function of additional tubes, in receivers as well as transmitters, is to amplify currents wherever they are needed, thereby providing greater transmitting range and clearer reception.

15.13. Vacuum Tube Amplifier

One of the most important functions of the vacuum tube is its use as an *amplifier* of radiofrequency or audiofrequency currents, as shown in Fig. 15Z(a).

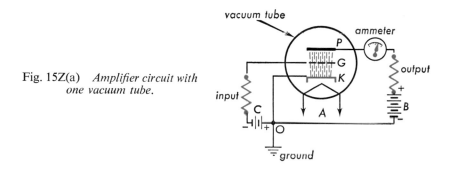

Fig. 15Z(a) *Amplifier circuit with one vacuum tube.*

The *input* resistor represents some part of any circuit in which a weak but varying current is flowing; the *output* resistor represents another circuit to which a stronger current of the same form is delivered. In some cases these are resistors as shown, but in others they are the primary and secondary windings of separate transformers. The source of the additional energy is the *B*-battery plate supply.

In amplifying any given signal current, a faithful reproduction of the *wave form* must be carried out, otherwise *distortion* will result; musical sounds from a radio will be harsh or pictures from a television receiver will be blurred. To amplify without distortion, a tube that has a long straight section in its *characteristic curve* must be used (see Fig. 15U), and it should be operated at the center of this straight portion. Such an operation is shown by the graph in Fig. 15Z(b). To make the tube operate at *M*, a small battery, called a *C* battery or *C* bias, is inserted in the grid circuit to maintain the grid at a negative potential. For the curve and tube shown, this requires −5 volts, while for other types of tubes it might well require greater or smaller potentials.

When no input signal potentials are imposed, the grid is held at −5 volts and a steady current of 150 milliamperes flows through the plate and output

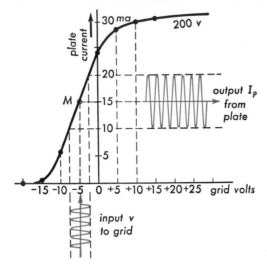

Fig. 15Z(b) *Graph showing amplifier operation.*

circuit. If now an alternating current like a radio frequency of constant amplitude is impressed across the input terminals, the grid potential will rise and fall in the same way, and an undistorted but amplified current will flow in the plate and output circuit. The time variations in grid potential are shown at the bottom in Fig. 15Z(b), and the corresponding plate current oscillations at the right. If the input radio frequency is voice-modulated, the amplified current will also be voice-modulated without distortion. It should be noted that if the impressed grid voltage variations are too large, say -20 to $+10$ volts, the amplified currents will reach the curved portions of the curve above and below, and *distortion* of the *wave form* will result. As long as the tube is operated on the straight portion of the curve, the plate current is directly proportional to the impressed grid potential, and faithful amplification takes place.

QUESTIONS

1. What is an oscillatory circuit? What are its three principal elements?

2. Upon what does the natural frequency of a circuit depend? What is the formula for the frequency?

3. What is electrical resonance? Under what conditions does it arise?

4. What are Hertzian waves? What is a Hertzian dipole?

5. What are electromagnetic waves? What is their nature? With what speed do they travel?

6. What are damped oscillations? How can damping be reduced?

7. Draw from memory a schematic wiring diagram of a full-wave rectifier showing the vacuum tube, transformer, and load.

8. Make a schematic diagram of a full-wave rectifier and filter circuit consisting of a transformer, vacuum tube, choke coil, two capacitors, and a load.

9. Draw from memory a wiring diagram of a one-tube radio receiver as shown in Fig. 15Q, but using a cathode type of vacuum tube in place of a De Forest audion.

10. Diagram the circuit of a vacuum tube oscillator for generating high radio frequencies of constant amplitude.

11. Draw from memory an amplifier circuit with one vacuum tube.

12. Make from memory a drawing of a radio transmitter employing a microphone and one vacuum tube.

PROBLEMS

1. Calculate the frequency and period of an oscillating circuit containing two 4-μf capacitors and an inductance of 5.2 μh if all three are connected in parallel.

2. What inductance connected to a capacitor of 0.05 μf will give the circuit a natural frequency of 6 megacycles/sec? *Ans.* 0.0141 μh.

3. What capacitance, if connected in parallel to an inductance of 10 μh, will give an oscillating circuit a frequency of 500 kilocycles/sec?

4. Determine the frequency of an oscillating circuit composed of two capacitors and one inductor, all connected in parallel: $C_1 = 5$ μf, $C_2 = 25$ μf, and $L = 10$ μh. *Ans.* 9.19 Kc/sec.

5. Two capacitors of 50 μf each are first connected in series and then the combination is connected across an inductor of 8 μh. Calculate the period of the oscillating circuit.

6. Calculate all of the possible frequencies that can be obtained by combining two or more of the following to form an oscillating circuit: $C_1 = 2$ μf, $C_2 = 4$ μf, $L = 8$ μh. *Ans.* 39.8, 28.1, 23.0, 48.8 Kc/sec.

7. A 6-mh inductor is connected in parallel with a 60-μf capacitor. What is the natural frequency of this circuit?

8. A 60-μh inductor is connected across a 75-μf capacitor. What is the frequency of the fifth harmonic of the oscillating circuit thus formed? *Ans.* 11.86 Kc/sec.

9. What capacitance connected in parallel to an inductance of 0.3 μh will produce an oscillating circuit with a fundamental frequency of 2 megacycles/sec?

10. What inductance connected in parallel to a capacitance of 20 μf will produce an oscillating circuit with a fundamental frequency of 30 kilocycles/sec? *Ans.* 1.41 μh.

11. Calculate the frequency of an oscillating circuit composed of a 2-μf capacitor and a 2-μh inductor.

12. What capacitance in parallel with an inductance of 0.1 μh will have a frequency of 1 megacycle/sec? *Ans.* 0.253 μf.

13. A capacitance of 0.05 μf is connected to an inductance of 6×10^{-8} henry. Find (a) the frequency and (b) the wavelength of the electromagnetic waves emitted.

14. What capacitance should be used with an inductance of 0.09 μh to produce electromagnetic waves having a wavelength of 10 cm? *Ans.* 0.0313 $\mu\mu$f.

15. What inductance should be used with a capacitance of 0.06 μf to produce electromagnetic waves of wavelength 50 cm?

The Solid State and Semiconductors

One of the most active and important branches of science and technology today is called *solid state physics*. This is a field of research concerned with a study of the physical properties of solids, and includes such subjects as the detailed configurations of atoms in solids, the behavior of free and bound electrons in a crystal lattice, the electrical conductivity of pure substances and those containing impurities, the magnetic and mechanical properties of solids at low and high temperatures. Since solid state physics has grown to such huge proportions (and so has the number of people active in the field) only a small part of the subject will be given in this chapter.

16.1. The Crystal Lattice

When a liquid cools and solidifies, the atoms tend to form a three-dimensional array called a *crystal lattice*. The model in Fig. 16A illustrates a cubic structure, the simplest known type of array in which the atoms take the positions at the corners of cubes.

Common table salt with its two kinds of atoms, *sodium* and *chlorine*, always

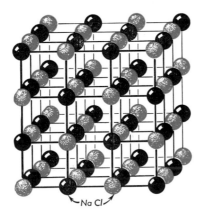

Fig. 16A *Atomic model for sodium chloride (NaCl is common table salt).*

forms such a cubic lattice, the individual atoms alternating in kind in each of the three directions, Na, Cl, Na, Cl, Na, etc.

A second model shown in Fig. 16B is a more complicated structure, and

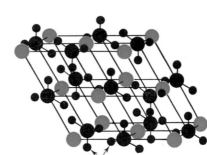

Fig. 16B *Atomic model for Ice-*
land spar, or calcite, CaCO₃.

$CaCO_3$

represents a chemical compound known as *calcite*. Calcite, which chemically is calcium carbonate, $CaCO_3$, is a clear transparent crystal found in nature. Note how the calcium and carbon atoms form the corners of parallelograms, each carbon atom surrounded by three oxygen atoms.

Although there are some twenty or more basic structures known to exist, two of the simplest and most common forms, face-centered and body-centered cubic crystals, are shown in Fig. 16C. These show the least number of atoms

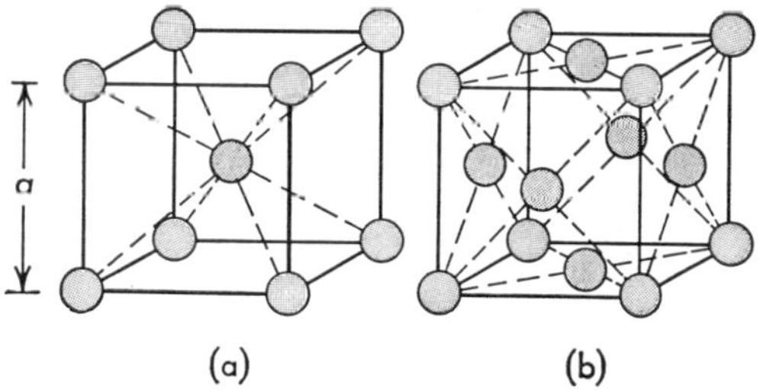

Fig. 16C *Unit cells for two com-*
mon crystal forms: (a) a body-
centered cube; and (b) a face-
centered cube.

(a)   (b)

necessary to illustrate the general structure of any given class of crystals and are called *unit cells*.

A number of chemical elements that form face-centered or body-centered cubic structures are listed in Table 16A. The size of the structure is specified in each case by the dimension *a* shown in Fig. 1C, and is given in angstroms: 1 angstrom = 1×10^{-10} m.

While some elements or compounds always seem to form the same crystal pattern on solidifying, others are known to take on any one of a number of different forms. For example, two of the forms of iron are given in Table 16A. While most of the different forms of crystals known today have been found in nature, there are some that have been produced in the laboratory only. Diamond, one of several known crystal forms of carbon atoms, is a closely packed

Table 16A. Some Elements and Their
Cubic Crystal Structures

FACE-CENTERED		BODY-CENTERED	
Element	a (angstroms)	Element	a (angstroms)
Aluminum	4.0493	Barium	5.0250
Copper	3.6149	α Iron	2.8665
Gold	4.0786	Rubidium	5.6300
γ Iron	3.5910	Sodium	4.2906
Lead	4.9505	Strontium	4.8500
Lithium	4.404	Titanium	3.3060
Nickel	3.5239	Tungsten	3.1147
Platinum	4.0862	Uranium	3.4740
Silver	4.0862	Zirconium	3.6200

crystal that has, until recently, defied laboratory reproduction. Small diamonds for use in the manufacturing of high-speed cutting tools are now produced in quantities by the General Electric Company. It was in the company's laboratories that the production of real diamonds was accomplished by applying extremely high temperatures and pressures to small graphite slugs.

The important features of the diamond structure are shown in Fig. 16D, and

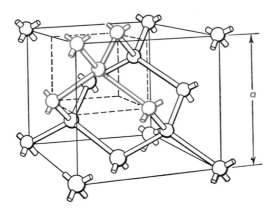

Fig. 16D   *Diamond crystal structure. For carbon, silicon, and germanium a = 3.56, 5.43 and 5.66 A, respectively. Nearest neighbor spacing is $a\sqrt{3}/4$, and maintains tetragonal symmetry. (After Shockley.)*

this is the characteristic structure for the elements C, Si, Ge, Sn, and Pb in the fourth column of the Periodic Table. See Appendix VII. Note that while the outermost atoms form a large cube, all atoms are bound by chemical bonds to their four nearest neighbors. Furthermore, each set of four nearest neighbors lie at corners of a cube, with atoms diagonally opposite each other on the square sides.

For C, Si, and Ge, $a = 3.56$, 5.43, and 5.66 A, respectively, while the nearest neighbor spacing is $a\sqrt{3}/4$.

Metals in general, when they cool down from the molten state, solidify into ultramicroscopic crystals that pack closely together to form a three-dimensional mosaic. This is well illustrated by the electron microscope photographs reproduced in Fig. 16E. Note the clear-cut cubic structure of pure aluminum.

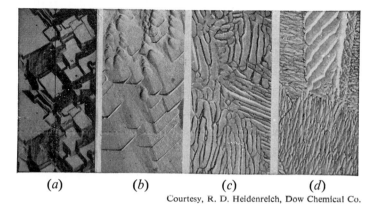

(*a*)   (*b*)   (*c*)   (*d*)

Courtesy, R. D. Heidenreich, Dow Chemical Co.

Fig. 16E *Electron microscope photographs showing the crystalline-like structure of metals: (a) pure aluminum 5600×, (b) magnesium-aluminum alloy 13,000×, (c) polished steel 14,000×, and (d) polished copper 14,000×.*

While each atom or molecule in a solid is confined to a definite space within the body lattice, it is in a state of vibration within that space. As the temperature decreases, this motion becomes slower and slower until at absolute zero, −273°C, all molecular motion ceases. By molecular motion is meant the motion of the molecule as a whole. At absolute zero, however, the atoms of certain solids are still vibrating. This residual energy of vibration is known to be an inherent property of some solids, which cannot be utilized or taken away.*

16.2. Conductors and Nonconductors

The crystal structure of the common highly conducting metals, like copper, silver, gold, and platinum, is such that the outermost electrons are shared by all the atoms. The complete inner shells of electrons are bound to their individual nucleus, but the outermost electrons in uncompleted shells, the so-called *valence electrons*, are free to wander throughout the substance. This diffusion process is similar to the random motion of molecules in a gas.

In contrast with *good conductors* there are *good insulators*, called *nonconductors*, which have practically no free electrons. In substances like quartz, mica, and sulfur, all of the electrons remain bound to their respective atoms. The resistivity of a good conductor is as low as 10^{-7} or 10^{-8} ohm-meters, compared to values as high as 10^{16} ohm-meters for an insulator like quartz.

* An introductory book on crystal structure is Alan Holden and Phylis Singer, *Crystals and Crystal Growing*, Doubleday, New York.

16.3. Semiconductors

There are a large number of solids that are neither good conductors of electricity nor good insulators. Since the resistance range between these two groups of materials is about 10^{23}, a majority of known substances lie in between. These substances are called *semiconductors*. In them, electrons are capable of being moved only by the application of relatively strong electric fields of hundreds of thousands of volts per meter.

Of the large number of semiconductors known to science, certain ones are of considerable importance. Typical examples are the crystalline forms of the elements listed in the fourth column of the periodic table (see Appendix VII). Two important ones are *silicon* and *germanium*.

Figure 16F is a diagram of a small germanium crystal about 5 mm square and

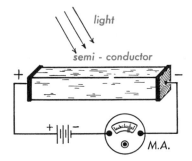

Fig. 16F *Circuit diagram for demonstrating the photoconductivity of a semiconductor like germanium or silicon.*

2 cm long, connected to a battery and a milliammeter by wires, and completing an electric circuit. When light is allowed to fall on the crystal, its electrical resistance decreases and the current rises. This response to light is instantaneous and is called *photoconductivity*.

If the germanium crystal is heated, the current again rises, indicating a decrease in electrical resistance. This *heating effect* is not instantaneous, however, since it takes a long time for the current to return to its original value, that is, for the temperature of the crystal to return to room temperature. The resistance of metallic conductors behaves in just the opposite way; their resistance increases with a rise in temperature.

In order to explain the light and heat effects described above, we must refer to the crystal lattice of semiconductors. Silicon and germanium atoms each have what the chemists call four *valence electrons*, that is, four electrons that enter into the chemical binding in solids. The atomic pattern of atoms in both crystals is a tetrahedral structure, as shown in Fig. 16D. In Fig. 16G each atom shares one of its electrons with each neighbor, the neighbor in turn shares one of its four with it. Such a sharing of electrons between two atoms is called a *covalent bond*.

Because of the difficulty of drawing a three-dimensional tetrahedral lattice

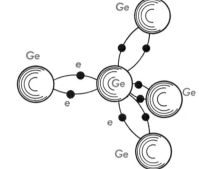

Fig. 16G   *Each atom in a germa-*
nium crystal is bound at the center
of a tetrahedron formed by its four
nearest neighbors.

structure, it is convenient to flatten the diagram out and represent the bonding as a square lattice, as shown in Fig. 16H.

At temperatures close to absolute zero, all electrons in a crystal are tied up

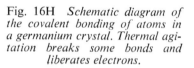

Fig. 16H   *Schematic diagram of*
the covalent bonding of atoms in
a germanium crystal. Thermal agi-
tation breaks some bonds and
liberates electrons.

strongly by these chemical bonds. When the crystal is raised to room temperature, however, the thermal vibrations of the atoms are sufficient to break some of the bonds and free some of the electrons to wander throughout the crystal. Where an electron has broken free, as shown at the upper right and lower left in Fig. 16H, a hole has been created and the process is referred to as *dissociation*. Since that part of the crystal was neutral beforehand, it now lacks an electron, and the vacant hole is equivalent to a net positive charge.

Owing also to thermal agitation, a *bound* electron next to a hole can move across to fill the gap, the net motion of the negative charge from one bonded position to another being in effect equivalent to the motion of a hole in the opposite direction. The motion of a hole is, therefore, equivalent to the motion of a positive charge. This action is shown at the lower center in Fig. 16H.

16.4. N-Type and P-Type Crystals

The widespread use of semiconductors in such devices as solar cells and transistors has resulted from the technical development of pure semiconductor crystals impregnated with minute quantities of certain impurities. The crystals most commonly used for these purposes are those mentioned in the preceding section, *silicon* and *germanium*.

If crystals are formed with *arsenic* as an impurity, the arsenic atoms, with five valence electrons each, provide a crystal lattice with extra electrons. Such a crystal as shown in Fig. 16I is therefore one in which each arsenic atom donates

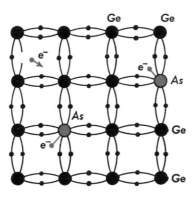

Fig. 16I *N-type crystal lattice with arsenic atoms as an impurity.*

one free electron to the system. With arsenic present in quantities of one to a million, *donor atoms* are on the average about 100 atoms apart, or there are about 10^{17} arsenic atoms and 10^{17} free electrons per cubic centimeter. In a good conductor like copper there are approximately 10^{23} free electrons per cm^3.

By thermal agitation a few bound electrons are also "shaken loose" and an equal number of holes thereby created. The ratio of the number of *donor electrons* to the number of unbound electrons or *holes* is approximately 10,000 to 1. Since by far the majority of free-to-move charge carriers are electrons, the lattice is called an *N-type crystal*.

If crystals are grown with *aluminum* as an impurity, the aluminum atoms, with only three valence electrons each, form a crystal lattice with an electron deficiency, that is, with holes. See Fig. 16J. Such a crystal is therefore one in

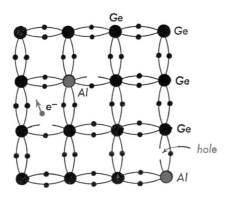

Fig. 16J *P-type crystal lattice with aluminum atoms as an impurity.*

which each aluminum atom provides one hole to the system that is free to accept an electron.

By thermal agitation a few bound electrons are also shaken loose and an equal number of holes are produced. Since by far the majority of the charge carriers are holes, and these act like positive charges, this type of lattice is called a *P-type crystal*.

Since the donor electrons in an N-type crystal, and the holes in a P-type crystal, far outnumber the electrons or holes contributed by thermal agitation, the donor charges are called *majority carriers*, and those caused by dissociation are called *minority carriers*.

Neither of these crystals, by itself, has a net charge. The surplus of free negatives in an N-type crystal is compensated for by the positive charges on the arsenic nuclei, while the surplus of holes in the P-type crystal is compensated for by the deficiency in positive nuclear charge of the aluminum nuclei.

When light falls on a crystal as in Fig. 16F, the light is absorbed within a few atomic layers. The absorbed energy breaks some of the electron bonds and creates holes. This process is called *photoionization*. The potentials applied at the ends of the crystal cause the electrons to move to the left, and the holes to the right. This flow of charge constitutes a current.

When the light is shut off, free electrons fall at random into holes as the crystal cools. The falling of an electron into a hole is called *recombination*.

16.5. The PN Junction, or Diode

When two semiconductors of the P and N types are brought into contact, as shown in Fig. 16K, they form what is called a *PN junction*, or *diode*. In the region

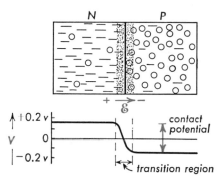

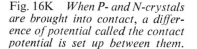

Fig. 16K   *When P- and N-crystals are brought into contact, a difference of potential called the contact potential is set up between them.*

of contact, a cloud of free electrons in the N crystal diffuse across the boundary to the right, much the same as gas atoms diffuse through a porous ceramic material.

Since electrons leave the N crystal, that side of the junction acquires a positive potential, while the P crystal gains electrons and acquires a negative potential. As this diffusion of electrons continues, the potential difference between the two sides rises. The filling of holes in the P crystal gives rise to more holes developing in the N crystal, and this action is the same as though holes had diffused across the boundary to the left.

As the N crystal becomes more positive, and the P crystal more negative, electrons will be attracted back toward the left, and an equilibrium condition will develop in which equal numbers of electrons will be crossing the boundary in opposite directions.

The electrostatic potential of different points across the crystal boundary under equilibrium conditions, as just outlined, is shown graphically below the PN junction in Fig. 16K. Notice that the sudden drop of the potential in the region of the junction gives rise to a large electric field ε.

Although various methods for producing diodes have been developed commercially, some mixing of the two semiconductor materials is purposely brought about at the junction. This area of mixing, called the *transition region*, accounts for the smooth way in which the potential curve changes from point to point through this region.

For a typical PN junction the transition region is about 6×10^{-5} mm thick, and the potential difference, called the *contact potential*, may have any value from a small fraction of a volt to 1 or 2 volts depending on the two materials in contact. These values indicate an electric field ε of several million volts per meter. This is called the *diffusion field* ε, shown in red at the center of the diagram.

If we now fuse metal plates to the ends of a PN junction and connect them to a milliammeter, the device becomes effective as a solar battery or cell. See Fig. 16L. When the entire crystal is maintained at a constant temperature in a

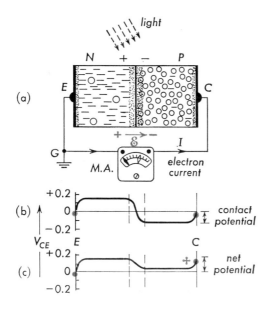

Fig. 16L   (a) *PN-junction used as a solar battery cell.* (b) *Potential graph when cell is in the dark, and* (c) *when it is in light.*

darkened room, no current will be observed through the milliammeter. The reason for this is that *reverse contact potentials*, due to electron diffusion, are set up between the crystal ends and the metallic electrodes, so that no potential difference exists between *E* and *C*. See diagram (b).

If we now shine light on the PN junction, the light is absorbed, freeing additional electrons and creating holes. By virtue of the strong electric field ℰ in the transition region, electrons now move to the left and holes to the right, and we have a current. Such a current is readily measured by a milliammeter M.A.

The drift of electrons to the left lowers its potential while available free holes move to the right and raise its potential. The lead connections have not changed their contact potentials; so we now have a net useful potential difference as shown in diagram (c). Acting as the terminals of a battery cell the end plates send an electron current *I* through the external circuit.

Since the useful ionization process arising from the absorption of light occurs only in the surface layers of atoms, solar batteries are made with very thin crystals deposited on some insulating material which serves as a rigid backing.

The electrostatic potential graphs in diagrams (b) and (c), are drawn assuming 0.15 volt contact potentials at the ends. These are typical but their values will depend upon what metals are used.

16.6. PN-Junction, or Diode, Rectifier

When a PN junction is connected to a battery for the purpose of sending a current through it, there are two ways in which the voltage can be applied. If the *P* crystal is connected to the positive terminal of the battery as shown in Fig. 16M, electrons will be pulled to the right and holes to the left. Both of these constitute a current. Note carefully that holes are being pulled from where

Fig. 16M *Schematic diagram of a PN-junction, or diode, with an applied voltage (forward bias). Graphs showing potentials across the junction at different voltages* V_{CE}.

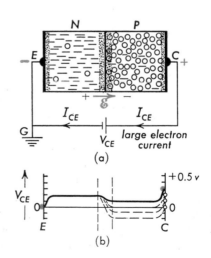

there are lots of holes and electrons from where there are lots of electrons, and we obtain a large current.

If we reverse the PN junction as in Fig. 16N, we will be trying to pull holes

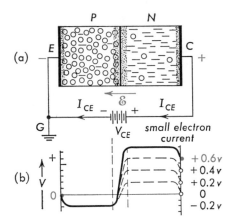

Fig. 16N *Schematic diagram of a PN-junction, or diode, with an applied voltage (reverse bias). Graphs show potentials across the junction at different voltage V.*

from where there are but few holes, and electrons from where there are but few electrons, and experimentally we obtain a relatively small current.

If we now apply different voltages, in turn, across a PN junction and measure the current for each value, we obtain a graph of the type shown in Fig. 16O.

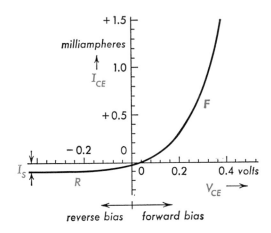

Fig. 16O *Typical characteristic curve for a PN-junction diode.*

This so-called *characteristic curve* shows that in one direction the current increases rapidly with small increases in voltage, whereas in the opposite direction the current is small and remains small even for relatively large values.

This is why a PN junction acts like a *rectifier* of alternating currents. When P is positive with respect to N, a large current flows and the junction has what is called a *forward bias*. When P is negative with respect to N, there is a small current and the junction is said to have a *reverse bias*.

With a small forward bias V_{CE} on the PN junction we are operating in the F region of the characteristic curve. Since this applied potential difference is in opposition to the diffusion field potential, the resultant drop in potential across the transition region is flattened out as shown in Fig. 16M(b). As the forward bias is increased, the current rises and the resultant potential curve flattens still more.

With a reverse bias V_{CE} on the PN junction, we are operating in the R region of the characteristic curve and there is a small current through the transition region. Even though there is a high positive charge on the N-crystal side of the junction, there are so few electrons to pull across the boundary from the P crystal that saturation is produced by a relatively small reverse bias. Reverse-bias potentials of many volts cannot increase this current and it will remain constant.

16.7. Transistor Triode

A transistor is composed of three semiconductor elements, two of the N-type crystals and one of the P-type crystals. See Fig. 16P. This combination is referred to as an *NPN-transistor*. We can also have a *PNP-type transistor*.

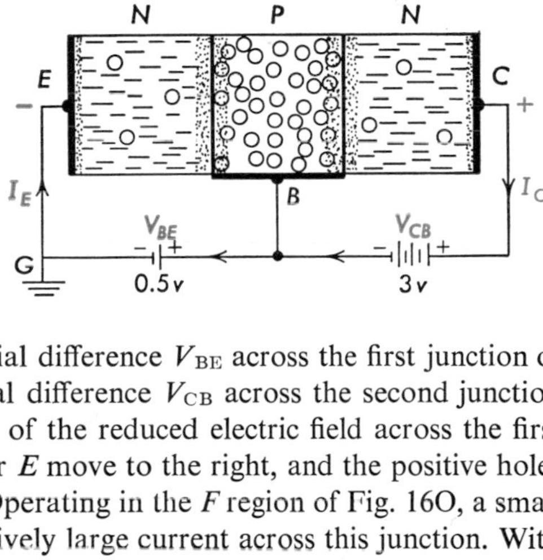

Fig. 16P *Schematic diagram of an NPN transistor, or triode, showing typical forward and reverse bias voltages.*

Suppose we now apply a potential difference V_{BE} across the first junction of the NPN transistor and a potential difference V_{CB} across the second junction as shown in the diagram. Because of the reduced electric field across the first junction, electrons from the emitter E move to the right, and the positive holes from the base B move to the left. Operating in the F region of Fig. 16O, a small voltage V_{BE} will give rise to a relatively large current across this junction. With a reverse bias on the second junction, electrons from the P crystal try to move to the right, and holes from the collector C to the left. Operating in the R region of the characteristic curve, little or no transport of charge takes place. Since there are few holes in the collector and few electrons in the base, then an increase of V_{CB} even to a large value will not produce an appreciable current. Because of this blocking action of the second junction a large current will flow between the emitter E and the base B.

By making the center element of the transistor very thin, as shown in Fig. 16Q an important change takes place. The electrons diffusing across the first junction

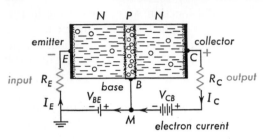

Fig. 16Q *Circuit diagram showing how a transistor or triode is used as a voltage amplifier.*

have little time to find a hole and be neutralized. Most of them are quickly attracted into the collector crystal by the high positive potential.

Because of the high resistance vertically along the thin P crystal, few electrons will get through to the base B. Hence the current across the two junctions is approximately the same.

Since a few electrons do recombine with holes in the center element, however, the current I_C is not quite as large as the current I_E, so that the ratio is slightly less than unity:

$$\frac{I_C}{I_E} = \alpha \tag{16a}$$

A typical value of α, for a well-designed transistor, would be 0.98, or 98%.

16.8. Transistor Structure

Junction diodes and triodes consist of semiconductors having relatively thin regions of change from N-type material to P-type material. There is a deliberate attempt in the manufacturing process to produce nonuniform distributions of impurities.

The physical structure of one type of NPN transistor is shown in Fig. 16R.

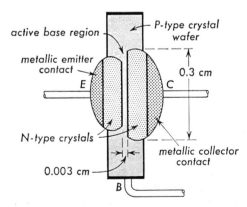

Fig. 16R *Cross-section diagram of an NPN transistor, or triode. The N crystals under pressure and temperature are fused into the P-type crystal wafer.*

From the dimensions given it will first be observed how very small it is. Note secondly how very thin the base region is. In some transistors the base is no more than 10,000 A thick while the transition region is not over 500 A. Thin as transition regions are, there are many atomic layers, since interatomic distances are from 2 to 4 A.

In some manufacturing processes the end crystals, with their metal electrodes already soldered or welded on, are heated and fused into the base wafer by pressure.

16.9. Single-Stage Transistor Amplifier

An amplifier circuit using a transistor in place of a vacuum tube is shown in Fig. 16Q. Since the current I_C is very nearly equal to the current I_E, little electron current flows through BM. With the input resistance R_E small, and the output resistance R_C large, the nearly equal currents through them signify a small $I_E R_E$ drop across the input and a large $I_C R_C$ drop across the output. Hence, there is a voltage gain and a power gain.

The ratio of the output power $I_C^2 R_C$ to the input power $I_E^2 R_E$ is called the power gain:

$$A = \frac{I_C^2 R_C}{I_E^2 R_E} \tag{16b}$$

A typical power gain would be $A = 40$. If we now apply a weak alternating current as an input power, the a.c. output power in the collector circuit could be 40 times larger.

Simple schematic diagrams of transistors, as used in circuit diagrams, are shown in Fig. 16S.

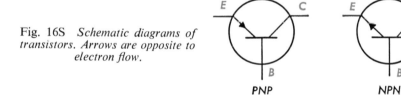

Fig. 16S *Schematic diagrams of transistors. Arrows are opposite to electron flow.*

PNP   NPN

The arrow is always shown on the emitter lead and its direction indicates the flow of positive current through the emitter material. The arrow therefore distinguishes between NPN and PNP transistors.

The basic principles of a transistor triode are analogous to those of a vacuum-tube triode, the base functioning very much as a control grid. A small alternating voltage is applied between B and E, and an amplified current I_C flows from the collector C.

As an example of the application of a single transistor as a voltage amplifier, consider the circuit diagram in Fig. 16T. This is a diagram of a *call system* in which by speaking into the first loud-speaker S_1 the amplified voice can be heard emerging from a distant loud-speaker S_2.

When the switch Sw is closed, an electron current flows from the battery $(-)$, through the switch, up through the NPN transistor and through R_L to

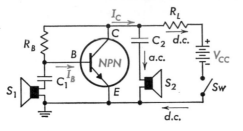

Fig. 16T   *Circuit diagram of a one-way call system employing one transistor.*

the battery (+). The function of R_B is to adjust the voltage across CE so that most of the potential difference is across CB, the collector, and a relatively small voltage is across BE, the emitter. See Fig. 16Q. R_B is called the *biasing resistor* and is carefully selected so as to adjust the transistor for maximum amplification.

When the sound vibrations of the voice enter S_1, they set the cone vibrating. In some types of speakers a small coil is mounted on the apex of the cone, and as the coil vibrates back and forth in the strong magnetic field of an Alnico permanent magnet, a minute "voice current" is generated. The speaker functions in this way as a microphone.

Since the generated voice current is an alternating current of varying amplitude and frequency, some of it passes through C_1 and is superimposed upon the input direct current to the emitter. Together these two currents constitute a varying d.c. as an input load.

The amplified signal from the collector circuit is also a varying d.c. The d.c. portion of this current returns through R_L to the battery, while the a.c. portion passes through C_2 and the speaker where it is converted into sound. Notice that C_1 and C_2 prevent d.c. battery current from flowing through S_1 and S_2, respectively. R_L limits the amount of direct current through the transistor and is called the *load resistor*.

To understand the basic principles involved in transistor operation, typical characteristic curves for a transistor triode are given in Fig. 16U. The upper set of curves (a) applies to the *output* and show how the collector current I_C varies with different voltages across CE. The lower curve (b) applies to the *input* and shows how the current I_B into the base varies with different emitter voltages across BE.

As a sample operation we select an NPN triode with the characteristics shown in Fig. 16U, and we choose a battery voltage $V_{CC} = 10$ v. To obtain good amplification it is then good practice to provide the appropriate load resistor R_L so that the voltage across CE is about $\frac{1}{2}V_{CC}$. In this example, therefore, we select $V_{CE} = 5$ v. These two points for V_{CC} and V_{CE} are shown on the horizontal voltage scale in diagram (A).

Upon drawing a vertical line at 5 volts, an appropriate base current curve is selected, and the one chosen here is $I_B = 0.20$ ma. A straight line is then drawn through P and V_{CC}, as the output current-voltage operating path, and a hori-

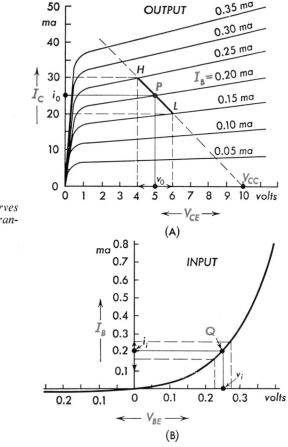

Fig. 16U   *Characteristic curves for the operation of a typical transistor triode.*

zontal line extended from P to establish the corresponding collector current $I_C = 25$ ma.

We can now determine the value of the appropriate load resistor R_L, since the IR drop across it should be approximately 5 v. This leaves the other 5 v for the transistor. Applying Ohm's law, $V = IR$, we can write

$$5 \text{ v} = 25 \text{ ma} \times R_L$$

from which we obtain

$$R_L = 200 \text{ ohms}$$

A normal biasing resistor R_B of 50 K to 100 K ohms will alter this so little its effect is negligible. See R_2, R_4, and R_8 in Table 16A.

With the selection of $I_B = 0.2$ ma for the input to the base, a horizontal line can be drawn through this point, as shown in diagram (B) of Fig. 16U. Where this line intersects the curve at Q a vertical line is drawn to find $V_{BE} = 0.25$ v.

The values thus far determined are the d.c. operating voltages and current, and we now apply an alternating potential difference to BE through the capacitor C_1. Suppose this alternating potential has a maximum amplitude of 0.025 v as shown by the double arrow at the bottom in diagram (B). This produces a current swing of I_B from approximately 0.15 ma to 0.25 ma.

Referring these values to the output operating path PV_{CC} on the upper graphs, the output current I_C swings from 20 ma to 30 ma while V_{CE} swings from 4 v to 6 v.

The reason the output operates along the slanted line PV_{CC} may be explained as follows. As an input voltage rises, V_{BE} diagram (B), the base current I_B rises. A rise in I_B, diagram (A), causes a rise in collector current I_C. This increased current through the load resistor R_L produces a bigger IR drop across R_L leaving less of the battery voltage V_{CC} across the resistor V_{CE}.

The voltage amplification of a transistor triode can be defined in several ways, but if we take as a definition *the ratio of the output voltage to the input voltage*, we obtain from Fig. 16U

$$A_V = \frac{v_o}{v_i} = \frac{1.0 \text{ v}}{0.025 \text{ v}} = 40$$

The current amplification may be defined as *the ratio of the output current to the input current*. Using the values from Fig. 16U as an example,

$$A_I = \frac{i_o}{i_i} = \frac{5 \text{ ma}}{0.05 \text{ ma}} = 100$$

The transistor is not a source of energy; it simply converts d.c. energy obtained from the battery to a.c. energy by the control signal applied at the base. These amplification values just illustrated are theoretical only and are not actually realized in practice. For one thing, the input capacitor C_1 and the output capacitor C_2 limit markedly the input and output frequencies. These and internal power losses limit the *power amplification* to approximately 40, as stated by Eq. (16b).

16.10. Three-Stage Transistor Amplifier

A circuit diagram for a three-stage amplifier is shown in Fig. 16V. This application is called an *intercommunication system*, in which each of two identical speakers is used as a microphone or as a loud-speaker. When the reversing switch Sw$_2$ is in the UP position, speaker S_1 is used as a microphone and S_2 as a loud-speaker, whereas with the switch in the DOWN position their functions are reversed.

The first transistor circuit forms the *input stage* in which the voice current is received from the speaker acting as a microphone. The second transistor circuit forms the *driver stage* because it receives the amplified signal current from the first stage. It is called the driver stage because it "drives" the third stage. This final transistor circuit is the *output stage* which powers the speaker acting now as a speaker.

Capacitor C_2 has a relatively small capacitance and shunts part of the a.c.

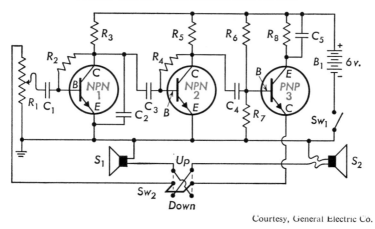

Fig. 16V   *Complete circuit diagram of a two-way intercommunication circuit employing three transistors.*

away from the second stage. This is done in order to prevent the high-frequency portion of the sound signal from reaching the final stage of amplification and the loud-speaker. There are limits to the frequencies that a loud-speaker can handle without distortion. The lower a.c. signals with frequencies in the middle of the audible range are passed by the larger capacitance C_3 to *EB*, the input of the second stage.

The output of the second stage passes the amplified a.c. signal through C_4 to the final stage. Since the emitter of the PNP transistor is at the top, the amplified voice signal emerges via the collector *C* and passes through the switch Sw_2 to the remote speaker S_2. This speaker ignores the d.c. portion of the current and reacts to the variations as if they were a.c.

Resistor R_8 limits the amount of d.c. current through PNP 3, while C_5 carries the a.c. output signal from the second stage to the emitter of the third stage. If C_5 were not provided, part of the power needed for driving the remote speaker would be lost in R_8. Table 16B gives a list of appropriate resistor and capacitor values.

**Table 16B. Resistance and Capacitance Values
for a Three-Stage Amplifier Circuit**

R_2	100 K ohms	C_1	0.22 µf
R_3	3.3 K ohms	C_2	0.05 µf
R_4	47 K ohms	C_3	0.22 µf
R_5	1.5 K ohms	C_4	6 µf (electrolytic)
R_6	470 ohms	C_5	50 µf (electrolytic)
R_7	1.5 K ohms	S_1	20 ohms
R_8	47 ohms	S_2	20 ohms

K = 1000

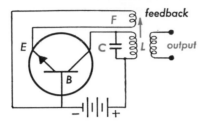

Fig. 16W   *Oscillator circuit using an NPN-transistor.*

Figure 16W shows a simple oscillator circuit employing an NPN transistor. The oscillation frequency is determined by L and C in the collector circuit, and feedback of a weak oscillating signal to the emitter is accomplished by the small coil F which picks up induced voltages from L.

QUESTIONS AND PROBLEMS

1. (a) What is a semiconductor? (b) What is a P-type crystal? (c) What is a PN junction?

2. (a) What is a solar battery? (b) What are the principles of the solar battery?

3. (a) What is a transistor? (b) What is a PNP transistor?

4. Explain the photoconductivity of a semiconductor like germanium.

5. Since an N-type semiconductor crystal has more free electrons than holes, why is it electrically uncharged?

6. Since a P-type semiconductor crystal has more holes than free electrons, why is it electrically uncharged?

7. Make a circuit diagram of a transistor amplifier using a PNP transistor. Show the $(+)$ and $(-)$ terminals of any batteries.

8. Make a diagram showing what instruments you would connect to a PNP transistor to obtain readings for plotting curves like those in Fig. 16U. How would you obtain different voltages and read the currents?

9. If each of the three stages of the circuit diagram in Fig. 16V has an amplification factor of 12, what is the over-all amplification?

10. Make a circuit diagram of a two-stage intercommunication system using an NPN transistor for the first stage and a PNP transistor for the second.

11. Make a circuit diagram of a three-stage intercommunication system using three NPN transistors.

12. If a 12-v battery is used for a transistor circuit of the kind shown in Fig. 16T, and a base current $I_B = 0.25$ ma is selected for its operation, find (a) the desirable voltage across CE, (b) the corresponding collector current I_C, (c) the load resistance to be used, (d) the base voltage V_{BE}, (e) the theoretical voltage amplification, and (f) the theoretical current amplification. Use the graphs in Fig. 16U.

17

Moving Frames of Reference

As an introduction to Einstein's theory of relativity let us consider the relative motions of bodies as seen from different frames of reference. We begin with the motion of a body in a medium which is itself moving. The drift of an airplane in a wind or the drift of a boat on a moving body of water is a good example.

17.1. The Airplane Problem

The pilot of a plane wishes to fly to a city directly to the north. If the plane has a cruising speed of 100 mi/hr and a steady wind is blowing from the west with a velocity of 50 mi/hr, at what angle should the pilot head his plane into the wind? See Fig. 17A(a). Because this type of problem is often solved incorrectly, its correct solution should be noted with care. The procedure to be followed is appropriately called the *domino method of vector addition.*

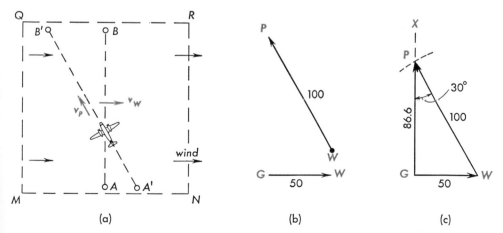

Fig. 17A *To fly a desired course, the wind velocity must be taken into account.*

With both the direction and magnitude of the wind velocity known, draw the first vector toward the east, 50 units long. See diagram (b). Label it with the moving body, the wind W at the head, and the ground G to which the velocity is referred, at the tail. Since only the magnitude of the airplane velocity is known, draw a temporary arrow 100 units long, at an arbitrary angle, and label it with the moving body, the plane P, at the head, and the air or wind W, to which the velocity is referred, at the tail.

The next step is to combine the two vectors with their like labels W together, as follows. After drawing the vector GW, draw a perpendicular line GX upward. With a compass of radius 100 units and the center at W, draw a short arc intersecting the vertical line at P.

The vector WP is then completed, and the length of the side GP measured. The vector GP of 86.6 pointing north represents the velocity of the plane P with respect to the ground G, while the angle of 30°, measured from the triangle, gives the direction in which the plane must be headed.

Diagram (a) in Fig. 17A shows how the plane, heading in a direction 30° west of north, and flying through the air with a velocity of 100 mi/hr, follows the northward land course from A to B with a *ground speed* of 86.6 mi/hr. In the air mass ($MNRQ$), which is a moving frame of reference, the plane flies from A' to B'.

17.2. Relative Times for Two Planes

Because of its direct bearing upon Einstein's theory of relativity, the following problem should be of primary interest to every student. Suppose two pilots with identical planes are to fly to different cities equally far away. As shown in Fig. 17B, one pilot flies north to city Y, at right angles to the wind, and returns,

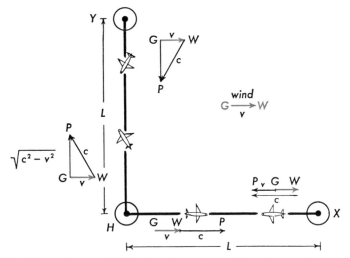

Fig. 17B   *Comparisons of flight time for a plane flying downwind, upwind, and crosswind.*

while the other flies east to city X, parallel to the wind, and returns. We now wish to find whether the two total flight times are the same or whether they are different.

Let c represent the cruising speed of both planes, and v the velocity of the wind. Starting out for city Y, the first pilot, cruising at velocity c, sets his course west of north so that his flight path, with respect to the ground, is due north, as shown by the left-hand vector diagram. To return to home base H, he sets his course west of south so that his flight path is due south, as shown by the upper velocity triangle. Since his ground speed GP is the same each way, by the Pythagorean theorem for a right triangle, we find that

$$GP = \sqrt{c^2 - v^2}$$

Starting out for city X, the second pilot, with the same air cruising velocity c, sets his course due east. Since his plane is flying *with* the wind, the ground speed GP is just the arithmetic sum of the two speeds:

$$GP = c + v$$

To return to home base H, the pilot sets his course due west. Flying against the wind, the plane has a reduced ground speed equal to the arithmetic difference between the two speeds:

$$GP = c - v$$

To find the total flight time $t_\perp$ of the first plane, we divide the total distance $2L$ by the velocity $\sqrt{c^2 - v^2}$:

$$t_\perp = \frac{2L}{\sqrt{c^2 - v^2}} \qquad (17a)$$

To find the total flight time $t_\parallel$ of the second plane, the two different velocities $c + v$ going, and $c - v$ returning must be used. Therefore, we have

$$t_\parallel = \frac{L}{c + v} + \frac{L}{c - v}$$

Placing these two terms over a common denominator, we obtain

$$t_\parallel = \frac{2Lc}{c^2 - v^2} \qquad (17b)$$

One way to compare these flight times is to divide the perpendicular flight time by the parallel flight time, and obtain

$$\frac{t_\perp}{t_\parallel} = \frac{2L}{\sqrt{c^2 - v^2}} \div \frac{2Lc}{c^2 - v^2}$$

Inverting the divisor and multiplying, we find

$$\frac{t_\perp}{t_\parallel} = \frac{2L}{\sqrt{c^2 - v^2}} \times \frac{c^2 - v^2}{2Lc} = \frac{\sqrt{c^2 - v^2}}{c}$$

from which we obtain the simplified result,

$$\frac{t_\perp}{t_\parallel} = \sqrt{1 - \frac{v^2}{c^2}} \qquad\qquad (17c)$$

This equation shows that flying with no wind blowing, $v = 0$, the ratio of the two flight times is unity, which means that they are equal, as one would expect. When the wind is blowing, however, the ratio is less than unity, and the plane flying parallel to the wind requires the greater time. If the wind velocity increases to nearly that of the cruising speed of the planes, the ratio tends toward zero. If the wind velocity exceeds the cruising speed of the planes, the ratio becomes imaginary: the first plane is blown off its course and the second plane cannot get back to home base.

It will be left as a problem for the student to show that the time of flight over both of these two courses is greater when the wind is blowing than when it is calm.

17.3. Frames of Reference

If two observers are moving with respect to one another, and both observers make measurements of any event, they could both be expected to come to the same conclusions as to what took place. In Fig. 17C we see two sets of rec-

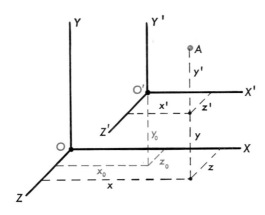

Fig. 17C *The coordinate distances of a point A measured in two different frames of reference (Cartesian coordinates).*

tangular, or *Cartesian*, coordinates called *frames of reference*. In the space common to these two frames is a single point marked A, and it is to this point that measurements are to be made. The first observer O' in the upper right-hand frame measures the distance to A and finds the coordinates of this distance to be x', y', and z'. The second observer O in the lower left-hand frame measures the distance to A and finds coordinate distances x, y, and z.

Suppose that observer O' now wishes to make measurements from which he can determine x, y, and z as observed by O. To do this, he measures the distance from O' to O and finds the coordinates to be x_0, y_0, and z_0. He can now write down the following equalities:

$$x = x' + x_0$$
$$y = y' + y_0$$
$$z = z' + z_0$$
(17d)

Suppose observer O wishes to make his own determinations of x', y', and z' as observed by O'. To do this, he measures his distance from O to O' and finds the coordinates to be x_0, y_0, and z_0. He then writes down the equations

$$x' = x - x_0$$
$$y' = y - y_0$$
$$z' = z - z_0$$
(17e)

Note that the two sets of equations are the same and that they permit either observer to transform measurements made in his frame of reference to those made in the other. It is for this reason that such equations are called *trans-formation equations*.

17.4. Distance Measurements in Moving Frames of Reference

To transform measurements from one moving frame of reference to another, we will confine the motion to the line joining the two origins. It is along this straight line that x-axes are set up for two frames, as shown in Fig. 17D. Here

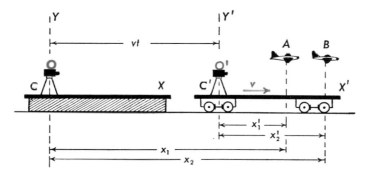

Fig. 17D   *Distance measurements of two airplanes, A and B, from two frames of reference.*

one observer O is located on a stationary platform C and the other observer O' on a flat car C' capable of moving freely along a track. Observer O and reference frame C will be assumed at rest, while observer O' and car C' will be assumed moving along together with constant velocity v.

Both observers start their stop clocks at the instant O' is opposite O. At some later time, both observers simultaneously note the time and snap pictures of two airplanes, A and B, directly over the track. From their stereophotographs, observer O finds the distances to be x_1 and x_2, while observer O' finds the distances to be x'_1 and x'_2. Each observer now decides to use the transformation equations to find the other observer's distance measurements. Observer O uses Eq. (17e) and writes

$$x'_1 = x_1 - vt \quad \text{and} \quad x'_2 = x_2 - vt \tag{17f}$$

and observer O' uses Eq. (17d) and writes

$$x_1 = x'_1 + vt \quad \text{and} \quad x_2 = x'_2 + vt$$

In each case they arrive at the same equations. Each observer finds that upon taking the difference between his own measured distances,

$$x_2 - x_1 = x'_2 - x'_1$$

Each observer concludes, therefore, that the distance between A and B is the same whether viewed from one frame or the other. The transformation of measurements from one moving frame of reference to another by means of the above equations is referred to as a *Galilean-Newtonian transformation*. Under such transformations, distances are said to be *invariant*, that is, they are the same.

We say, therefore, that in any Galilean-Newtonian system an object measured and found to have a length l in one reference frame will be found to have the same length l' when measured in any other reference frame:

$$l' = l \tag{17g}$$

Note that this equality of results does not depend upon the value of v or of t, but does depend upon all observations being made at the same instant, that is, *simultaneously*. In other words, anything happening in one frame at an instant t' is observed from the other frame as occurring at the same instant t:

$$t' = t \tag{17h}$$

17.5. Velocity Measurements in Moving Frames of Reference

We have seen in the preceding section that, under a Galilean-Newtonian transformation, the straight-line distance between two points is invariant. The question next arises whether the *velocity* of a moving body should be invariant, i.e., Should the velocity be the same when observed from different frames which are themselves moving with different velocities? To find the answer to this question, we again make use of two observers, one stationary, and the other on a flatcar as shown in Fig. 17E.

Both observers start their stopclocks at the instant the moving observer O' is opposite the stationary observer O. At some later time, both observers simultaneously note the time t_1 as they snap stereo pictures of an airplane A_1 flying along and over the track. A short time later, both observers again note the time t_2 as they simultaneously snap pictures of the same plane at A_2. From their respective photographs each observer determines the plane's distance for each of the two times. Applying the transformation equations, Eq. (17f), they can write

$$x'_1 = x_1 - vt_1 \quad \text{and} \quad x'_2 = x_2 - vt_2$$

To find the average velocity of the plane, observer O' takes his measured

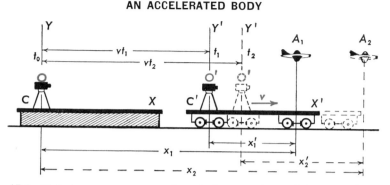

Fig. 17E   *Velocity measurements of an airplane as observed from two frames of reference.*

distance it has traveled, $x'_2 - x'_1$, and divides by the elapsed time $t'_2 - t'_1$. Taking the difference, he obtains

$$x'_2 - x'_1 = (x_2 - x_1) - (vt_2 - vt_1)$$

and dividing each term by either of the two equal time differences, he finds

$$\frac{x'_2 - x'_1}{t'_2 - t'_1} = \frac{x_2 - x_1}{t_2 - t_1} - \frac{v(t_2 - t_1)}{t_2 - t_1}$$

The term on the left represents the average velocity u' of the plane as observed by O', while the middle term represents the average velocity u observed by O.

$$u' = \frac{x'_2 - x'_1}{t'_2 - t'_1} \quad \text{and} \quad u = \frac{x_2 - x_1}{t_2 - t_1}$$

Observer O' can therefore write

$$\boxed{u' = u - v} \tag{17i}$$

Exactly the same equations can be written by observer O. Thus the two velocities measured by the two observers are not the same; they differ by the relative velocity of the two observers. We say that, *under a Galilean-Newtonian transformation, velocities are not invariant.*

17.6. An Accelerated Body Viewed from a Frame Moving with Constant Velocity

Suppose that the plane in Fig. 17E is increasing its speed, and by appropriate means of observation observers O and O' determine its velocity when it passes point A_1, and again when it passes A_2. Using the transformation equation just derived for velocities, Eq. (17i), both observers can write down

$$u'_1 = u_1 - v \quad \text{and} \quad u'_2 = u_2 - v$$

Taking the difference between the two velocities u'_2 and u'_1, each would obtain

$$u'_2 - u'_1 = u_2 - u_1 \tag{17j}$$

Since each observer measured the time t between positions A_1 and A_2, each can divide his velocity difference by t to obtain the acceleration

$$\frac{u'_2 - u'_1}{t} = \frac{u_2 - u_1}{t}$$

or

$$a' = a \tag{17k}$$

In other words, the acceleration seen by both observers is exactly the same.

The conclusion to be drawn from this result is that *acceleration is invariant*. If we now assume the mass of a body is invariant, we can assume that the two observers could apply Newton's second law of motion, $F = ma$, and find the same force exerted by the plane's jet engines.

17.7. Falling Body Observed from a Constant-Velocity Frame

Suppose you as an observer O', riding in a train, plane, bus, or ship with constant velocity, hold out a coin in front of you and drop it. The coin will appear to fall straight downward with the acceleration $g = 9.8$ m/sec².

If an observer outside on the ground were to observe the coin's motion, he would find it traverses a parabolic path. While the two observed paths would appear to be different, both observers would measure the same time of fall, and each could apply the transformation equations to determine what the other observer saw.

A simple demonstration of such an experiment is shown in Fig. 17F. A small

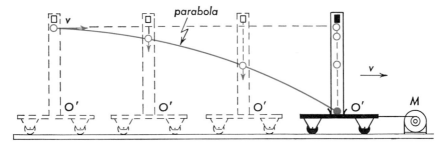

Fig. 17F   *A falling body experiment, showing the trajectory as viewed from a stationary frame and from a moving frame.*

car with a vertical stand is pulled along the table top at constant speed by means of an electric motor and drum. While the car is moving, an electric switch is opened; this deactivates an electromagnet and releases a steel marble for free fall. The ball falls in a small cup at the base of the stand, thus illustrating that to an observer O' on the car the trajectory is straight downward.

To a stationary observer O the released ball has an initial horizontal velocity v, and traverses the parabolic path as a projectile. If we apply the transformation Eq. (17i) to this example, the horizontal velocity u of the ball, as seen by observer O, is equal to the velocity v of the car, so that $u = v$. This gives $u' = O$ as observed by O', and the ball falls vertically downward.

17.8. Falling Bodies Observed from an Accelerated Frame

Suppose you, as an observer O' riding in an accelerated vehicle, drop an object and observe its free fall. To an observer on the ground outside the object will appear to follow a parabolic path, but to you it will appear to fall in a straight-line path making an angle with the vertical. It will behave as though in addition to a downward acceleration g it also has a backward horizontal acceleration a.

An informative experiment demonstrating these observations is shown in Fig. 17G. A small car with a vertical stand attached is accelerated along the

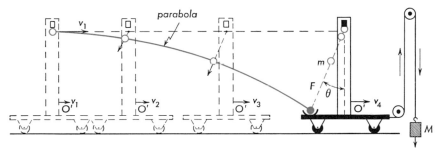

Fig. 17G   *A falling body experiment, showing the trajectory as viewed from a stationary frame and from an accelerated frame.*

table top by a cord passing over two pulleys to a large mass M. Shortly after the car starts, an electric switch is opened, deactivating an electromagnet, releasing a steel marble for free fall. The marble falls in a small cup at one side of the base of the stand.

To a stationary observer O, the released marble has an initial horizontal velocity v_1 (the velocity of the car at the instant of release), and it traverses the parabolic path as shown. To the accelerated observer O' there appears to be a force F giving the marble an acceleration a' along a straight-line path making an angle θ with the vertical. This force is the vector resultant of the downward force of gravity F_1 and a horizontal force F_2. These forces, as well as their corresponding accelerations, g, a, and a', are shown in Fig. 17H:

$$F_1 = mg, \quad F_2 = ma, \quad \text{and} \quad F = ma'$$

Since there is no apparent reason for the horizontal force F_2 it is called a *fictitious force*. To the observer O this force does not exist and the marble obeys Newton's laws of motion.

17.9. Motion in a Rotating Frame

Imagine that you, as an observer O', are on a merry-go-round and while it is turning with constant speed you place a marble on the floor at your feet. Since

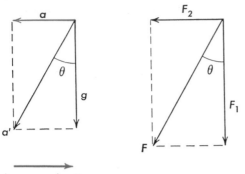

Fig. 17H *A falling body in an accelerated frame of reference appears to have two simultaneous accelerations, one an acceleration g vertically downward due to gravity, and the other an acceleration a opposite in direction to that in which the frame is accelerated.*

frame acceleration

it is free to move, the marble will roll away from your feet and out along a curved path toward the periphery.

If a stone is fastened to a string and you hold the other end in your hand, it will not hang vertically but with the string making an angle with the vertical. At the center of rotation it will hang straight down, but at points away from the center the centrifugal component of force will increase the radius of its circular path.

If a large gun, fixed in position on the earth's surface, fires a shell at a distant ground target, the projectile deflects to the right in the northern hemisphere and to the left in the southern. The explanation of this phenomenon, that it is due to the earth's rotation, was first given by the French scientist Coriolis about the middle of the last century.

Consider a demonstration in which a gun G, mounted on a turntable as shown in Fig. 17I, is aimed at targets T and P. The target P is off the table and in a

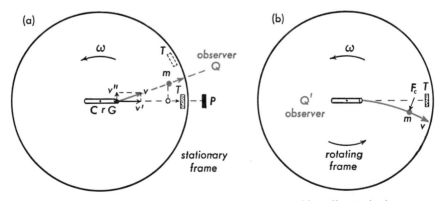

Fig. 17I *The projectile from the gun on the rotating table will miss both targets, because of the Coriolis force.*

stationary frame of reference, while target T is on the table and in a rotating frame of reference. If the table is not turning, the projectile leaving the gun with a velocity v' will pass through T and hit the target P.

If the table is rotating with an angular velocity ω, and the gun is fired at the instant the two targets are in line, the projectile will pass to the right of T (shown dotted) and to the left of P.

Since the gun rotates with the table, the tip end of the gun barrel, at a distance r from the center of rotation C, will give the projectile an additional velocity v'' at right angles to v', and equal in magnitude to the product ωr:

$$v'' = \omega r$$

Since the projectile is given two simultaneous velocities v' and v'', with respect to the fixed frame of reference, a stationary observer will observe the mass m moving with a velocity v along the straight-line path GQ.

To an observer on the rotating table there appears to be a force F_c at right angles to v' deflecting the projectile to the right of the target T at which it is aimed. This apparent force, called the *Coriolis force*, is demonstrated in Fig. 17J.

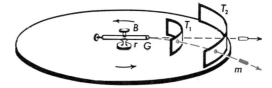

Fig. 17J *Apparatus for demonstrating the curved path of a projectile and the Coriolis acceleration.*

A rotating table made of plywood with a diameter of 4 ft contains a small spring gun and two metal frame targets, T_1 and T_2. The push button B for a trigger is located on the axis, and thin paper is glued over the target frames. When the table is set rotating at a relatively low angular speed, the trigger B is depressed, firing the small wood or aluminum projectile. When the table is stopped, the gun muzzle and the holes in the paper targets clearly indicate the curved path.

A derivation of the formula for the Coriolis acceleration and Coriolis force follows from Fig. 17K. A mass m is given a velocity v in a direction which is radially outward from C toward the stationary observer O. When it passes the

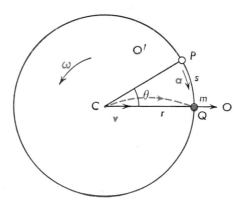

Fig. 17K *Geometry for the derivation of the Coriolis acceleration $a = 2vw$.*

point Q the table has turned through an angle θ. To the rotating observer O' the projectile has moved away from the point P a distance S. From classical mechanics

$$s = \tfrac{1}{2}at^2$$

Since $s = r\theta$, and $\theta = \omega t$, we substitute and find

$$r\omega t = \tfrac{1}{2}at^2$$

from which

$$a = \frac{2r\omega}{t}$$

Since the velocity v is just r/t, we obtain

$$a = 2v\omega \tag{17l}$$

as the Coriolis acceleration. By Newton's second law of motion, $F = ma$, the Coriolis force becomes

$$F = 2mv\omega \tag{17m}$$

17.10. Classical Relativity; Newtonian Mechanics

From the treatment of the kinematics of motion in this chapter, we have seen how an observer in one reference frame can make measurements of something happening in another reference frame and, by applying simple transformation equations, find the measurements that apply to the second frame.

We have seen that, if the relative velocity of the two frames is constant, the distance between two points, or the length of an object measured from either frame, is found to be the same. If the *acceleration* of an object is measured from both frames, the two results again come out the same, but if the *velocity* is measured, the two values are different. In other words,

Distances and accelerations are invariant and do not depend upon any relative velocity of an observer, while velocities are variant.

Even though velocities may be different for two observers, the transformation equations make the two velocities compatible. If measurements of the motion of a body in one frame of reference are transformed to find the measurements that would apply in another frame, we say that we are applying the Newtonian transformation equations. These transformation equations are

$$
\begin{aligned}
x' &= x - vt \\
u' &= u - v \\
l' &= l \\
a' &= a \\
t' &= t
\end{aligned}
\tag{17n}
$$

A set of coordinates that is fixed relative to an observer is called the *observer's frame of reference.*

If a frame of reference is accelerated, a body initially at rest or moving in that frame appears to have a fictitious force acting upon it.

Since, therefore, Newton's laws of motion do not apply in an accelerated frame of reference, a frame that is not accelerated is called a *Galilean-Newtonian* or *inertial frame of reference*.

In setting up the Galilean-Newton transformation equations, it was assumed that all measurements were simultaneous, i.e., that the light with which one sees an object travels with an infinite speed. If the relative velocities of moving frames are small compared with the speed of light (186,000 mi/sec), the speed of light can be assumed to be infinite, and the principles developed in this and the preceding chapters are valid. Newtonian mechanics applies, therefore, to velocities that are low compared with the speed of light, while a modified system called *relativistic mechanics* will apply to velocities comparable to the speed of light. (See Chapter 19.)

PROBLEMS

1. A motorboat capable of 12 mi/hr is headed straight across a river. If the water flows at the rate of 2 mi/hr, what will be the velocity of the boat with respect to the starting point on the bank?

2. To an observer on a ship sailing east at 24 knots, it appears that a 16-knot wind is blowing from the northwest. Find the true wind velocity. *Ans.* 37.1 knots at 17.7° south of east.

3. A pilot with a plane having a cruising speed of 150 mi/hr leaves an airport and sets his course at 30° north of east. After flying for 1 hr, he discovers he is 110 mi directly east of his starting point. What is the average wind velocity that blew him off his course?

4. Two pilots flying identical aircraft with a cruising speed of 200 mi/hr fly to equally distant cities 400 mi away, and return to their home base. City X lies to the east and city Y lies to the north as shown in Fig. 17B. Aloft, there is a 70-mi/hr wind blowing from the west. Find (a) the time of flight going to each city, (b) the time of flight returning, (c) the difference in total times of flight, and (d) the ratio of the two total flight times. *Ans.* (a) 1.48 hr to X, 2.14 hr to Y; (b) 3.08 hr from X, 2.14 hr from Y; (c) 0.28 hr or 16.8 min; (d) 1.067.

5. Two swimmers, starting simultaneously from the same anchored marker in a river, swim to distant markers and return. One marker X is directly downstream 300 ft, and the other, Y, is directly across-stream 300 ft. If both swimmers can make 2 mi/hr in the water, and the river flows at 1 ft/sec, (a) how long will it take each swimmer to reach his distant marker, (b) what is each swimmer's time to return, (c) what is the difference between their total times, and (d) what is the ratio of these times as determined from Eq. (17c)?

6. Two pilots flying identical aircraft with a cruising speed of 280 mi/hr fly to equally distant cities 300 mi away, and return to their home base. City X lies to the west, and city Y lies to the south of home base. Aloft there is an 80-mi/hr wind blowing from the west. Find (a) the time of flight going to each city, (b) each time of flight returning, (c) the difference in total time of flight times, and (d) the ratio of total flight times using Eq. (17c). *Ans.* (a) 1.50 hr to X, 1.12 hr to Y; (b) 0.833 hr from X, 1.12 hr from Y; (c) 0.093 hr; (d) 1.04.

7. If the small car in Fig. 17G has an acceleration of 2.5 m/sec², and the steel ball is dropped from a height of 0.5 m, find (a) the angle of descent, and (b) the lateral displacement of the ball as observed by the moving observer.

8. A merry-go-round is rotating with an angular velocity of 5 rpm. A 2-Kg ball starts rolling radially outward with a velocity of 4 m/sec. Calculate (a) the

Coriolis acceleration, (b) the Coriolis force, and (c) the average radius of its path on the floor. *Ans.* (a) 4.18 m/sec², (b) 8.36 newtons, (c) 3.82 m.

9. A bicycle and rider must lean over at an angle of 16° to the vertical as they round a curve of 100-ft radius. What is their speed?

10. A small spring gun, as shown in Fig. 17J, fires a 100-gm projectile with a velocity of 15 m/sec. If the table turns at 3 rps, find (a) the Coriolis acceleration, (b) the Coriolis force, and (c) the average radius of the deflected path. *Ans.* (a) 565 m/sec², (b) 56.5 newtons, (c) 0.398 m.

11. A stone attached to a wire 2 m long travels in a horizontal circle, the wire describing the surface of a cone. Find the angular speed in rps at which the wire will stand out at an angle of 10° with the horizontal.

Interferometers and Lasers

In this chapter we will first take up the *Michelson interferometer*, and then Fizeau's experiment by which the speed of light is found to be altered by a moving stream of water. This experiment leads directly to the principles of the famous *Michelson-Morley experiment* on ether drift, which in turn leads to Einstein's theory of relativity.

We will then consider the *Fabry-Perot interferometer* as it applies to the hyperfine structure of spectrum lines, and finally to the principles of the recently developed instruments called *lasers*.

18.1. The Michelson Interferometer

The form of the Michelson interferometer generally found in the science laboratory is that shown in Fig. 18A. The optical parts consist of two mirrors, M_1 and M_2, and two parallel plates of glass, G_1 and G_2. Oftentimes the rear side

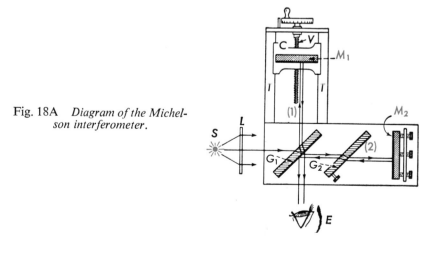

Fig. 18A   *Diagram of the Michelson interferometer.*

of plate G_1 is lightly silvered (shown heavy in the figure) so that light coming from the source S is divided into (1) a reflected and (2) a transmitted beam of equal intensity. The light returning from M_1 passes through G_1 a third time before reaching the eye. The light returning from M_2 is reflected from G_1 and into the eye. The purpose of plate G_2 is to render the total path in glass equal for the two rays.

The mirror M_1 is mounted on a well-machined guide and can be moved along slowly by means of a screw V. When mirror M_2 is made exactly perpendicular to M_1 by screws on its back face, interference fringes similar to those found with a double slit may be seen, or photographed, at E. Photographs of typical fringes when mirror M_1 is at different distances are shown in Fig. 18B.

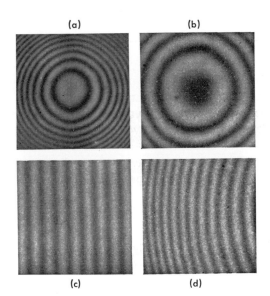

Fig. 18B   *Interference fringe patterns as photographed with a Michelson interferometer.*

When monochromatic light is used as a source and the mirrors are in exact adjustment, circular fringes are observed as shown in photographs (a) and (b). If the mirrors are not exactly at right angles to each other, fringes like those in photographs (c) and (d) are obtained. If circular fringes are observed, and mirror M_1 is slowly moved along by the turn of the screw V, the circular fringe pattern will expand or contract; if it expands, new fringes will appear as a dot at the center, widen, and expand into a circle; and if it contracts, fringes grow smaller, become a dot, and vanish at the center. If straight or curved fringes are observed, the motion of M_1 causes the fringes to drift across the field at right angles to the fringe lines.

The expansion, contraction, or drift in the pattern for a distance of one fringe corresponds to M_1 moving a distance of exactly $\frac{1}{2}$ wavelength of light. When M_1 moves back a distance of $\frac{1}{2}\lambda$, the total light path (1) increases a whole wavelength. If M_1 moves 1λ, the pattern will move two fringes because the total light path (1) has changed by 2λ. Any bright fringe that one observes is caused by both beams coming together in phase. When the one path is changed by $\frac{1}{2}$λ, 1λ, $\frac{3}{2}$λ, etc., the two beams arriving at the same field points will again be in phase.

By counting the number of fringes required to move the mirror M_1 a given distance, the wavelength of light can be calculated. This would appear to be the most direct and accurate method for the wavelength measurements of different light sources. Knowing the accurate wavelength of any light source, one can then use the interferometer to accurately measure distances.

It is by means of the Michelson interferometer that the standard meter was determined in terms of the wavelength of orange light, λ = 6057.80 A of krypton, element 36.*

1 meter — 1,650,763.73 wavelengths
(for orange light of krypton)

18.2. Velocity of Light in Moving Matter

In 1859, the French physicist Armand Fizeau measured the velocity of light in a moving stream of water and found that the light was carried along by the stream. A schematic diagram of his apparatus is shown in Fig. 18C.

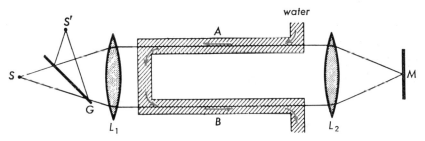

Fig. 18C   *Fizeau's experiment for measuring the velocity of light in a moving medium.*

Light from a monochromatic source S is separated into two beams by means of a lens L_1. These two beams pass through tubes A and B which contain water flowing rapidly in opposite directions. After reflection from M the beams traverse opposite tubes, so that upon arrival at L_1 one has traversed streams A and B in the direction of flow, while the other has traversed both tubes but always against the flow. Part of each beam is reflected by the half-silvered plate G and the two brought together at S' where interference fringes are produced.

If the light travels faster by one path than the other, the time will be different and the fringes formed at S' will have shifted. Using tubes 1.5 m long and a water speed of 7.0 m/sec, Fizeau found a shift of 0.46 of a fringe upon a reversal of the water stream. This shift corresponds to a decrease in the speed of light

---

* Adopted as the international legal standard of length on October 14, 1960, by the General Conference on Weights and Measures in Paris, France.

in one direction, and an increase in the other, of about half the speed of the water. In other words, the moving water has a dragging effect upon the light waves.

In 1818, the French physicist Augustin Fresnel* derived a formula for this dragging effect, based upon the existence of what was then called the *ether*. His formula gives v' the increase in the velocity of light in any medium, due to the motion of the medium as

$$v' = v\left(1 - \frac{1}{\mu^2}\right) \tag{18a}$$

where v is the velocity of the medium, and μ is the index of refraction. For water, with an index of 1.33, $v' = 0.43v$ in reasonably good agreement with Fresnel's observations.

18.3. The Michelson-Morley Experiment

This, the most famous experiment in optics, was first performed by Michelson and Morley in 1881, in an effort to detect the motion of the earth through space. If the transmission of light through space requires an ether, that is, a medium for it to move in, then light should be dragged along by this ether as the earth moves along through space. In order to detect such a drift, the Michelson interferometer appeared to be the most sensitive instrument to use.

In principle the ether drift test consists simply of observing whether there is any shift of the interference fringes of light in the Michelson interferometer when the entire instrument is turned through an angle of 90°. Let us assume that the interferometer and the earth are at rest and that the ether is moving by with a velocity v as shown in Fig. 18D. If no ether drag is effective, light paths

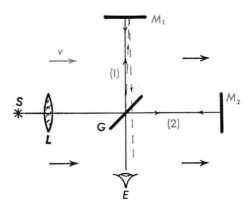

Fig. 18D *The Michelson interferometer arrangement for detecting an ether drift.*

---

* Augustin Fresnel (1788–1827), French physicist. Although starting his career as a civil engineer, Fresnel became interested in optics at the age of 26. His mathematical development of the wave theory of light and its complete validation of experiment have marked this man as an outstanding genius of the nineteenth century. His true scientific attitude is illustrated by a statement from one of his memoirs, "All the compliments that I have received from Arago, Laplace, and Biot never gave me so much pleasure as the discovery of a theoretic truth, or the confirmation of a calculation by experiment."

(1) and (2) will be out and back, as shown by the solid lines and arrows, and a set of interference bands like those shown in Fig. 18B will be observed.

If we now suppose the light to be dragged along by an ether, the time it takes for light to traverse path (1) at right angles to the ether stream, and the time it takes to traverse path (2), first with and then against the stream, will both be increased.

The times to traverse paths (1) and (2) are given by Eqs. (17a) and (17b) as

$$t_\perp = \frac{2L}{\sqrt{c^2 - v^2}} \qquad t_\| = \frac{2Lc}{c^2 - v^2} \tag{18b}$$

where c is the velocity of light *in vacuo*, and v is the very much slower drift velocity.

A little study of these equations will show that, while both times have been increased a slight amount, the increase is twice as large in the direction of motion. Furthermore, the ratio of the two times is given by Eq. (17c), as

$$\boxed{\frac{t_\perp}{t_\|} = \sqrt{1 - \frac{v^2}{c^2}}} \tag{18c}$$

An ether drift should, therefore, cause a shift in the fringes observed in the interferometer. Since neither the earth's motion, nor the ether, can be stopped, in an effort to observe this shift, a rotation of the interferometer through 90° should have a similar effect. By interchanging paths (1) and (2), the time difference $t_\| - t_\perp$ is reversed, and any fringe shift should be doubled.

Michelson and Morley made the light paths as much as 11 m long by reflecting the light back and forth between 16 mirrors, as shown in Fig. 18E. To prevent distortion by the turning of the instrument, the entire apparatus was mounted on a concrete block floating in mercury, and observations of the fringes were made as it rotated slowly and continuously about a vertical axis.

If we assume the ether velocity v to be 18.6 mi/sec (the speed of the earth in its orbit around the sun), and the speed of light c to be 186,000 mi/sec, a shift of $\frac{1}{2}$ a fringe should have been observed. No shift as great as $\frac{1}{10}$ of this was observed. Such a negative result was so surprising and so disappointing that others have repeated the experiment.

The most exacting work was done by D. C. Miller who used Michelson and Morley's arrangement, but with optical paths of 64 m in place of 11 m. While Miller thought he found evidence for a shift of $\frac{1}{80}$ of a fringe, the latest analysis of Miller's data makes it probable that no shift exists.

18.4. The Fabry-Perot Interferometer

Unlike the Michelson interferometer that produces interference fringe patterns with two coherent beams of light, the Fabry-Perot interferometer produces interference with a large number of coherent beams. The principles of this device are shown in Fig. 18F.

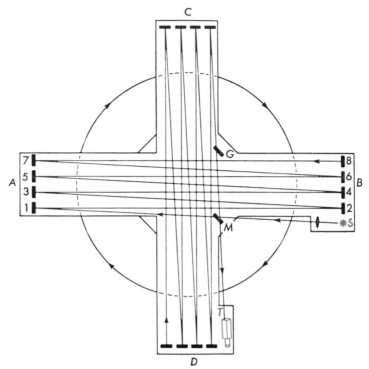

Fig. 18E *Miller's elaborate arrangement of the Michelson-Morley experiment to detect ether drift.*

Two optically flat glass or quartz plates, each partially silvered on one face only, are mounted in rigid frames. By means of fine screws the plates are adjusted until their two silvered surfaces are parallel to a high degree of precision. Light from an extended source S, upon passing through the interferometer undergoes reflection back and forth, and the emerging parallel rays are brought together to interfere in the focal plane of a lens.

In Fig. 18F a ray of light from the point P is shown incident on the first surface at an angle θ. Part of this light is reflected and part is transmitted. Part

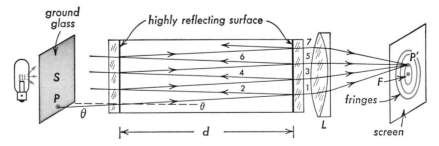

Fig. 18F *Diagram of the paths of light rays in a Fabry-Perot interferometer.*

of the transmitted ray 1 is reflected at the second surface and part is transmitted. Repeating this behavior at each mirrored surface it can be seen that rays 1, 3, 5, 7, etc., emerging as parallel rays have traveled successively greater distances. These numbers specify the number of times each ray traverses the gap of width d.

The *path difference* between successive rays 1, 3, 5, 7, etc., will be shown at the end of this section to be just $2d \cos \theta$. If this distance is exactly equal to a whole number of wavelengths, the emergent rays when brought together at the point P' will all be in phase and produce a bright spot. For such a bright spot we can write

$$2d \cos \theta = n\lambda \qquad (18d)$$

where n is a whole number and λ is the wavelength.

For all rays from all points of the source incident at the same angle θ, identical phase relations will exist, and the lens will bring all sets of parallel rays to a focus at points lying on a circle in the focal plane of the lens. For those rays in which the angle θ is such that the path difference between successively reflected rays is $(n + \frac{1}{2})\lambda$, the waves in alternate rays will be out of step and destructively interfere to produce darkness. Hence the existence of bright and dark concentric rings on the screen indicates coherence as well as interference.

The interference ring patterns in Fig. 18G illustrate the difference between

(a) (b)

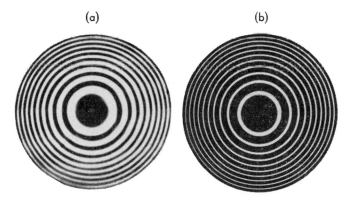

Fig. 18G *Comparison of the types of fringes produced with (a) the Michelson interferometer and (b) the Fabry-Perot interferometer.*

the kinds of fringes observed when (a) two beams of light are brought together, as in the Michelson interferometer and (b) when a large number of beams are brought together as with the Fabry-Perot interferometer.

Note that in Fig. 18F, for rays entering the interferometer parallel to the axis, that is, with $\theta = 0$, the multiply reflected rays will be brought together at the focal point F, and if these waves arrive in phase, a bright spot will be produced there, and

$$2d = n\lambda \qquad (18e)$$

The number n is called the *order of interference*, or in microwave terminology, the *principal oscillation mode* of the cavity. For an interferometer with $d = 10$ cm, and green light $\lambda = 5000$ A, $n = 400,000$ for the light emerging parallel to the axis.

To derive Eq. (18d) for the bright fringes of a Fabry-Perot interferometer, we refer to Fig. 18H. As ray 1 leaves the point A, ray 3 must travel the extra

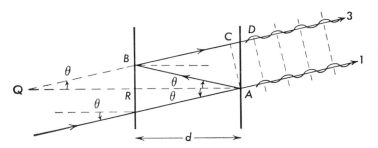

Fig. 18H   *Geometry for calculating the path difference between two successive rays of a Fabry-Perot interferometer.*

distance A to B to C to emerge in phase with ray 1. By extending the line BC back to the left, until it intersects the normal AR at the point Q, an isosceles triangle QBA is formed. Since this makes the line QB equal to AB, the total path difference

$$AB + BC = QC$$

Since $QR = AR$, and $AR = d$, we have

$$QA = 2d$$

Since the base of the right triangle QAC is the path difference QC, we obtain

$$QC = 2d \cos \theta$$

For bright fringes, QC must equal $n\lambda$, and this leads directly to Eq. (18d).

Eq. (18d) shows that the highest order of interference occurs when $\theta = 0$, that is, for the most central fringe. As θ increases, $\cos \theta$ decreases, and the path difference QC decreases. If, for example, the central fringe is for $n = 400,000$, the next two fringes out will be $n = 399,999$ and $n = 399,998$, respectively.

18.5. Hyperfine Structure of Spectrum Lines

One of the principal uses of the Fabry-Perot interferometer has been that of resolving what is called the *hyperfine structure* of spectrum lines. Through the study of line structure it has been possible to determine the spin angular momenta and magnetic moments of many atomic nuclei. See Chapter 12.

The ability to resolve lines differing ever so slightly from one another in wavelength is attributed to the great length of the path difference between consecutive rays and the large number of rays producing the interference fringes. Each wavelength produces its own set of fringes as given by Eq. (18d), and because bright fringes are very much sharper than the dark spaces between, several slightly different wavelengths can be seen as separate bright ring systems.

Since the spectrum of most light sources contains many lines, the ring system from a source to be studied is usually focused on the jaw faces of the slit of a prism spectrograph, and photographs are made with the slit wide open or with the slit about a millimeter or two wide.

Figure 18I is a reproduction of a section of the visible spectrum of mercury.

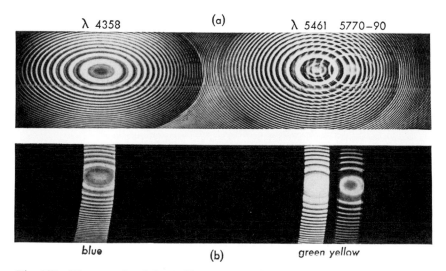

Fig. 18I   *Photographs of the visible spectrum of mercury made with a Fabry-Perot interferometer and a prism spectrograph. Spectrograph slit (a) wide open, and (b) 1 millimeter wide.*

In this relatively simple spectrum the overlapping of interference ring systems in (a) is confusing and the use of a slit to separate interference patterns as in (b) is clearly an improvement. In a complex spectrum like that of lanthanum, $Z = 57$, the slit must be narrowed still further to clearly separate most of the lines and see their ring structure. See Fig. 18J.

By tilting the Fabry-Perot interferometer so the center of the interference ring

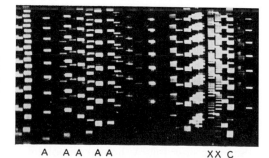

Fig. 18J   *Photograph of a section of the lanthanum spectrum made with a Fabry-Perot interferometer and a prism spectrograph. (After D. E. Anderson.)*

pattern is just off the lower end of the spectrograph slit, only half of each ring strip is observed. Occasionally it will be found that a line which appears sharp and single in an ordinary spectroscope will yield ring systems that are complex. Examples are found in the lines marked X in the photograph, which upon close examination reveal at least seven components. Those marked A are single sharp lines. Line C is broad and probably composed of several unresolved lines.

Two line patterns selected from the Fabry-Perot spectrum of tantalum, $Z = 73$, and tungsten, $Z = 74$, have been enlarged and reproduced in Fig. 18K.

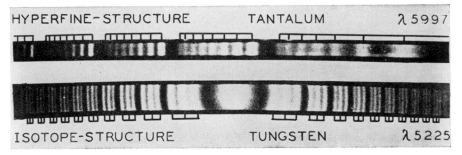

Fig. 18K *Photographs of a section of the Fabry-Perot interferometer pattern of a single line from the spectrum of tantulum and tungsten.*

The upper pattern shows an eight-component line in the orange part of the spectrum at $\lambda = 5997$ A, the structure being due to the fact that tantalum atoms of atomic mass 181 have a nuclear spin of $\dfrac{7}{2}\dfrac{h}{2\pi}$. See Table 32A.

The lower pattern shows a three-component line illustrating what is called *isotope structure*. Tungsten has five isotopes with the following relative abundance.

Isotope	180	182	183	184	186
Abundance, %	0.2	26.4	14.4	30.6	28.4

The even isotope 180 is quite rare and produces too faint a line to be seen.

The odd isotope 183 is known to produce a pattern of several lines, and because this pattern is spread over the three single lines produced by the three more abundant even isotopes, they are too faint to show in the reproduction. The innermost fringe of the three lines in each order is due to isotope 182.

To evaluate the differences in wavelength between the component lines of an otherwise single spectrum line we make use of the general equation, Eq. (18d). Any given fringe of a wavelength λ_1 is formed at such an angle that

$$2d \cos \theta_1 = n\lambda_1 \qquad (18f)$$

The next fringe out for this same wavelength is given by

$$2d \cos \theta_2 = (n - 1)\lambda_1 \qquad (18g)$$

Suppose now that λ_1 has a component line λ_2 which has a slightly different wavelength than λ_1, and that this component in order n falls on order $n - 1$ of λ_1. We can therefore write

$$2d \cos \theta_2 = n\lambda_2 \tag{18h}$$

If we now let the small difference in wavelength $\lambda_1 - \lambda_2$ be written as $\Delta\lambda$, that is,

$$\Delta\lambda = \lambda_1 - \lambda_2$$

we may substitute $\lambda_1 - \Delta\lambda$ for λ_2 in Eq. (18h) and obtain

$$2d \cos \theta_2 = n(\lambda_1 - \Delta\lambda) \tag{18i}$$

Equating right-hand sides of Eqs. (18g) and (18i) we find

$$\lambda_1 = n \, \Delta\lambda$$

and upon substituting the value of n from Eq. (18f), and solving for $\Delta\lambda$ we obtain

$$\Delta\lambda = \frac{\lambda_1{}^2}{2d \cos \theta_2}$$

Since θ_2 is practically zero, we can write $\cos \theta_2 = 1$, and obtain the very useful equation

$$\Delta\lambda = \frac{\lambda_1{}^2}{2d} \tag{18j}$$

This is the wavelength interval between the centers of successive fringes and is independent of n. Knowing d and λ_1, we can calculate the wavelength difference of component lines lying in this small range from measured positions.

Example. The line pattern for $\lambda = 5225$ A in Fig. 18K was made with a Fabry-Perot plate spacing of 5 cm. Find the wavelength interval between successive ring patterns of the same isotope.

Solution. The given quantities are $\lambda = 5225$ A, and $d = 5$ cm. First change 5 cm to angstroms, and then substitute in Eq. (18j)

$$\Delta\lambda = \frac{(5225 \text{ A})^2}{2 \times 5 \times 10^8 \text{ A}} = 0.0273 \text{ A}$$

The interval between orders is 0.0273 A, and the three components differ from each other by about one-quarter of this. (See Problem 5 at the end of this chapter.)

18.6. Lasers

The term *laser* derives its name from the description, *Light Amplification by Stimulated Emission and Radiation.* In principle the laser is a device that produces an intense, concentrated, and highly parallel beam of light. So parallel would be the beam from a visible light laser 1 foot in diameter that at the moon the beam would be no more than a mile wide.

Historically the laser is the outgrowth of the *maser*, a device using Microwaves instead of Light waves. The first successful maser was built by C. H. Townes at Columbia University in 1953. During the next seven years great strides were made in developing intense microwave beams, the principal contributions being made by the Bell Telephone Research Laboratories and the Lincoln Laboratories at the Massachusetts Institute of Technology.

The first successful laser, using a large synthetic ruby crystal, was built by T. H. Maiman of Hughes Aircraft Company Laboratories in the summer of 1960. Hundreds of extensive researches on laser development have been carried on since that time, and because such devices appear to have great potential in so many different fields of research and development, a brief account of their basic principles will be presented here.

Lasers are of three general kinds, those using solids, those using liquids, and those using gases. For the case of liquid or gas lasers, a Fabry-Perot interferometer, with silvered end plates like the ones shown in Fig. 18F, is filled with

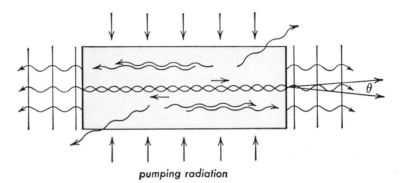

pumping radiation

Fig. 18L   *Diagram illustrating the coherent stimulation of light waves in a solid-state laser such as a ruby crystal.*

a fluid. In the solid laser, the ends of a crystal are polished and silvered as shown in Fig. 18L. Since the first successful laser was made with a large single crystal of ruby, this device will be explained as representative of solid state lasers.

The atomic lattice structure of a ruby crystal has the properties of absorbing light of certain frequencies v_0 and of holding this absorbed energy for a period of time. Then by bouncing light of a different frequency v_1 back and forth between the silvered ends, the excited atoms may be stimulated to emit their stored energy as light of the same frequency v_1 and in exact phase with the original light waves. As these intensified waves bounce back and forth, they stimulate others, thus amplifying the original beam intensity.

Because the light waves emerging from the end of the laser are all in phase, the beam is said to be coherent, and the angular spread of the light is given by the relation

$$\theta = \frac{2.44\lambda}{d}$$

where θ is in radians, λ is the wavelength of light, and d is the diameter of the emergent beam. This equation arises from the treatment of the diffraction of light by a circular aperture. See §6.6, and Eq. (6k).

18.7. Optical Pumping

A convenient method of describing laser action is to refer to an energy-level diagram of the electronic states involved in light absorption and emission. An energy-level diagram for the electrons in the atomic lattice of a ruby crystal is given in Fig. 18M. Here there are three sets of levels, the *normal state*, the

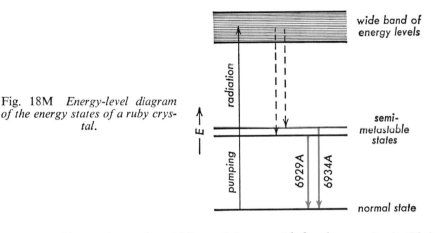

Fig. 18M *Energy-level diagram of the energy states of a ruby crystal.*

semi-metastable states near the middle, and the top *wide-band energy levels.* (Pink ruby is a crystal lattice of aluminum and oxygen atoms, Al_2O_3, with a small amount of chromium ions as an impurity, 0.04 % of Cr^{3+}.)

Semi-metastable states are states in which electron transitions to lower levels are not entirely forbidden, as they are in ionized calcium, but there is a delay in the electrons' jumping down to a lower level. For metastable states see Fig. 13K.

When a beam of white light enters a ruby crystal, strong absorption occurs in the blue and green part of the spectrum, and the transmission of only the red region is what gives the ruby its red color.

On shining a strong beam of blue-green light into a ruby crystal the absorption that takes place raises many electrons to the wide-band levels, as shown by the up arrow at the left in Fig. 18M. Because of internal atomic activity the electrons quickly drop down to the intermediate levels, not by the emission of photons, but by the conversion of energy into vibrational kinetic energy of the atoms forming the crystal lattice.

Once in the intermediate levels the electrons remain there for some time and randomly jump back to the normal state emitting visible red light. This *fluorescent light*, as it is called, enhances the red color of the ruby.

Since an incident beam of blue-green light steadily increases the number of electrons in the semi-metastable states, the process is called *optical pumping*. To greatly increase the electron populations in the middle levels, very intense light sources as well as efficient light-gathering systems are frequently used.

One of many systems developed for doing this is shown in Fig. 18N. By placing

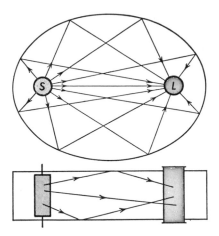

Fig. 18N *Elliptical reflector for concentrating light from a source S on a laser L.*

the exciting light source at one focus of an elliptical reflector and the laser at the other focus, high efficiency can be obtained.

In some lasers, where a steady light beam is required, the random emission of light within the gas, liquid or solid mass will by chance find some of the waves emitted along the axis. See Fig. 18L. This light, because it may bounce back and forth many times between the highly reflective ends, will stimulate coherent emission parallel to itself and to the axis. These two waves traveling back and forth in step with each other now induce other emission, thereby giving rise to a rapid growth or chain reaction. Thus by continuous pumping action from a separate light source, a large part of the stored energy is converted into a coherent beam of light of a different wavelength. Under these conditions the device is said to "laze" spontaneously.

If the silvered ends are not highly reflecting, too much light escapes from the ends and spontaneous lazing cannot occur. Under these conditions a beam of light of the stimulating frequency from another laser can be sent into the crystal where it is intensified. By modulating the input beam the greatly intensified output beam will be modulated accordingly. This modulation capability is one of the important properties that gives lasers their promising future applications to the field of communications.

18.8. The Helium-Neon Gas Laser

The first successful optical gas laser was set into operation by Javan, Bennett, and Heriott in 1961. Since that time many different gas lasers, using ten or more gas systems and several different excitation methods, have been made to operate.

Because it was the first to function successfully, a helium-neon laser will be described here.

The gas laser shown in Fig. 18O is composed of a glass tube nearly a meter

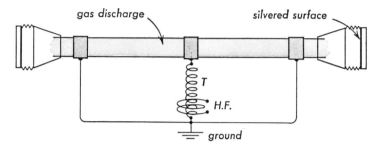

Fig. 18O *Diagram of a helium-neon gas laser showing Fabry-Perot plates at the ends of the discharge tube, and a high-voltage, high-frequency electrical source for excitation.*

long containing a mixture of 1 mm Hg pressure of helium and $\frac{1}{10}$ mm Hg pressure of neon gas. The highly reflecting Fabry-Perot plates on the ends are so sealed as to prevent leakage and so mounted as to be adjustable to a high degree of parallelism.

A high-voltage, high-frequency potential difference, such as that obtained from a Tesla coil, is applied by means of three metal bands around the outside of the tube.

Although there are ten times as many helium atoms present as there are neon atoms, the orange-red color of the gaseous discharge is characteristic of neon. The visible spectrum of helium contains strong lines in the red, yellow, green and blue, so the discharge in helium alone appears as white light. The spectrum of neon, on the other hand, has so many strong lines in the yellow, orange and red, and so few in the green, blue, and violet that its gaseous discharge appears orange-red. The neon spectrum also reveals a large number of strong lines in the near infrared spectrum.

Simplified energy-level diagrams for helium and neon are shown in Fig. 18P for the purpose of explaining the atomic processes involved in laser action. The normal state of helium is a $^1S_0$ level arising from two electrons in $1s$ orbits. The excitation of one electron to a $2s$ orbit finds the atom in a $^3S_1$ or a $^1S_0$ state. Both of these are metastable, since transitions to the normal state are forbidden by selection rules. See Eq. (13h).

Neon, with $Z = 10$, has ten electrons and in the normal state is represented by the configuration $1s^2\, 2s^2\, 2p^6$. When one of the $2p$ electrons is excited to a $3s$, $3p$, $3d$, $4s$, $4p$, $4d$, $4f$, $5s$, etc., orbit, triplet and singlet levels arise.

A subshell such as $2p^5$, lacking but one electron to form a closed subshell, behaves as though it were a subshell containing but one $2p$ electron. The number and designations of the levels produced are therefore the same as with two electrons.

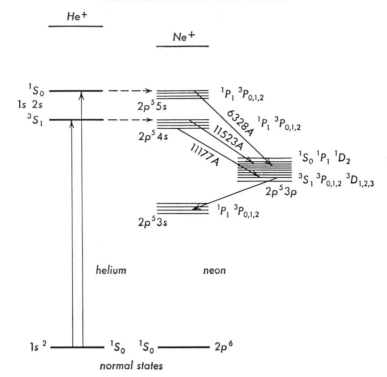

Fig. 18P   *Correlation diagram of the energy levels of helium and neon involved in the helium-neon gas laser.*

As free electrons collide with helium atoms during discharge, one of the two bound electrons may be excited to $2s$ orbits, that is, to the $^3S_1$ or $^1S_0$ states. Since downward transitions are forbidden by radiation selection rules, these are *metastable states*, and the number of excited atoms increases. We, therefore, have optical pumping, out of the normal $^1S_0$ state and into the metastable states $^3S_1$ and $^1S_0$.

When a metastable helium atom collides with a neon atom in its normal state, there is a high probability that the excitation energy will be transferred to the neon, raising it to one of the $^1P_1$ or $^3P_0$, $^3P_1$, $^3P_2$ levels of $2p^5\,4s$ or $2p^5\,5s$. The small excess energy is converted to kinetic energy of the colliding atoms.

In this process, called a *collision of the second kind*, each helium atom returns to the normal state as each colliding neon atom is excited to the upper level of corresponding energy.

The probability of a neon atom being raised to the $2p^5\,3s$ or $2p^5\,3p$ levels by collisions of the second kind are extremely small because of the energy mismatch. The collision transfer, therefore, selectively increases the population of the upper levels of neon. Since selection rules permit transitions from these levels downward to the ten levels of $2p^5\ 3p$, and these in turn to the four levels of $2p^5\ 3s$, stimulated emission can speed up this process by laser action.

Light waves emitted within the laser at wavelengths such as 6328 A, 11,177 A, and 11,523 A, will occasionally be emitted parallel to the tube axis. Bouncing back and forth between the parallel end plates, these waves will stimulate emission of the same frequency from other excited neon atoms, and the initial wave, with the stimulated wave, moves parallel to the axis and in phase.

Most of the amplified radiation emerging from the ends of the helium-neon gas laser are in the near infrared region of the spectrum, between 10,000 A and 35,000 A. The most intense amplified wavelength in the visible spectrum being the red line at $\lambda = 6328$ A. The strongest amplified line in the infrared is at $\lambda = 11,523$ A.

PROBLEMS

1. A Fabry-Perot interferometer has its two silvered faces accurately spaced 1 mm apart. If green light of wavelength 5000 A is sent through, and a 1-meter focal length lens is used on the far side to produce an interference ring pattern at the focal plane, find (a) the principal oscillation mode n, (b) the diameter of the $n - 1$ circular fringe, and (c) the equivalent wavelength spacing between adjacent rings.

2. A Fabry-Perot interferometer has its two silvered faces accurately spaced 1.8 cm apart. If blue light of wavelength 4500 A is sent through, and a 1-meter focal length lens is used on the far side to produce an interference ring pattern at the focal plane, find (a) the principal oscillation mode n, and (b) the equivalent wavelength interval in angstroms between adjacent rings. *Ans.* (a) 80,000, (b) 0.05625 A.

3. A ruby laser produces a beam of red light of wavelength 6943 A, with a circular cross section $\frac{1}{8}$ inch in diameter. Find the diameter of this beam at a distance of ten miles.

4. A ruby laser produces a beam of red light of wavelength 6943 A, with a circular cross section 1 cm in diameter. Find the diameter of this beam at a distance of 1000 kilometers. *Ans.* 169 m.

5. Use a millimeter scale and measure the distances between line components in the fringe pattern for $\lambda = 5997$ in Fig. 18K. Assuming that the Fabry-Perot plate separation was 7 cm, calculate (a) the wavelength interval between successive ring patterns and (b) the wavelength intervals between the strongest component and the other seven components.

Relativity

The mention of the word "relativity" suggests the name of Albert Einstein,* the scientist to whom we are indebted for the now famous theory. To begin with, Einstein was a realist, and his theory rests upon physical facts which have been verified by repeated observations of well-planned experiments. Reference is made in particular to the Michelson-Morley experiment, described in Chapter 18.

Beginning in Chapter 17 we have seen that moving observers in each of two different frames of reference may make their own simultaneous measurements of some single event and by the use of transformation equations, determine the observations made by the other observer.

A frame of reference moving with constant velocity has been called an *inertial frame* because Newton's laws are obeyed within it. The term inertial is used to distinguish such a frame from an accelerated frame in which a fictitious force appears to be ever present.

In transforming measurements from one inertial frame to another, distances and accelerations are found to have the same values whether viewed from one frame or the other. For these reasons *distance* and *acceleration* are said to be

---

* Albert Einstein (1879–1955), German-Swiss physicist, was born of Jewish parents at Ulm, Württemberg, on March 14, 1879. His boyhood was spent in Munich where his father, a dealer in chemicals, had settled in 1880. When the family moved to Italy in 1894, young Albert went to Switzerland to study. There he worked his way through school, finally taking his Ph.D. degree at the University of Zürich in 1902. He was appointed extraordinary professor of theoretical physics at the University of Zürich in 1909, and in 1913 he was called to Berlin as director of the Kaiser-Wilhelm Institute for Physics. While at this post, he was elected a member of the Prussian Academy of Sciences and a member of the Royal Society of London. In 1921 he received the Nobel Prize in physics and, in 1925, the Copley Medal of the Royal Society. From 1933 to 1945 he was with the Institute for Advanced Studies in Princeton, where he died. He is best known for his theory of relativity, the theory and explanation of Brownian motion, the theory of the photoelectric effect, and the quantum theory of radiant heat energy. Twice married, Einstein had several children. For his friends he was a quiet, sincere, and modest man who loved his pipe and violin and disliked formality.

invariant. See §17.6. Observed velocities, on the other hand, differ from one frame to another and are not invariant.

In deriving the Galilean-Newtonian transformation equations in Chapter 17 it was assumed that light travels with an infinite velocity. This means that measurements could be made simultaneously from different frames. Since light has a finite velocity, light itself should not be expected to be invariant; the velocity of light should be different when measured from different inertial frames.

It should then be possible by accurately observing a single event from two inertial frames, to determine this difference in the speed of light and find whether or not the two frames are moving through the ether at different rates.

This was the purpose of the Michelson-Morley experiment, and it failed.

19.1. The Lorentz-Fitzgerald Contraction

From the time Michelson and Morley announced the negative results of their ether-drift experiment, scientists tried to explain why the experiment failed. An ingenious explanation was first advanced by Fitzgerald in 1890. If objects moving through space have to push against the immovable ether, he suggested, they would be compressed in the direction of motion. This compression would, therefore, shorten the Michelson interferometer arms holding the mirrors and might exactly compensate for an existing ether drift.

Lorentz,* the famous Dutch physicist, studied this problem from an atomic point of view. All matter, he proposed, is made up of atoms, and atoms are made up of charged particles that produce electric and magnetic fields. These fields must exert forces on the electromagnetic ether, thus causing the atoms and molecules in moving matter to be pushed closer together. Starting with the well-known principles of electricity and magnetism, Lorentz derived the following formula for the length of any object:

$$l' = l\sqrt{1 - \frac{v^2}{c^2}} \qquad (19a)$$

where v is the velocity of the object through the ether, l is its length when at rest in the ether, and c is the velocity of light. Note the resemblance of this equation to Eq. (18c). Suppose that an object is at rest, so that $v = 0$. Upon substitution of $v = 0$ in Eq. (19a), we find $l' = l$, which says the object's length l' will be just equal to its rest length l. If a rod were moving lengthwise with $\frac{3}{4}$ the

* Hendrik A. Lorentz (1853–1928), Dutch physicist, was born at Arnheim on July 18, 1853. He was educated at the University of Leyden, where he was appointed professor of theoretical physics at the age of 25. Those who knew him never lost the opportunity of mentioning his charming personality and kindly disposition. Of his numerous contributions to science he is best known for (1) a set of four algebraic equations, known as the Lorentz transformation, equations which later came out of Einstein's special theory of relativity, and (2) for his theoretical explanation of the Zeeman effect. In 1922 he was awarded, jointly with Zeeman, the Nobel Prize in physics.

speed of light, however, the substitution of $v = \frac{3}{4}c$ into the equation gives $l' = 0.66l$. This indicates that the moving rod is only $\frac{2}{3}$ as long as when it is at rest. (See Fig. 19A.)

l

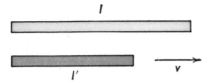

Fig. 19A   *Diagram of the Lorentz-Fitzgerald contraction.*

l' v

It is interesting to see what this equation reveals if a rod could move length-wise with the speed of light, i.e., with $v = c$. The result is $l' = 0$. This means that any object moving with the speed of light would be compressed to zero length. The velocity of light, therefore, becomes an upper limit for the velocity of any moving object.

The above formula, when applied to the cross arms of the interferometer used in the Michelson-Morley experiment (see Fig. 18E), shows that the arms are shortened by just the right amount to compensate for the expected drift. See Eq. (18c). This shortening of an object cannot be measured, for, if one attempts to measure the length of a moving object, the measuring stick must move with the same velocity, and an equal length of the stick shortens by the same amount.

19.2. Einstein's Special Theory of Relativity

Einstein interpreted the failure of the Michelson-Morley experiment to mean that the velocity of light is invariant, that time and distance are relative, and that Galilean-Newtonian mechanics must be modified accordingly. Out of these modifications came the theory of relativity.

Relativity is divided into two parts. One part is called the *special*, or *restricted*, *theory of relativity*, and the other is called the *general theory*. The special theory, developed by Einstein in 1905, deals with observers and their reference frames moving with constant velocities. The mathematics of the special theory is simple enough, and we will consider several of the relationships that are necessary for the satisfactory explanations of atomic phenomena.

The general theory, proposed by Einstein in 1915, deals with motions of bodies in accelerated frames of reference. The mathematics of the general theory is quite difficult, and the experimental evidence for its validity is not as well founded as for the special theory.

Since the Michelson-Morley experiment fails to provide a fixed frame of reference in space, Einstein's theory assumes that all such experiments will fail, and that at relatively high speeds the laws of Newton are not valid. Einstein's special theory of relativity shows that the laws of physics can be restated so that they will apply to any frame of reference, and that at low relative velocities these laws reduce to Newton's laws of motion. The first postulate for setting up these equations is:

The laws of physics apply equally well for all observers as long as they are moving with constant velocities.

The second postulate follows from the assumption that the velocity of light is *invariant*.

The velocity of light in free space has the same value regardless of the motion of the source and the motion of the observer.

To see the meaning of this second statement, consider a reference frame and observer O at rest as shown in Fig. 19B. A source of light S is set up, and by

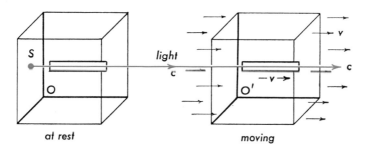

at rest   *moving*

Fig. 19B   *The velocity of light is the same to all observers, that is, it is invariant.*

means of an experiment the velocity of light is measured and found to be 3×10^8 m/sec. Another observer O', moving with a velocity v with respect to O, allows the light from the same source S to pass through his apparatus. Upon measuring the velocity of this same light in his frame, he too finds 3×10^8 m/sec.

For these two identical results to be consistent, Einstein derived new transformation equations. To do this he assumed that *distance* and *time* are relative, i.e., they are not *invariant*. His transformation equations may be written down and compared with the nonrelativistic equations as follows:

$$x' = x - vt \qquad\qquad x' = \gamma(x - vt) \tag{19b}$$

$$t' = t \qquad\qquad t' = \gamma\left(t - \frac{vx}{c^2}\right) \tag{19c}$$

Nonrelativistic   **Relativistic**

where γ is given by

$$\gamma = \frac{1}{\sqrt{1 - v^2/c^2}} \tag{19d}$$

Note the similarity of these two sets of equations. By setting $v = 0$ in Eq. (19d), we get $\gamma = 1$, and the relativistic equations reduce to the nonrelativistic, or Newtonian, equations. The value of γ is just the ratio of the two *travel times* for the light paths in the Michelson-Morley experiment.

Classical laws such as Newton's laws of motion can be used in most applications of kinematics and dynamics to the motions of macroscopic bodies; but at speeds above 10% the speed of light, the relativistic equations should be used.

Since Eqs. (19b), (19c), and (19d) were first derived by Lorentz in 1895, their use in any problem is referred to as a *Lorentz transformation*. Lorentz arrived at the equations by assuming a contraction of moving objects in an ether, whereas Einstein in 1905 derived the equations by assuming the speed of light as *invariant*.

19.3. Relativistic Velocity Transformation

Suppose the velocity of a body in an observer's frame of reference is u, and we wish to calculate the velocity u' of that same body as measured by an observer moving with a velocity v. For this calculation we use the velocity transformation equations, Eq. (12m):

$$u' = \frac{x'_2 - x'_1}{t'_2 - t'_1} \quad \text{and} \quad u = \frac{x_2 - x_1}{t_2 - t_1} \tag{19e}$$

We now make use of the relativistic transformation equations, Eqs. (19b) and (19c), and substitute unprimed terms for each primed term in the first relation. This gives

$$u' = \frac{\gamma(x_2 - vt_2) - \gamma(x_1 - vt_1)}{\gamma(t_2 - vx_2/c^2) - \gamma(t_1 - vx_1/c^2)} \tag{19f}$$

Upon canceling the γ's, collecting like terms, substituting from the right-hand equation of Eq. (19e), and simplifying, we obtain*

$$\boxed{u' = \frac{u - v}{1 - uv/c^2}} \tag{19g}$$

Relativistic

This is the velocity transformation equation in the theory of relativity. It shows that velocity, as in Newtonian mechanics, too, is not *invariant*. Different observers find different velocities.

$$u' = u - v \tag{19h}$$

Nonrelativistic

Example 1. An observer on the earth (assumed to be an inertial frame of reference) sees a space ship A receding from him at 2×10^8 m/sec and overtaking a space ship B receding at 1.5×10^8 m/sec. Find the relative velocity of (a) space ship B as observed by A, (b) space ship A as observed by B, and (c) space ship B relative to space ship A as observed by O.

Solution. This example is shown schematically in Fig. 19C. The given quan-

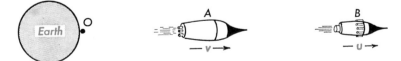

Fig. 19C   *Diagram of two space ships receding from the earth with constant velocities.*

---

* The algebraic steps from Eq. (19f) to Eq. (19g) are left as a student exercise. See Problem 14.

tities for (a) are $v = 2 \times 10^8$ m/sec; $u = 1.5 \times 10^8$ m/sec; and $c = 3 \times 10^8$ m/sec. Upon substitution in Eq. (19g) we obtain

Part (a)

$$u' = \frac{1.5 \times 10^8 \text{ m/sec} - 2.0 \times 10^8 \text{ m/sec}}{1 - 1.5 \times 10^8 \times 2.0 \times 10^8/(3 \times 10^8)^2}$$

$$u' = \frac{-0.5 \times 10^8}{1 - 3 \times 10^{16}/9 \times 10^{16}} = -0.75 \times 10^8 \text{ m/sec}$$

For (b) we reverse the velocity symbols; $u = 2 \times 10^8$ m/sec and $v = 1.5 \times 10^8$ m/sec. Upon substitution in Eq. (19g), we obtain

Part (b)

$$u' = \frac{2.0 \times 10^8 - 1.5 \times 10^8}{1 - 2.0 \times 10^8 \times 1.5 \times 10^8/(3 \times 10^8)^2}$$

$$u' = \frac{0.5 \times 10^8}{1 - 3 \times 10^{16}/9 \times 10^{16}} = +0.75 \times 10^8 \text{ m/sec}$$

For (c) we take just the difference between the two velocities observed by O.

$$u' = 1.5 \times 10^8 - 2.0 \times 10^8 = -0.5 \times 10^8 \text{ m/sec}$$

Example 2. Suppose space ship B in Example 1 (Fig. 19C) is replaced by a beam of light moving from left to right, which observer O measures and finds to be $c = 3 \times 10^8$ m/sec. What will the velocity of this same light be, as observed by space ship A?

Solution. The given quantities are $u = c$, $v = 2.0 \times 10^8$ m/sec, and $c = 3 \times 10^8$ m/sec. Upon first replacing u by c in Eq. (19g) and solving for u', we obtain

$$u' = \frac{u - v}{1 - uv/c^2} = \frac{c - v}{1 - cv/c^2} = \frac{c - v}{1 = v/c} = \frac{c - v}{\dfrac{c - v}{c}} = c$$

Hence the observer in A finds the velocity of light to be c regardless of his velocity. Hence the velocity of light is the same to all observers; it is *invariant*.

19.4. Relativistic Mass

Einstein's special theory of relativity shows that if the mass of an object is measured by two different observers, one moving with respect to the other, the results are different. Mass, therefore, is not invariant. Although the derivation will not be presented here, the special theory gives, for the transformation equation,

$$m = \gamma m_0 \tag{19i}$$

or

$$\boxed{m = \frac{m_0}{\sqrt{1 - v^2/c^2}}} \tag{19j}$$

Relativistic

where m_0 is the mass of an object at rest in the observer's reference frame, and m is its mass when it is moving with a velocity v. A schematic diagram of a practical situation is shown in Fig. 19D in which the rest mass m_0 is not moving

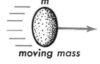

Fig. 19D *Schematic diagram illustrating the relativistic increase in mass, and the Lorentz-Fitzgerald contraction, due to motion.*

with respect to you, the observer, while at the right the same mass m is shown moving with a velocity v.

Table 19A gives the values of the relativistic mass of objects for a large range of velocities.

Table 19A. Relativistic Mass for Different Velocities

Velocity ratio v/c in per cent	1 %	10 %	50 %	90 %	99 %	99.9 %
Relative mass m/m_0	1.000	1.005	1.15	2.3	7.1	22.3

At 10 % the speed of light (18,630 mi/sec) the mass of a body is only $\frac{1}{2}$ of 1 % greater than its rest mass. At 50 % the speed of light the mass m has increased 15 %, while at 99.9 % the speed of light, it has jumped to over 22 times its rest mass. These values are in excellent agreement with experiments on high-speed atomic particles, a subject that will be considered in detail in later chapters.

It is important to note that, as the speed of any given mass increases, the mass rises slowly at first, and then much more rapidly as it approaches the speed of light. No object, however, can move with the speed of light, for by Eq. (19j) its mass would become infinite.

For low velocities v, Eqs. (19d) and (19j) are hard to evaluate, and the following approximation formula should be used:

$$\frac{1}{\sqrt{1 - v^2/c^2}} \cong 1 + \frac{1}{2}\frac{v^2}{c^2} \qquad (19k)^*$$

19.5. Einstein's Mass-Energy Relation

Just as sound, heat, and light are forms of energy, Einstein's special theory of relativity shows that mass is a form of energy. The expression giving the relation between mass and energy is an equation familiar to everyone. It is

$$\boxed{E = mc^2} \qquad (19l)$$

Relativistic

* The right side of this equation represents the first two terms of a mathematical series expansion, and for relatively low velocities v, the third and all succeeding terms are negligibly small:

$$1 + \frac{1}{2}\frac{v^2}{c^2} + \frac{3}{8}\frac{v^4}{c^4} + \frac{5}{16}\frac{v^6}{c^6} + \cdots$$

where m is the mass, c is the velocity of light, and E is the energy equivalence of the mass. The validity of this equation is now well established by hundreds of experiments involving atomic nuclei as well as the general subject called *atomic energy.*

If an object has a rest mass of m_0, it has stored within it a total energy m_0c^2. If the same mass is moving with a velocity v, its mass has increased to m and the total stored energy is mc^2. These two masses are related by Eq. (19j).

When a force F is applied to accelerate a given mass, the amount of work done is given by

$$W = F \times s$$

As a result of this *work done*, the object, whose rest mass is m_0, is moving with a velocity and has kinetic energy E_k:

$$F \times s = E_k \tag{19m}$$

Applying the law of conservation of energy, we can write

$$m_0c^2 + E_k = mc^2$$

Transposing, we obtain for the kinetic energy of a moving mass, the relation

$$E_k = mc^2 - m_0c^2 \tag{19n}$$

Another form for this equation is obtained by substituting Eq. (19j) for m:

$$E_k = \frac{m_0c^2}{\sqrt{1 - v^2/c^2}} - m_0c^2 \tag{19o}$$

or

$$E_k = m_0c^2 \left[\frac{1}{\sqrt{1 - v^2/c^2}} - 1 \right] \tag{19p}$$

In the abbreviated notation,

$$\boxed{E_k = m_0c^2(\gamma - 1)} \tag{19q}$$

PROBLEMS

(Assume the speed of light to be 3×10^8 m/sec, or 186,300 mi/sec, in all of the following problems.)

1. Find the length of a meter stick moving lengthwise at a speed of 2.8×10^8 m/sec. Assume a Lorentz-Fitzgerald contraction.

2. If a space ship 50 m long were to pass the earth traveling at 2.4×10^8 m/sec, what would be its apparent length, assuming a Lorentz-Fitzgerald contraction? *Ans.* 30 m.

3. The brakeman on a freight train traveling 60 mi/hr is walking forward on top of one of the box cars at 4 mi/hr. A train just ahead of this one is traveling at 80 mi/hr. What is the velocity of the brakeman with respect to the engineer in the train ahead, to three significant figures, (a) in Newtonian mechanics, and (b) in special relativity?

4. A man on the ground observes a plane taking off on the airport runway at 120 m/sec. A car traveling 32 m/sec follows it down the runway. Find the relative velocity of the plane as seen by the driver of the car. *Ans.* 88 m/sec.

5. An observer on the earth sees one space ship traveling at 2.55×10^8 m/sec overtaking another space ship traveling at 2.25×10^8 m/sec. What is the relative velocity of (a) the second ship as seen by the first, (b) the first ship as seen by the second, and (c) the relative velocity as seen from the earth?

6. An observer on the earth sees a space ship, receding from the earth at 2.0×10^8 m/sec, launch a projectile ahead of it. As seen from the earth this projectile has a speed of 2.25×10^8 m/sec. What is the velocity of the projectile with respect to the space ship as seen from (a) the space ship, and (b) the earth? *Ans.* (a) 0.5×10^8 m/sec, (b) 2.25×10^8 m/sec.

7. An earth observer sees a space ship approaching the earth at $\frac{1}{2}$ the speed of light. It launches an exploration vehicle which from the earth appears to be approaching at $\frac{3}{5}$ the velocity of light. What is the velocity of the vehicle with respect to the space ship as seen from (a) the space ship, and (b) the earth?

8. Two space ships are observed from the earth to be approaching each other, each with a velocity of $\frac{2}{3}$ the speed of light. With what velocity is each space ship approaching the other, as seen from either ship? *Ans.* 2.77×10^8 m/sec, or 0.92 c.

9. Atomic particles in the form of a beam have a velocity of 92% the speed of light. What is their relativistic mass as compared with their rest mass?

10. Atomic particles in the form of a beam have a velocity of 95% speed of light. What is their relativistic mass compared with their rest mass? *Ans.* $m/m_0 = 3.20$.

11. An atomic particle has a rest mass of 2.5×10^{-25} Kg. Find its total mass energy when it is (a) at rest, and (b) when it has a velocity of 0.90 the speed of light.

12. Two atomic particles, each with a rest mass of 2.0×10^{-25} Kg, approach each other in head-on collision. If each has an initial velocity of 2.0×10^8 m/sec, what is (a) the velocity of one atom as seen from the other, and (b) the relativistic mass of one as seen by the other? *Ans.* (a) 2.77×10^8 m/sec, (b) 5.20×10^{-25} Kg.

13. If an atomic mass of 4.2×10^{-25} Kg were converted into energy, and all of it imparted as kinetic energy to another atomic particle with a rest mass of 2.0×10^{-25} Kg, what would be the atom's velocity?

14. Starting with Eq. (19f), carry out the algebraic steps necessary to obtain Eq. (19g).

20

Electron Optics

There exists a remarkable similarity between optical systems of prisms and lenses, as they act upon light rays, and electric and magnetic fields as they act upon streams of electrons. It is the purpose of this chapter to consider some of these similarities and to treat several practical applications of *electron optics*. To begin with, it is convenient to present one of the standard methods of producing a beam of electrons, and to give the formula for calculating electron velocity.

20.1. An Electron Accelerator

A schematic diagram of an electron accelerator is shown in Fig. 20A. The source of electrons is a caesium-oxide-coated cathode K, heated by a filament F.

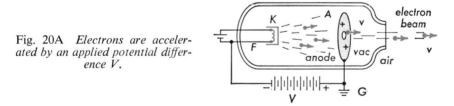

Fig. 20A *Electrons are acceler-ated by an applied potential differ-ence V.*

The cathode and filament are connected to the negative terminal, and the circular disk at the center to the positive terminal of a high-voltage battery V. Starting from rest at the cathode, the electrons are accelerated along the electric lines of force, acquiring at the anode A a velocity v.

By connecting the (+) terminal to the ground, the anode is brought to the potential of the surrounding walls of the room and the electrons are not attracted back toward A, but continue on with constant velocity. With a thin aluminum foil at the end and a high applied voltage V, electrons may be projected into the air beyond.

One of the results of J. J. Thomson's experiments with cathode rays was the discovery that the velocity of electrons depends upon the *potential* applied between the anode and the cathode. The higher the voltage, the higher is the electron velocity. Since the energy required to carry an electric charge q through a difference of potential V is given by Vq (see Eq. (2k)), the kinetic energy acquired by an electron of charge e falling through a difference of potential V will be $V \times e$. If we equate this product to the kinetic energy $\frac{1}{2}mv^2$,

$$Ve = \tfrac{1}{2}m_0v^2 \qquad (20a)$$

where V is the applied accelerating *potential* in volts, m_0 is the rest mass of the electron in Kg, v the velocity in meters per second, and e the charge on the particle in coulombs.

Example. Calculate the velocity of electrons accelerated by a potential difference of 10,000 volts. (The electronic charge $e = 1.60 \times 10^{-19}$ coul, and $m = 9.1 \times 10^{-31}$ Kg.)

Solution. Direct substitution in the above equation gives

$$10,000 \times 1.6 \times 10^{-19} = \tfrac{1}{2}(9.1 \times 10^{-31}) \times v^2$$

from which

$$v = \sqrt{\frac{10,000 \times 1.6 \times 10^{-19} \times 2}{9.1 \times 10^{-31}}} = 0.59 \times 10^8 \text{ m/sec}$$

This is approximately $\frac{1}{5}$ the velocity of light. (Velocity of light $c = 3 \times 10^8$ m/sec.

Instead of calculating the velocity of electrons in meters per second, it is customary to refer to their kinetic energy in terms of the applied voltage. For example, in the problem above the energy gained by the electrons is said to be 10,000 *electron volts* (abbr. 10,000 ev, or 10 Kev). They are also sometimes referred to as 10,000 volt electrons.

If voltages greater than 255,000 volts are used in Eq. (20a), the calculated velocities will be greater than the velocity of light. Consequently, for voltages of about 20,000 volts or more, the relativistic formula should be used. From the formula for the kinetic energy of a high-speed mass, Eq. (19q), we obtain

$$Ve = m_0c^2(\gamma - 1) \qquad (20b)$$

where

$$\gamma = \frac{1}{\sqrt{1 - v^2/c^2}} \qquad (20c)$$

20.2. Refraction of Electrons

When a moving electron, entering an electric field, makes an angle with the electric lines of force, it is bent in its path according to *Bethe's law of refraction* (see Fig. 20B). A correlation of this law with Snell's law in optics is indicated by the following parallel equations:

SNELL'S LAW	BETHE'S LAW	
$\dfrac{\sin i}{\sin r} = \dfrac{v_1}{v_2}$	$\dfrac{\sin \alpha}{\sin \beta} = \dfrac{v_2}{v_1}$	(20d)

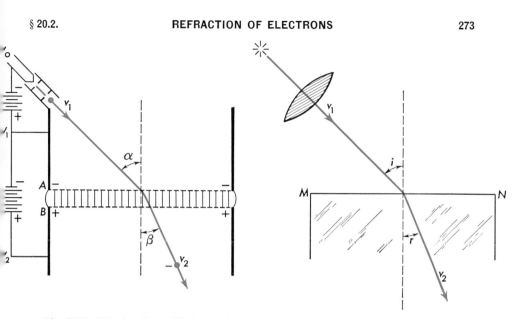

Fig. 20B   *The bending of light in refraction is analogous to the bending of the path of an electron.*

Note the reverse order of the velocities v_1 and v_2. When a ray of light enters a more dense medium, it is slowed down and at the same time bent toward the normal. Electrons, on the other hand, are deflected toward the normal when, in crossing a potential layer, they are speeded up. If the grid potentials are reversed, the electrons will be retarded in crossing the potential layer and they will be deflected away from the normal. In other words, *reverse the direction of the electrons, keeping their speed the same, and they will retrace their paths exactly.* Such a behavior is analogous to the very useful principle in geometrical optics that *all light rays are retraceable.*

To carry the refraction analogy a little further, consider the bending of electron paths by electrically charged bodies as shown in Fig. 20C. Attraction

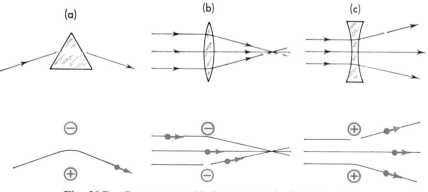

Fig. 20C   *Comparison of light optics with electron optics.*

by the positively charged wire and repulsion by the negative produce a prism-like action in case (a). A negatively charged metal ring produces a converging lenslike action in case (b); a positively charged ring produces a diverging lens action in case (c).

Since, by Eq. (20a), $v^2 \propto V$ and $v \propto \sqrt{V}$, the velocities v_1 and v_2 in Bethe's law of electron refraction may be replaced by $\sqrt{V_1}$ and $\sqrt{V_2}$, respectively, giving

$$\frac{\sin \alpha}{\sin \beta} = \frac{\sqrt{V_2}}{\sqrt{V_1}} \tag{20e}$$

V_1 and V_2 are the potentials of the two grids A and B in Fig. 20B, taken with respect to the cathode source of electrons in the electron gun as zero.

20.3. Electron Lenses

An electron lens, known as a double-aperture system, is shown in Fig. 20D; it is to be compared in its action to parallel rays of light incident on a converging

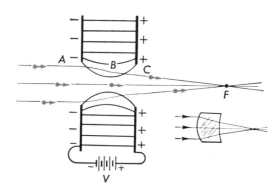

Fig. 20D *Double-aperture electron lens and its optical analogue.*

glass lens as shown at the lower right. While both are converging systems, the essential difference between the two is that whereas light rays are bent only at the two surfaces, electrons are refracted continuously as they pass through the potential layers.

The focal length of a glass lens is fixed in value by the radius of curvature of its two faces and the refractive index for the light used, but the focal length of an electron lens can be varied at will by altering v, the velocity of the electrons, and V, the voltage applied to the system. In this respect, the latter can be compared to the crystalline lens of the eye where the focal length can be changed by altering the lens curvature.

In the diagram, refraction for the upper path is greatest near A and, although it changes sign at some point near B, the gain in velocity due to the electric field produces a lesser deviation over the second half of the path, thereby causing convergence. If the electrons are reversed in direction on the right, they will retrace their paths and emerge parallel at the left. If the electric field is reversed in direction, however, the electron paths will not be the same but the system will still act as a converging lens.

A second type of electron lens, known as a double-cylinder system, is shown in Fig. 20E. In passing through the potential gap, the electric field has a con-

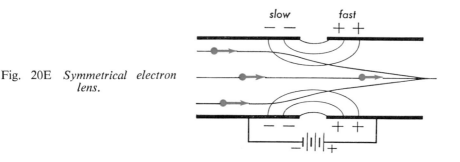

Fig. 20E *Symmetrical electron lens.*

verging action for the first half of the distance and a diverging action during the second half. Because they spend a greater time in the first half of the converging field, and the force on a charged particle is independent of velocity, the impulse (force × time) is greater for the convergence interval than it is for the divergence interval.

By making the second cylinder larger than the first, as in Fig. 20F, the electric

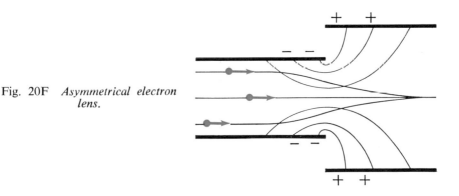

Fig. 20F *Asymmetrical electron lens.*

lines of force spread out more in the second cylinder. Such spreading weakens the field in the larger cylinder and reduces the divergent action to bring the electrons to a shorter focus.

20.4. An Electron Gun

A narrow beam of high-speed electrons, all having as nearly as possible the same velocity, has many practical applications in the field of electronics and atomic research. A device for producing such beams is called an "electron gun" (see Fig. 20G).

Electrons from a small filament-heated cathode *K* are accelerated by a

Fig. 20G   *Electron gun.*

difference of potential V applied to the cylinders of an electrostatic lens system, A_1 and A_2. The purpose of the guard ring maintained at the potential of the cathode is to improve the properties of the lens action of the first aperture and thereby collect a maximum number of emitted electrons into the collimated beam.

The function of the second lens is to converge the bundle toward a focus and then introduce enough divergence to straighten the beam out into a narrow pencil. The velocity of the emergent beam is given by Eq. (20a) where V is the over-all voltage from cathode K to anode A_2.

20.5. The Cathode-Ray Oscilloscope

One of the simplest applications of an electron gun is to be found in every *cathode-ray oscilloscope*, an instrument whose purpose is to reveal the detailed variations in rapidly changing electric currents, potentials, or pulses (see Fig. 20H). In appearance this device looks like J. J. Thomson's cathode-ray tube

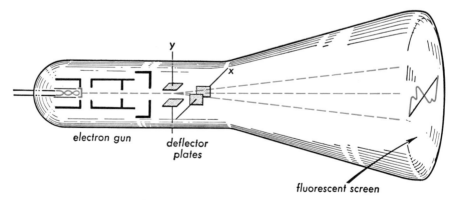

Fig. 20H   *Cathode-ray oscilloscope.*

(see Fig. 4F), and is actually the important element in one type of television receiver.

A cathode-ray oscilloscope is a vacuum tube containing *an electron gun* at one end, two pairs of *deflector plates* (or magnetic coils) near the middle and a *fluorescent screen* at the other end. When an alternating voltage is applied to the *x-plates*, the electron beam bends back and forth from side to side and, when applied to the *y-plates*, it bends up and down. The luminous spot produced where the beam strikes the fluorescent screen traces out a horizontal line in the first instance and a vertical line in the second.

It is customary to apply a *saw-tooth potential difference* to the *x*-plates (see Fig. 20I(a)) and the unknown potential difference to be studied, (b), to the

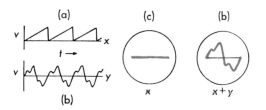

Fig. 20I   *Potentials applied to a cathode-ray tube, and graphs appearing on a fluorescent screen.*

y-plates. The saw-tooth potential difference supplied by a special radio tube circuit, called a "sweep circuit," causes the beam spot to move from left to right across the screen at constant speed and then jump quickly back from right to left to repeat the motion, (c). When the vertical deflections occur at the same time, the spot draws out a graph of the varying potential difference as in diagram (d). By varying the sweep circuit frequency until it matches the frequency of the studied signal, repeated graphs will be drawn out, one on top of the other, and persistence of vision and the fluorescent screen will present a stationary graph.

Green fluorescent screens are used for visual observation, since the eye is most sensitive to this color; blue screens are used for photographic purposes, since films and plates are most sensitive to blue.

The oscilloscope has many practical applications, and is to be found in every research laboratory as well as in every radio and television and repair shop. Its principal function is to analyze, or diagnose, rapidly changing potential differences whose frequencies may be as low as a fraction of a cycle per second or as high as thousands of megacycles per second. Periodic or transient potential differences as small as a fraction of a microvolt may also be studied by first amplifying them with standard vacuum-tube or transistor circuits. See Chapters 15 and 16.

Another valuable feature of the oscilloscope is its ability to measure time intervals between electrical impulses less than a microsecond apart. One microsecond is equal to one-millionth of a second.

20.6. Infrared Telescope

A telescope for seeing objects in the dark illuminated by infrared light is diagramed in Fig. 20J. An image of the object to be observed is focused on the photo-cathode of the vacuum tube by means of an ordinary glass lens *L*. The caesium-oxide-coated cathode under infrared illumination emits photoelectrons

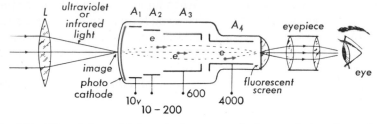

Fig. 20J   *Ultraviolet and infrared telescope employing an electron-image tube.*

which, accelerated to the right by $A_1A_2A_3$ and A_4, are brought to a focus on the green fluorescent screen at the right. The visible light they produce there by their impact is then observed by means of a magnifying eyepiece.

Electron focusing is accomplished by varying the potential difference applied to the second anode A_2; the infrared light image is focused by moving the lens L; and the visible light image is focused by moving the eyepiece.

20.7. Magnetic Lenses

When electrons cross a magnetic field and their paths make an angle with the magnetic lines, they are deflected in spirallike paths which, if properly controlled, may bring them to a focus. Such focusing properties of magnetic fields, illustrated by the cross section of a flat coil in Fig. 20K, were first demonstrated and proved mathematically by Busch in 1926. It can be shown that the focal length of such a lens, the magnetic field strength, and the electron velocity fit into well-known formulas in optics.

By encasing a flat coil in a hollow iron ring, the magnetic field becomes more concentrated and the refraction of electrons becomes more abrupt as they pass through the field. As a consequence the refraction more nearly resembles that

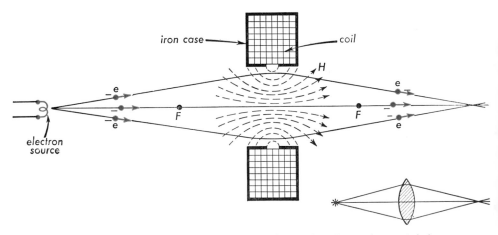

Fig. 20K   *Magnetic lens for electrons. (Optical analogue, lower right.)*

of optical lenses. Still greater concentration is brought about by providing a small narrow gap on the inside of the iron casing, as shown in the diagram.

If electron paths diverge too far from the principal axis of a coil lens, aberrations of the kind found with light and glass lenses arise. For this reason *diaphragms* are often used to confine electron beams to the center of the coil, as an *iris diaphragm* is used to confine light rays to the center of a lens.

20.8. Electron Microscope

The electron microscope, like the optical microscope, is an instrument used principally in the research laboratory for magnifying small objects to such an extent that their minutest parts may be observed and studied in detail. The importance of this device in the field of medical research cannot be over-estimated. To illustrate, many viruses known to medical science as being responsible for certain human diseases lie beyond the range of the optical microscope. With the electron microscope, magnifications of from 10 to 100 times that of the finest optical microscopes make many of these viruses, and some of their detailed structure, visible to the eye. While the highest magnification obtained with the best optical microscope is about 2000×, electron microscopes have already been made that give magnifications as high as 100,000×.

A schematic diagram of an electron microscope employing magnetic lenses is shown in Fig. 20L. At the bottom, a source of electrons is concentrated on the

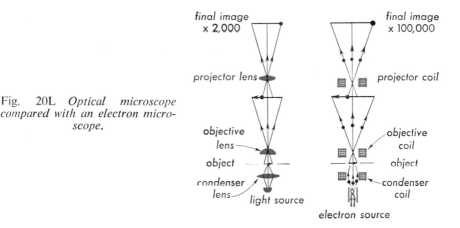

Fig. 20L *Optical microscope compared with an electron microscope.*

object (small arrow) by a condenser coil. Passing through or around the object, these electrons focus a magnified image of the object just below the projector coil. Only a small central section of these electrons pass into the projector coil, to be brought to focus in a further magnified image at the top. There the image of only a small section of the object can be seen directly on a fluorescent screen or can be photographed with ordinary photographic plates.

Electrons, like light waves, are stopped by metallic films; only when the films are extremely thin can transmitted rays be employed. For opaque objects the light, or electrons, may be reflected from the surface and only surface structures may be observed.*

20.9. The Electrocardiograph

When two small metal electrodes are placed at random on the skin of the human body, small and continuously changing potential differences are found to exist between them. When these varying *potentials* are carefully measured and compared with muscular activity, many of them can be correlated directly with certain activities of the body organs.

Each muscular action within the body is preceded and accompanied by an electrical impulse, the magnitude of which depends upon many factors. Such impulses accompanying the beating of the heart, for example, are exceedingly large (of the order of 1 millivolt); they are of such importance to the medical profession that special instruments have been devised and successively used to graphically record their intricate and minute variations. Such an instrument is called an *electrocardiograph*.

A schematic wiring diagram of one type of electrocardiograph is shown in Fig. 20M. *RA*, *LA*, and *LL* represent three metal electrodes that are taped to

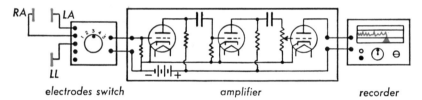

electrodes switch   amplifier   recorder

Fig. 20M   *Schematic diagram for an electrocardiograph used by physicians and surgeons.*

the patient's *right arm*, *left arm*, and *left leg*, respectively. Varying potential differences created between any pair of electrodes are selected by the switch box and, after being amplified, are applied to the vertical deflector plates of an oscilloscope or to the ink stylus of a recorder. The latter instrument draws out the potential changes on a moving strip of paper to make what is called an *electrocardiogram*.

Small sections of electrocardiograms are shown in Fig. 20N. Curve (a) is characteristic of a normal heart action. After a relaxation period, *T—P*, an electrical pulse *P*, associated with the activity of the right and left atrium, is recorded. This activity stimulates the right and left ventricles, and their response results in the *QRS* section of the curve. Between *S* and *T*, both ventricles are still activated, the pulse *T* indicating a restoration to the relaxed condition.

Although an electrocardiogram is only a record of a series of complex

---

* For a more complete treatment of electron optics, see *Electron Optics* by V. K. Zworykin and others, John Wiley and Sons.

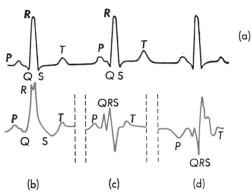

Fig. 20N (a) Electrocardiogram
of a normal heart. (b), (c), and (d):
Waves showing three types of abnormalities.

(a)

(b)   (c)   (d)

electrical events, it does indicate the time relations and magnitudes of certain
muscular activity. In (b), (c), and (d), single-period sections of abnormal
electrocardiograms are shown as samples of the thousands of different curves
found in practice. If, as an illustration, the electrical conduction to one ventricle
is poor, the *QRS* section of the curve is delayed as well as modified. If, on the
other hand, there is a larger than normal heart muscle, the *QRS* curve widens,
etc.

20.10. Electroencephalography

When small metal electrodes are placed on a patient's scalp, potential differences of the order of from 1 to 50 microvolts are found to exist between any
given pair. The predominant frequencies of these varying potentials are lower
than 100 cycles per second. They are called "brain waves." Instruments used to
record these waves are essentially the same as those used in electrocardiography
(see Fig. 20M) and are called *electroencephalographs*.

At the present time it is common practice to use a dozen or more electrodes
on one patient in order to localize with precision tumors and other abnormalities. Potential waves from any three or more selected points are often recorded
simultaneously to provide additional information by correlation. Reproductions
of two electroencephalograms (abbr. EEG), are shown in Fig. 20O. It will be
noted that there is a marked difference between (a) the waves from a normal
patient's record, and (b) those from a patient in epileptic seizure.

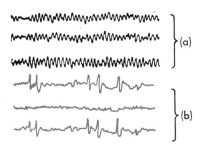

Fig. 20O *Electroencephalograms*
(*EEG*): (a) *normal*, (b) *epileptic
seizure.*

(a)

(b)

When EEG recordings are analyzed into their component waves, as is done with complex sound waves, various frequencies can be identified and associated with recognized abnormalities. Very low frequencies up to about 8 cyc/sec are called *delta waves;* those around 10 cyc/sec, *alpha waves;* and those in the range 10 to 60 cyc/sec, *beta waves.* Of the many brain disorders detectable from EEG records, the most familiar are: *epilepsy; cerebral thrombosis; encephalitis* and *meningitis;* and *brain tumor.*

QUESTIONS AND PROBLEMS

1. What is Bethe's law? In what way is it like Snell's law in optics? In what way is it different?

2. What kind of electric field will deviate an electron beam the way a glass prism deviates a light beam?

3. What are electron lenses? How are they made?

4. Make a diagram of an electron gun, using two hollow tubes of different diameter. Show the electric field and electron paths. Reverse the potentials, and assume a parallel beam again coming in from the left.

5. Make a diagram of a cathode ray oscilloscope. Label the principal parts. Briefly describe its action.

6. Make a diagram of your idea of an electron microscope. Label the principal elements.

7. What voltage applied to an electron gun will give electrons a velocity of 2×10^7 cm/sec?

8. If 5000 volts are applied to an electron gun, what will be the velocity of the electrons? *Ans.* 4.19×10^7 m/sec.

9. Make a table of electron velocities for the following accelerating voltages: $V = 5, 25, 90, 150, 300,$ and 600 volts.

10. What voltage applied to an electron gun will produce electrons having a speed of 2000 mi/hr? *Ans.* 2.25×10^{-6} v.

11. Electrons accelerated by a potential $V_1 = 1200$ volts enter an electric field between two grids as shown in Fig. 20B. If the angle of incidence is 40° and the potential across the grids is 600 volts, find the angle of refraction.

12. If in Problem 11 the voltage across the grids is reversed, find the angle of refraction. *Ans.* 65.3°.

13. Electrons accelerated by a potential of 250 volts enter an electric field as shown in Fig. 20B. If the angle of incidence is 50°, and the angle of refraction is 30°, find the potential difference between the grids.

14. If the angles are reversed in Problem 13, find the potential difference between the grids. *Ans.* 106 v.

15. Electrons from an electron gun enter a uniform magnetic field perpendicular to the lines of induction. If their circular path has a diameter of 40 cm and a potential of 300 volts is applied to the gun, find the magnetic induction B.

16. Electrons are to be accelerated by an electron gun and then allowed to enter a uniform magnetic field. If the magnetic induction is 8×10^{-4} weber/m², and the circular path is to have a radius of 5 cm, what voltage should be applied to the gun? *Ans.* 140.5 v.

17. Electrons accelerated by a potential $V = 750$ volts enter a uniform magnetic field in which the magnetic induction is 8×10^{-4} weber/m². Moving at right angles to the field, what is the radius of their circular path?

18. Calculate the velocity of electrons accelerated by a potential of 500,000 volts. *Ans.* 2.59×10^8 m/sec.

Radio, Radar, Television, and Microwaves

This chapter is concerned with electromagnetic waves at the long-wavelength end of the electromagnetic spectrum. The basic relation involved is the well-known wave equation

$$c = \nu\lambda \qquad (21a)$$

where ν is the frequency, λ the wavelength, and c the speed of the waves. In a vacuum, c is the same for all electromagnetic waves and is equal to the speed of light:

$$c = 3 \times 10^8 \, \frac{m}{sec} \qquad (21b)$$

A chart of the complete electromagnetic spectrum extending from the shortest known waves, the γ rays, to the longest known waves of radio is given in Fig. 21A. The long-wavelength end of this chart is seen to be divided into equally spaced bands with the designations shown in Table 21A.

Table 21A. Radio Wave Bands

BAND	DESIGNATION	ν	λ
VLF	very low frequency	3 Kc	100 Km
LF	low frequency	30 Kc	10 Km
MF	medium frequency	300 Kc	1 Km
HF	high frequency	3,000 Kc	100 m
		30 Mc	10 m
VHF	very high frequency	300 Mc	1 m
UHF	ultra high frequency	3,000 Mc	10 cm
SHF	super high frequency	30,000 Mc	1 cm
EHF	extremely high frequency	300,000 Mc	0.1 cm

The first four bands in this table are used for AM and FM radio communications of all kinds. The VHF and UHF bands are used for television, while

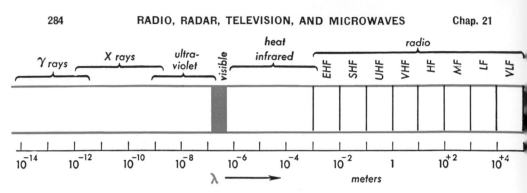

Fig. 21A   *Electromagnetic spectrum showing wavelengths of radio bands.*

the SHF and EHF bands, called *microwaves*, are used principally for experiments of all kinds.

$$1000 \frac{\text{cycles}}{\text{sec}} = 1 \frac{\text{kilocycle}}{\text{sec}} = 1 \text{ Kc}$$

$$1,000,000 \frac{\text{cycles}}{\text{sec}} = 1 \frac{\text{megacycle}}{\text{sec}} = 1 \text{ Mc}$$

21.1. Long-Distance Reception

It has long been known that certain bands of radio waves travel farther at night than they do in the daytime. During daylight hours programs heard on standard broadcast bands, 550 to 1600 kilocycles/sec, are generally received at distances up to 10 to 100 mi. At night, however, signals can be heard at 10 to 100 times these distances, but with intensities often varying in an erratic manner. Such nocturnal variations are due to changing atmospheric conditions and are known as "fading." Fading is more pronounced at greater distances but improves considerably after the first hour or two after sundown, only to become worse again shortly before sunrise.

Short radio waves in the range 1.5 to 40 megacycles travel great distances day or night but are strongly susceptible to changing atmospheric conditions. At one time of day, for example, communications between two greatly distant stations may be carried on over 20 m waves but not over 40 m waves. Later during the day the reverse may be true.

It is now known that the great distances spanned by radio are due to the reflection (actually refraction) of waves by a layer of electrically charged atoms and molecules in the upper atmosphere. The possible existence of such a layer was first postulated by O. Heaviside and A. E. Kennelly in 1902 to account for the propagation of waves around the earth's curved surface. The existence of such a layer was first demonstrated experimentally in 1925 by G. Breit, M. Tuve, and others, and its elevation shown to vary from 50 to 150 mi.

As illustrated in Fig. 21B, waves radiated upward from a transmitter *T* are bent back toward the earth where they are again reflected upward. The diagram indicates how the entire earth's surface might be covered with radio waves, the

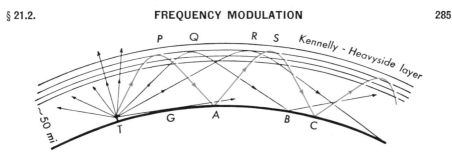

Fig. 21B   *Radio waves are refracted (reflected) by the ionized gas layers high in the earth's atmosphere.*

intensity of which should decrease with distance from the transmitter. At short distances of only a few miles a "ground wave," *T* to *G* in the figure, is heard as a strong steady signal day and night.

For some reason the Kennelly-Heaviside layer does not reflect the ultrahigh frequency waves used in *television* and *radar*. For this reason both are restricted to short-range operation, the waves traveling only in straight lines like visible light. To receive radar or television signals, therefore, the receiver antenna must be within sight of the transmitting antenna. To cover a large area the transmitting antenna should be located high in the air atop a building, hill, or mountain peak.

21.2. Frequency Modulation

A majority of the standard broadcasting stations, as well as amateur and commercial radio stations, still employ *amplitude modulation* (abbr. AM). The audiofrequency currents from the microphone are used to vary the amplitude of the continuous oscillations and waves from the transmitter.

In frequency modulation (abbr. FM), the audiofrequency currents from the microphone are used to vary the frequency of the carrier wave yet keep its amplitude constant. Louder sounds with AM mean greater variations in amplitude while with FM they mean greater changes in frequency. The two are compared graphically in Fig. 21C.

While just as good reproduction of voice and music can be accomplished with AM as is possible with FM, the latter has the advantage of eliminating "static."

Fig. 21C   *Graphs comparing amplitude modulation (AM) with frequency modulation (FM) of radio waves.*

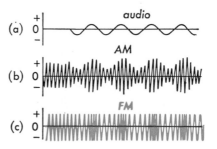

Static is the term applied to those noises in radio reception directly attributed to local atmospheric disturbances. Frequent minor lightning discharges in the air induce undesirable current pulses in any radio receiving antenna. When the receiver is designed to receive AM, these current pulses are amplified along with the broadcast and are heard as sharp cracking sounds or sometimes as a kind of "frying" noise. Where the receiver is designed for FM, however, the sudden current pulses picked up by the antenna do not change the frequency in any way and do not get through the *frequency discriminator circuit* to the audio amplifier. The resulting quietness to radio reception is remarkable and must be heard on a well-designed receiver to be fully appreciated.

21.3. Width of Radio Bands

The above discussion should not imply that the frequency of a carrier wave is unaltered by AM. Theory, confirmed by experimentation, shows that if a carrier wave of N cycles per second is amplitude modulated by a sound wave of n cycles per second, two new frequencies are produced, one $N - n$ and the other $N + n$. The first one is the beat note or difference frequency, whereas the other is a summation frequency not easily visualized from a graph like Fig. 21C.

When, for example, a carrier of 1 million cycles is modulated by a sound of 5000 cycles, a receiver tuned to the carrier must also respond to 995,000 cycles and 1,005,000 cycles, and to hear all other sound frequencies between zero and 5000 cycles must respond to all frequencies between these two limits. Hence a frequency range of 10,000 cycles ($N \pm 5000$ cycles) is required to transmit sounds from zero to 5000 cycles.

Transmitting stations must have their carrier frequencies 10,000 cycles apart to avoid interfering with each other; and receivers, when tuned to a given carrier, must not respond to frequencies of more than 5000 cycles up or down. To obtain higher quality reception 10,000 and even 15,000 cycles should be used. The latter would require a band width of 30,000 cycles and broadcasting stations would have to be at least 30,000 cycles apart. The desirability of high fidelity suggests shifting of all standard broadcast bands into the short-wave region (see Fig. 21A) where there is ample space for numerous stations 30,000 cycles apart.

21.4. The Scanning Process in Television

For years the sending of pictures by wire or radio has been an everyday occurrence. The fundamental principle involved in this process, and illustrated in Fig. 21D, is known as *scanning*. Every picture to be transmitted is scanned by an *exploring spot* which, starting at the top, moves in straight lines over the entire picture. The spot first moves from A to B, then from C to D, then E to F, etc., until the entire picture has been covered. Each time the spot reaches the right-hand side, it jumps back to the left and starts on the next line.

The exploring spot in any scanning device is so constructed that it generates

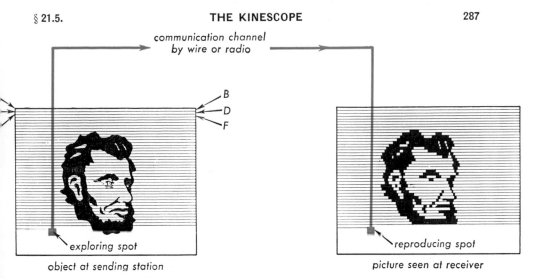

Fig. 21D   *Illustration of the process of picture scanning.*

an electric current proportional to the brightness of its instantaneous position. Such a pulsating current, called the *video signal*, is transmitted over wires or radio waves to the receiving station. There in a specially designed instrument a *reproducing spot*, whose brightness is proportional to the video signal amplitude, moves over a viewing screen in a path similar to that of the exploring spot. In this way the reproducing spot reconstructs the original picture.

It will be realized that the smaller the scanning and reproducing spots and the greater the number of lines, the better will be the details of the scanned picture being reproduced at the receiving end. The diagram shown here includes only 50 lines per picture as compared with 525 lines used in some standard (black and white) broadcasts.

If a single picture is to be sent by wire, as is generally the case in the *telephotographic newspaper service*, the scanning process requires from 10 to 20 min. In television, however, it is standard practice to scan and transmit 30 distinct and separate pictures in every second of time. At the receiving station these pictures are rapidly flashed one after the other upon a viewing screen. All are still pictures differing progressively one from the next so that, due to persistence of vision, the motions seem smooth and continuous, just as with moving pictures.

To avoid spurious shadows and images, the process of *interlacing* is employed. By this process each picture is scanned twice, first by running the exploring spot over the odd numbered lines 1, 3, 5, 7, etc., and then over the even numbered lines 2, 4, 6, 8, etc.

21.5. The Kinescope

In many respects the construction of a television receiver and its operation are similar to those of an ordinary radio receiver. The carrier wave from a nearby transmitter after being tuned in, detected, and amplified with conven-

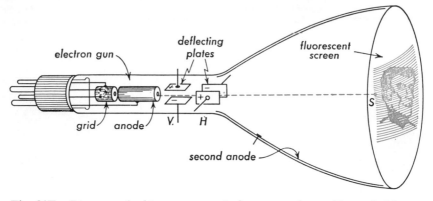

Fig. 21E   *Diagram of a kinescope, a typical vacuum tube used in a television re-ceiver to reproduce pictures on a fluorescent screen.*

tional radio tube circuits, is fed as a video signal into a kinescope in place of a loud-speaker. A kinescope is a large vacuum tube used for scanning and viewing the transmitted pictures.

A kinescope using electrostatic deflection plates for scanning is shown in Fig. 21E. Electrons from an electron gun at the left travel down the length of the tube to where, impinging upon a fluorescent screen, they produce a bright luminescent spot *S*. The purpose of the deflecting plates *V* and *H* is to deflect the electron beam with the identical frequency and scanning motion of the transmitting station. Two special oscillator tubes and circuits in the receiver supply saw-tooth potentials (see Fig. 21F) to these plates; the high-frequency

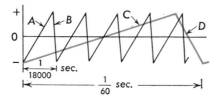

Fig. 21F   *Saw-tooth potentials ap-plied to plates of the television tube to produce horizontal scanning A and B, and vertical scanning C and D. A horizontal deflection, B hori-zontal return, C vertical deflection, and D vertical return.*

potentials to the *H-plates* for horizontal scanning and the lower frequency potentials to the *V-plates* for vertical scanning.

The proper fluctuations in the intensity of the luminescent spot are brought about by applying the video signal shown in Fig. 21G to the *grid* of the electron gun. This grid controls the flow of electrons through to the anode in the same

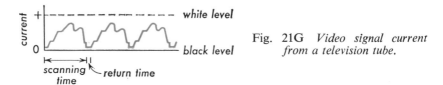

Fig. 21G   *Video signal current from a television tube.*

way that the grid controls the current to the plate in an ordinary three-element radio tube.

For a small fraction of a second, between successive pictures being scanned for transmission, current pulses of a certain type and frequency are sent out from the sending station as part of the video signal. These, picked up by the receiver, act as a triggerlike mechanism to bring the reproducing spot to the top left of the screen at the proper time to start the next picture. In other words the transmitter sends out signals that enable the receiver to automatically keep "in step" with the pictures as they are sent.*

21.6. Radar

Radar was one of the most important electronic developments of World War II and may be defined as the art of determining by means of *radio echoes* the presence, distance, direction, and velocity of distant aircraft, ships, land masses, cities, and other objects. RADAR derives its name from the longer title "RAdio Detection And Ranging."

Basically, a complete radar station consists of a *transmitter*, a *receiver*, and an *indicator*. As shown by a schematic diagram in Fig. 21H, the transmitter sends out high-frequency radio waves which, traveling outward with the velocity of light, are reflected from a distant object. The small portion of the reflected waves returning toward the station is picked up and amplified by the receiver. The signal is then fed into any one of a number of indicating devices, some of which are so complete as to give continuously the instantaneous *distance*, *direction*, and *relative velocity* of the object.

With one type of air-borne unit, ground objects can be observed on the screen of a kinescope even though fog or clouds intervene. Such systems are extremely useful in reducing the flying hazards that are always present during inclement weather.

The wavelengths of the waves used in radar are in the *microwave* region of

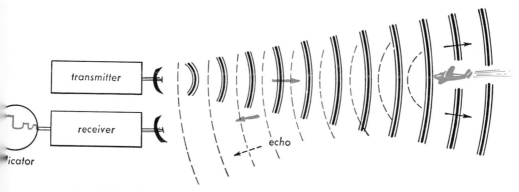

Fig. 21H   *Illustration of the principles of radar detecting and ranging.*

* *Television* by V. K. Zworykin and G. A. Morton, Wiley, New York.

the electromagnetic spectrum. They have wavelengths in the range of 1 cm to 10 cm.

21.7. The Pulsed System

Since a powerful transmitter must operate side by side with a supersensitive receiver, some provision is always made whereby the power from the transmitter is blocked out of the receiver. In most radar equipment this is accomplished by means of intermittent transmission, commonly called the *pulsed system*. According to this system, the transmitter is turned on for only a fraction of a second to send out a train of waves while the receiver is made very insensitive. When the transmitter goes off, the receiver is turned on to full sensitivity to receive the faint echo signal returning. When the receiver goes off, the transmitter comes on again to send out another wave train and repeat the above process hundreds of times per second.

A graph of the received pulses from an object 9 mi away is shown in Fig. 21I.

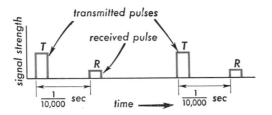

Fig. 21I *Graph of transmitted pulses showing received pulse, or echo, from an object 9 mi distant.*

Since the velocity of radio waves is 186,300 mi/sec, the same as light, the time interval between each transmitted pulse T and its echo R returning is a direct measure of the distance. If a frequency of 30,000 megacycles is used, the wavelength is 1 cm and each pulse will contain thousands of waves. Rectified by the receiver, an entire wave train appears as a voltage pulse as in the graph.

One type of *indicator* used for determining this time interval is a *cathode-ray oscilloscope* of the type shown in Fig. 20H. While the scanning spot is kept at constant intensity, a saw-tooth potential difference is applied to the horizontal sweep to make it move with constant speed across the fluorescent screen.

Electrical circuits are so arranged that the spot starts at the left just prior to the transmitter's emission of a pulse. When, a fraction of a second later, a pulse is initiated, a small part of the energy is applied as a vertical deflection of the spot, thereby producing a trace T as shown in Fig. 21J. When the returning echo signal arrives at the receiver, it too is applied as a vertical deflection, and a peak like the one at R_1 is produced. Upon reaching the right-hand end of the screen, the spot is extinguished and returned to the left, where it is again turned on and the above process repeated. As the spot retraces the same line many times every second, persistence of vision gives rise to the appearance of a steady trace.

If several different objects reflect waves of sufficient intensity to be picked up by the receiver, several peaks R_1, R_2, etc., will be seen on the indicator screen.

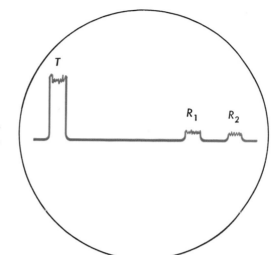

Fig. 21J   *Trace of spot on cathode-ray tube as used in radar ranging.*

In radar parlance, each such peak on the trace is called a *pip*, and its distance along the horizontal line from *T* is a direct measure of the time required for the signal to go out and return and is therefore a measure of the range of the object that caused it. Various methods of accurately measuring the distance interval have been developed.

If an object is coming toward or receding from a radar station, the frequency of the waves reflected from it will be increased or decreased respectively as in the Doppler effect. Hence by measuring the frequency change between the waves going out and those coming back, the velocity of approach or recession becomes known.

21.8. Wave Guides

The term "wave guide" is generally applied to a special class of metallic conductors having the property of conducting high-frequency oscillations from one place to another. To be more specific, it is a radio transmission line by which power generated at an oscillator can be transmitted to some utility point with little or no loss along the line.

Two types of wave guide commonly used at present are shown in Fig. 21K. The first, called a *coaxial cable* or *concentric line*, consists of a wire conductor

Fig. 21K   *Wave guides commonly used in radar and television transmission lines: (a) coaxial cable, and (b) hollow conductor.*

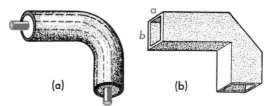

insulated from, and running lengthwise through, the center of a tubular conductor. Power from any high-frequency source, when connected to the central wire and tubular sheath, is propagated as waves through the dielectric between the two conductors.

The second is a hollow rectangular pipe called a *wave guide*. Power introduced as electromagnetic waves at one end is guided by the conducting walls to the other end. Each conductor is shown with a 90° bend to show that waves can be guided around corners.

While there is no limit to the frequency transmitted by coaxial lines, there is a lower limit for hollow wave guides. This lower limit, called the *cut-off frequency* or critical frequency, is the limiting case of a so-called *dominant mode* of vibration inside the guide and is analogous in some respects to the *fundamental vibration* of a given air column in sound. The dominant mode occurs in a rectangular pipe when the wider of the two dimensions b is $\frac{1}{2}$ a wavelength. The narrow dimension of the latter is not critical, but in practice is made to be about $\frac{1}{2}b$.

Since wave guides are comparable in cross section to the waves they propagate, and a coaxial cable will transmit any frequency no matter how low, the latter is generally used for waves longer than 10 cm, whereas hollow pipes are used with waves shorter than 10 cm. The power capacity of a hollow pipe, transmitting at its dominant mode, is greater than a coaxial cable of the same size.

Figure 21L shows an arrangement in which the high-frequency oscillations

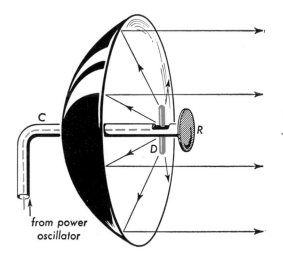

Fig. 21L   *Radar antenna system for producing a nearly parallel beam.*

from an oscillator tube source (not shown) are fed through a coaxial cable C to a single dipole, or Hertzian doublet, D. Radiated waves from the doublet are reflected into a parallel beam by the mirror. Since the over-all length of a dipole must be equal to $\frac{1}{2}$ a wavelength, the two small rods for 10 cm waves would each be 2.5 cm, or 1 in., long.

Two commonly used methods of feeding waves into one end of a wave guide, or withdrawing them from the other end, are shown in Fig. 21M. *Probe coupling* illustrated at (a) and (d) shows the center wire of a coaxial cable protruding

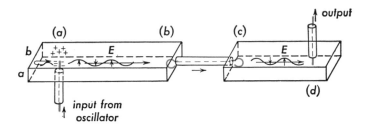

Fig. 21M   *Probe and loop coupling between wave guides and coaxial cables.*

into a rectangular guide a short distance from the end. *Loop coupling* illustrated at (b) and (c) shows a single loop of wire protruding through the end wall of the guide. The size and position of the probe and loop are critically determined for each separate installation.

In probe coupling as at (a) the protruding wire becomes alternately positive and negative. By attraction and repulsion of electrons the metal surface above the probe acquires an opposite charge thereby giving rise to a periodically changing electric field E parallel to the probe. The surging charges in the probe simultaneously produce magnetic field lines B around the probe, perpendicular to the electric lines, hence the propagation through the guide of electromagnetic waves with E and B perpendicular to each other (see Fig. 15G). The probe location is such that waves reflected from the end will be "in step" with those already moving down the guide. The reverse process takes place at (d) where the probe acts like a radio receiver antenna set into oscillation by the incoming waves.

In loop coupling (b) the surging charge in the loop constitutes a changing current that sets up magnetic lines of force threading through it. This changing magnetic field induces electric currents in the metal walls such that the bunching of charges above and below gives rise to a simultaneously reversing electric field E. The net result is the same as with probe coupling; electromagnetic waves are propagated through the guide.

21.9. Wave Propagation in Guides

In sound, the resonance of an air column is described as the result of two waves traveling in opposite directions and stationary nodes and loops are formed. The propagation of electromagnetic waves in a wave guide, on the other hand, may be described in terms of two wave trains crisscrossing each other as they reflect back and forth between the walls of the conductor. For one of these modes, the so-called *dominant mode*, see Fig. 21N.

The dotted lines in the "top" view represent wave fronts of the field E, *dots*

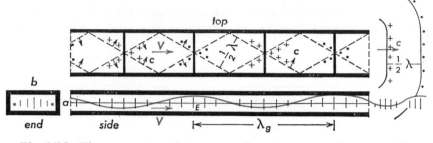

Fig. 21N *The propagation of microwaves through a rectangular wave guide.*

for the wave crests where the field is *up* out of the page, and *crosses* for the troughs where the field is *down* into the page. These two waves, traveling normal to their wave fronts with the velocity of light c, form resultant waves traveling lengthwise down the tube with a velocity V. The velocity V is always greater than the velocity of light and the wave length λ_g in the guide is always greater than the wave length λ outside. $V = \nu\lambda_g$.

As the b-dimension of the guide is reduced, the oblique wave fronts stay the same normal distance apart while the wave length λ_g and the velocity V increase, becoming infinite when $b = \frac{1}{2}\lambda$ (see Fig. 21O). Under this latter condition the

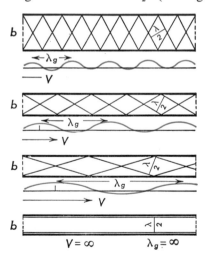

Fig. 21O *As a wave guide is narrowed down, the wavelength and wave velocity inside increases, becoming infinite when the critical width $b = \frac{1}{2}$.*

two component wave fronts are parallel to the wave guide and the electric field at all points in the guide is rising and falling together. This is the resonant condition of the dominant mode, and the guide width b is at its minimum allowable value.

21.10. Scanning

In a majority of radar installations the receiver uses the same antenna as the transmitter, thereby requiring a rapid switching mechanism that connects the transmitter to the antenna, when a pulse is to be radiated, and then to the receiver to pick up a possible echo signal (see Fig. 21P). Such dual use of one antenna saves space and weight and eliminates the mechanical difficulty of

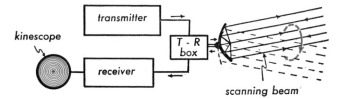

Fig. 21P *Block diagram indicating a T-R box for alternately connecting the antenna reflector to the transmitter and receiver. Conical or spiral scanning is also indicated.*

making two directive antennas point in exactly the same direction while they are moved about.

In certain types of radar installation the transmitter beam is made to sweep back and forth across the ground or sky with a scanning motion similar to that used in television. In one system a spiral scanning motion adapts itself to total coverage of a given area. When the beam crosses the path of any reflecting object, an echo signal returns to the receiver where it is amplified and applied to the cathode beam of a kinescope. As the transmitter beam carries out its scanning motion, the cathode ray spot on the kinescope is made to traverse a similar path. At that instant, when an echo signal returns, the spot brightens, and its location on the screen locates the relative position of the object in the scanning field.

Because the reflector is rotated mechanically, the radar scanning speed is considerably slower than in television. Its usefulness, however, is unquestionable, for high above the clouds in a plane it is possible to observe the positions and shapes of many landmarks on the ground below, as well as other planes and turbulent air of violent storm centers ahead.

21.11. The Magnetron

The magnetron is a vacuum tube known to radar engineers as a powerful source of electromagnetic waves. A tube no larger than a man's hand, for example, is capable of delivering 100 kilowatts in the form of 1 cm waves. A typical arrangement is shown in Fig. 21Q where the vacuum tube itself, in the

Fig. 21Q *How a magnetron power tube is mounted in the strong field of a permanent magnet.*

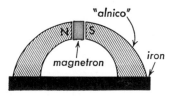

shape of a pill box, is mounted between the poles of a small permanent magnet.

Diagrams showing the resonant cavities inside the tube are given in Fig. 21R. Electrons emitted by the hot cathode K are accelerated toward the ring-like

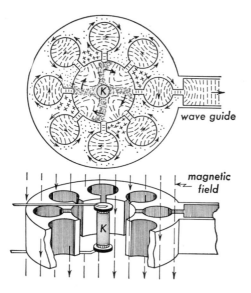

wave guide

Fig. 21R   *Details of magnetron tube showing resonant cavities and their action on electrons and electric lines of force.*

magnetic field

plate by a potential of several thousand volts. The magnetic field urges the electrons to follow a spirallike path, and at the same time causes them to bunch into clouds shaped like the arms of a windmill. These radial arms of spiraling electrons sweep clockwise around the main central cavity several billion times per second.

As the electron arms whirl around they repel the free electrons in the interstices between the small cavities, causing them to move alternately clockwise, then counterclockwise around their respective openings. These oscillatory surges of negative charge make the metal interstices alternately positive, then negative, and at the same time set up electric and magnetic lines of force in the cavities and slots as shown. These fields extending into the central cavity slow down some electrons and speed up others to bunch them as stated above.

The high-frequency oscillating fields in each of the eight cavities constitute electrical energy which can be tapped through any one of the cavities by an opening leading into a wave guide as shown at the upper right. By pulsing the cathode to plate voltage thousands of times per second, bursts of microwaves are sent along the wave guide channel to some antenna system. Since the dominant mode of a cylindrical cavity occurs when the diameter is about 0.7λ the smaller the diameter of the magetron cavities, the shorter is the wavelength of the generated waves.*

* *Principles of Radar*, by M.I.T. Radar School Staff, McGraw-Hill Book Co., Inc.

QUESTIONS AND PROBLEMS

1. What is RADAR? What do the letters stand for?

2. What is a pulsed radar system? Why is the operation intermittent?

3. How is the distance of an object determined? How is the velocity of a moving object determined?

4. What is a wave guide? How many kinds are there?

5. What is the process called scanning? What is a video signal?

6. What is meant by "interlacing"?

7. Radar waves of frequency 5×10^3 Mc/sec are reflected from a paraboloidal metal reflector. Calculate the over-all length of the dipole used at its focal plane.

8. The dipole of a radar transmitter has an over-all length of 2.5 cm. Calculate the frequency in Mc/sec. *Ans.* 600 Mc/sec.

9. A television transmitter broadcasts on a frequency of 94 Mc/sec. Find the length of the dipole of the transmitting antenna.

10. A television receiver uses a straight dipole with an over-all length of 1.8 m. What is its natural frequency? *Ans.* 83.3 Mc/sec.

11. A rectangular wave guide has a vertical opening of width $b = 2.2$ cm. If waves of frequency 0.8×10^{10} vib/sec travel through the guide as shown in Fig. 21K, find the wavelength λ_g and the wave velocity V.

12. A rectangular wave guide as shown in Fig. 21K has a vertical dimension $b = 1.5$ cm. If waves of frequency 1.5×10^{10} vib/sec travel through the guide, find the wavelength λ_g and the wave velocity V. *Ans.* (a) 2.68 cm, (b) 4.02×10^{10} cm/sec.

13. The dipole of a radar transmitter has an over-all length of 2.40 cm. Calculate the frequency in Mc/sec.

14. A television receiver uses a straight dipole with an over-all length of 2.5 m. What is its natural frequency? *Ans.* 60 Mc/sec.

15. A cross section of a rectangular wave guide has an opening of 6 mm by 14 mm. If waves of frequency 18,000 Mc/sec are sent through the guide, find (a) the wavelength λ, and (b) the wave velocity V.

Photon Collisions and Atomic Waves

In the preceding chapters we have seen that light waves consist of small finite bundles of energy, called *quanta* or *photons*, and that they too, like atomic particles, may be made to collide with atoms of one kind or another. This was the case both in the *photoelectric effect* (Chapter 10) and in the production of X rays (Chapter 14). The first part of the present chapter deals with the *corpuscular nature of light*, and the last part with the *wave nature of atomic particles*.

This last statement suggests a sort of Dr. Jekyll and Mr. Hyde existence for light waves as well as for atoms. Under some conditions, light and atoms may both act as though they were waves, whereas under other conditions they may both act like small particles.

22.1. Photoelectric Effect with X Rays

When a beam of X rays is allowed to shine on the surface of a thin sheet of metal like gold, several different phenomena may be observed to take place. Acting like waves, the X rays may be scattered at different angles to produce a diffraction pattern (see Fig. 14F), or acting like particles, they may collide with atoms and eject electrons as in the photoelectric effect (see Chapter 10).

Even though a beam of X rays may contain waves all of the same frequency, not all of the ejected photoelectrons acquire the same velocity, but they are divided into several well-defined groups. These different groups are illustrated schematically by the lengths of the arrows in Fig. 22A.

Careful measurements of the velocities of the photoelectrons, first made by Robinson and his collaborators in 1914, have shown that each velocity group is to be associated with the various shells of electrons within the atoms. The slowest electrons, all with the same velocity v_K, are ejected from the K-shell, the next faster group with a velocity v_L from the L-shell, the next group with a velocity v_M from the M-shell, etc.

The closer an electron is to the nucleus (see diagram (b)), the greater is the

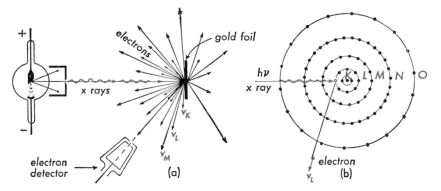

Fig. 22A　*Photoelectric effect produced with X rays gives rise to electrons with several different velocities. (b) Detail of an X ray ejecting an L electron from a heavy atom.*

attracting force and the greater is the force and energy necessary to liberate it from the atom. The velocity of the electrons in each group is given by Einstein's photoelectric equation,

$$h\nu = W + \tfrac{1}{2}mv^2 \tag{22a}$$

where W, the work function, is the energy necessary to free an electron from any one of the different electron shells (see §10.6).

While all of the incident X-ray photons have the same energy $h\nu$, more energy W will be used in liberating a K electron than there will be in liberating an L electron. This being the case, a photon liberating a K electron will have less energy left over for the electron than would another photon liberating an L electron from a similar atom. This experiment shows as well as one could wish that electrons exist in shells within the atom.

It should be pointed out that the energy W, used up in ejecting a photoelectron, is not lost by the atom but is later given out again in the form of X rays of various frequencies. In atoms where a K electron has been ejected, an L electron may jump into the vacated K-shell, with the simultaneous emission of a K X ray. This may be followed immediately by an M electron's jumping into the vacated L-shell and the emission of an L X ray.

22.2. The Compton Effect

While making a spectroscopic study of scattered X rays in 1923, A. H. Compton* discovered a new phenomenon, now known as the Compton effect. After considerable controversy with other experimenters, Compton proved quite conclusively that an X ray may collide with an electron and bounce off

---

* Arthur H. Compton (1892–1962), American physicist, born in Wooster, Ohio, on September 10, 1892. He received the degree of Doctor of Philosophy at Princeton University in 1920. In 1923 he discovered the change in wavelength of X rays when scattered by carbon, the phenomenon now known as the Compton effect. In recognition of this important discovery, in 1927 he was awarded the Nobel Prize in physics jointly with C. T. R. Wilson of England.

with reduced energy in another direction. This is analogous to the collision of two billiard balls.

Compton's historic experiment is illustrated schematically in Fig. 22B. X rays

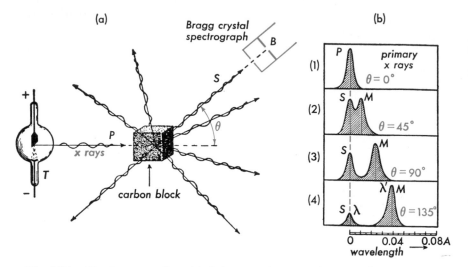

Fig. 22B   *The Compton effect. (a) Schematic diagram of Compton's experiment. (b) Graphs of the X-ray spectrum lines observed with a Bragg crystal spectrograph.*

from a tube *T* were made to strike one face of a small carbon block and scatter out in various directions. With an X-ray spectrograph at one side of the block he measured the wavelength of the X rays *S* scattered in a direction θ. These wavelengths he then compared with those of the incident beam *P*.

The comparisons are illustrated by graphs in diagram (b). The top curve (1) represents the wavelength λ of the X rays in the beam *P*, before striking the block. The other three curves, (2), (3), and (4), represent the two wavelengths λ and λ', observed when the spectrograph is located at the angles $\theta = 45°, \theta = 90°$, and $\theta = 135°$, respectively. These graphs show that some of the scattered X rays have changed their wavelengths whereas others have not. They further show the important result that as the angle increases, the change in wavelength of the modified rays *M* increases.

To explain the modified wavelengths *M*, Compton invoked the quantum theory of light and proposed that a single X-ray photon, acting as a material particle, may collide with a free electron and recoil as though it were a perfectly elastic sphere (Fig. 22C(a)). Applying the law of conservation of energy to the collision, Compton assumed that the energy $\frac{1}{2}mv^2$ imparted to the recoiling electron must be supplied by the incident X-ray quantum $h\nu$. Having lost energy, the X ray moves off in some new direction with a lower frequency ν' and energy $h\nu'$. By applying conservation of energy, we obtain

$$h\nu = h\nu' + \tfrac{1}{2}mv^2 \tag{22b}$$

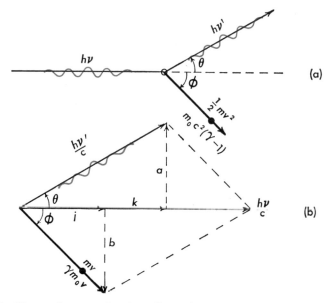

Fig. 22C   *Vector diagrams for the collision between an X-ray photon and a free*
electron. The Compton effect.

Since in most cases the velocity of the recoiling electron is so near the velocity
of light c, the relativistic equations must be applied. Using Eq. (19q) for the
kinetic energy of a moving mass, we write

$$hv = hv' + m_0c^2(\gamma - 1)$$ (22c)

where

$$\gamma = \frac{1}{\sqrt{1 - v^2/c^2}}$$ (22d)

and m_0 is the *rest mass* of the electron.

As with two perfectly elastic balls, Compton also applied the law of conservation of momentum and derived an equation from which he could calculate the
change in wavelength λ' of the scattered X ray. These calculated changes were
found to agree exactly with those observed by the experiment.

The fact that a beam of light has the equivalence of a momentum mv, and can
exert a pressure on a wall on which it falls, has long been known. According to
the quantum theory the momentum of a single photon is given by the energy hv
divided by the velocity of light c,

$$\text{momentum of a photon} = \frac{hv}{c}$$ (22e)

Compton's experiment is considered a proof of this equation.

Since momentum is a vector quantity, we construct a vector diagram as shown in Fig. 22C(b). The momentum of the X ray before impact is hv/c, while its momentum after impact is hv'/c, and the momentum of the electron is mv.

For electrons close to the speed of light the momentum mv is written in the relativistic form $\gamma m_0 v$. See Eq. (19i). By resolving the two momenta into two components, we obtain a and k as components of hv'/c and b and j as components of $\gamma m_0 v$.

Conservation of momentum requires the vector sum of j plus k to equal the initial momentum hv/c, and the vectors a and b to cancel each other. As equations,

$$\frac{hv}{c} = \frac{hv'}{c} \cos \theta + \gamma m_0 v \cos \phi \qquad (22\text{f})$$

and

$$\frac{hv'}{c} \sin \theta = \gamma m_0 v \sin \phi \qquad (22\text{g})$$

By combining Eqs. (22c), (22f), and (22g), and changing from frequencies v and v' to wavelengths by λ and λ', respectively, Compton derived the equation

$$\lambda' - \lambda = \frac{h}{m_0 c} (1 - \cos \theta) \qquad (22\text{h})$$

The quantity $h/m_0 c$ is called the *Compton wavelength*, and is equal to 2.43×10^{-12} meters. For those X rays that are scattered at an angle of 90°, the observed or calculated change in wavelength is just this amount and is the same for all X-ray wavelengths incident on the scatterer.

Compton's success is to be attributed to the exact agreement he found between the wavelength shift calculated from his application of the quantum theory and the values measured by experiment.

The first discoveries of the recoil electrons from the Compton effect were made by C. T. R. Wilson, and by Bothe and Becker. The existence of these collision products is readily shown by sending a beam of X rays through a Wilson cloud chamber just prior to its expansion (Fig. 22D).

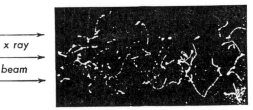

x ray

beam

Fig. 22D *Recoil electrons from X rays passing through the air in a Wilson cloud chamber. The Compton effect. (After C. T. R. Wilson.)*

When an X ray collides with a free electron the Compton effect can be expected, since the recoiling photon and electron are able to conserve energy and momentum. But when an X ray collides with an electron bound to an atom, the photoelectric effect takes place, since the atom can now recoil and conserve energy and momentum with the electron.

22.3. De Broglie's Electron Waves

In 1924 De Broglie, a French theoretical physicist, derived an equation predicting that all atomic particles have associated with them waves of a definite wavelength. In other words, a beam of electrons or atoms should, under the proper experimental conditions, act like a train of light waves or a beam of photons. The wavelength of these waves, as predicted by De Broglie, depends upon the mass and velocity of the particles according to the following relations:

$$\lambda = \frac{h}{mv} \qquad (22i)$$

This is known as *De Broglie's wave equation.* For an electron moving at high speed, the denominator mv is large and the wavelength is small. In other words, the faster an electron moves, the shorter the wavelength associated with it. See Fig. 22E.

Fig. 22E *Schematic diagram of a De Broglie wave.*

To acquire some concept of the relative wavelengths of electrons moving with different velocities, several values have been computed from De Broglie's equation (see Table 22A). The velocities are listed in miles per second in column

Table 22A. Wavelengths Associated with Electrons Moving with Different Velocities According to De Broglie's Wave Equation

$V \left(\begin{array}{c} \text{applied} \\ \text{voltage} \end{array} \right)$	$v \left(\begin{array}{c} \text{velocity} \\ \text{in mi/sec} \end{array} \right)$	$\frac{v}{c} \left(\begin{array}{c} \text{velocity} \\ \text{in percent} \end{array} \right)$	$\lambda \left(\begin{array}{c} \text{wavelength} \\ \text{in angstroms} \end{array} \right)$
1	370	0.20	12.23
10	1,100	0.62	3.87
100	3,700	1.98	1.22
1,000	18,000	6.26	0.38
10,000	36,000	19.50*	0.12
100,000	100,000	54.80*	0.03
1,000,000	175,000	94.10*	0.01

* These values take into account the increase in mass of the electron due to the theory of relativity. See Eq. (20b).

2 and in percent of the velocity of light in column 3. The potential differences listed in column 1 are the voltages required by Eq. (20a) to give an electron any

one of the velocities listed in columns 2 and 3. It will be noted that the wavelengths at the bottom correspond closely to those for X rays and γ rays.

22.4. The Davisson-Germer Experiment

The first experimental proof of the wave nature of atomic particles was demonstrated in 1927 by two American physicists, C. J. Davisson and his collaborator, L. H. Germer. Their experiment is illustrated schematically in Fig. 22F. Electrons from a hot filament are accelerated toward an anode, where,

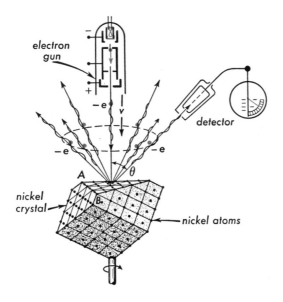

Fig. 22F *The Davisson-Germer experiment. Electrons striking the surface layers of a crystal are diffracted at different angles just as if they were waves with a very short wavelength.*

upon passing through a system of pinholes, they emerge as a narrow beam as indicated. This source acts as an electron gun from which electrons of any desired velocity may be obtained by applying the proper potential difference V.

Upon striking one of the polished faces of a nickel crystal, the electrons, acting like waves, are diffracted off in certain preferred directions. These preferred directions are located by means of a detector in which the electrons are collected and their accumulated charge measured. The detector is mounted so that it may be turned to any angle θ, and the crystal is mounted so it may be turned about an axis parallel to the incident beam.

With the electron-beam incident perpendicular to the crystal surface shown in Fig. 22F, the preferred direction of diffraction for 54-volt electrons was found to be 50°. Under these conditions the surface rows of atoms parallel to AB act like the rulings of a diffraction grating, producing the first-order spectrum of 54-volt electrons at $\theta = 50°$. This is illustrated in a cross-section detail in Fig. 22G. The waves reflected from one row of atoms M must travel one whole wavelength farther than the waves from the adjacent row N.

Fig. 22G   *Diagram of electron diffraction from the surface layer of a nickel crystal. The regular spacing of the atoms makes the crystal act like a diffraction grating.*

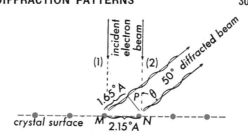

22.5. Electron Diffraction Patterns

Experiments analogous to Von Laue's X-ray diffraction experiments (see Fig. 14F) were first performed in 1928 by the English physicist G. P. Thomson, and independently by the Japanese physicist Kikuchi. A schematic diagram of their experimental apparatus is given in Fig. 22H. Electrons of known velocity

Fig. 22H   *Experimental arrangement for observing the diffraction of electron waves by thin films or crystals.*

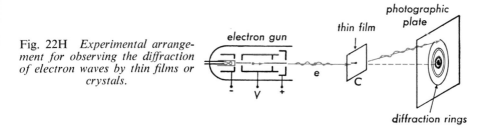

from an electron gun are projected at the front face of a thin metal film or crystal at *C*. A short distance farther on, the diffracted electrons strike a photographic plate where they produce patterns of the type reproduced in Fig. 22I.

Kikuchi's photograph (b) was made by projecting 68,000-volt electrons through a thin mica crystal. In this instance we have the exact analogue to the X-ray diffraction patterns of Friedrich, Knipping, and Von Laue (see Figs. 14F and 14G). The electrons, in passing through the crystal, are diffracted by the

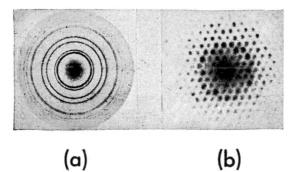

(a)   **(b)**

Fig. 22I   *Photographs of electron diffraction patterns demonstrating the wave nature of electrons. (a) 36,000 volt electrons from a thin silver foil (after G. P. Thomson). (b) 68,000 volt electrons from a thin mica crystal (after Kikuchi).*

atom centers in such a way that the various crystal planes act like mirrors to reflect them the same as they do with X rays of an equivalent wavelength. Because high-speed electrons had to be used to penetrate the crystal, the diffraction spots are closer together; the reason for this is that the equivalent electron wavelength, $\lambda = 0.047$ A, is about $\frac{1}{50}$ of the crystal spacing.

22.6. Electron Waves Within the Atom

The most recent development in the theory of atomic structure shows that the Bohr picture of the atom with sharply defined electron orbits is not correct. The new theory does not discard the Bohr theory entirely, but only modifies it to the extent that the electron does not behave as though it were a particle. The electron behaves as if it were made up of waves (sometimes called De Broglie waves) of the type described in preceding sections.

The new theory of the hydrogen atom was worked out independently by the two German theoretical physicists W. Heisenberg and E. Schrödinger in 1925 and was later modified and improved by the English theoretical physicist P. Dirac in 1928.* Schrödinger, making use of De Broglie's idea of electron waves, pictures the single electron in the hydrogen atom as moving around the nucleus as a kind of *wave packet*. This wave packet, as it is called, is formed in somewhat the same way that standing waves are set up and maintained in sound waves.

To set up these standing waves, according to Schrödinger, the length of the path of an electron around the hydrogen nucleus must be a whole number of wavelengths. Since the circumference of a circle is $2\pi r$, and the De Broglie wavelength $\lambda = h/mv$, the conditions to be satisfied by the new theory are

$$n \frac{h}{mv} = 2\pi r \tag{22j}$$

where $n = 1, 2, 3$, etc. This is exactly the condition proposed by Bohr in his orbital theory presented in Eq. (11b), since transposing the equation we obtain $mvr = nh/2\pi$. It is not surprising, therefore, that the new theory also gives exactly the Bohr equation, Eq. (11g), for the frequencies of the hydrogen spectrum.

One method of representing the electron in the atom is to picture an electron wave as one having a considerable length, so that it extends to standing waves. These may be illustrated schematically as shown in Fig. 22J. In the first figure there are two radial nodes; in the second, four radial nodes; and in the third, six radial nodes and one spherical node. In this representation the electron is not thought of as a particle located at some point within the atom but as though its mass and charge were spread out symmetrically throughout the space immediately surrounding the nucleus of the atom. While the Bohr circular orbits

---

* For their contributions to the new theory of atomic structure, Heisenberg was granted the Nobel Prize in physics for the year 1932, while Schrödinger and Dirac were jointly granted the prize one year later.

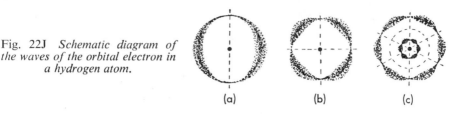

Fig. 22J *Schematic diagram of the waves of the orbital electron in a hydrogen atom.*

(a)　　　(b)　　　(c)

were confined to a plane, the wave model allows the electron distribution to be three-dimensional.

Even though the new theory of the hydrogen atom is an improvement upon the older Bohr orbit theory, and gives a more satisfactory explanation of all known phenomena, it is more difficult to form a mental picture of what an atom might look like. Indeed, the modern theoretical physicist goes so far as to say that the question, "What does an atom look like?" has no meaning, much less an answer. There are others, however, who still maintain that only those things that can be pictured are the things that are understood and that all mental thought processes are made in terms of things we detect by sight or touch. For this reason an interpretation is often given to the theory and its resultant equations that the amplitude of the electron waves within an atom represents the distribution of the electronic charge and mass. At the nodes where the motion is practically zero, there is assumed to be little or no charge, while at the antinodes there is a maximum amount of charge.

Photographs representing a few of the possible states of the single electron in hydrogen are shown in Fig. 22K. These are not pictures of real atoms but are made to represent them. They are made by photographing a specially designed mechanical top. Where the electronic charge density is large, the figure is white; where it is practically zero at the nodes, it is dark. The three-dimensional distributions can be visualized by imagining each figure to be rotating about a vertical axis, as illustrated by the white line in the second figure. This particular figure in three dimensions would have a shape similar to a smoke ring.

22.7. The New Atomic Picture

Although the Bohr atom has been replaced by the more satisfactory model of a *nucleus surrounded by electron waves*, it is still customary, for convenience only, to talk about electron *shells* and *orbits*. The reason for this is that there is a close analogy between the old and the new models. When the Bohr-Stoner scheme of the building-up of the elements is extended to the new theory of electron waves, the electrons are found to distribute their charge in such a way that something analogous to shells is formed.

This is illustrated by the graph for a rubidium atom, atomic number 37, in Fig. 22L. The shaded area above represents the distribution of the charge of 37 electrons on the new theory, and the lower orbital model represents the

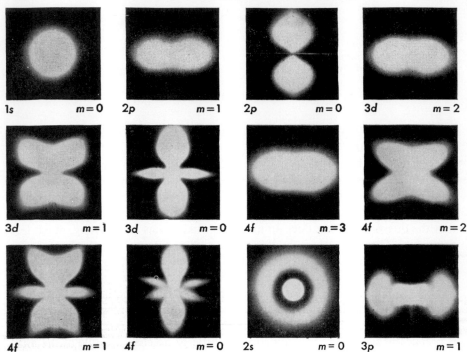

Fig. 22K *Electron wave density figures representing the single electron states of the hydrogen atom.*

electron shells on the old theory. The new model is represented by a graph because it is spherically symmetrical in space, while the old model is represented by orbits because it is confined to one plane. Proceeding out from the nucleus it is seen that the charge rises to several maxima at distances corresponding closely to the discrete K, L, M, N, and O shells of the orbital model. In other words, the new atom also has a shell-like structure.

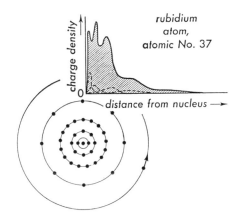

Fig. 22L *Diagrams comparing the new and the old theories of atomic structure, according to wave mechanics and the Bohr-Stoner model.*

22.8. Heisenberg's Uncertainty Principle

The quantum theory description of a light beam as being made up of discrete packages of energy $h\nu$, called photons, would seem to rest upon our ability to determine for a given photon both the *position* and the *momentum* that it possesses at a given instant. These are usually thought of as measurable quantities of a material particle. It was shown by Heisenberg, however, that for particles of atomic magnitude it is in principle impossible to determine both position and momentum simultaneously with perfect accuracy. If an experiment is designed to measure one of them exactly, the other will become uncertain, and vice versa.

An experiment can measure both position and momentum but only within certain limits of accuracy. These limits are specified by the *uncertainty principle* (sometimes called the *principle of indeterminacy*), according to which

$$\Delta x \cdot \Delta p = h \tag{22k}$$

Here Δx and Δp represent the variations of the value of position and the corresponding momentum of a particle which must be expected if we try to measure both at once, i.e., the uncertainties of these quantities.

The uncertainty principle is applicable to photons, as well as to all material particles from electrons up to sizable bodies dealt with in ordinary mechanics. For the latter, the very small magnitude of h renders Δx and Δp entirely negligible compared to the ordinary experimental errors encountered in the measuring of its position x and its momentum p.

When p is very small, as it is for an electron or a photon, the uncertainty may become a large percent of the momentum itself, or else the uncertainty in the position is relatively large.

According to Bohr the uncertainty principle of Heisenberg's provides complementary descriptions of the same phenomenon. That is, to obtain the complete picture of any event we need both the wave and corpuscular properties of matter, but because of the uncertainty principle it is impossible to design an experiment that will show both of them in all detail at the same time. Any one experiment will reveal the details of either the wave or corpuscular character, according to the purpose for which the experiment is designed.

The interference fringes in Young's double-slit experiment, shown in Fig. 6I, constitute one of the simplest manifestations of the wave character of light. Identical fringe patterns, however, can be obtained by sending a beam of electrons or protons through the same slits. On the wave theory a small part of the incident wave goes through one slit and another small part goes through the other, and these when they come together produce constructive and destructive interference at different areas on the screen. See Fig. 22M.

If on the corpuscular theory, however, a photon $h\nu$, or a particle m, goes through one slit, how can it be affected by the other slit, and hence go to one of the appropriate fringe areas on the screen, and never to one of the zero points?

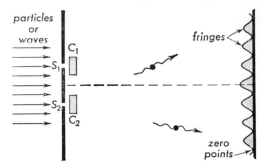

Fig. 22M   *Young's double-slit experiment modified to demonstrate both wave and corpuscular properties of matter.*

To answer this question let us suppose small sensitive detectors, such as scintillation counters (see Figs. 25H and 25I), are placed behind the two slits as shown in Fig. 22M. These detectors C_1 and C_2 can register each photon or particle that goes through one slit or the other. But, in so doing, the fringe pattern will have been destroyed because of the deflections suffered by the corpuscles in producing scintillations. Because of a change in direction the momentum p is changed so that its new momentum is uncertain.

Hence, when we use slit detectors to make Δx small, we introduce a large change in momentum p, and destroy the fringe pattern. When we do without the slit counters to restore the fringe pattern making Δp small, Δx becomes large because we do not know what slit the corpuscle went through.

Some philosphers regard the uncertainty principle as one of the profound principles of nature. Physicists, on the other hand, are inclined to believe it to be an expression of our inability thus far to formulate a better theory of radiation and matter.

PROBLEMS

1. Calculate the momentum of an X ray having a wavelength of 0.40×10^{-10} meter.

2. What would be the De Broglie wavelength of a 2000-Kg car moving along the highway at 30 m/sec? *Ans.* 1.1×10^{-38} m.

3. Find the wavelength associated with an electron moving with $\frac{1}{20}$ the velocity of light.

4. Calculate the wavelength associated with a proton moving with $\frac{1}{20}$ the velocity of light. *Ans.* 2.64×10^{-14} m.

5. Find the wavelength of an α-particle accelerated by a potential difference of 25,000 volts.

6. The equivalent wavelength of a moving electron is 0.25×10^{-10} m. What voltage applied between two grids will bring it to rest? *Ans.* 2400 volts.

7. If 5000 volts are applied as an accelerating potential difference to an electron gun, what is the equivalent wavelength of the electrons?

8. The electron beam in a television receiver tube is accelerated by 12,000 volts. What is the De Broglie wavelength of the electrons? *Ans.* 0.112 A.

9. Briefly explain the Davisson-Germer experiment. What was observed and what were the conclusions?

10. If the electron beam in a TV picture tube is accelerated by 10,000 volts, what is the De Broglie wavelength? *Ans.* 1.23×10^{-11} m.

11. Calculate the momentum of each photon in a beam of visible orange light, $\lambda = 6000$ A.

12. Find the momentum of each photon in a beam of violet light, $\lambda = 4000$ A. *Ans.* 1.66×10^{-27} Kg m/sec.

13. Find the speed of an electron whose wavelength is equivalent to 0.5 A.

14. X rays of wavelength 0.82 A fall on a metal plate. Find the wavelength associated with the photoelectrons emitted. Neglect the work function of the metal. *Ans.* 0.0997 A.

23

Radioactivity

Radioactivity may be defined as a spontaneous disintegration of the nucleus of one or more atoms. The phenomenon was discovered originally by Becquerel* in 1896 and is confined almost entirely to the heaviest elements in the periodic table, elements 83 to 102. What Becquerel discovered was that uranium, element 92, gave out some kind of rays that would penetrate through several thicknesses of thick black paper and affect a photographic plate on the other side. When the same phenomenon was confirmed several months later by Pierre and Marie Curie,† these rays became known as Becquerel rays.

23.1. Discovery of Radium

Unlike the discovery of many new phenomena, the discovery of radium by Pierre and Madame Curie in 1898 was brought about intentionally by a set of carefully planned experiments. Having found that pitchblende was active in emitting Becquerel rays, the Curies chemically treated a ton of this ore in the

* Antoine Henri Becquerel (1852–1908), French physicist. Born in Paris on December 15, 1852, Antoine succeeded to his father's chair at the Museum of Natural History in 1892. In 1896 he discovered radioactivity, the phenomenon for which he is most famous. The invisible but penetrating rays emitted by uranium and other radioactive elements are now called Becquerel rays. For these researches he was granted the Nobel Prize in physics in 1903.
† Pierre Curie (1859–1906) and Marie Curie (1867–1936), French physicists. Pierre Curie was educated at the Sorbonne where he later became professor of physics. Although he experimented on piezoelectricity and other subjects, he is chiefly noted for his work on radioactivity performed jointly with his wife, Marie Sklodowska, whom he married in 1895. Marie was born in Poland on November 7, 1867, where she received her early scientific training from her father. Becoming involved in a student's revolutionary organization, she left Poland for Paris where she took a degree at the university. Two years after the discovery of radioactivity by Becquerel, Pierre and Madame Curie isolated polonium and radium from pitchblende by a long and laborious physical-chemical process. In 1903 they were awarded the Davy Medal of the Royal Society, and (jointly with Becquerel) the Nobel Prize in physics. Professor Curie, who was elected to the Academy of Sciences in 1905, was run over and killed by a carriage in 1906. Succeeding him as professor at the university, Madame Curie in 1911 was awarded the Nobel Prize in chemistry. She has the rare distinction of having had a share in the awards of two Nobel Prizes.

hope of isolating from it the substance or element responsible for the activity. The first concentrated radioactive substance isolated was called *polonium* by Madame Curie, a name chosen in honor of her native country, Poland. Five months later came the isolation of a minute quantity of *radium*, a substance that was a powerful source of Becquerel rays. Continued experiments by the Curies, and others, soon led to the isolation of many other substances now recognized as radioactive elements. Some of the more common of these are *ionium, radon,* and *thorium*.

23.2. The Properties of Becquerel Rays

It is to the experimental genius of Rutherford* that we owe the complete unraveling of the mystery surrounding the nature of Becquerel rays. As the result of an extensive series of experiments, Rutherford and his co-workers discovered that these penetrating rays are of three quite different kinds. A simplified experiment demonstrating this is illustrated in Fig. 23A. A small

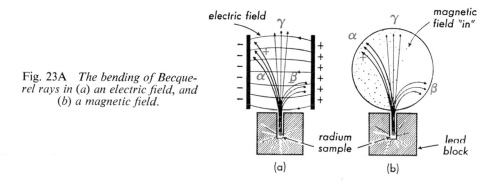

Fig. 23A   *The bending of Becque-*
rel rays in (a) an electric field, and
(b) a magnetic field.

sample of radium is dropped to the bottom of a small drill hole made in a block of lead. This produces a narrow beam of rays emerging from the top of the block, since rays entering the walls of lead are absorbed before reaching the surface. When electrically charged plates are placed at the side of this beam as shown in diagram (a), the paths of some rays are bent to the left, some to the right, and some are not bent at all. A magnetic field as shown in diagram (b) exhibits the same effect. Paths bending to the left indicate positively charged particles called *α rays* or *α particles*, those bending to the right indicate negatively

* Lord Rutherford (1871–1937), British physicist, was born in New Zealand where he attended the university. In 1898 he became Macdonald professor of physics at McGill University, Montreal, Canada, and in 1907 professor of physics at Manchester University. In 1919 he became professor and director of experimental physics at the University of Cambridge, and in addition held a professorship at the Royal Institution in London. He is most famous for his brilliant researches establishing the existence and nature of radioactive transformations and the electrical structure of the atom. For this work and until the time of his death in 1937 he was acclaimed by many as the greatest living experimental physicist. He was awarded the Nobel Prize in chemistry in 1908, and was knighted in 1914.

charged particles called *β rays* or *β particles*, and those going straight ahead indicate no charge and are called *γ rays* or photons.

Rutherford, by a series of experiments, was able to show that each *α* ray is in reality a *doubly ionized helium atom*, i.e., a helium atom with both of its electrons removed. Such a particle is nothing more than a bare helium nucleus with double the positive charge of a hydrogen nucleus or proton, and a mass number or atomic weight four times as great. The *β rays* he found are ordinary electrons with a mass of $\frac{1}{1836}$ the mass of a *proton* or $\frac{1}{7360}$ the mass of an *α particle*, while *γ rays* are electromagnetic waves of about the same or a little higher frequency than X rays. Although *γ* rays all travel with exactly the velocity of X rays and visible light, *α* rays are ejected with a speed of from $\frac{1}{10}$ to $\frac{1}{100}$ the velocity of light; *β* particles move faster than *α* particles, some of them traveling with 99% the velocity of light.

23.3. Identification of Alpha Particles

The first conclusive evidence that alpha particles are helium nuclei was obtained by Rutherford and Royds in England in 1909. A special glass tube as shown in Fig. 23B was used for this purpose. A thin-walled glass tube *A* contain-

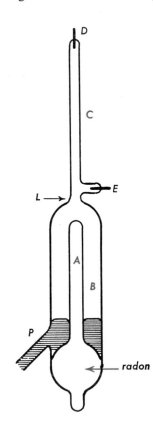

Fig. 23B *Special discharge tube used in proof that alpha particles are helium nuclei.*

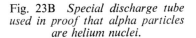

ing radon gas was sealed inside a thick-walled tube B containing mercury. At the top was a glass capillary tube C with two metal electrodes D and E sealed through the glass as shown.

Tubes B and C were evacuated through the side tube P, and then mercury was raised to the level shown. After maintaining this condition for some time, some of the alpha rays emitted by the radon gas atoms passed through the thin glass walls of tube A to be collected in tube B. Here these particles acquire electrons and become neutral atoms. After standing for six days the mercury was raised to the level L, forcing the helium into the capillary. A high voltage was applied to the electrodes, and the light from the discharge was observed with a spectroscope. The line spectrum observed was found to be identical with that obtained when a discharge was sent through another tube containing regular helium gas.

23.4. Ionizing Power

When Becquerel rays penetrate matter in the gaseous, liquid, or solid state, they do not continue to move indefinitely, but are brought to rest slowly by ionizing atoms all along their path. Being ejected from their radioactive source with tremendously high speeds, all three types of rays collide with electrons and knock them free from atoms. They are, therefore, *ionizing agents*. The relative number of ionized atoms created along the path of an α particle, however, is much greater than the number created by a β particle or γ ray. If, in traveling the same distance in a given material, a γ ray produces one ionized atom, a β particle will, on the average, produce approximately 100, and an α particle will produce about 10,000. Thus α particles are powerful ionizing agents, while γ rays are not.

As stated above, an α particle, ejected from a radioactive atom, is but a nucleus of a helium atom and lacks the two electrons necessary to make of it a neutral atom. As this particle speeds through matter, it picks up and loses electrons at a rapid rate. No sooner does an electron become attached than it is swept off again by other atoms. Finally, upon coming to rest, however, each α particle collects and retains two electrons, becoming a *normal helium atom*.

23.5. Penetrating Power

At each collision with an atom, Becquerel rays lose, on the average, only a small part of their initial energy. Usually an α particle or β particle will make several thousand collisions before being brought to rest. At each collision, some of the kinetic energy is expended in ionizing the atom encountered while giving that same atom a certain amount of kinetic energy. Since α particles produce the greatest number of ions in a given path, they penetrate the shortest distance and therefore have the poorest penetrating power. The penetrating powers of the three kinds of rays are roughly inversely proportional to their ionizing power.

	α	β	γ
Relative ionizing power	10,000	100	1
Relative penetrating power........	1	100	10,000

23.6. Methods of Detecting Becquerel Rays

There are several well-known methods for detecting and measuring radio-activity; the most common of these are

Electroscopes	Geiger-Müller counters
Electrometers	Scintillation counters
Cloud chambers	Ionization chambers
Bubble chambers	Photographic emulsions
Semicoductors	Spark chambers

We have already seen how X rays passing through an electroscope cause the charge to disappear and the gold leaf to fall. This same action may be demonstrated with α, β, and γ rays. The stronger the source of rays or the nearer the sample is brought to the electroscope, the more rapid is the discharge. Experiments show that, if the walls of the electroscope are too thick, only the γ rays get through to produce ionization on the inside. For this reason specially designed electroscopes made with thin windows of light material like aluminum are used for measuring α and β rays.

The Braun type of electrometer is convenient for demonstration purposes. This device uses a lightweight metal pointer, pivoted a trifle above its center of gravity and insulated from its ring support by a nonconductor I as shown in Fig. 23C. In principle, the electrometer operates exactly like an electroscope; when charged positively or negatively, the needle rises and the pointer indicates the acquired potential on a scale.

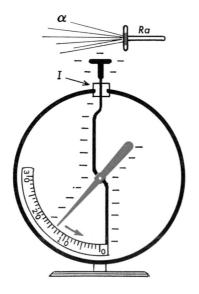

Fig. 23C *Diagram of a Braun-type electrometer—used here to detect radioactivity.*

If a radioactive source is brought close to the knob or terminal of a charged electrometer, the needle slowly returns to the vertical, or no charge, position. The α particles produce many ions in the air close to the source; oppositely charged ions are drawn toward the knob and there neutralize the charge.

23.7. The Wilson Cloud Chamber

In 1912 C. T. R. Wilson devised a method by which one may actually observe the paths of α and β particles. As will be seen in the following chapters, this method is used extensively in modern atomic physics as a means of studying many different atomic processes. The device by which this is accomplished consists of an expansion chamber in which water vapor is made to condense upon ions produced by the high-speed particles that have previously passed through it.

To begin with, the conditions under which water in the vapor state will condense into fogdrops are quite critical. These conditions are (1) there must be water vapor present, (2) there must be dust particles or ions on which the drops can form, and (3) the temperature and pressure must be brought to a definite value. That water drops will condense only upon ions or dust particles can be demonstrated with an ordinary glass jar containing a little water as shown in Fig. 23D. If allowed to stand for a short period of time, some of the

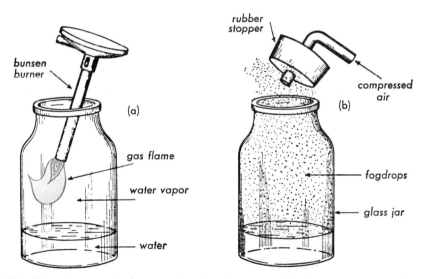

Fig. 23D   *Experiment demonstrating the formation of fog drops on ions in a glass jar.*

water will evaporate and fill the bottle with vapor. Ions are next formed in the bottle by momentarily inserting a small gas flame as shown in diagram (a). Compressed air is then injected into the bottle through a tube, so that when the stopper is quickly removed the sudden expansion will produce a dense fog as shown in diagram (b). If the flame is not first inserted to produce ions, no appreciable fog can be formed.

The purpose of the compressed air and subsequent expansion of the chamber is to lower the temperature, thus causing the air to become supersaturated with water vapor. Under these conditions the vapor will condense on all ionized molecules present.

When an α or β particle shoots through the air, positive and negative ions are formed all along its path. The removal, by collision, of each electron from a neutral atom or molecule leaves a positively charged ion. The electron that attaches almost immediately to another neutral atom or molecule forms a negatively charged ion. If immediately after an α particle has gone through a cloud chamber an expansion takes place, fogdrops will form on the newly created ions, revealing clearly the path the particle has taken. As illustrated by the photographs in Fig. 23E, such α-ray tracks are straight and quite dense,

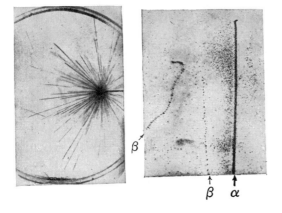

Fig. 23E (a) *α-ray tracks from radium as seen in a Wilson cloud chamber*. (b) *One α-ray and two β-ray tracks*. (*After C. T. R. Wilson*, Proceedings of the Royal Society of London, *Vol. 87, 1912, p. 292.*)

whereas the β-ray tracks are crooked and sparsely lined with drops. The β rays, being very light particles, are easily deflected by collision, while the relatively heavy α particles "plow" right through thousands of atoms with only an occasional deflection.

Gamma rays are never observed in a cloud chamber, since they produce so few ions. In passing through several feet of air a single γ ray will, on the average, produce only one or two ions. This is not enough to produce a recognizable cloud track. If a very strong source of γ rays is available, however, their presence can be observed in a cloud chamber by the chance collisions some of them have made with electrons. These recoiling electrons are called "Compton electrons" and were explained in §22.2, and shown in Fig. 22D.

A diagram of a simple type of Wilson cloud chamber is shown in Fig. 23F. The arrangement is made from an ordinary flat-bottomed flask with a rubber bulb attached to the neck. A tiny deposit of radium or polonium is inserted in the end of a thin-walled glass tube as indicated. When the rubber bulb is squeezed to compress the air in the top, and then released to cause an expansion,

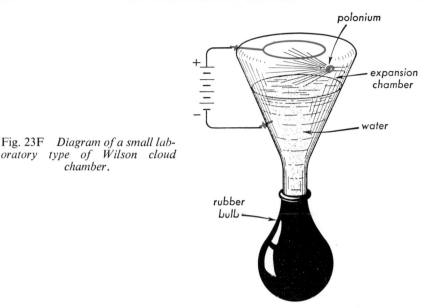

Fig. 23F   *Diagram of a small laboratory   type   of   Wilson   cloud   chamber.*

fogdrops will form on the ions created by the α particles. The battery and the wires leading to the wire ring in the top of the chamber and the water below are for the purpose of quickly removing ions previously formed in the chamber. This clears the field of view for newly formed tracks.

Another type of cloud chamber, one that is readily made in any laboratory workshop, is shown in Fig. 23G. This is the so-called *diffusion cloud chamber*, a

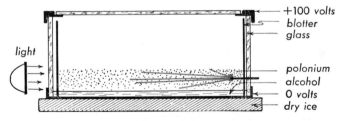

Fig. 23G   *Cross-section diagram of a diffusion cloud chamber.*

device that is continuously sensitive to track formation. A glass cylinder separates and insulates a shallow metal pan below from a metal ring and glass disk above. The bottom pan contains alcohol and rests on a slab of Dry Ice (solid CO_2).

A blotter, extending 80% of the way around the walls, rests with its lower edge in the alcohol. Evaporated alcohol around the warm upper edge of the blotter mixes with the air and slowly settles as it cools. In the dotted region, alcohol vapor is saturated and small droplets will form on any ions present.

Alpha particles shooting out through this space create positive and negative ions, and hence tracks are observed.

A potential difference of 100 volts or so between the bottom pan and the top ring will clear the field of ions so that newly formed tracks are not masked by a dense fog of droplets. A strong source of light shining through the 20% open space through the blotter wall illuminates the tracks, thus making them visible.

23.8. Range

The range of an α particle is defined as the distance such a particle will travel through dry air at normal atmospheric pressure. In a partial vacuum where there are fewer air molecules per centimeter to bump into, the distance traveled before coming to rest will be greater, whereas in air under higher than normal atmospheric pressure there are more molecules per centimeter and the distance will be diminished. Experiments show that some radioactive elements eject α particles with a higher speed than others. The higher the initial speed, the greater is the range. The range of the α particles from *radium* is 3.39 cm, whereas the range of those from *thorium C'* is 8.62 cm.

The ranges of α particles in general have been determined in three different ways: (1) by the Wilson cloud chamber, (2) by the number of ions produced along the path, and (3) by scintillations produced on a fluorescent screen.

In the Wilson cloud chamber photograph of Fig. 23H, α particles of two

Fig. 23H *Wilson cloud-chamber tracks from thorium C and C'. (After Rutherford, Chadwick, and Ellis.)*

(Courtesy of Cambridge University Press.

different ranges are observed. The radioactive sample used to obtain this picture was a mixture of *thorium C* and *thorium C'*. The shorter tracks with a 4.79 cm range are due to the α particles from *thorium C* which disintegrate to become *thorium C''*, and the longer tracks of 8.62 cm range are due to the

α particles from *thorium C'* which disintegrate to become lead (see the last chart in Appendix III).

When one measures the number of ions produced along the path of an α particle, curves similar to those in Fig. 23I are obtained. At the end of each

Fig. 23I   *Graphs of the relative number of ions produced at various points along the path of an α particle (a) from thorium C, and (b) from thorium C'.*

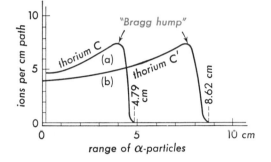

range of α-particles

track the number is seen to reach a maximum and then drop to zero within a very short distance. This maximum on the graph is called the "Bragg hump" in honor of W. H. Bragg, who discovered the phenomenon. The maximum number of ions is therefore produced just before the particles are stopped, i.e., where they are moving at relatively low speeds. The point at which the ion density drops rapidly to zero gives the range as shown in the figure. Although the experimental method by which ion density is usually determined will not be presented here, another method will be. This is the well-known method of counting fogdrops. An enlarged photograph of a cloud track, as reproduced in Fig. 23J, reveals the individual fogdrops separated sufficiently to enable the

Fig. 23J   *Enlarged photograph of a cloud-chamber track showing individual fog drops. (After Brode.)*

number of drops per centimeter of path to be counted. In measurements of this kind, it is assumed that each ion produces one fogdrop.

The third method used in measuring the range is illustrated in Fig. 23K. When each α particle strikes a fluorescent screen, a tiny flash of light is produced. These flashes, called *scintillations*, are observed by means of a microscope. When the sample is moved farther and farther away, by pulling the rod R back, a point is reached where scintillations are no longer observed. The distance d, where the α particles just fail to reach the screen, is a direct measure of the range.

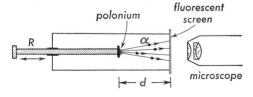

Fig. 23K *Experimental method of measuring the range of α particles.*

23.9. Range of Alpha Particles—Experiment

A simple laboratory experiment for measuring the range of α particles is shown in Fig. 23L. It involves the use of a radioactive source obtained from any

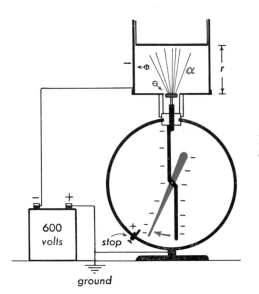

Fig. 23L *An experimental arrangement for measuring the range of α particles.*

physics supply house, a Braun-type electrometer, and an ionization chamber made from ordinary sheet metal. The ionization chamber, about the size of the most common tin can, has a hole in the bottom and a sliding disk plunger for a top. The radioactive source on the Braun electrometer, and a small metal clamp is used to ground the needle to the frame when it swings about 3 cm from its zero position. A 600-volt radio battery of small capacity is a convenient source of high potential.

The plunger distance r is first set at about 8 cm. In this position the needle rises steadily, hits a small metal clamp stop, and quickly drops back to zero. The needle rises again and drops back quickly at a regular rate, due to the continual collection of charge from the ions in the chamber above. With a stop watch the time t required for any given number n of discharges of the needle is recorded. The plunger is then lowered 5 mm and the discharge time again measured. This process is repeated for distances diminishing by 5-mm steps, and the discharge rate determined.

The quantity n/t for each setting of the plunger distance r is a direct measure of the ion current, that is, to the number of ions produced per second. If n/t is plotted against r, a curve similar to that shown in Fig. 23M is obtained. The

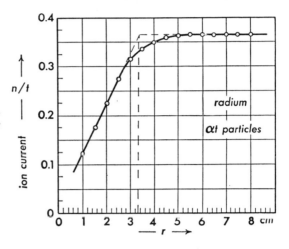

Fig. 23M   *Graph of the ion current from the* α *particle range experiment (see Fig. 23L).*

drop in ion current at distances less than the *range* is due to the fact that α particles hit the plunger before they reach the end of their range in air and are prevented from creating their full quota of ions. The leveling-off of the current at larger values of r signifies that all α particles reach the end of their range in air and create their full ion quota. From the intersection of the dotted lines the α-particle range is found to be approximately 3.3 cm, in good agreement with the accurately known value of 3.39 cm.

The ranges of α particles from some of the natural radioactive elements are given in Chapter 24, Tables 24A and 24B.

QUESTIONS

1. Who discovered radioactivity? What were the circumstances?
2. Who discovered radium, polonium, and thorium?
3. What names are given to the different radioactive rays? Who unraveled this mystery?
4. What are α rays, β rays, and γ rays?
5. What are the relative ionizing powers and penetrating powers of the different rays?
6. What is a Wilson cloud chamber? How does it work?
7. What kinds of particles do not leave tracks in a Wilson cloud chamber? Why not?
8. Define or briefly explain each of the following: (a) scintillations, (b) the Bragg hump, (c) penetrating power, (d) ionizing power, and (e) an ionized atom.
9. What is meant by the range of an α particle?
10. From your knowledge of atomic structure explain how an α particle loses energy as it passes through matter, such as a gas.
11. Draw a graph showing how the specific ionization along the path of an α particle varies with distance. Indicate the range of the particle on the plot.

12. Why does the α particle-range graph of Fig. 23M round off at the top instead of showing a sharp break in the curve?

13. What effect would β and γ rays from a radium source have upon the measured ion current and the α-particle range measurements described in §23.9?

14. What is the purpose of the Dry Ice in the operation of a diffusion cloud chamber? (See Fig. 23G.)

Disintegration and Transmutation

We have seen in the preceding chapter how the heaviest elements in the periodic table spontaneously emit three kinds of radiation, *alpha particles* which are positively charged helium nuclei, *beta particles* which are negatively charged electrons, and *gamma rays* which are electromagnetic waves of extremely high frequency called photons. Such emission is called *radioactivity.*

24.1. Transmutation by Spontaneous Disintegration

A careful study of radioactivity indicates that α, β, and γ rays originate from within the nucleus of the atom and are the result of a nuclear disintegration. When a radium atom disintegrates by ejecting an α particle, the nucleus loses a net positive charge of 2. Since the number of positive charges on the nucleus determines the exact number of electrons outside of the atom, and this in turn determines the chemical nature of an atom, the loss of an α particle, with two positive charges, leaves a new chemical element. Thus a *radium atom*, for example, in disintegrating, changes into a new atom called *radon*. We say that there has been a *transmutation*. Not only does a nucleus lose a double charge by emitting an α particle and thereby *drops down two places in atomic number*, but it also loses a weight of four units and thus *drops down four units in atomic weight*, or four atomic mass units.

It is common practice among physicists to designate all atomic nuclei in an abbreviated form. The nucleus of radium, for example, is written $_{88}Ra^{226}$. The subscript to the left of the chemical symbol gives the *atomic number*, i.e., the number of positive charges on the nucleus, and the superscript on the right gives the atomic *mass number*, or weight.

The disintegration of radioactive nuclei may be written in the form of simple equations, as follows. For radium

$$_{88}Ra^{226} \rightarrow {}_{86}Rn^{222} + \alpha \tag{24a}$$

or

$$_{88}Ra^{226} \rightarrow {}_{86}Rn^{222} + {}_{2}He^{4} \tag{24b}$$

As another example, for polonium

$$_{84}Po^{210} \rightarrow {}_{82}Pb^{206} + {}_2He^4 \tag{24c}$$

When a nucleus like *radium B* disintegrates by ejecting a β particle (an electron) to become *radium C*, the nuclear positive charge *increases by one unit*. Such a transmutation yields a new element one atomic number higher in the chemical table. Since an electron weighs only $\frac{1}{1836}$ part of a hydrogen atom or proton, the charge in mass due to a β particle leaving a nucleus is too small to change the atomic mass number. Although the loss in weight is measurable, it changes the atomic weight so slightly that for most purposes of discussion it can be, and is, neglected. For radium B

$$_{82}RaB^{214} \rightarrow {}_{83}RaC^{214} + {}_{-1}e^0 + \gamma \text{ ray} \tag{24d}$$

In each equation the sum of the subscripts on the right side of the equation is equal to the subscript on the left. The same is true for the superscripts. The designation $_2He^4$ represents the α particle, and $_{-1}e^0$ represents the β particle. In nearly all radioactive disintegrations where a β particle is emitted, one finds a γ ray also. In such cases, as shown by the example in Eq. (24d), *radium B* ejects a β particle and a γ ray to become *radium C*, a nucleus higher in atomic number by unity, but with the same mass number.

A γ ray, like the β ray particle, changes the weight of a nucleus by a negligible amount, and, since it has no charge, it does not alter either the atomic number or the mass number.

24.2. Half-Life

The half-life of a radioactive element is the time required for half of a given quantity of that element to disintegrate into a new element. For example, it takes 1600 years for $\frac{1}{2}$ of a given quantity of radium to change into radon. In another 1600 years, $\frac{1}{2}$ of the remainder will have disintegrated, leaving $\frac{1}{4}$ of the original amount. The half-life of radium is therefore said to be 1600 years.

The rate at which a given quantity of a radioactive element disintegrates, that is, *decays*, is found by observing the activity of a given sample over a period of time and plotting a graph of the type shown in Fig. 24A. Here, for *polonium*,

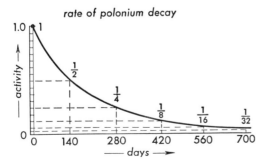

Fig. 24A *Decay curve for the radioactive element polonium. Polonium has a half-life of 140 days.*

the activity drops to $\frac{1}{2}$ of its original value in 140 days. In another 140 days it again drops to half value, etc. The term activity may be defined as the number of rays given off per second of time, or as the number of ionized atoms produced each second by the rays.

The only difference between the decay curve of one element and that of another is the horizontal time scale to which they are plotted. To turn Fig. 24A into a decay curve for radium, the times 140, 280, 420 days, etc., need only be changed to read 1600, 3200, 4800 years etc., respectively. Since, therefore, all radioactive decay curves follow the same law, one does not have to wait for half of a given sample to disintegrate to be able to calculate how long it will be before half will have changed. This would require too many years of waiting for some elements.

A decay curve for the radioactivity of any given substance is best shown on what is called a *semilog graph*. Such a graph involves the use of cross-section paper in which the spacings on the horizontal scale are uniform, while on the vertical scale they are proportional to the logarithms of numbers. Such cross-section paper can be purchased, or made by using the B or C scale on the slipstick of a slide rule.

In Fig. 24B, the fractions $\frac{1}{1}$, $\frac{1}{2}$, $\frac{1}{4}$, $\frac{1}{8}$, $\frac{1}{16}$, etc., are plotted on semilog graph

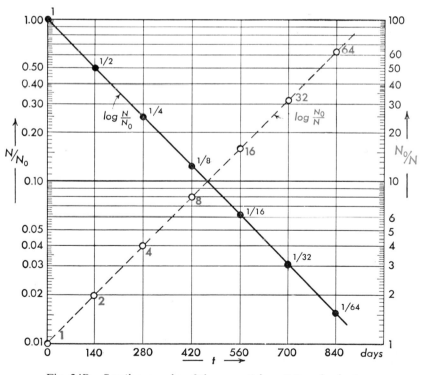

Fig. 24B   *Semilog graphs of the α particle activity of polonium.*

paper with the polonium *time scale* plotted horizontally. Note that a straight line is the result. If a semilog graph is plotted for the activity of any other radioactive element, it too will be a straight line, but with a different slope.

Suppose that for a given sample of radioactive material the initial number of particles given off per minute is N_0. After a time t, the number of atoms yet to disintegrate will have decreased and the reduced number of particles given off per minute will be N. Hence, the ratio N/N_0 will decrease with time t. If the logarithm of the inverse ratio N_0/N is plotted against t, as shown by the dotted line in Fig. 24B, it too will produce a straight line. The latter shows that the log N_0/N is proportional to t. We can therefore write

$$\log_e \frac{N_0}{N} = \lambda t \qquad (24e)$$

where λ is the proportionality constant and is called the *decay constant*.

To find the half-life of an element from the semilog graph we find the point where $N = \frac{1}{2}N_0$, and call the corresponding time t, the *half-life T*. For this graph point, Eq. (24e) gives

$$\log_e \frac{N_0}{\frac{1}{2}N_0} = \lambda T \qquad (24f)$$

or

$$\log_e 2 = \lambda T \qquad (24g)$$

Looking up the Naperian logarithm of 2, we find 0.693. Hence

$$0.693 = \lambda T \qquad (24h)$$

or

$$\boxed{T = \frac{0.693}{\lambda}} \qquad (24i)$$

This is the relation between the decay constant λ and the half-life T of any element.

Example 1. A count-rate meter is used to measure the activity of a given sample. At one instant the meter shows 4750 counts per minute (abbr. cpm). Five minutes later it shows 2700 cpm. Find (a) the decay constant, and (b) the half-life.

Solution. To find the decay constant λ, use Eq. (24e). We first find

$$\frac{N_0}{N} = \frac{4750}{2700} = 1.760$$

To find the logarithm to the base e of any number we can look up the logarithm to the base 10 of the number, and multiply by 2.3026. From a table of ordinary logarithms, we find

$$\log_{10} 1.760 = 0.2455$$

Multiplying this by 2.3026, we obtain

$$\log_e 1.760 = 0.5653$$

Using Eq. (24e), we obtain

(a) $$\lambda = \frac{0.5653}{5\,\text{min}} = 0.1131\ \text{min}^{-1}$$

Substituting in Eq. (24i), we obtain

(b) $$T = \frac{0.693}{0.1131} = 6.1\ \text{min}$$

The mean life is defined as the average time an atom or particle exists in a particular form, and is given by the reciprocal of the decay constant.

$$\tau = \frac{1}{\lambda}$$

As shown by Eq. (24i) the half-life T is just slightly over $\frac{2}{3}$ the mean life τ:

$$T = 0.693\tau$$

24.3. Radioactive Series

It was Rutherford and his colleagues who discovered that when one radio-active atom disintegrates by ejecting an α or β particle, the remaining atom is still radioactive and may sooner or later eject another particle to become a still different atom. This process they found to continue through a series of elements, ending up finally with a type of atom that is stable and not radioactive. It is now known that nearly all natural disintegration processes, occurring among the heaviest elements of the periodic table, finally end up with *stable lead atoms*, atomic number 82.

There are at least four known radioactive series or chains of elements, one starting with uranium-238, a second with thorium-232, a third with uranium-235, and a fourth with plutonium-241. The first and fourth of these series are given in Tables 24A and 24B. All four series are given in a graphical tabulation in Appendix III.

When a uranium atom of mass number 238 and atomic number 92 disintegrates by ejecting an α particle, the remainder is a new atom, *uranium X_1*, of mass number 234 and atomic number 90. When a uranium X_1 atom disintegrates by ejecting a β particle to become *uranium X_2*, the mass number remains unchanged at 234, while the atomic number increases by 1 to become 91. This increase of 1 positive charge is attributed to the loss of 1 negative charge. These processes of successive disintegration continue until *lead*, a *stable* atom, is the end result.

As explained in §5.4, all atoms with the same atomic number but different mass number are called isotopes of the same element. For example, $_{82}\text{RaB}^{214}$, $_{82}\text{ThB}^{212}$, $_{82}\text{AcB}^{211}$, $_{82}\text{RaD}^{210}$, $_{82}\text{Pb}^{209}$, $_{82}\text{Pb}^{208}$, $_{82}\text{Pb}^{207}$, and $_{82}\text{Pb}^{206}$ are isotopes of the same chemical element, lead (see Appendix III). Even though the first five of these are radioactive, i.e., unstable, the other three are stable. Chemically they behave exactly alike and are separated only with difficulty. The isotopes 214, 210, and 206 belong to the uranium-238 series; 212 and 208 belong to the

Table 24A. Uranium-238 Series

ELEMENT	SYMBOL	ATOMIC NUMBER	MASS NUMBER	PARTICLE EJECTED	RANGE IN AIR	HALF-LIFE
Uranium I.......	UI	92	238	α	2.70 cm	2×10^9 yr
Uranium X$_1$......	UX$_1$	90	234	β	....	24.5 days
Uranium X$_2$......	UX$_2$	91	234	β	....	1.14 min
Uranium II......	UII	92	234	α	3.28	3×10^5 yr
Ionium..........	Io	90	230	α	3.19	83,000 yr
Radium.........	Ra	88	226	α	3.39	1600 yr
Radon..........	Rn	86	222	α	4.12	3.82 days
Radium A.......	RaA	84	218	α	4.72	3.05 min
Radium B.......	RaB	82	214	β	....	26.8 min
Radium C.......	RaC	83	214	α, β	....	19.7 min
Radium C'.......	RaC'	84	214	α	6.97	10^{-6} sec
Radium C''......	RaC''	81	210	β	....	1.32 min
Radium D.......	RaD	82	210	β	....	22 yr
Radium E.......	RaE	83	210	β	....	5 days
Polonium........	Po	84	210	α	3.92	140 days
Lead............	Pb	82	206	stable	....	infinite

thorium series; 211 and 207 belong to the uranium-235 series; and 209 belongs to the neptunium series.

24.4. Daughter Products

When an element disintegrates by emitting α or β rays, it produces a new chemical element. Such "offspring" atoms are referred to as the *daughter element*, an element which itself may or may not be radioactive. Consider as an

Table 24B. Neptunium Series†

ELEMENT	SYMBOL	ATOMIC NUMBER	MASS NUMBER	PARTICLE EJECTED	RANGE IN AIR	HALF-LIFE
Plutonium....	Pu	94	241	β	..	...
Americium....	Am	95	241	α	4.1*	500 yr
Neptunium...	Np	93	237	α	3.3*	2.25×10^6 yr
Protoactinium	Pa	91	233	β	..	27.4 days
Uranium.....	U	92	233	α	3.3*	1.63×10^5 yr*
Thorium.....	Th	90	229	α	3.3	7×10^3 yr
Radium......	Ra	88	225	β	..	14.8 days
Actinium.....	Ac	89	225	α	4.4	10 days
Francium.....	Fa	87	221	α	5.0	4.8 min
Astatine.....	At	85	217	α	5.8	0.018 sec
Bismuth......	Bi	83	213	$\beta(94\%)$ $\alpha(4\%)$	4.6	47 min
Polonium.....	Po	84	213	α	7.7	10^{-6} sec
Lead........	Pb	82	209	β	..	3.3 hr
Bismuth......	Bi	83	209	stable	..	infinite

* See *Radioactivity and Nuclear Physics* by J. M. Cork, D. Van Nostrand, Princeton.
† See F. Hagemann, L. I. Katzin, M. H. Studier, A. Ghiorso, and G. T. Seaborg. *Physical Review*, vol. 72, 252, August, 1947.

example the radioactive element radon, $Z = 86$. Of the several known isotopes of this element (see Appendix III), isotope 220 is a derivative of the thorium series beginning with $_{90}Th^{232}$.

Thoron-220 is a gas and is α-active, that is, it disintegrates by giving off α particles. The daughter product, polonium-216, called ThA, is α-active, and has an extremely short half-life of 0.158 sec. The daughter product of this element, however, has the relatively long half-life of 10.6 hours. The reactions involved here are as follows:

$$_{86}Tn^{220} \rightarrow {_{84}}ThA^{216} + {_2}He^4 \qquad (T = 54.5 \text{ sec})$$
$$_{84}ThA^{216} \rightarrow {_{82}}ThB^{212} + {_2}He^4 \qquad (T = 0.158 \text{ sec})$$
$$_{82}ThB^{212} \rightarrow {_{83}}ThC^{212} + {_{-1}}e^0 \qquad (T = 10.6 \text{ hr})$$

If a given quantity of thoron gas is confined to a closed vessel, the Tn nuclei that disintegrate will tend to accumulate as ThA. Not much of this daughter product can accumulate, however, since with the short half-life of 0.158 sec, most of the nuclei quickly disintegrate into ThB. In the period of ten to fifteen minutes, few of these ThB disintegrate, and they do accumulate. A graph of this decrease of parent element Tn and the accumulation of daughter element ThA and its daughter element ThB is shown in Fig. 24C.

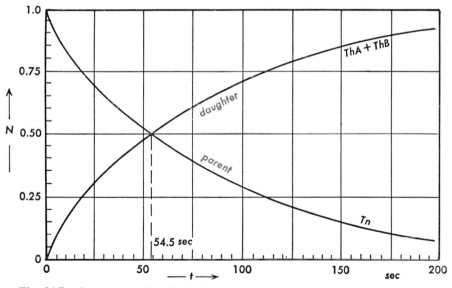

Fig. 24C   *Decay curve for $_{86}Tn^{220}$ (thoron) and the growth curve for the daughter products, $_8ThA^{216} + {_{82}}ThB^{212}$.*

24.5. Half-Life Experiment

A relatively simple experiment for measuring the half-life of a radioactive element is shown in Fig. 24D. Thoron gas $_{86}Tn^{220}$ from a small glass vessel A, injected into an ionization chamber C, is detected by the ion current I as it flows through the electrometer E into the ground G.

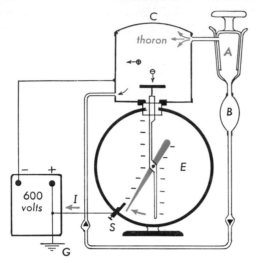

Fig. 24D   *Simple apparatus for determining the half-life of a radio-active gas like thoron, $_{86}Tn^{220}$.*

Immediately upon squeezing the rubber bulb *B* and injecting the thoron, the electroscope needle rises and falls, discharging intermittently through the stop *S*. The time *t* is recorded from a seconds clock each time a discharge occurs, and these are recorded against the discharge number *n*. A typical set of recorded

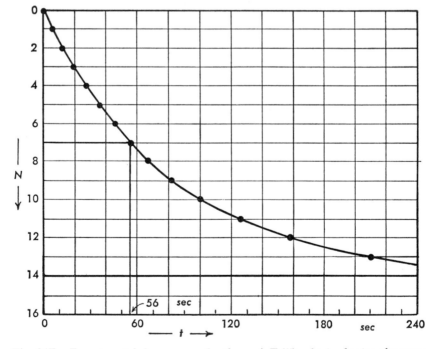

Fig. 24E   *Experimental decay curve for thoron ($_{86}Tn^{220}$), obtained using the apparatus shown in Fig. 24D.*

times are plotted in Fig. 24E. The leveling-out of the graph after some time signifies that little thoron gas is left, and a reasonable horizontal line is drawn at $n = 14$ as the last discharge that could be obtained. Coming down, therefore, to the point $N = 7$, where the activity was at half value, the time $t = 56$ sec is read from the graph as the half-life of $_{86}Tn^{220}$.

The decay curve of Fig. 24E can be plotted as a semilog graph by plotting the difference between the line $N = 14$ and the curve value at each 10-second interval, and plotting these as shown in Fig. 24F. When the best possible straight line is drawn through the points, the half-value point comes at $t = 56$ sec.

Mathematically, this value can be computed from the graph and Eq. (24e) as follows. At time $t = 0$ we obtain $N_0 = 14$. Choosing any other point far down on the graph, such as the one at $t = 150$ sec, we find $N = 2.2$. Dividing one value by the other gives $N_0/N = 6.36$. Looking up this number in a table of

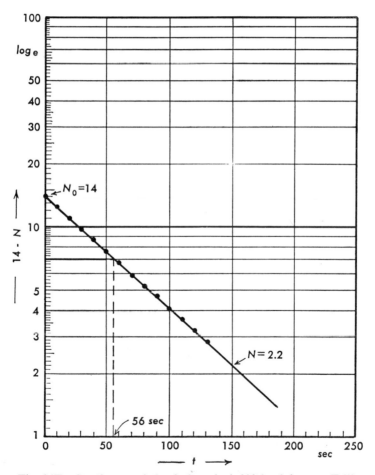

Fig. 24F   *Semilog graph for finding the half-life of thoron, $_{86}Tn^{220}$.*

Naperian logarithms, or using the method employed in Example 1, we find $\log_e 6.36 = 18.50$. Substitution of this value in Eq. (24e), along with $t = 150$ sec, gives as the value for the decay constant:

$$\lambda = \frac{1.850}{150} = 0.0123$$

When this is substituted in Eq. (24i), we obtain

$$T = \frac{0.693}{0.0123} = 56.3 \text{ sec}$$

as the half-life.

24.6. Different Forms of Energy

Einstein, in working out the theory of relativity, arrived at a number of simple equations concerning the nature of the physical world. One of these equations, having to do with the increase in mass of a moving object, was presented in §19.5. It is important at this point to consider this mass-energy equation again:

$$E = mc^2 \tag{24j}$$

where m is the mass, c is the velocity of light, and E is the energy equivalence.

From this relation we can predict that mass can be turned into energy, or energy into mass. In other words, mass is a form of energy, for if a quantity of mass m could be annihilated, a definite amount of energy E would become available in some other form. To illustrate this, suppose that a 1-gm mass could be completely annihilated and the liberated energy given to some other body in the form of kinetic energy:

$$E = 1 \times 10^{-3} \text{ Kg} \times 9 \times 10^{16} \text{ m}^2/\text{sec}^2$$
$$= 9 \times 10^{13} \text{ joules}$$

In the English system of units this is equivalent to 7×10^{13} ft-lb, or enough energy to propel the largest battleship around the world.

The annihilation of mass then is a source of undreamed-of energy. In the following chapters we will see that disintegration is one means whereby mass can be annihilated or created through planned laboratory experiments.

In the following chapters we will see that if an atom, a part of an atom, or an electron is annihilated, the energy may either be transformed into kinetic energy and given to another atomic particle in the form of a velocity, or it may appear as a γ ray of specified frequency v and energy hv. To find the equivalence between mass energy, γ-ray energy, and kinetic energy, all of the following quantities are equated to each other:

$$\boxed{E = mc^2 = hv = \tfrac{1}{2}m_0v^2 = Ve} \tag{24k}$$

It is customary among physicists to express each of these energies in terms of V in volts. Thus one speaks of a one-million volt γ ray, a three-million volt electron, or a 12.5-million volt proton, etc. This terminology is used for convenience only, and denotes the value of V in the above equation which, with the electronic charge substituted for e, gives the energy of the γ-ray photon, or the energy of the moving atomic particle. For all energies, m in the second term is the relativistic mass given by Eq. (19j), and the fourth term is the kinetic energy of any particle whose rest mass is m_0. At very high velocities the fourth term must be replaced by the relativistic kinetic energy $m_0 c^2 (\gamma - 1)$. See Eq. (19q).

When an atomic nucleus ejects an α or β particle, the mass of that nucleus diminishes not only by the rest mass of the ejected particle, but by an amount called the *annihilation energy*. A small part of the nuclear mass is annihilated and given to the ejected particle as kinetic energy. While the annihilated mass varies from isotope to isotope, it is usually less than 1 % of one atomic mass unit.

As a convenient figure to use in disintegration problems, we will calculate the energy equivalent to the annihilation of unit atomic mass, namely, 1.66×10^{-27} Kg. Using the second and last terms of Eq. (24k), we find

$$\boxed{Ve = mc^2} \tag{24l}$$

$$V = \frac{1.66 \times 10^{-27} \times 9 \times 10^{16}}{1.60 \times 10^{-19}}$$

$$= 931,000,000 \text{ volts}$$

The annihilation energy of 1 amu = 931 million electron volts. Abbreviated,

$$1 \text{ amu} - 931 \text{ Mev} \tag{24m}$$

When a γ ray is ejected from a nucleus, it carries with it an energy $h\nu$. This energy may be expressed by (a), the mass equivalence as given by the second and third terms of Eq. (24k), or by the voltage equivalence as given by the third and fifth terms. It is customary to use the latter relation and write

$$\boxed{Ve = h\nu} \tag{24n}$$

Example 2. A γ ray has a wavelength of 4.5×10^{-13} m. Find its energy equivalence in million electron volts.

Solution. Using the wave equation $c = \nu\lambda$, and substituting c/λ for ν in Eq. (24n), we obtain

$$V = \frac{hc}{e\lambda}$$

Substituting, we find that

$$V = \frac{6.62 \times 10^{-34} \times 3 \times 10^8}{1.60 \times 10^{-19} \times 4.5 \times 10^{-13}}$$

$$= 2.75 \times 10^6$$

$$V = 2.75 \text{ Mev}$$

It is common practice to express the initial energy of α particles by specifying either their range in air or their energy equivalence in Mev. This applies to

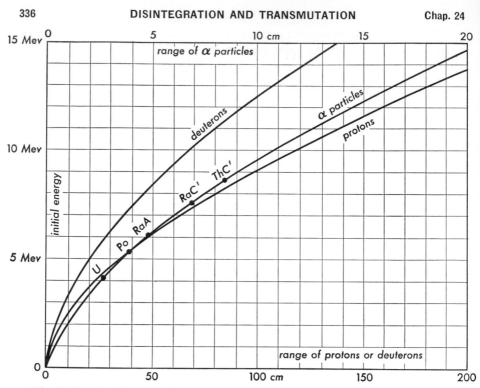

Fig. 24G *Graphs giving the range of protons, deuterons, and α particles for different initial energies in million electron volts.*

other high-speed particles as well, particles such as protons and deuterons which are the nuclei of the two hydrogen isotopes. A graph like the one shown in Fig. 24G is useful for converting the observed range of an α particle to its equivalent energy in Mev, or vice versa.

QUESTIONS AND PROBLEMS

1. What is spontaneous disintegration?

2. What is transmutation? What change takes place in the nucleus of an atom when an α particle is emitted? What change takes place when a β particle is emitted?

3. What is meant by the half-life of a radioactive isotope?

4. What is meant by the range of radioactive rays? Which of the three kinds of rays should have the greatest range?

5. What is the abbreviated designation for atomic nuclei?

6. Write down the nuclear reaction, in the abbreviated form, for the emission of an α particle by an ionium nucleus (atomic number 90). See Appendix III.

7. Write down the nuclear reaction for the emission of an α particle by radon-222 (atomic number 86). See Appendix III.

8. Write down the nuclear reaction for α particle emission by uranium-238. See Appendix III.

9. Write the nuclear reaction for the emission of a β particle by actinium-227. See Appendix III.

10. Write the nuclear reaction for the emission of a β particle by radium B. See Appendix III.

11. If the activity of a radioactive sample drops to $\frac{1}{16}$ of its initial value in 1 hr and 20 min, what is its half-life?

12. If the activity of a radioactive sample drops to $\frac{1}{32}$ of its initial value in 7.5 hr, find its half-life. *Ans.* 1.5 hr.

13. How long will it take a sample of radon to decrease to 5% if its half-life is 3.82 days? Find your answer by plotting a decay curve.

14. How long will it take a sample of radium D to decrease to 10% if its half-life is 22 years? *Ans.* 73.1 yr.

15. If the activity of a radioactive sample drops to $\frac{1}{32}$ of its initial value in 6 hr and 45 min, find its half-life.

16. The activity of a radioactive sample drops to $\frac{1}{32}$ of its initial value in 54 hr. Find its half-life. *Ans.* 10.8 hr.

17. How long will it take for the activity of a sample of radon-222 to decrease to 5%? Determine your answer by graphing the time decay curve.

18. How long will it take for the activity of a sample of francium-221 to decrease to $\frac{1}{10}$ of 1%? *Ans.* 8.0 hr.

19. What radioactive atoms are isotopes of bismuth? (See Appendix III.) Give symbols, atomic numbers, and mass numbers.

20. If 5 lb of lead could be completely annihilated, how many joules would be produced? *Ans.* 20.4×10^{16} joules.

21. Calculate the mass equivalent to an energy of (1) 18 Mev, (b) 50 Mev, and (c) 2.5 Mev.

22. If the following masses could be annihilated, how much energy, in Mev, is created? (a) 0.0216 atomic mass unit, (b) 0.0589 atomic mass unit. *Ans.* (a) 20.1, (b) 54.8.

23. At one instant a count-rate meter shows a radioactive sample emitting 2000 counts per minute. Ten minutes later the activity has dropped to 1520 counts per min. Find (a) the decay constant, and (b) the half-life.

24. In recording the activity of a given radioactive sample, a count-rate meter indicates 6520 counts per minute. Two minutes later the meter shows the activity has dropped to 4840 counts per min. Calculate (a) the decay constant, and (b) the half-life. *Ans.* (a) 0.149 min^{-1}, (b) 4.65 min.

Beta and Gamma Rays

Beta rays emitted by the natural radioactive elements have been studied by many people. It was Becquerel who first found them to be comparable to cathode rays. Today we know them as electrons. One of the greatest mysteries of atomic physics has been the origin of these particles. The great mass of experimental evidence shows that most of them come from the nucleus, yet they do not exist in the same form inside any of the known nuclei. Furthermore, unlike α particles and γ rays, they do not emerge from the nuclei with the same energy and range, but with a wide band of velocities.

25.1. Beta Ray Spectrograph

This is an instrument by which one can experimentally determine the velocity of β particles from a radioactive source. Figure 25A is a diagram of one such

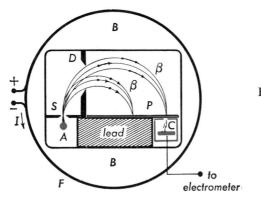

Fig. 25A *Diagram of a β-ray spectrograph.*

instrument, developed in principle by Robinson in England. Beta particles from a small radioactive source A, consisting of a fine wire with a deposit of material on its surface, are allowed to pass through a slit S. With the entire vacuum tube located in a uniform magnetic field, the β particles follow circular paths and are brought to focus on a photographic plate P or at the open slit of an ionization chamber C. Each of the three semicircular paths in any one group shown in the diagram has the same radius and diameter. The focusing action indicated can be demonstrated by using a compass and drawing several semicircles with the same radius, but slightly displaced centers. The velocity of any group of electrons is given by Eq. (4e), as

$$Bev = \frac{mv^2}{r} \qquad (25a)$$

where r is the path radius, B the magnetic induction, and e, m, and v the electron's charge, mass, and velocity, respectively.

If the field B is constant, and a photographic plate P is located as shown in the figure, its development after a time t will result in a photograph like the drawing in Fig. 25B. In addition to a darkened background all along the plate,

Fig. 25B   *Diagram of a β-ray spectrogram.*

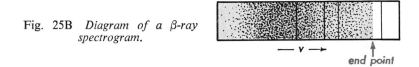

one finds several lines parallel to the slit. The background indicates β particles present with all different velocities, while the lines signify groups with discrete and definite velocities. Not all spectrograms with such a continuous background contain these lines, however.

In some β-ray spectrographs, the β particles are collected in an ionization chamber C as shown in Fig. 25A. Different velocities are determined by slowly and continuously changing the magnetic induction B. This is accomplished by changing the current I in the field coils F. If measurements of electrometer current are recorded for different values of B, a graph similar to that shown in Fig. 25C can be plotted.

The measured ion current is plotted as the number of β particles, and the coil current I is plotted as the energy the particles would have to enter the slit at C. This curve, made using RaE as a source, which is the radioactive bismuth isotope $_{83}Bi^{210}$, shows β particles with a maximum energy at 0.15 Mev, and an *end-point* energy of 1.17 Mev. The end point, or cut-off, at the high-energy end of this curve, signifies a velocity limit or maximum. As shown by Table 25A, such end-point energies have different values for different isotopes.

25.2. The Neutrino Postulate

In 1931 Pauli, of Germany, suggested that in β-ray emission all nuclei of the same isotope emit the same amount of energy, and that this is the *end-point energy* shown in Fig. 25C (see also Table 25A). To account for the observed fact

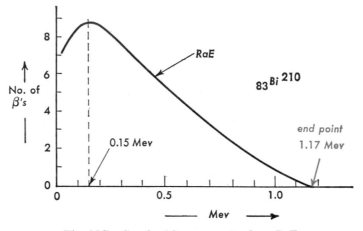

Fig. 25C   *Graph of β-ray energies from RaE.*

that some β particles emerge with much less energy than others, he made the following postulate. The emission of a β particle by any nucleus is accompanied by a companion particle having a variable energy E. While the β particle is an

Table 25A. End-Point Energies for β Rays

ISOTOPE	DESIGNATION	END POINT E	HALF-LIFE
$_{82}Pb^{214}$	Ra B	0.72 Mev	26.8 m
$_{83}Bi^{210}$	Ra E	1.17 Mev	4.8 d
$_{87}Fr^{223}$	Ac K	1.20 Mev	21.0 m
$_{88}Ra^{225}$	—	0.320 Mev	15 d
$_{89}Ac^{228}$	Ms Th$_2$	1.55 Mev	6.1 h
$_{90}Th^{231}$	UY	0.21 Mev	25.6 y
$_{90}Th^{234}$	UX$_1$	0.193 Mev	24 d

ordinary negatively charged electron, the companion particle, now called a *neutrino*, has no charge. In some respects a neutrino is like a photon; it has no rest mass, it has energy E, and it travels with the speed of light c. Hence, the reaction for β emission by a nucleus like $_{83}Bi^{210}$, can be written

$$_{83}Bi^{210} \rightarrow {}_{84}Po^{210} + \beta^- + \nu \tag{25b}$$

where ν represents the neutrino.

The neutrino postulate has served another important function in atomic processes; it permits the retention of the law of conservation of momentum. All even-numbered nuclei, *even A*, are known to have an angular momentum given by an integral quantum number, $I = 0, 1, 2, 3, 4, \ldots$, all odd-numbered nuclei, odd-A, are known to have an angular momentum given by a half-integral quantum number, $I = \frac{1}{2}, \frac{3}{5}, \frac{5}{2}, \frac{7}{2}, \ldots$. Furthermore, all electrons are known to have a spin angular momentum given by a half-integral quantum number, $s = \frac{1}{2}$. By assigning the neutrino a spin angular momentum equal in magnitude to that of the electron, the two spins can either cancel each other as shown in Fig. 25D, or add together, thus keeping the nuclear spin integral or half-integral valued as the case may be.

Fig. 25D *Each β particle from a nucleus is accompanied by a neutrino.*

nucleus

Only recently have experiments been performed that appear to establish the existence of these phantom particles, neutrinos. Since nuclei are composed of neutrons and protons, the two ejected particles are created by the nucleus, a neutron transforming into a proton as it emits an electron of mass m_0, charge $-e$, and spin $\frac{1}{2}\hbar$, and a neutrino of energy E, no charge, and spin $\frac{1}{2}\hbar$. See Eq. (12h).

25.3. Conservation of Nuclear Energy

The simultaneous emission of a β particle and a neutrino from the nucleus of an atom requires energy. The disintegrating nucleus gives up not only the mass of the two particles but some additional mass which it converts into kinetic energy. By adding up all this energy in the form of mass, the total loss in the mass of an atom can be calculated.

Consider as an example the β emission of $_{83}Bi^{210}$ as represented by Eq. (25b). The masses of the two atoms involved are known to be

$$
\begin{array}{ll}
_{83}Bi^{210} & 209.984110 \text{ amu} \\
_{84}Po^{210} & 209.982866 \text{ amu} \\
\hline
\Delta m = & 0.001244 \text{ amu}
\end{array}
$$

To convert this mass into Mev, we multiply Δm by the value 931 Mev given by Eq. (24m), which gives

$$E = 1.17 \text{ Mev}$$

This is just the end-point energy derived from experiment and shown in Fig. 25C. The mass of the ejected electron need not be included here, since the two masses above are for the neutral atoms. $_{83}Bi^{210}$ has 83 orbital electrons included in its mass of 209.984110 amu, while $_{84}Po^{210}$ includes 84 electrons. When $_{83}Bi^{210}$ ejects an electron from the nucleus, the daughter product $_{84}Po^{210}$ picks up a stray electron to become a neutral atom.

25.4. Gamma Rays

For many radioactive elements the emission of an α or β particle from a nucleus is immediately followed by the emission of a γ ray. It has been shown by crystal diffraction spectrographs that γ rays are electromagnetic waves and that they consist of sharp lines of discrete wavelengths. Just as visible light, and ultraviolet and infrared radiation as well as X rays are known to be emitted from the outer structure of the atom by an electron's jumping from one energy level to another, so γ rays are believed to arise from a transition of a nucleon from one energy state to another within the nucleus. (Neutrons and protons as constituents of nuclei are called *nucleons.*)

There is good evidence that γ rays are emitted by the daughter element, that is, that they are preceded by particle emission. A good example is to be found in the case of RaD. This nucleus is the lead isotope $_{82}Pb^{210}$, and with a half-life of 22 years it emits a β particle. The reaction is

$$_{82}Pb^{210} \rightarrow _{83}Bi^{210} + \beta^- + \nu + \gamma \ \text{ray} \qquad (25c)$$

These emissions are represented on a nuclear energy level diagram in Fig. 25E.

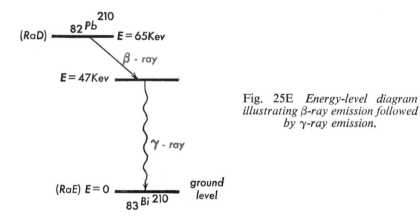

Fig. 25E *Energy-level diagram illustrating β-ray emission followed by γ-ray emission.*

When a $_{82}Pb^{210}$ nucleus emits a β particle and a neutrino, the end-point energy is found to be 18,000 electron volts (abbr. 18 Kev). This leaves a $_{83}Bi^{210}$ nucleus in what is called an *excited state* or energy level. A transition down to the ground level is accompanied by the emission of a γ ray with an energy $h\nu$. This energy is equivalent to 47 Kev. Although it is not shown in this diagram, this nucleus, too, is radioactive and emits another β ray to become $_{84}Po^{210}$. See Eq. (25b).

25.5. Internal Conversion

When a γ ray is emitted by a disintegrating nucleus, it must of necessity pass some of the outer electrons on its way out of the atom. Upon passing close to one of the electrons, a photoelectric action may occur whereby the γ ray is absorbed and an electron is emitted. Energetically this process must follow the photoelectric Eq. (10a),

$$h\nu = W + \tfrac{1}{2}mv^2 \qquad (25d)$$

Part of the γ-ray energy $h\nu$ is used to pull the electron away from the atom, and the remainder is imparted to it as kinetic energy. As a rule this kinetic energy is relatively large, and $\frac{1}{2}mv^2$ must be replaced by the relativistic form $m_0c^2(\gamma - 1)$. See Eq. (19q).

Since the energy required to remove an electron from an atom will differ from one shell to another, the electrons can be expected to have any one of several discrete energies. The definite sharp lines shown in Fig. 25B supply the evidence for this assumption. Furthermore, the experimentally determined energies check exactly with those determined from X rays as shown in Fig. 22A. The process just described is illustrated schematically in Fig. 25F, and is called internal conversion (abbr. IC).

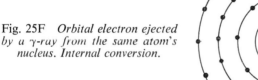

Fig. 25F *Orbital electron ejected by a γ-ray from the same atom's nucleus. Internal conversion.*

If by internal conversion an electron is ejected from the innermost K shell, leaving a vacancy there, another electron from the L shell or M shell falls in to take its place and, in so doing, emits an X ray. Confirmation of such a process is assured since the measured wavelengths of such X rays are identical with those emitted by the same element in an X-ray tube.

25.6. Scintillation Counters

A scintillation counter is a sensitive device used in nuclear physics studies for the detection and measurement of high-energy atomic radiation. In principle it is based upon the earliest discoveries in radioactivity that α particles upon striking a fluorescent material, like zinc sulfide, produce a tiny flash of light. See Fig. 23K. These flashes, called *scintillations*, can be seen by the dark adapted eye, or they can be detected by a photomultiplier tube and amplified.

It is now well known that when high-energy-charged atomic particles pass through certain transparent materials, fluorescent light is produced all along the path. See Fig. 25G. As the fast-moving particle collides with atoms and molecules, electrons are raised to excited energy levels and in returning to their ground states emit light. For many crystals and plastics this *fluorescent light* is blue or violet in color, while for others it is ultraviolet or infrared.

A typical scintillation counter tube is shown in Fig. 25H. A block of fluorescent material is mounted on the flat end of a special photomultiplier tube and

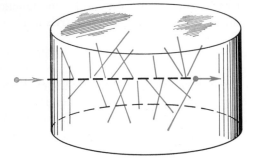

Fig. 25G *Diagram of the fluo-rescent light developed by an atomic particle traversing a trans-parent crystal or plastic fluor.*

then encased in a thin-walled, light-tight aluminum shield. When a particle traverses the fluor, the light ejects electrons from the photocathode by the photoelectric effect. The charge multiplication built up by the eight or more dynodes makes a sizable voltage pulse that activates some electronic counting device. See Fig. 25I.

When γ rays are to be detected, the fluorescent materials frequently used are crystals of sodium iodide, NaI, and caesium iodide, CsI. For high-energy β rays, such plastics as polystyrene, impregnated with anthracene, are used. These are inexpensive and very effective and can be quite large in size. For α particles with their relatively low penetrating power, a thin layer of zinc sulfide deposited on the photomultiplier tube or a plastic surface is commonly used.

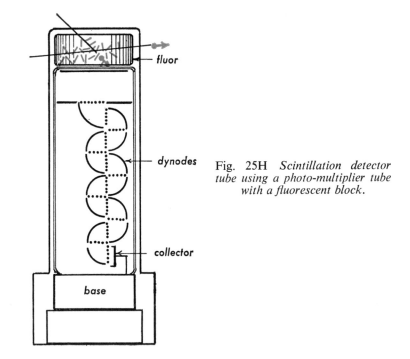

Fig. 25H *Scintillation detector tube using a photo-multiplier tube with a fluorescent block.*

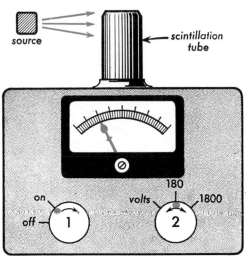

Fig. 25I *One type of scintillation counter with a count-rate meter, two ranges, and an on-off switch.*

The principal advantages of scintillation counters over other detectors of nuclear radiation are these. (1) They operate in air or in a vacuum; (2) they deliver an electrical impulse which is proportioned to the energy lost by the traversing particle; and (3) they can count at amazingly high speeds.

While the duration of a single pulse from a NaI crystal counter will last about one microsecond, the pulse time from an anthracene counter can be as short as one thousandth of a microsecond (10^{-9} sec).

25.7. Cerenkov Radiation

In 1934 Cerenkov, in Russia, discovered that fast-moving electrons, such as β particles from radioactive materials, will produce light within a transparent medium if their velocity is greater than the speed of light in that medium. In a medium like glass or plastic, the speed of light is about $\frac{2}{3}$ to $\frac{3}{4}$ the speed of light in a vacuum, yet many β particles are ejected at speeds greater than this and some of them close to $c = 3 \times 10^8$ m/sec.

The phenomenon of Cerenkov radiation is analogous to (a) the production of the V-shaped wave from a ship when the ship travels through water at a speed greater than the wave velocity or (b) the shock waves from a missile traveling through the air at a speed greater than sound.

In Fig. 25J a particle is shown generating a conical wave as it travels with a velocity V through a medium of refractive index μ. While the conical wave front makes an angle with the particle's direction, the light travels outward at right angles to the wave front.

The principle of Cerenkov radiation is frequently used to detect high-energy atomic particles. A Cerenkov counter has the same general construction as shown in Fig. 25H, except that the fluor is replaced by a transparent medium,

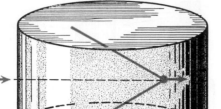

Fig. 25J   *The conical wave from a high-speed atomic particle in a transparent medium: Cerenkov radiation.*

liquid or solid, and with a refractive index specified for the particular particle speeds to be detected.

25.8. Semiconductor Detectors

Recent experiments show that semiconducting materials like those used in transistors are useful detectors of α rays and other heavy atomic particles. One type of semiconductor detector used is shown in Fig. 25K. A thin layer of gold

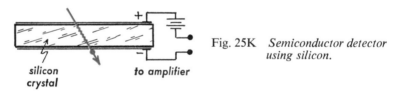

silicon crystal   to amplifier

Fig. 25K   *Semiconductor detector using silicon.*

metal is deposited on the two surfaces of a thin wafer of very pure silicon and the two conducting surfaces are connected to a 1- to 20-volt battery and amplifier. When a charged particle enters the crystal, free electrons and holes develop, increasing its conductivity. The sudden current pulse thereby created is amplified and then measured or counted.

25.9. Radiation Absorption

As high-energy-charged particles travel through matter in the solid, liquid, or gaseous state, their energy is gradually dissipated in a number of different ways, principally by the excitation and ionization of atoms and molecules. Gamma rays on the other hand, have a far greater range because they have no electric charge. Unless their energy is over 1 Mev, their absorption is due to the photoelectric effect or the Compton effect. In either case, ions are formed in the process, and, in relation to those produced by charged particles, are far apart.

A diagram representing the absorption of γ rays is presented in Fig. 25L. If I_0 represents the intensity of the beam as it enters the medium, there will be some depth d_1 at which the intensity will have dropped to $\frac{1}{2}I_0$. Imagine now that we divide the medium into layers, each of thickness d_1. The beam intensity $\frac{1}{2}I_0$

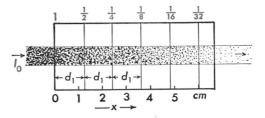

Fig. 25L *Diagram illustrating the absorption of γ rays by matter.*

entering the second layer will be reduced to $\frac{1}{2}$ of this initial value and emerge to enter the third layer with an intensity $\frac{1}{4}I_0$. Upon traversing the third layer, the entering beam $\frac{1}{4}I_0$ will be reduced to half value, or to $\frac{1}{8}I_0$, etc.

If we now plot a graph of the beam intensity I against the number of absorbing layers, or the depth x, we obtain the curve shown in Fig. 25M. While the

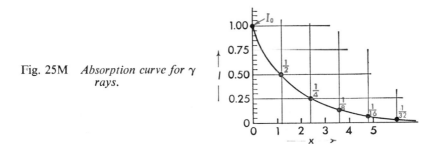

Fig. 25M *Absorption curve for γ rays.*

absorption curves for different γ-ray energies, and different absorbers, are not identical, they are alike in that they follow the same law. Just as in the case of the half-lives of radioactive isotopes (see Fig. 24B), if we plot the absorption on a semilog graph, we obtain a straight line. From this straight line we can write

$$\log_e \frac{I_0}{I} = \mu x \qquad (25e)$$

The proportionality constant μ is often called the *absorption coefficient* and represents the fraction of the beam absorbed from the beam per centimeter path. The calculation of μ can be made from the experimental data by selecting any two values of I from the straight-line section of such a semilog graph.

25.10. Radiation Absorption

If a beam of β and γ rays, emitted by a radioactive source, are allowed to enter an absorbing medium, the β rays are absorbed within a much shorter distance than are the γ rays. A schematic diagram of the relative absorptions is shown in Fig. 25N.

Suppose we perform an experiment, using a small sample of radium as a radiation source, a scintillation counter as a detector, and aluminum or lead

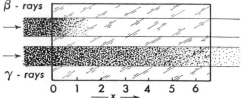

Fig. 25N *Schematic diagram showing the relative absorption of β and γ rays.*

sheets as absorbers. See Fig. 25O. Radiation from a radium source *S*, in passing through absorbers at *A*, enters the fluor block *C*, and the signal developed is amplified by the photomultiplier tube *PM*.

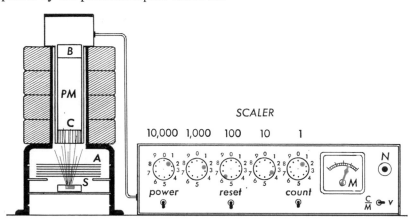

Fig. 25O   *Scintillation tube with scaler counter for radiation absorption measurements.*

The number of counts per minute (C/M), detected by the scintillation counter, is shown here being recorded with an electronic amplifier called a *scaler*.

The five circles represent the end view of electronic tubes, each having ten small pins. Only one pin at a time in each tube will glow orange-red in color, the light arising from a neon gas discharge around it. Each pulse from the scintillator will cause the glow to jump to the next pin clockwise in tube "1." When the tenth pulse arrives and the glow jumps from pin 9 to 0, the glow in the next tube "10" will jump from pin 0 to pin 1. The next time around for tube "1," the glow in tube "10" will jump to 2, thus indicating a total count of 20. This process continues, thus activating the third tube for hundreds, the fourth for thousands, etc. The reading showing in the diagram is 12,631.

By depressing the RESET switch, all five tubes return to their zero positions. To determine the number of counts per minute from any setting, the COUNT switch is raised at the same time the button on a stop-watch is pressed. At the end of one minute, by the watch, the COUNT switch is depressed, stopping the counting process. The total counts are then read directly from the tubes.

Such scalers as these are usually equipped with a dual-purpose milliammeter as shown at the right. When the accompanying switch is thrown to the *V* position, the knob N can be turned to adjust and set the total voltage applied to the photomultiplier tube. When it is thrown to the C/M position, the meter pointer will read directly the counts per minute. As one observes this meter for a fixed set of counting conditions, the pointer will fluctuate about some median position, and the observer must estimate the average value. It is for this reason that, for accuracy, the tube readings are preferred.

With no absorbers in place, and the source far removed from the scintillation counter, stray radiation from nearby objects and cosmic rays can and should always be measured and recorded as "background C/M." After the source is inserted in the holder, absorbers are inserted one at a time between the source and the counter, and the C/M determined for each. The results of this experiment, using first a series of lead absorbers and then a series of aluminum absorbers, are shown on a semilog graph in Fig. 25P.

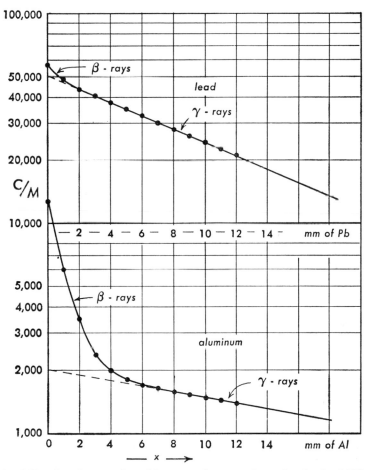

Fig. 25P   *Semilog graphs of β ray and γ ray absorption by lead (Pb) and aluminum (Al).*

Since β rays from a radioactive source have a range of velocities, the upper part of each graph is curved. At the point where each curve straightens out, the β rays are completely absorbed and only the more penetrating γ rays are left. The straight section signifies that γ ray absorption follows Eq. (25e), and that absorption coefficients can be determined for both lead and aluminum.

PROBLEMS

1. Write down the reaction for the disintegration of RaD. (See Appendix III.)

2. Write down the reaction for the disintegration of AcK. (See Appendix III.)

3. Atoms of MsTh$_2$ emit β particles with an end-point energy of 1.55 Mev. (a) Write down the reaction and (b) find the loss in nuclear mass due to this particle and the neutrino. (See Table 25A.)

4. Atoms of UY emit β particles with an end-point energy of 0.21 Mev. (a) Write down the reaction and (b) find the loss in nuclear mass due to the emission of this particle and the neutrino. (See Table 25A.) *Ans.* (b) 0.00023 amu.

5. Atoms of RaD have a mass of 209.984177 amu, and emit β particles followed by a γ ray as shown in Fig. 25E. (a) Write down the reaction, and (b) find what the atomic mass is of the daughter product.

6. An electron has a kinetic energy equivalent of 2 Mev. What is its velocity relative to the velocity of light, v/c? (See Eq. 20b). *Ans.* 97.9%.

7. An electron has a velocity of 99.8% the speed of light ($v/c = 0.998$). Calculate its equivalent energy in Mev.

8. Calculate the absorption coefficient of aluminum for γ rays as given by the experimental graph in Fig. 25P. *Ans.* 0.032 mm^{-1}.

9. Calculate the absorption coefficient of lead for γ rays as given by the experimental graph in Fig. 25P.

10. Atoms of the radioactive isotope $_{90}$Th^{234} have a mass of 234.04357 and a half-life of 24 days. Sixty-five percent of these nuclei emit β particles with an energy of 193 Kev, while 35% emit 100-Kev β particles, followed by a γ ray. Either way they become the same daughter element $_{91}$Pa^{234}. (a) Make a single energy-level diagram for these changes. (b) Write down the two different reactions. (c) Find the γ-ray energy in amu. (d) What is the mass of the daughter atom? *Ans.* (c) 0.0010 amu. (d) 234.04336 amu.

11. Atoms of the radioactive isotope $_{88}$Ra^{235} have a mass of 225.023518 amu and a half-life of 15 days. Sixty-three percent of these nuclei emit a β particle with an energy of 280 Kev, followed by a γ ray, while 37% emit a 320-Kev β particle with no γ ray. Either way they become the same daughter element $_{89}$Ac^{225}. (a) Make a single energy-level diagram for these two disintegration modes. (b) Write down the two reactions. (c) Find the γ-ray energy in amu. (d) What is the mass of the daughter atom?

Atomic Collisions and Nuclear Disintegration

The continual search of the scientists for some knowledge of the ultimate particles into which all matter may be subdivided has led within the last sixty years to the discovery of still smaller particles than molecules and atoms: protons, neutrons, mesons, neutrinos, etc. These discoveries are but the first steps toward solving the age-old mystery of why all solids, large or small, do not fall apart.

26.1. Rutherford's Scattering Experiments

As early as 1903 P. Leonard sent cathode rays through thin films of metal and measured their penetration and absorption in matter. He concluded from his experiments that the mass associated with solid matter is not distributed uniformly throughout the body, but is concentrated upon myriads of tiny isolated centers which he called *dynamids*. It was for these experiments that he was awarded the Nobel Prize in physics in 1905.

During the following decade Sir Ernest Rutherford, and his collaborators H. Geiger and E. Marsden, performed a series of ingenious experiments on the scattering of α particles, the results of which implied that the positive charge and mass of every atom are confined to a particle smaller than 10^{-12} cm in diameter. Historically this marks the beginning of the idea of a nuclear atom proposed formally by Niels Bohr several years later. A schematic diagram of the scattering experiments is given in Fig. 26A.

High-speed α particles from the radioactive element radon, confined to a narrow beam by a hole in a lead block were made to strike a very thin gold foil F. While most of the α particles go straight through the foil as if there were nothing there, some of them collide with atoms of the foil and bounce off at some angle. The latter phenomenon is known as *Rutherford scattering*.

The observations and measurements made in the experiment consisted of counting the number of particles scattered off at different angles θ. This was done

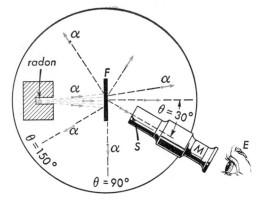

Fig. 26A   *Diagram of the Rutherford scattering experiments.*

by the scintillation method of observation. Each α particle striking the fluorescent screen S produces a tiny flash of light, called a *scintillation*, and is observed
as such by the microscope M. With the microscope fixed in one position the
number of scintillations observed within a period of several minutes was
counted; then the microscope was turned to another angle, and the number was
again counted for an equal period of time.

In the schematic diagram of Fig. 26B, α particles are shown passing through a

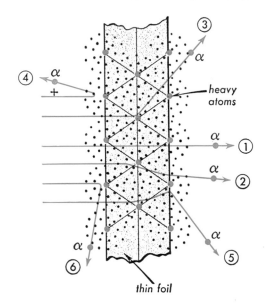

Fig. 26B   *Schematic diagram of α
particles being scattered by the
atomic nuclei in a thin metallic
film.*

foil three atomic layers thick. Although the nuclear atom was not known at the
time the experiments were performed, each atom is drawn in the figure with the
positively charged nucleus at the center and surrounded by a number of electrons. Since most of the film is *free space*, the majority of the α particles go
through with little or no deflection as indicated by ray (1). Other α's like (2)

passing relatively close to an atom nucleus are deflected at an angle of a few degrees. Occasionally, however, an almost *head-on collision* occurs as shown by (4) and the incoming α particle is turned back toward the source.

As an α particle approaches an atom, as represented by ray (6) in Fig. 26B, it is repelled by the heavy positively charged nucleus and deflected in such a way as to make it follow a curved path. The magnitude of the repulsive force is at all times given by Coulomb's law, see Eq. (2a):

$$F = k \frac{QQ'}{r^2} \tag{26a}$$

Whatever the force of repulsion may be at one distance r, it becomes 4 times as great at $\frac{1}{2}$ the distance, 9 times as great at $\frac{1}{3}$ the distance, 16 times as great at $\frac{1}{4}$ the distance, etc. We see, therefore, that at very close range the mutual repulsion of the two particles increases very rapidly and finally becomes so great that the lighter α particle is turned away. The repelling force, still acting, gives the particle a push, causing it to recede with the same velocity as that with which it approached. The actual trajectory is in every case a hyperbolic orbit with the nucleus at the focus. See Fig. 26C.

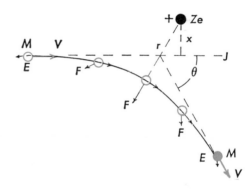

Fig. 26C   *Diagram of the deflection of an α particle by a nucleus:
Rutherford scattering.*

26.2. The Rutherford Scattering Experiments

To see how the principles of mechanics, and Coulomb's law, Eq. (26a), are applied to the deflection of α particles passing by an atomic nucleus, we refer to Fig. 26C. The α particle of mass M, charge E, moving with a velocity V along the line MJ would, in the absence of Coulomb's law, pass within a distance x of a relatively heavy nucleus of charge Ze. Owing to the mutual repulsion of the two positive charges, the force F, expressed by

$$F = k \frac{Ze \cdot E}{r^2} \tag{26b}$$

and acting on the α particle at all points, gives rise to a hyperbolic trajectory as shown.

Approaching along one asymptote, and receding along the other, the α particle is deflected through a total angle θ.

By applying these principles to find the numbers of particles scattered at different angles Rutherford obtained the following formula:

$$N = N_0 \frac{nt(Ze)^2 E^2}{2^4 r^2 (\frac{1}{2}MV^2)^2 \sin^4 \frac{1}{2}\theta} \tag{26c}$$

N = number of α's striking the screen
n = number of atoms per unit volume
t = foil thickness
$\frac{1}{2}MV^2$ = kinetic energy of α particle
θ = angle scattered
N_0 = number of α's striking the foil

While this formula looks quite complicated, it can be separated into parts as follows:

1. Foil thickness	$N \propto t$	
2. Kinetic energy	$N \propto 1/(\frac{1}{2}MV^2)^2$	(26d)
3. Scattering angle	$N \propto 1/\sin^4 \frac{1}{2}\theta$	
4. Nuclear charge	$N \propto (Ze)^2$	

Repeated experiments with different films made of light and heavy elements, like copper, silver, and gold, showed that the relative number of the wide-angled deflections increases with atomic weight. From all of these results and numerous calculations, Rutherford came to the following conclusions.

1. An increase in film thickness t increases proportionately the number of target nuclei to be hit, and hence the number of particles scattered out.
2. The greater the energy or velocity of the incident particle, the smaller will be the angle through which it will be deflected.
3. A single impact can deflect a particle through a large angle and by head-on collision reverse its direction.
4. Increased deflections resulting from foils of increasing atomic number Z are the result of stronger Coulomb forces arising from increased nuclear charge Ze.
5. Deflections resulting from nuclear collisions are perfectly elastic and obey the laws of conservation of mechanical energy and momentum.
6. All of the positive charge of an atom is confined to a particle smaller than 10^{-12} cm in diameter.
7. Practically all of the weight of an atom is confined to this same particle.
8. The amount of positive charge in atomic units is approximately equal to half the atomic weight.

Although an α particle (mass number 4) is light compared with an atom of a metal like gold (mass number 197), it is 7000 times heavier than a single electron. For this reason the electrons surrounding the atomic nucleus are pushed to either side as the α particle goes speeding through, and they have little effect upon the shape of the trajectory.

A graph representing the force of repulsion between an α particle and a

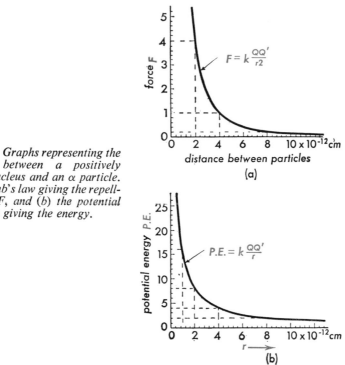

Fig. 26D *Graphs representing the repulsion between a positively charged nucleus and an α particle. (a) Coulomb's law giving the repelling force F, and (b) the potential curve giving the energy.*

positively charged nucleus is illustrated in Fig. 26D. Diagram (a) shows the rapid increase in force as the distance decreases, while diagram (b) shows the rapid rise in potential energy. The potential energy between two electric charges is given by Eq. (2l):

$$E_p = k \frac{QQ'}{r} \tag{26e}$$

The reason for giving this equation, and the potential energy curve in Fig. 26C, is that an interesting mechanical model for demonstrating Rutherford scattering can be derived from it. Such a model is illustrated in Fig. 26E, where the circular

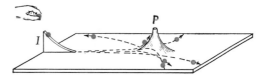

Fig. 26E *Mechanical model of an atomic nucleus for demonstrating Rutherford scattering.*

peak at the right represents the nucleus of an atom and has a form generated by rotating curve (b) of Fig. 26D, about its vertical axis at $r = 0$.

Marbles, representing α particles, roll down a chute and along a practically level plane where they approach the potential hill. Approaching the hill at various angles, the marbles roll up to a certain height and then off to one side or the other. The paths they follow, if watched from above, are *hyperbolic* in shape. Approaching the hill in a head-on collision, the ball rolls up to a certain point, stops, then rolls back again. Thus the potential energy of the α particle close to the nucleus is analogous to the potential energy of a marble on the hillside, and the electrostatic force of repulsion is analogous to the component of the downward pull of gravity.

26.3. Elastic Collisions Between Atoms

Collisions between free atomic particles were first studied by Rutherford with apparatus as shown in Fig. 26F. A long glass tube, containing a small sample of radioactive material *R*, was first thoroughly evacuated by means of a vacuum pump and then filled with a gas of known constitution. Alpha particles from the radioactive source were then permitted to travel through the gas to the other end of the tube where, upon passing through a thin aluminum foil to a fluorescent screen *S*, they could be observed as scintillations in the field of view of a microscope *M*. This is exactly the arrangement used by Rutherford in measuring the range of α particles from different radioactive elements (see Fig. 23K).

With air in the tube *T* and *radium C'* as a source of α particles, scintillations could be observed with the screen as far back as 7 cm. With hydrogen in the tube it was found that the distance *d* could be greatly increased. Inserting thicker and thicker aluminum foils between *F* and *S* in front of the fluorescent screen, the range of the particles was calculated to be equivalent to 28 cm of air.

The conclusion Rutherford drew from this result was that an α particle occasionally collides with a hydrogen atom, much as a large ball collides with a lighter one, imparting to it a greater velocity and hence a greater penetrating power. Of the many recoil hydrogen atoms from collisions with α particles, some undergo head-on collisions and go shooting off in a forward direction with over half again (more accurately 1.6 times) the speed of the incident α particle.

The enormous increase in range is due principally to a higher velocity. Each hydrogen nucleus, called a *proton*, is stripped of its orbital electron, and, with but a single positive charge, produces fewer ions per centimeter of path than an

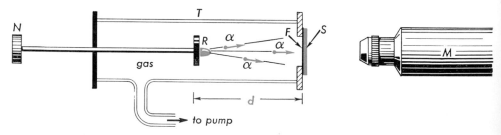

Fig. 26F   *Rutherford's apparatus used in observing atomic collisions between α particles from radium and the atoms of a gas like hydrogen, helium, nitrogen, oxygen, etc.*

α particle with its double positive charge. Curiously enough, protons or α particles having the same velocity have about the same range. The reason for this is that, whereas an α particle has double the charge of a proton and produces more ions per centimeter of its path, which tends to slow it down more rapidly, it also has four times the mass and therefore four times the energy. The range curves in Fig. 24G show that for *protons* and α *particles* with the same kinetic energy, the protons have about ten times the greater range.

(a)   (b)   (c)

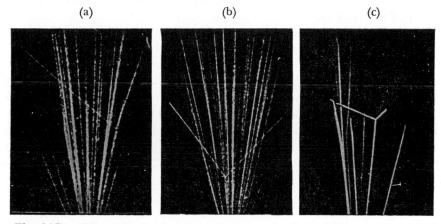

Fig. 26G   *Wilson cloud-chamber photographs of collisions between α particles and (a) a hydrogen atom, (b) a helium atom, and (c) an oxygen atom. (After Rutherford, Chadwick, and Ellis.)*

A more convincing study of such atomic collisions can be made with a Wilson cloud chamber. In the thousands of cloud-chamber photographs of the ion tracks made by α particles from radioactive elements, one occasionally observes forked tracks of the type reproduced in Fig. 26G. When each of these pictures was taken, the cloud chamber contained different gases. For photograph (a) the cloud chamber contained hydrogen, for (b) it contained helium, and for (c) it contained oxygen. Schematic diagrams of these same collisions are illustrated in Fig. 26H. Note in (b) that the angle between the two recoiling particles is 90°, a right angle.

Fig. 26H   *Diagrams of collisions between α particles and other nuclei of different mass.*

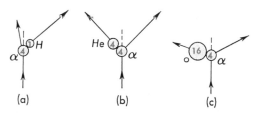

(a)   (b)   (c)

Most elastic collisions between atoms are not head-on collisions, but ones in which the incident particle strikes the other a glancing blow. When an α particle having a mass of 4 units collides with a hydrogen atom of 1 unit, the α particle is deviated only a little from its path, whereas the hydrogen atom nearly always recoils off at quite a large angle. This is in agreement with the laws of conservation of energy and momentum applied to two perfectly elastic spheres.

When an α particle collides with a helium atom, both particles have the same mass of 4 units each, and the two always glance off at right angles to each other. The laws of mechanics show that for a head-on collision between perfectly elastic spheres of equal mass, one moving and one at rest, the incident particle is stopped by the collision and the second body goes on in the forward direction with all of the velocity.

When an α particle collides with an oxygen atom having a mass of 16 units, the oxygen atom recoils to one side with a relatively low velocity, and the α particle glances off to the other side with a high or low velocity depending upon the angle of recoil. The oxygen atom with its greater mass and charge ionizes more particles per centimeter path and therefore leaves a heavier track.

Atomic collisions involving such high velocities take place between the heavy nuclei of the atoms and are little affected by the light orbital electrons. If a nucleus is hit hard by a collision, it will be partially or wholly denuded of its electrons. When it comes to rest, it will soon pick up enough electrons to make it a neutral atom again.

26.4. The Discovery of Nuclear Disintegration

Upon repeating the range experiments illustration in Fig. 26F, with a heavy gas in the tube T, Rutherford made a new and startling discovery in 1919. When nitrogen gas (atomic weight 14) was admitted to the tube, scintillations could be observed at a distance of 40 cm or more from the source. No such long-range particles had ever been observed before. What were these long-range particles? They could not be electrons or γ rays, for these were not capable of producing visible scintillations. Rutherford allowed the new rays to pass through a magnetic field and discovered from their deflection that they had the mass and charge of protons. In other words, the long-range particles were hydrogen nuclei.

Rutherford was not long in coming forward with the correct explanation of the phenomenon. An α particle, near the beginning of its range where its velocity is high, may make a head-on collision with a nitrogen nucleus and be captured. This capture is then followed immediately by a disintegration in which a proton is ejected with high speed. The process is illustrated in Fig. 26I, and the transformation can be represented by the following simple reaction:

$$\boxed{{}_2\text{He}^4 + {}_7\text{N}^{14} = ({}_9\text{F}^{18}) = {}_8\text{O}^{17} + {}_1\text{H}^1} \tag{26f}$$

When the α particle, with a charge of $+2$ and mass 4, collides with the nitrogen nucleus with a charge of $+7$ and mass 14, they form a single particle with a charge of $+9$ and mass 18. Since an atom with a nuclear charge of $+9$ would be expected to have all the chemical properties of *fluorine*, atomic number 9, the newly formed nucleus is labeled ${}_9\text{F}^{18}$.

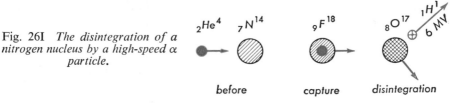

Fig. 26I *The disintegration of a nitrogen nucleus by a high-speed α particle.*

before   capture   disintegration

An examination of the table of isotopes, however (see Appendix III), shows that no such isotope exists in nature. The reason becomes apparent when it is realized that such a combination of particles is not stable. A fluorine nucleus of mass 18 is unstable and disintegrates by discharging a proton, a particle with a charge of +1 and a mass of 1. This leaves behind a residual nucleus with a charge of +8, and a mass of 17. Under atomic number 8 in the same Table III, an oxygen isotope of mass 17 is seen to have been found in nature.

Thus the above disintegration process started with two stable nuclei, *helium* and *nitrogen*, and out of them were created two new stable nuclei, *oxygen* and *hydrogen*. This is called a *transmutation* of elements. Because the intermediate step indicates but a momentary existence of a fluorine nucleus, $_9F^{18}$, this step is often omitted from any discussion of the above process and the disintegration reaction simply written

$$_2He^4 + _7N^{14} = _8O^{17} + _1H^1$$

Such transformation reactions are like equations and must balance: first, the total amount of charge must remain the same, and second, the mass numbers must balance. The first of these is accomplished by having the sum of the subscripts on one side of the reaction equal to the sum of the subscripts on the other side, and the second by having the sum of the superscripts the same on both sides. In every known atom the subscript, representing the nuclear charge, is the sole factor determining the chemical element to which the atom belongs.

26.5. Chadwick's Identification of the Neutron

In 1932 Chadwick, in England, performed an experiment for which he was awarded the Nobel Prize in physics in 1935. As diagramed in Fig. 26J his experi-

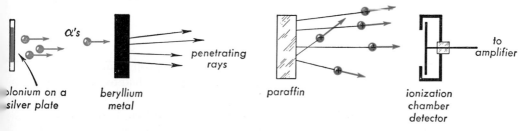

α's

penetrating rays

to amplifier

polonium on a silver plate   beryllium metal   paraffin   ionization chamber detector

Fig. 26J *The experiment by which Chadwick discovered the neutron. The penetrating rays are neutrons and α particles.*

ment consisted of bombarding a beryllium target with α particles. Penetrating particles emerging from the beryllium were permitted to impinge upon a block of paraffin from which protons were found to emerge with high speed. From energy calculations he was able to show that the penetrating rays were uncharged particles with the mass of protons; these he called *neutrons*. The disintegration taking place in the metal target is the following (see Fig. 26K):

$$_2\text{He}^4 + {}_4\text{Be}^9 = {}_6\text{C}^{12} + {}_0\text{n}^1 \tag{26g}$$

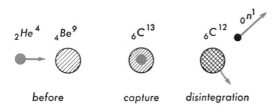

Fig. 26K　*The neutron process discovered by Chadwick. See Fig. 26J.*

The α particle, $_2\text{He}^4$, makes a collision and unites with a beryllium nucleus, $_4\text{Be}^9$, causing a disintegration; whereupon a neutron, $_0\text{n}^1$, is expelled with high velocity. The residual particle with a charge of $+6$ and mass of 12 units is a stable carbon nucleus such as found in nature.

The penetrating rays from the beryllium block in Fig. 26J are mostly neutrons which, in bombarding the paraffin block, collide elastically with hydrogen atoms, knocking them out on the other side. An elastic head-on collision between two particles of the same weight, like a neutron and proton, finds the entire velocity of one transferred to the other; the neutron is stopped and the proton goes on. The protons, having a positive charge, can be observed by their tracks in a Wilson cloud chamber, whereas neutrons cannot.

The reason fast neutrons have such a high penetrating power is that they are not slowed down by ionizing atoms as they pass close by them. A proton, electron, or α particle has a charge and can ionize atoms by attracting or repelling electrons from a distance, but a neutron without a charge cannot do this. It must make a direct collision with another particle to be slowed down or stopped.

Neutrons can now be produced in such intense beams that they are often used in place of X rays wherever radiations of high penetrating power are desired.

26.6. The Nucleus Contains Neutrons and Protons

Since the time of Chadwick's identification of the neutron as an elementary particle, our ideas concerning the nucleus of the atom have had to be modified. We now believe that the nucleus contains two kinds of particles, neutrons and protons. Each neutron has a mass of one unit and no charge, whereas each proton has a mass of one unit and a positive charge of one unit. This differs from the older idea that the nucleus contains protons equal in number to the atomic weight and enough electrons to neutralize the surplus charge in excess of the amount specified by the atomic number.

Since only the proton has a charge, any given nucleus of atomic number Z and mass number M is now believed to have Z protons and $M-Z$ neutrons, and in a neutral atom the number of protons is believed to be equal to the number of orbital electrons. The nuclear particles of a few of the elements of the periodic table are given in Table 26A, as examples.

Table 26A. Neutrons and Protons in the Nuclei of a Few Elements

ATOM	PROTONS	NEUTRONS
$_1H^1$	1	0
$_1H^2$	1	1
$_2He^4$	2	2
$_3Li^6$	3	3
$_3L^7$	3	4
$_4Be^9$	4	5
$_4Be^{10}$	4	6
$_5B^{11}$	5	6
$_7N^{13}$	7	6
$_8O^{16}$	8	8
$_{11}Na^{23}$	11	12
$_{29}Cu^{65}$	29	36
$_{80}Hg^{200}$	80	120
$_{92}U^{238}$	92	146

Schematic diagrams of the nucleus of five different atoms are given in Fig. 26L.

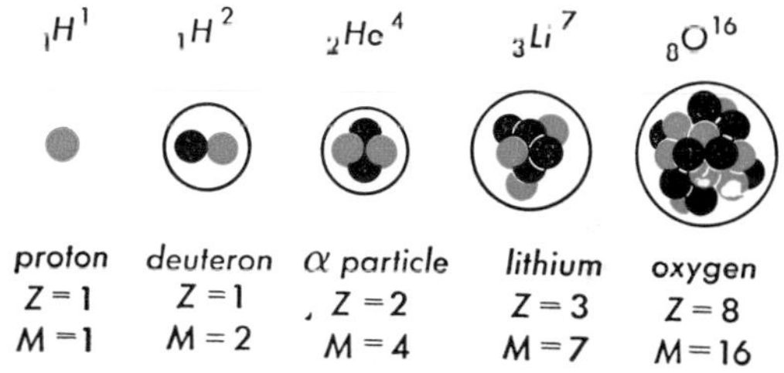

Fig. 26L *The number of protons and neutrons in the nuclei of hydrogen, deuterium, helium, lithium, and oxygen.*

$_1H^1$	$_1H^2$	$_2He^4$	$_3Li^7$	$_8O^{16}$
proton	deuteron	α particle	lithium	oxygen
$Z=1$	$Z=1$	$Z=2$	$Z=3$	$Z=8$
$M=1$	$M=2$	$M=4$	$M=7$	$M=16$

26.7. Atomic Masses Are Not Whole Numbers

When an α particle collides with the nucleus of an atom and produces a disintegration as shown in Fig. 26I, the total energy before collision must be equal to the total energy after collision. To verify this, all forms of energy involved in the process must be included: (1) *the kinetic energy of all particles*, (2) *the energy of a γ ray if one is involved*, and (3) *the mass energy*. The latter is necessary since disintegration experiments show that the total mass of the two colliding particles is not in general equal to the total mass after disintegration. To test this change, it is necessary that we know the exact masses of all atoms individually.

Recent mass spectrographic measurements by Aston, Bainbridge, and others (see Chapter 5) show that the masses of atoms are not exactly whole number values as previously suspected and given in Appendix III. A list of the most recent mass determinations of some of the lighter elements of the periodic table is given in Appendix IV. These values are all based upon the carbon isotope 12 as having a mass of exactly 12.000000.

26.8. Conservation of Energy in Nuclear Disintegrations

To illustrate the law of conservation of energy as it applies to nuclear disintegrations, consider Rutherford's first experiment, shown in Figs. 26F and 26I, where α particles from radium C′ passing through nitrogen gas make collisions with, and disintegrate, nitrogen nuclei. The total energy of any two particles before impact will be the sum of the masses of the two nuclei, $_2\mathrm{He}^4 + _7\mathrm{N}^{14}$, plus their kinetic energy E_1. The total energy after impact will be the sum of the masses of the two nuclei, $_8\mathrm{O}^{17}$ and $_1\mathrm{H}^1$, plus their kinetic energy E_2. If we insert accurately known masses, the reaction becomes

$$_2\mathrm{He}^{4.002604} + _7\mathrm{N}^{14.003074} + E_1 = _8\mathrm{O}^{16.999133} + _1\mathrm{H}^{1.007825} + E_2 \qquad (26h)$$

It is customary to express the energies E_1 and E_2 in *mass units* or in million electron volts, Mev. The kinetic energy before impact is confined to the α particle from radium C′, which has been measured and found to be equivalent to 7.7 Mev. Dividing this value by 931, from Eq. (24m), gives the equivalent of 0.00827 mass unit. Adding mass-energy for both sides of Eq. (26f), we must obtain the same total.

Using mass values given in Appendix IV we obtain,

$$
\begin{array}{ll}
_2\mathrm{He}^4 = 4.002604 & _8\mathrm{O}^{17} = 16.999133 \\
_7\mathrm{N}^{14} = 14.003074 & _1\mathrm{H}^1 = 1.007825 \\
E_1 = \underline{0.008270} & E_2 = \underline{?} \\
\text{Total} = 18.013948 & 18.013948
\end{array} \qquad (26i)
$$

Simple addition and subtraction show that, to yield the proper sum for the right-hand column, E_2 must be equal to 0.006990 mass unit. Multiplying by 931 gives, this time, 6.51 Mev as the energy liberated. This is the liberated energy utilized in the "explosion" which drives the proton and oxygen nuclei apart.

In general, when an atomic nucleus disintegrates by splitting up into two particles, the annihilation energy is divided between them. Experiments show that this division takes place according to the ordinary laws of mechanics, that *the kinetic energies of the two particles are approximately inversely proportional to their respective masses.*

When, in the above example, the available energy is divided between an oxygen nucleus of mass 17 and a proton of mass 1, the $_8\mathrm{O}^{17}$ nucleus acquires an energy of 0.36 Mev, and the proton an energy of 6.15 Mev. A proton with this kinetic energy and velocity has a range of 49 cm in air. Recent repetitions of Rutherford's experiment give measured ranges of 48 cm and an energy of 6 Mev, in good agreement with the calculation.

Graphs showing the ranges of protons, deuterons, and α particles for different energies are drawn in Fig. 24G.

26.9. The Cockcroft-Walton Experiment

Believing that the disintegration of atomic nuclei might be accomplished by using other than α particles as projectiles, Rutherford instigated in 1930 the construction of a high-voltage, direct-current generator at the Cavendish Laboratory. The purpose of this *million-volt* source of potential difference was to accelerate hydrogen nuclei, *protons*, to high speeds and then cause them to strike known substances. In this way he hoped to produce new and various kinds of disintegrations.

Becoming impatient with the relatively slow progress of the project, however, Rutherford suggested to Cockcroft and Walton that lower voltages be tried in the meantime to see if, by chance, disintegrations might occur. In 1932 Cockcroft and Walton announced that they had successfully disintegrated lithium atoms with protons accelerated by relatively low voltages. Their apparatus is schematically represented in Fig. 26M.

Electrons from a hot filament F, passing through hydrogen gas in the region of A, ionize many hydrogen atoms. These protons with their positive charge are then accelerated toward the other end of the tube by a potential difference V of 150,000 volts. Upon passing through the opening C and a window W, they emerge from the acceleration chamber as a narrow beam of protons.

This tube, acting as a "proton gun," is aimed at a target consisting of lithium metal. Cockcroft and Walton observed and measured α particles emanating from the metal with a range of 8 cm, an energy equivalent to 8.5 Mev. Considering the relatively low energy of the bombarding protons of only 0.15 Mev, this is a tremendous release in atomic energy. The transmutation taking place here is written as follows:

$$_1H^1 + {}_3Li^7 + E_1 = {}_2He^4 + {}_2He^4 + E_2 \tag{26j}$$

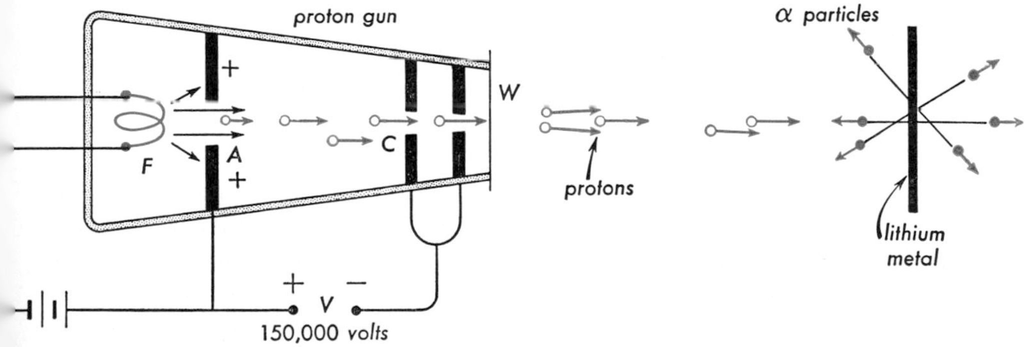

Fig. 26M   *Schematic diagram of the Cockcroft-Walton experiment. Lithium is disintegrated by 150,000-volt protons.*

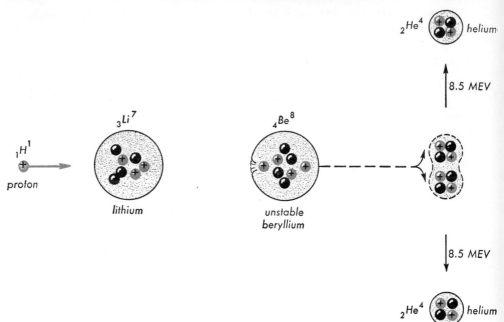

Fig. 26N   *Disintegration of a lithium nucleus by a proton of 0.15 Mev energy:*
the Cockcroft-Walton experiment.

This reaction, illustrated in Fig. 26N, shows a proton, $_1H^1$, of energy $E_1 = 0.15$ Mev, entering a lithium nucleus, $_3Li^7$, to form a new but unstable beryllium nucleus, $_4Be^8$. Being unstable, this compact structure of eight particles splits up into two α particles which are driven apart with great violence. Since the measured energy of each α particle is equivalent to 8.5 Mev, each disintegration involves the liberation of 17.0 Mev energy. The source of energy is to be found in the annihilation of a part of the total atomic mass.

The loss in mass can be calculated from the table of atomic weights, given in Appendix IV. Listing the involved masses in two columns and adding, we obtain

$$\begin{array}{ll}
_1H^1 = 1.007825 & \\
_3Li^7 = 7.016005 & _2He^4 = 4.002604 \\
E_1 = \underline{0.000161} & _2He^4 = \underline{4.002604} \\
8.023991 & 8.005208
\end{array}$$

E_1 is the mass equivalent to the energy of the incident proton and is obtained by dividing 0.15 Mev by 931 (see Eq. (24m)). The difference between the two sums, $8.023991 - 8.005208 = 0.018783$ mass unit, represents the loss in mass by the disintegration. When multiplied by 931, this gives 17.48 Mev as the liberated energy, a value in good agreement with the experimentally determined value of 17.0 Mev.

It might be thought that such a disintegration as the one described above could be used as a source of energy, but as yet it has not been feasible to use it so. While each nuclear collision and disintegration liberates at least one hundred times as much energy as that supplied to the proton, it takes many particles to make a few collisions. In other words, only a small percentage of the proton bullets hit the tiny nuclear targets as they pass through matter. Most of them are slowed down by electron collisions and the ionization of atoms. To a proton bullet the lithium nuclei, as targets in the lithium metal, present an area millions of times smaller than the space between them.

QUESTIONS AND PROBLEMS

1. If, in Rutherford's scattering experiments, the number of α particles observed at an angle $\theta = 60°$ is 36 per minute, how many per minute should be observed at 45° and 135°?

2. When the scintillation microscope used in a Rutherford scattering experiment is set at an angle, $\theta = 30°$, the number of α particles observed is 3000 per hour. How many per hour should be observed at 60° and at 150°? *Ans.* 216, and 16.

3. If, in Rutherford's experiments, scintillations are observed at an angle of 20° for a silver foil, 300 α particles are counted in 1 min. How many counts per minute would be observed if the foil were replaced by foils of equal thickness made of (a) aluminum, (b) copper, and (c) gold?

4. Complete the following disintegration reactions:

$$_1H^2 + {}_8O^{16} \rightarrow \; + {}_2He^4 \qquad {}_1H^2 + {}_5B^{10} \rightarrow \; + {}_0n^1$$
$$_2He^4 + {}_{13}Al^{27} \rightarrow \; + {}_1H^1 \qquad {}_1H^1 + {}_3Li^6 \rightarrow \; + {}_2He^3$$

5. From the known masses of the atoms in the reactions of Problem 4, find the energy liberated by each disintegration in Mev. Assume the incident energy of the lighter nucleus in each reaction to be 20 Mev.

6. From the graphs plotted in Fig. 24G, write down the ranges of the following particles: (a) 10 Mev protons, (b) 13 Mev protons, (c) 12 Mev α particles, (d) 10 Mev deuterons, and (e) 8 Mev α particles. *Ans.* (a) 116 cm, (b) 182 cm, (c) 14.1 cm, (d) 67 cm, (e) 7.3 cm.

7. If the following mass is annihilated, and 95 % of it goes into kinetic energy of an α particle, find the range of the particle in air at normal atmospheric pressure: (a) 0.011547 amu, and (b) 0.003725 amu.

8. Briefly explain the experiment by which Chadwick discovered the neutron. Make a diagram.

9. Give a brief description of the Rutherford experiment in which the first nuclear disintegrations were discovered.

10. Make a table of the following atoms giving the number of protons and neutrons in each nucleus: Ba^{138}, Ca^{44}, Fe^{56}, I^{127}, and Au^{197}.

11. The range of protons in air is found to have the following values: (a) 6 cm, (b) 15 cm, (c) 80 cm, and (d) 180 cm. Find their energies in Mev.

12. Deuterons from a radioactive source are found to have three definite ranges: (a) 10 cm, (b) 75 cm, and (c) 115 cm. Determine their energies from Fig. 24G. *Ans.* (a) 3.4 Mev, (b) 10.7 Mev, (c) 13.5 Mev.

13. Alpha particles from a radioactive material are found to have three ranges: (a) 2.8 cm, (b) 6.8 cm, and (c) 16.5 cm. Find their energies using Fig. 24G.

14. In a nuclear disintegration process, a mass of 0.00820 amu is annihilated and 95 % of the liberated energy is imparted as kinetic energy to an alpha particle. Find (a) its energy in Mev, and (b) its range. *Ans.* (a) 7.23 Mev, (b) 6.2 cm.

Cosmic Rays

Karl K. Darrow has described the subject of cosmic rays "as unique in modern physics for the minuteness of the phenomena, the delicacy of the observations, the adventurous excursions of the observers, the subtlety of the analysis, and the grandeur of the inferences." It is impossible for anyone to say when and by whom cosmic rays were first studied. From the time of the discovery of radioactivity by Becquerel (in 1896) and the discovery of radium by the Curies but a few months later, the radioactive rays from the ground, air and outer space have been investigated by many scientists. Extending over a period of some fifty years these investigations have led to some of the most interesting and important discoveries in the structure of atomic nuclei.

27.1. Early Experiments

It has long been known that a charged electroscope, if left standing for some little time, will discharge regardless of how well the gold leaf is insulated. Realizing that the rays from radioactive materials can be stopped by a sufficient thickness of heavy matter, Rutherford and Cooke (in Canada, 1903) surrounded an electroscope with a thick wall of brick and found very little decrease in the rate of discharge. McLennan and his co-workers (also in Canada) lowered an electroscope into a lake, hoping that the thick layer of water would screen off the rays. This experiment, like the other, failed.

In 1910 Glockel, with an electroscope, rose nearly 3 mi in a balloon in order to get away from the ground radiation, but to his astonishment he found that the rate of discharge did not decrease, but increased, the higher he went. The same effect was observed by Hess (in Austria, 1911) and Kolhörster (in Germany, 1914). Rising to heights as great as $5\frac{1}{2}$ mi, both of these observers independently found that the intensity of these unknown radiations became greater the higher they went.

Because in one of his scientific publications concerning these results Hess suggested the possibility that some kind of penetrating rays were entering the earth's atmosphere from outer space, he is usually credited with the discovery of cosmic rays. For this reason he was granted the Nobel Prize in physics for the year 1936.

27.2. Millikan and Bowen's Discovery

Soon after World War I (1922), R. A. Millikan, with the help of I. S. Bowen, constructed several small, self-recording string electroscopes. Making use of their war time experiences with sounding balloons, they sent these electroscopes high into the stratosphere by fastening each one to two sounding balloons. As shown in Fig. 27A, the string electroscope E consists of two gold-covered quartz

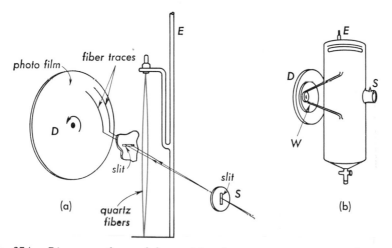

Fig. 27A　*Diagrams of one of the sensitive electroscopes sent up into the strato-sphere by Millikan and Bowen to measure cosmic rays. (a) Schematic diagram, (b) scale drawing of entire instrument 6 in. high.*

fibers insulated and mounted with their ends together. When they are charged, the fibers spread apart in mutual repulsion, and as they discharge they slowly come together.

Daylight, passing through a narrow vertical slit S in the instrument case, casts a shadow of the center section of the fibers on a rotating disk D which contains a photographic film. As the film turns slowly and the fibers come together, they leave a double trace, as indicated in the diagram. On the same film, a small oil manometer recorded the height of ascent and a small thermometer recorded the temperature. The film was driven by a watch W, the whole apparatus weighing only 7 oz. On one of the best record flights, only one of the balloons burst at a height of 10 mi and the other brought the instruments safely to earth.

Like the earlier results obtained by other experimenters, Millikan and Bowen found the ionization to increase with increasing altitude. After extending the observations of previous workers to higher altitudes, Millikan and Bowen became convinced, and announced their belief, that the rays were coming from interstellar space.

27.3. The Penetration of Cosmic Rays

In order to determine the nature of the new rays, Millikan and his co-workers, Otis, Cameron, and Bowen, in the fall of 1922 began an extensive study of the penetrating power of cosmic rays. Since cosmic rays penetrate our atmosphere of many miles of air, how far might they penetrate beyond?

Self-recording electroscopes were lowered to various depths in snow-fed lakes as illustrated schematically in Fig. 27B. Measurements taken at Arrowhead Lake

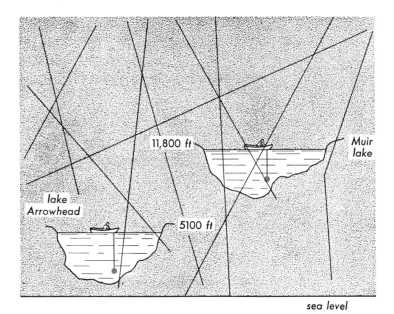

Fig. 27B *Illustrating the lowering of self-recording electroscopes into deep, snow-fed lakes to measure the absorption of cosmic rays by the water.*

in Southern California (at an elevation of 5100 ft) agreed approximately with those taken at Muir Lake near Mt. Whitney (at an elevation of 11,800 ft), provided one took into account the increased air path for the lower elevation. The extra mile and a quarter of air is equivalent in weight to 6 ft of water.

As cosmic rays penetrate deeper and deeper below the surface of water, their number decreases, until at a depth of 100 ft the intensity is reduced to about one ten-thousandth of that at the surface. With very sensitive electroscopes, cosmic radiation capable of penetrating 2000 ft of water has more recently been detected. This is a far greater penetrating power than that possessed by any known X rays or γ rays from radioactivity.

27.4. The Geiger-Müller Tube Counter

There are at least nine methods of observing and measuring cosmic rays. These are

(a) Geiger-Müller tube counters
(b) Wilson cloud chambers
(c) Ionization chambers
(d) Photographic emulsions
(e) Scintillation counters
(f) Bubble chambers
(g) Electroscopes
(h) Semiconductors
(i) Spark chambers

The Geiger-Müller tube, named after its inventors, is one of the simplest electrical instruments ever designed (see Fig. 27C). It consists of an open-ended

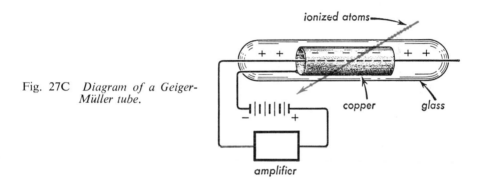

Fig. 27C   *Diagram of a Geiger-Müller tube.*

copper cylinder from 1 to 25 in. long, fitted inside a thin-walled glass cylinder with a fine tungsten wire stretched along the middle. After the tube has been partially evacuated (a pressure of from 5 to 10 cm of mercury is convenient), a potential difference of about 1000 volts is applied, the positive to the center wire and the negative to the cylinder.

When a single cosmic ray or high-speed particle from a radioactive source goes through a Geiger-Müller tube, ions are created by the freeing of electrons from air molecules. These freed electrons are attracted by the positively charged wire and move toward it, acquiring within a very short distance a high velocity of their own. Because of this velocity they, too, can ionize other atoms, thus freeing more electrons.

This multiplication of charges repeats itself in rapid succession, producing within a very short interval of time an *avalanche of electrons toward the central wire*. This sudden surge of charge is equivalent to a small current impulse along the electrical circuit. When this current has been intensified by an amplifier, it may be made to operate an electric switch, a radio loud-speaker, or any kind of electrical device.

Quite frequently, the impulses of a Geiger-Müller tube are made to operate a small counting device. Each cosmic ray particle passing through the tube is therefore counted automatically. The number of counts received per second depends upon the size of the counter tube. An average-sized tube 1 in. in diameter and several inches long, at sea level, gives from 50 to 100 counts per minute.

A common form of Geiger counter is shown in Fig. 27D. It consists of a

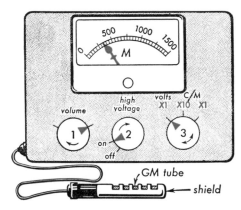

Fig. 27D *Geiger counter instrument complete with G-M tube and counting-rate meter.*

G-M counter tube about 10 cm long connected to a box containing vacuum tube circuits, a small loud-speaker, and a counting-rate meter *M*.

Dial (1) operates the speaker and permits the individual ray pulses to be heard. Dial (2) turns on the vacuum tubes as in any radio or TV receiver, and turning it clockwise increases the voltage supplied to the G-M tube. Dial (3) has three position points to which it can be turned. The position marked volts connects the meter *M* so that it shows the voltage on the G-M tube. When in the ×1 position, under C/M, the meter M directly reads counts per minute, and in the ×10 position the meter reading should be multiplied by 10.

27.5. Directional Effects

To observe the direction of the greatest cosmic ray intensity, a cosmic ray telescope is used. Such a telescope is made by connecting two or more Geiger-Müller tubes *in coincidence*, and mounting them on a common support some distance apart. Tubes in coincidence are so connected electrically that a current will flow in the accompanying electric circuit only when both tubes discharge at the same time.

When one tube is set above the other, as shown in Fig. 27E (a), a single cosmic ray, on going through both cylinders, will cause a current pulse and a count to be made. If, however, a particle goes through one and not the other, no count is recorded. Experiments at sea level show that, when the telescope is mounted

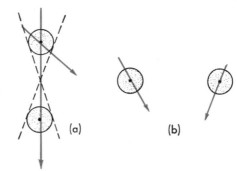

Fig. 27E   *Diagram of two Geiger counters. Connected in coincidence, they form a cosmic-ray telescope.*

in the horizontal position (b), few counts are made, whereas when it is mounted in a vertical direction many more counts are recorded. The interpretation to be made, therefore, is that cosmic rays come principally from overhead.

As a verification of the telescope method, a Wilson cloud chamber is frequently inserted between two Geiger counter tubes as shown in Fig. 27F, and a photograph of the path of each cosmic ray is taken. Thousands of such photographs are made automatically by having a single cosmic ray take its own picture. This is accomplished by allowing the sudden electric current from the counter tubes, produced by a ray in transit, to open and close a camera shutter, to cause the cloud chamber to expand, and to flash a light, illuminating the fog track that forms.

In the reproduction of Fig. 27F, either one of the two cosmic rays would have tripped the electrical devices and taken the picture. It should be noted that both rays passed right through a 0.5-in. lead plate without being deviated. Cloud chamber pictures are not photographs of cosmic rays, but of the path traversed by the rays.

27.6. The Altitude Effect

The results of airplane and balloon flights into the stratosphere have shown that the intensity of cosmic rays increases up to a height of from 12 to 15 mi, and then decreases again. In 1935 Stevens and Anderson,* for example, rose to a height of nearly 14 mi carrying with them, among other scientific instruments, Geiger-Müller tube counters. With these instruments they measured the cosmic-ray intensity at various altitudes on both their ascent and descent.

The compiled experimental results of various observers taken at different elevations are illustrated by the curves in Fig. 27G. Near the city of Omaha, at a magnetic latitude of 51°N, the maximum is found at a height of 15 mi where an intensity 170 times as great as that at sea level has been measured. From that altitude to the highest points that observations have been made, about 17 mi, there is a gradual decrease in total intensity.

* Capt. A. W. Stevens and O. A. Anderson, *National Geographic Magazine*, Vol. LXIX, 1936, p. 693.

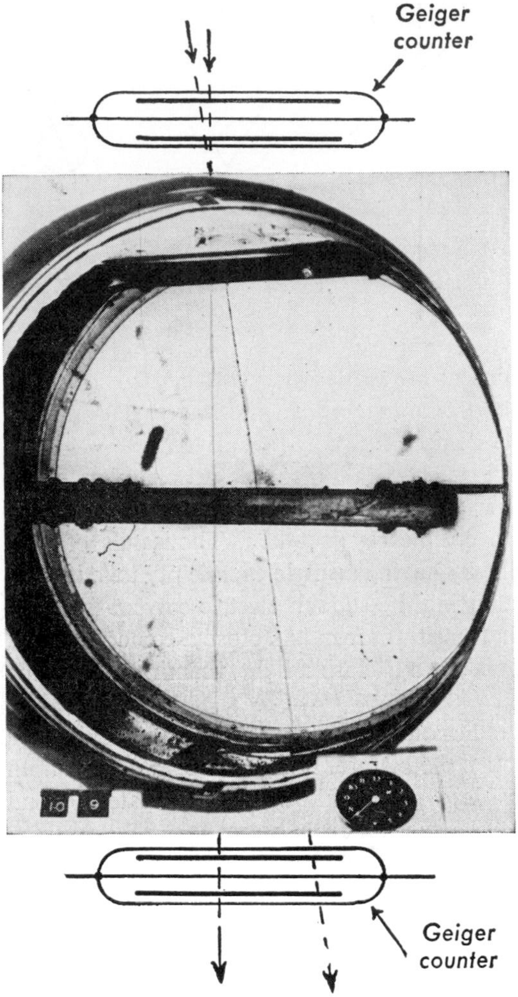

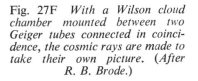

Fig. 27F *With a Wilson cloud chamber mounted between two Geiger tubes connected in coincidence, the cosmic rays are made to take their own picture. (After R. B. Brode.)*

The four different curves in the figure show, from altitude measurements made by observers all over the world, that in nearing the magnetic equator the cosmic-ray intensity decreases at high altitudes, as well as at sea level.

27.7. Primaries and Secondaries

Experimental observations show that the cosmic rays entering our atmosphere are almost entirely composed of positively charged atomic nuclei. Of these so-called *primary cosmic rays* 89% are protons, and another 10% are about 90% α particles and 10% heavier nuclei like carbon, nitrogen, oxygen, iron, etc. See Table 27A in § 27.13.

Upon entering the atmosphere, a high-energy primary particle soon collides with another atomic nucleus, splitting one or both particles into a number of

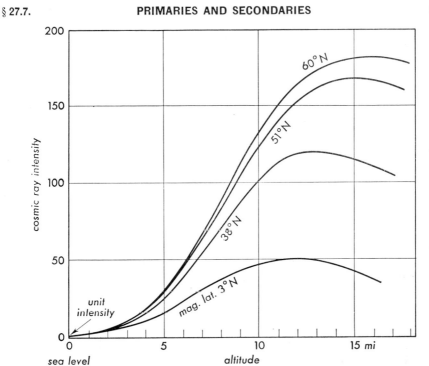

Fig. 27G　*The intensity of cosmic rays increases with altitude up to a height of 10 to 15 mi, and then decreases.*

smaller nuclear fragments, each one of which carries away some of the primary's energy. These high-speed particles in turn collide with other nuclei, further dividing their energy to produce other high-speed particles. All of these rays, with the exception of the primary particle, are called *secondary cosmic rays*.

One of the results of cosmic-ray collision processes is the creation of very high-frequency and highly penetrating gamma rays. These photons, too, are included in the classification, Secondary Cosmic Rays.

At a height of some 15 mi, about ten to fifteen times as many secondary cosmic rays exist as have entered the atmosphere as primaries. At this level, where more than $\frac{9}{10}$ of the earth's atmosphere still lies below, as many rays are observed moving in a horizontal direction as in the vertical. From the diagram presented in Fig. 27H it becomes quite clear why the primaries are difficult to distinguish from the far greater number of secondaries.

At lower altitudes the total intensity decreases, since many of the secondaries produced above are stopped by collision. In other words, so much energy is lost by successive collisions that the energy is gradually absorbed as heat motion by the air molecules. By the time sea level is reached, the remaining rays consist principally of a few high-speed secondaries and primaries. Even at sea level, some of these rays have enough energy left to penetrate several hundred, and even several thousand feet of earth and water.

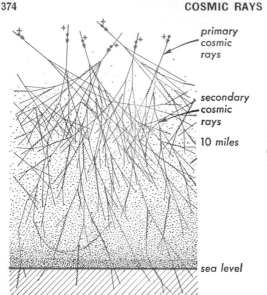

primary
cosmic
rays

secondary
cosmic
rays

10 miles

Fig. 27H *Schematic illustration
of secondary cosmic rays produced
from primaries entering the earth's
atmosphere.*

sea level

It should be borne in mind that, between collisions of the type indicated in
Fig. 27H, each cosmic-ray particle is continually being slowed down as it "plows
through" thousands of air molecules, knocking electrons free to produce ions.
These are the ions on which fogdrops form, revealing the path in a cloud
chamber.

Although γ rays also lose energy by collisions with atoms to produce Compton
electrons, the γ rays themselves do not leave visible tracks in a Wilson cloud
chamber. The reason for this is that collisions are few and far between, and the
resulting fogdrops are too far apart.

27.8. The Latitude Effect

By the year 1930, studies of cosmic-ray intensities by Millikan indicated that
the number of cosmic rays arriving at the earth's surface was constant at all
latitudes. This led him to the conclusion that the primary cosmic radiation
entering the atmosphere must be γ rays of very high energy. He reasoned that,
if they were charged atomic particles, they would be deflected by the earth's
magnetic field, and fewer would reach the earth near the magnetic equator.

In 1931 the Dutch physicist Clay, sailing from Amsterdam in the Northern
Hemisphere to Batavia, Dutch Guiana, in the Southern Hemisphere, carried
Geiger counters with him aboard ship. Measuring the cosmic-ray intensity daily
en route, he obtained the results shown by the graph in Fig. 27I. The curve
shows that as one proceeds from magnetic north to magnetic south, at sea level,
the cosmic-ray intensity remains quite constant, until a magnetic latitude of
about 42° is reached. In this region the intensity begins to drop appreciably,
reaches a minimum at the equator, and rises again to symmetrical intensities in
the Southern Hemisphere.

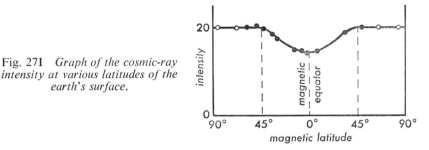

Fig. 27I   *Graph of the cosmic-ray intensity at various latitudes of the earth's surface.*

27.9. Effect of the Earth's Magnetic Field

The decrease in cosmic-ray intensity at the earth's magnetic equator (see Fig. 27I) is now explained as being due to the earth's magnetic field. This is illustrated in Fig. 27J. The paths of all charged particles crossing the earth's magnetic field are bent by a force that is perpendicular to the direction of the field.

Because of these forces, many charged particles from the sun and outer space are trapped by the field in two belts. The recent discovery of these belts by Van Allen and his colleagues at State University of Iowa was made from instruments carried into space by American earth-circling satellites. These belts

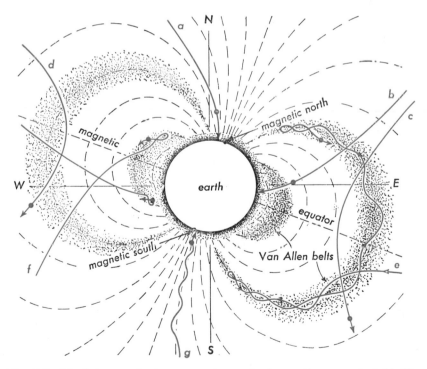

Fig. 27J   *The behavior of primary cosmic rays in the earth's magnetic field. The shaded arches are the Van Allen belts.*

surround the earth, except at the regions of the magnetic poles, the outer one being caused largely by the slower particles, protons and electrons, from the sun. The inner Van Allen belt is formed by more energetic particles from outer space, and is centered about 2500 mi above the magnetic equator.

If the primary energy is very large, little deflection will occur, and particles like *a* and *b* will reach our atmosphere and perhaps the ground. At somewhat lower energies they will be deflected back into space as *c* and *d*. At still lower energies they may follow paths like *e*, *f*, and *g*.

Particle *e* spirals around the field lines and follows them in toward the earth. As the field gets stronger, a point is reached where the particle is turned back, and spiraling around the field lines approaches the earth again on the other side. Such particles running back and forth are trapped; they account for the large number of ionized particles that form the Van Allen belts.

Slow particles, such as *g*, entering the earth's field parallel to the lines of force from far away, will be guided by the field and reach the earth's surface. A day or two after an active display of solar flares, so many particles become trapped in the outer Van Allen belt that they spill out into the earth's atmosphere, creating auroras. Since particles with only the highest of energies can get down to the earth's atmosphere near the magnetic equator, the cosmic ray latitude effect is well understood.

27.10. Discovery of the Positron

The positron, or positive electron, was discovered by Anderson in 1932 by photographing the tracks of cosmic rays in a Wilson cloud chamber. Under the influence of a strong magnetic field applied perpendicular to the face of the cloud chamber, positively charged particles should bend to the right and negatively charged particles should bend to the left. In order to be certain that those bent one way were not all coming from above, and those bent the other way were particles of the same kind and charge coming from below, Anderson inserted a block of lead in the chamber to slow down the particles. Under these conditions photographs similar to the one shown in (a) of Fig. 27K were obtained.

Here Anderson could be quite certain, from the curvature of the track on each side of the lead, that the particle entered from the side shown above, for in passing through the lead plate it could only have been slowed down and not speeded up. Knowing the direction of motion, the direction of the field, and the direction of bending, Anderson concluded that such a particle had a positive charge.

Comparing the track with well-known electron tracks and α particle tracks, he concluded that the new particle had about the same mass as the electron. Later experiments continued to give more positive proof of the existence of a positive electron. Now, very strong beams of positrons can be produced in the laboratory.

It should be pointed out here that near sea level most cosmic rays come from

Fig. 27K   *Wilson cloud-chamber photographs of pair production. An X ray coming close to the nucleus of an atom produces a pair of electrons, one positive and one negative.* (a) *Discovery of a positron (after Anderson),* (b) *three pairs of electrons produced by X rays (after Anderson),* (c) *pair produced in air by X rays (after Lauritson and Fowler),* (d) *pair produced in air by X rays from thorium C" (after Simons and Zuber).*

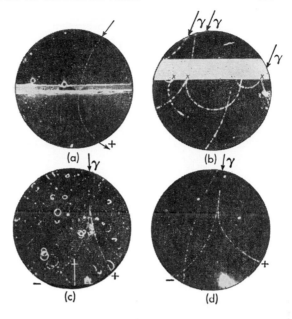

above, whereas a few come from other angles and the horizontal, and some even from below.

27.11. Creation of Electron Pairs

Soon after Anderson's discovery of the positron, several theoretical physicists attempted to calculate the conditions under which a positron might exist in nature. An extension of the quantum theory of the electron, proposed earlier by P. Dirac, led them to the prediction that if a high-energy photon, i.e., a high-frequency γ ray, were to come close enough to the nucleus of an atom, the electric field of the nucleus would be strong enough to annihilate the γ ray and create in its place a *pair of particles, an electron and a positron.* These two particles, the theory predicts, should have the same mass, and equal but opposite charges. A schematic diagram of pair production is given in Fig. 27L.

Blackett, Anderson, and others, looking for such pairs in a cloud chamber, soon found them exactly as predicted. Gamma rays from a radioactive element like *thorium C"*, in passing through matter, were observed to produce pairs of electrons. Three photographs of such incidents are shown in Fig. 27K. In (b) three different pairs are seen emerging from the points marked X on the lower side of a lead plate, and in (c) and (d) a pair is seen having been produced apparently in mid-air. As usual the γ rays that produced these pairs do not show up in the cloud chamber.

When an electron pair is created, *conservation of energy and momentum* requires the two particles to move almost straight forward. Without a magnetic field applied to the cloud chamber, the particles travel side by side in almost

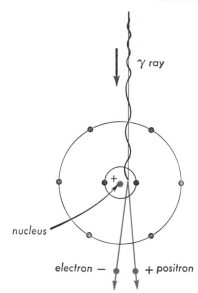

Fig. 27L　*Schematic diagram of electron pair production.*

parallel paths, but with a magnetic field the path of the positron bends to one side and that of the electron to the other.

The reason positrons were not discovered earlier in the history of physics is that they do not exist long in the free state. As soon as a positron meets with an electron, the two are annihilated. Because a positron can annihilate an electron it is called an *antiparticle*.

Experiments indicate that all electrons and positrons spin around an axis through their center of mass. There is good evidence that when a positron and an electron come close together, they frequently combine by revolving around each other like a double-star, with their spin axes parallel to one another. As such a pair they are called *positronium*.

Positronium is very short lived, for soon the two particles disintegrate completely; in their place γ rays are created. If the particles were spinning in the same direction, they would disintegrate into three γ rays of different energies, whereas, if they were spinning in opposite directions, they would produce two γ rays. Conservation of energy and momentum requires each of these latter rays to have an energy of $\frac{1}{2}$ Mev. (See Eq. (24k).)

27.12. Cosmic-Ray Showers

Out of hundreds and hundreds of cloud chamber photographs of cosmic rays, the experimenter is occasionally rewarded with a picture of a cosmic-ray shower. Instead of one or two tracks in the picture, he finds anywhere from half a dozen or more to several hundred. As shown by the photographs in Fig. 27M most of the tracks of a shower seem to come from one localized region, usually within a solid piece of matter like a lead plate or the wall of the cloud chamber.

An extensive study of showers by both the experimental and theoretical physicists has led to the conclusion that each shower is produced by a single, high-energy cosmic ray. A charged particle of very high energy upon entering a solid block of matter, where atoms are packed very close together, carries out the multiple collision process illustrated in Fig. 27H. Here, within a short distance of 1 or 2 cm of lead, enough atoms are encountered to yield many secondaries. These secondaries, emerging from the lower face of the metal, result in the observed photographs.

In a thin sheet of metal, relatively small showers are usually found, whereas with a thick metal block showers of many tracks are occasionally photographed. The first five photographs in Fig. 27M were taken without a magnetic field so that the tracks are all straight, while the last photo was taken with a magnetic field. The bending of three tracks to the right and three to the left indicates equal numbers of positrons and electrons.

In photograph (a), a single high-speed particle is seen to enter the lead plate from above and to produce some twenty or more secondary particles, each with enough energy to get through and into the air space below. In (b), two small showers of particles enter the top surface of the lead plate, whereas a single

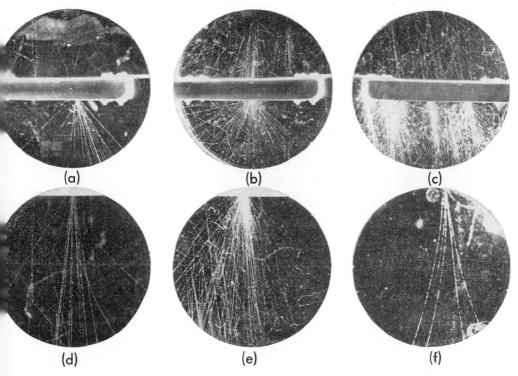

Fig. 27M   *Wilson cloud-chamber photographs of cosmic-ray showers. (The first three photographs are reproduced through the courtesy of R. B. Brode and the last three through the courtesy of C. D. Anderson and the* Physical Review.)

larger shower emerges from the bottom. Apparently one or two of the particles at the center of the one shower above have the necessary high energy to produce the lower shower, whereas the others of lower energy are stopped by the lead. Note particularly the fanning out of the rays below. In (c), a small shower of very high-energy particles enters the chamber from above, having been produced far above the cloud chamber in a shower-producing process, probably by a single particle of extremely high energy. As some of these secondaries pass through the lead, each produces a shower of its own.

Direct evidence that some showers originate with a single high-energy particle is shown in Fig. 27N. Here, in a cloud chamber with five equally spaced lead

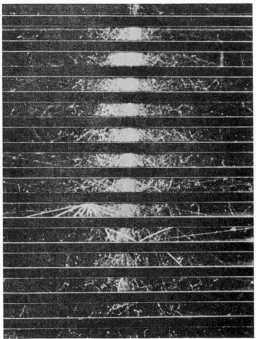

Fig. 27N   *Cloud-chamber photograph showing cascade shower of cosmic rays developed in 13 lead plates, each 1.3 cm thick.*

Courtesy, Wm. B. Fretter

plates, a relatively large shower is seen to have grown from but one or possibly two particles at the top. Not only does this avalanche grow in numbers with each traversal of a lead plate, but the relatively small spread of the tracks indicates how nearly each new particle recoils along with the others in the forward direction. In this picture, one observes in the small space of several inches the process that, in Fig. 27H, requires several miles of air.

27.13. Mesons

The presence in cosmic rays of charged particles having a mass several hundred times that of an electron, yet considerably lighter than a proton, was discovered by Anderson and Nedermeyer in 1938. These particles, now called *mesons*, are of several kinds, and experimental data taken in balloons and airplanes show that most of them are produced high in the atmosphere by the collisions of primary cosmic rays with air nuclei.

In these collisions positively and negatively charged π mesons are produced along with neutral π mesons, protons, and neutrons as shown in Fig. 27O.

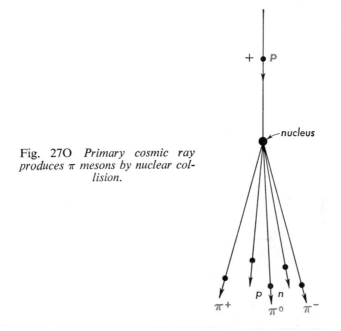

Fig. 27O *Primary cosmic ray produces π mesons by nuclear collision.*

The π mesons, each with a mass of about $275m_e$, along with other nucleons, recoil forward with speeds close to that of light. (m_e = mass of an electron.) The term nucleons is here applied to only those particles believed to exist in atomic nuclei and consist of protons and neutrons.

The possible existence of mesons and their spontaneous disintegration were first predicted by Yukawa* in 1935, and first photographed by Williams and Roberts in 1940. All charged π mesons seem to have a half-life of 2×10^{-8} sec, and each one decays into a charged μ particle called a *muon*, and a lightweight neutral particle called a neutrino. The charged muon in turn decays, with a half-life of 2×10^{-6} sec, into an electron and two neutrinos, as shown in Fig. 27P and Fig. 27Q. The neutrino ν_0 is an uncharged particle, postulated first by Pauli in order to explain nuclear phenomena in keeping with the fundamental laws of the conservation of energy and of momentum. (See §25.2.)

The uncharged π mesons are very unstable. With a half-life of less than 10^{-14} sec, they decay into two γ rays. In the upper atmosphere these γ rays create cascade showers of electrons by electron pair-production and bremsstrahlung.

* H. Yukawa, *Proceedings Physical and Math. Society*, Japan, 17, 48 (1935).

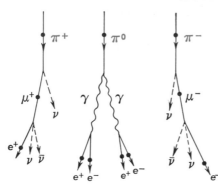

Fig. 27P   π *mesons disintegrate into* μ *mesons,* γ *rays, electrons, and neutrinos.*

See Fig. 27N. Many of the charged μ mesons, with their mass of about $210m_e$, traverse the atmosphere before decaying, and reach the surface of the earth. At sea level the charged cosmic rays are about 70% μ mesons and 29% electrons and positrons, with about 1% heavier particles like protons, deuterons, α particles, etc.

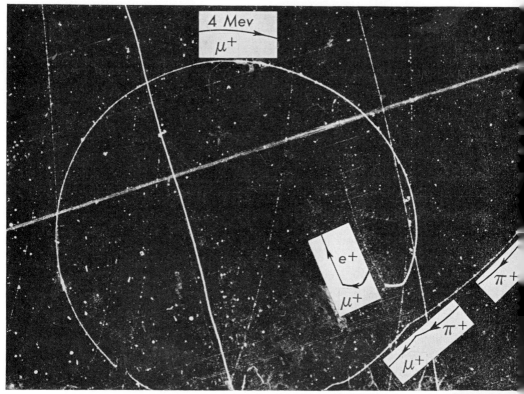

Courtesy, W. Po⋯

Fig. 27Q   *Wilson cloud chamber photographed in a magnetic field of 8000 gauss showing decay of a* π *meson into a* μ *meson and of the* μ *meson into a positron.*

In traversing solid matter, negatively charged π mesons frequently slow down to such a speed that, upon an encounter with a nucleus, they are attracted by the $+$ charge and captured. In this process the meson mass is transformed into energy exciting the nucleus to such a state that it literally explodes by shooting out a number of heavier particles like protons, deuterons, α particles, etc. Fig. 27R shows such a "star" event in a photographic emulsion.

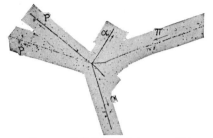

Fig. 27R   *"Star" in photographic emulsion showing the explosion of a nucleus resulting from the capture of a slow π meson.*

Courtesy, C. F. Powell's Laboratory, University of Bristol, England

Recent studies of cosmic ray tracks made in photographic emulsions and cloud chambers indicate the presence of particles with various other masses and charges (to be discussed in Chapter 35).

Most recent cosmic-ray observations at high altitudes and at sea level show the following:

Table 27A. Composition of Primary and Secondary Cosmic Rays

PRIMARIES		SEA LEVEL	
H	89%	μ Mesons	70%
He	9%	e^+ and e^-	29%
Li, Be, B	0.5%	Heavier particles	1%
C, N, O	0.5%		
Ne, Mg, Si	0.1%		
Fe	0.03%		

QUESTIONS

1. Make a diagram of a Geiger-Müller tube counter, and briefly explain how it is able to be used to detect atomic particles of high energy.

2. Who was awarded the Nobel Prize in physics for the discovery of cosmic rays? What was his experimental observation?

3. How does the intensity of cosmic rays vary with altitude?

4. Explain why the intensity of cosmic rays is a minimum near the equator. Make a diagram.

5. How was the positron discovered, and by whom?

6. Under what conditions are electron pairs created?

7. What is a cosmic ray telescope, and how is it made?

8. What are primary cosmic rays, and of what are they composed?

9. What are secondary cosmic rays, and of what are they composed?

10. Make a list of all the atomic particles found in cosmic rays.

11. Make diagrams to show the disintegration of π mesons.

Atomic Particle Accelerators

At the time Cockcroft and Walton were performing their first disintegration experiments (see Fig. 26M), E. O. Lawrence,* an American physicist, and his assistant S. L. Livingston were developing a new type of atomic accelerator which soon attracted the attention of the leading physicists the world over. So successful was this "atomic machine gun" in producing high-speed atomic projectiles for disintegration experiments that a new and larger instrument was soon constructed and put into operation. Because the principles of this accelerator involved the cyclic motion of charged atomic particles in a uniform magnetic field, the device was appropriately called a *cyclotron*. Today a cyclotron of considerable size occupies a most prominent position in many of the leading physics laboratories of the world.

One of the early cyclotrons built at the University of California, called the "sixty-inch," is an instrument capable of producing intense beams of protons, deuterons, or α particles having energies of 10, 20, and 40 Mev, respectively. The purpose of these high-speed particles, as is the case with all such instruments, is to subject various known substances to bombardment and thus produce disintegrations and transmutations of all kinds.

* Ernest O. Lawrence (1901–1958), American experimental physicist. After an early education in South Dakota, Lawrence obtained the A.B. degree at the University of South Dakota in 1922, the master's degree at Minnesota in 1923, and the Ph.D. at Yale University in 1925. After two years as National Research Fellow he became, at the early age of 26, Assistant Professor of Physics at Yale University. The following year he was appointed Associate Professor of Physics at the University of California, and in 1930 was made full Professor. Having built up the Radiation Laboratory at the same institution, he became its Director in 1936. In 1937 he was awarded the Comstock Prize of the National Academy of Sciences, the Cresson Medal of the Franklin Institute, and the Hughes Medal of the Royal Society of London. Lawrence was a member of the National Academy of Sciences and was noted principally for his invention and development of the cyclotron and its application to the production of induced radioactivity. For these discoveries he was granted the Nobel Prize in 1939. During World War II he directed one of the main research projects leading to the isolation of uranium-235 used in atomic bombs.

28.1. The Lawrence Cyclotron

Although the operation of a large cyclotron requires an elaborate outlay of apparatus and equipment, the principles upon which it operates are quite simple. As a means of explaining these principles, cross-section diagrams of a cyclotron are shown in Figs. 28A and 28B.

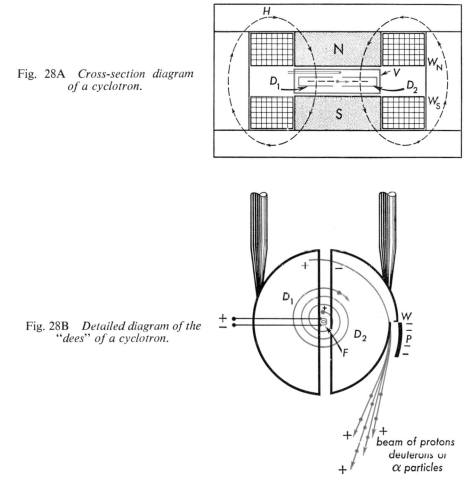

Fig. 28A　*Cross-section diagram of a cyclotron.*

Fig. 28B　*Detailed diagram of the "dees" of a cyclotron.*

beam of protons
deuterons or
α particles

The very heart of the instrument consists of two short, hollow, half cylinders, D_1 and D_2, mounted inside of a vacuum chamber V, between the poles of a powerful electromagnet, and connected on the outside to the two terminals of a high-frequency alternating-current generator. This generator is really a short-wave radio transmitter supplying energy to the "dees" (D_1 and D_2) instead of to the antenna.

When a trace of hydrogen gas is admitted to the evacuated chamber, the hot-wire filament F ionizes some of the hydrogen atoms, thereby producing the protons to be used as atomic bullets. At the particular instant when D_1 is charged positively and D_2 is charged negatively, a proton in the neighborhood of F will be accelerated toward D_2. Moving through the strong magnetic field of the huge magnet, this positively charged particle traverses a circular path as shown in the diagram.

If, after making a half-turn, the potential difference is reversed so that D_1 becomes negatively charged and D_2 positively charged, the proton will be attracted by one and repelled by the other, causing it to increase its speed. With added speed it therefore moves in the arc of a larger circle as shown.

After this second half-turn, the potential difference again reverses, making D_1 positive and D_2 negative, and again the proton speeds up. Thus, as the potentials reverse periodically, the proton travels faster and faster, moving in ever-expanding circles, until, reaching the outer edge, it passes through a narrow open window W.

Upon leaving W all protons must pass close to a negatively charged plate P where, by attraction, their paths are straightened out and they become a separated beam of projectiles. Whatever substance is to be bombarded is then placed in this beam, and the disintegrated fragments are studied by means of various detective devices.

The fundamental principle that makes the cyclotron work at all is the fact that *the time required for a charged particle to make one complete turn within the dees is the same for all speeds*. The faster a particle travels the larger is the circle it must traverse, thus keeping the time constant. Hence, with a constant frequency of the alternating current supply, some particles may be just starting their acceleration near the center while others farther out have already acquired higher speeds. The result is a pulsed stream of protons emerging from the window W.

If the alternating current voltage applied between the *dees* of the cyclotron is 200,000 volts, with each half-turn a particle obtains an added velocity equivalent to 200,000 volts. If a proton makes 25 complete revolutions before leaving the chamber at W, it will have acquired a velocity equivalent to 200,000 times 25 times 2, or 10,000,000 volts. Here, then, is a beam of 10 Mev protons acquired by the application of a potential difference only $\frac{1}{50}$ as great.

Deuterium is a gas composed of atoms of mass 2, the heavy isotope of hydrogen. (See Appendix III.) A deuterium atom without its one and only orbital electron is called a *deuteron*. When hydrogen in the evacuated chamber of the 60-in. cyclotron is replaced by *deuterium* and the magnetic field strength is doubled, a beam of high-energy deuterons is obtained. Having twice the mass, but the same unit positive charge as protons, these particles acquire twice as much energy. If helium gas is used in place of deuterium, many of the atoms become doubly ionized at the source and after acceleration emerge from the cyclotron window as alpha particles with an energy of about 40 Mev. By increasing or decreasing the frequency of the potential applied to the dees, and properly adjusting the magnetic field, protons, deuterons, or α particles of 10, 20, and 40 Mev, respectively, can be produced.

Some of the details of the cyclotron shown in Fig. 28A are as follows: The dimension, 60 in., refers to the diameter of the poles of the cyclotron magnet; this in turn limits the size of the dees and therefore the maximum energy available in the form of atomic projectiles. Most of the instrument's total weight of 200 tons lies in the solid iron core and the pole pieces located inside the field windings. The latter, consisting of many turns of thick copper wire, are encased in tanks W_S and W_N through which cooling fluid is continually circulated.

A photograph of an 11-Mev deuteron beam from the Harvard University cyclotron is shown in Fig. 28C. From the point where the particles emerge from

Courtesy, Harvard University Press and A. K. Solomon

Fig. 28C   *Photograph by Paul Donaldson of an 11-Mev deuteron beam from the Harvard cyclotron.*

the cyclotron window at the left center to where they come to rest in mid-air at the lower right, they ionize the air molecules and atoms, causing them to emit visible light.

28.2. Theory of the Cyclotron

The theory of the cyclotron involves simple classical laws describing the motion of a charged particle in a uniform magnetic field. By Eq. (4e), the force on a particle in a magnetic field is given by Bev, and this is equal to the centripetal force mv^2/r:

$$Bev = \frac{mv^2}{r} \tag{28a}$$

Here e is the charge on the particle in coulombs, m is its mass in kilograms, v is its velocity in meters per second, and r is the radius of its circular path in meters. To find the time required for any charged particle to make one complete circle the formula $s = vt$ from mechanics is employed. The distance traveled in one turn is represented by s, and T represents the time:

$$T = \frac{s}{v} = \frac{2\pi r}{v} \tag{28b}$$

By solving Eq. (28a) for v, and substituting Eq. (28b), we obtain

$$v = \frac{Ber}{m} \quad \text{and} \quad T = \frac{2\pi}{B} \cdot \frac{m}{e} \tag{28c}$$

The right-hand equation shows that the *period* T is independent of r and v and for like particles (that is, the same e and m) varies with the magnetic induction B. The alternating potential difference E applied to the cyclotron dees must therefore match in frequency the particles' motion due to the field B. It is customary in practice to apply a fixed frequency to the dees and adjust the current in the magnetic field coils until resonance occurs. In the 60-in. cyclotron, $B = 1.6$ webers/meter² or 16,000 gauss.

28.3. The Van de Graaff Generator

This machine, developed in 1931 by R. Van de Graaff at Princeton University, employs the principle of the electrostatic generator discovered many years ago.

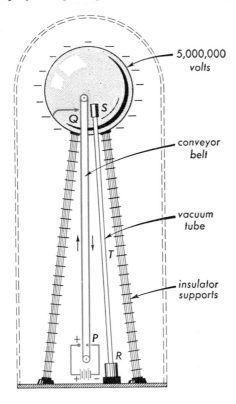

5,000,000 volts

conveyor belt

vacuum tube

insulator supports

Fig. 28D　*Diagram of a Van de Graaff generator of high voltage.*

A typical installation, as shown in Fig. 28D, consists of a large hollow sphere, supported on insulating columns and charged by a belt conveying electrical charges from a battery at ground potential and depositing it inside the sphere. The fabric conveyor belt, a foot or more in width and running over well-aligned rollers, travels about 60 mi/hr.

As the belt passes between the metallic surface and row of needle points at P, electrons from the points jump toward the positive electrode and are caught by the belt. Upon entering the sphere at the top, the electrons jump to the needle points Q where they go quickly to the outside surface of the sphere. The "spraying" of electrons *to* and *from* the points is assured by keeping the battery potential high (about 50,000 volts) to maintain a "brush discharge." As more and more electrons arrive at the sphere, its negative potential rises higher and higher until leakage into the surrounding air and through the insulators becomes equally fast.

Atomic particles to be accelerated are generated inside a vacuum tube source S inside the sphere. Starting at the top of a long straight vacuum tube T, electrons are accelerated downward toward ground potential, where, acquiring the full energy of the available voltage, they are allowed to bombard whatever target is being studied. Where installations are designed for accelerating protons, deuterons, or α particles, the battery potential is reversed and the sphere acquires a high positive potential with respect to the ground.

28.4. The Betatron

The *betatron*, invented in 1941 by D. W. Kerst at the University of Illinois, is an electron accelerator capable of producing electron beams of high energy as well as X rays of extremely high penetrating power. This ingenious device differs from the cyclotron in at least two fundamental respects: first, the electrons are accelerated by a rapidly changing magnetic field, and, second, the circular orbit of the particles has a constant radius.

A cross-section diagram of a 20-Mev betatron is shown in Fig. 28E. A glass vacuum tube in the shape of a *doughnut* and containing an *electron gun* is mounted between the poles of an electromagnet. An alternating current (180 cycles/sec) applied to the coils causes some of the magnetic lines of force to pass through the vacuum tube at the electron orbit and the remainder through the orbit center, as shown below. Electrons are injected only at the beginning of each quarter cycle when the field begins to increase in the "up" direction. The increasing field through the center of the orbit gives rise to an electromotive force, tangent to the orbit, speeding up the electrons; whereas the increasing field at the orbit is just sufficient to increase the centripetal force and keep the electrons from spiraling outward. The stability of such an orbit is brought about by properly shaping the pole faces of the magnet, adjusting the frequency and strength of the magnetic field, and injecting the electrons at the proper voltage at the appropriate time.

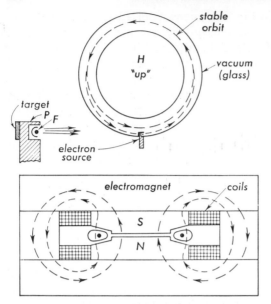

Fig. 28E   *Cross-section diagrams of an electron accelerator, called a betatron.*

During World War II, a 350-ton betatron was constructed by the General Electric Company and put into use as a source of extremely penetrating X rays. In this instrument electrons accelerated to 100-Mev energy, and impinging upon a target, give rise to X rays capable of penetrating many feet of solid iron and lead.

28.5. The Synchro-Cyclotron

The firm belief that new and fundamental discoveries in nuclear physics can be made with atomic projectiles having greater and greater energies has led scientists and engineers in various institutions to combine their efforts in groups to design and construct larger and larger atomic accelerators. One such instrument is the synchro-cyclotron shown in Fig. 28F.

The fundamental differences between this and the orthodox cyclotron are the use of one dee in place of two and the use of an applied alternating-current potential whose frequency is made to rise and fall periodically instead of remaining constant. The principles of operation are illustrated in Fig. 28G. Starting at the center, protons, deuterons, or α particles are made to move in circles of increasing radius, the acceleration taking place as they enter and leave the lips of the dee. At the outer edge they are deflected out of the field, as in the cyclotron, or allowed to strike a suitable target inside the vacuum chamber.

The theory of the synchro-cyclotron* is based upon "phase stability," a principle fundamental to nearly all high-energy accelerators. This relatively new

* The theory of phase stability of atomic accelerators of high energy was first developed in 1945 by V. Veksler (*Journal of Physics*, USSR, Vol. 9, p. 153 (1945)), and independently by E. M. McMillan (*Physical Review*, Vol. 68, p. 143 (1945)).

Fig. 28F *Photograph of the 184-in., 4000-ton, synchro-cyclotron at the University of California.*

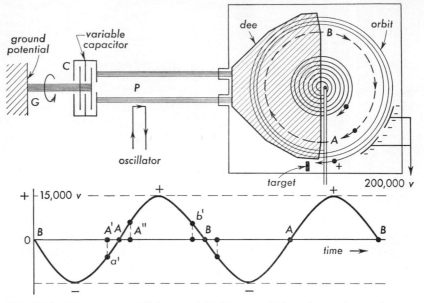

Fig. 28G   *Illustration of "phase stability" as applied to the synchro-cyclotron.*

idea can be derived from the well-founded theory that, as a particle is accelerated and approaches the speed of light, continued acceleration increases the mass, while the speed approaches a nearly constant value. In other words, as more and more energy is given to a particle, more and more of it is stored as *increased mass* and less and less as *increased speed*.

To apply this principle to an accelerator like the synchro-cyclotron, consider the conditions that exist when a positively charged particle in a uniform magnetic field is moving with constant speed in an orbit of constant radius and at the same time in synchronism with an applied high-frequency potential difference. Such an orbit is represented by the dotted circle in Fig. 28G, with the particle entering and leaving the lips of the dee at A and B when the potential (see graph below) is zero. By Eq. (28a), $Bev = mv^2/r$. Since, from mechanics, angular velocity $\omega = v/r$, and the frequency of revolution $f = \omega/2\pi$, direct substitution in Eq. (28a) gives $m = Be/2\pi f$. Multiplying both sides of this equation by the square of the speed of light c^2 gives

$$mc^2 = \frac{Bec^2}{2\pi f}$$

where mc^2 represents the total energy E of the particle,

$$E = \frac{Bec^2}{2\pi f} \tag{28d}$$

It may be shown from these equations that mc^2 includes the *rest mass* m_0c^2 of the particle, and that to increase E the magnetic induction can be *increased* as in the betatron and synchrotron (see §28.6), or *the frequency can be decreased* as in the synchro-cyclotron.

Returning now to the high-speed but stationary orbit *AB* in Fig. 28G, assume that a particle is a little early in entering the dee. Arriving there, as shown by *A′* in the graph below, the dee has a negative potential *a′* and the particle is accelerated by attraction. Being accelerated, the mass increases with little increase in speed. Owing principally to increased mass, the particle now describes a larger circle, and the next time around it has dropped back to arrive more nearly at the time of *zero potential.* Hence, if the magnetic induction *B* were to be increased or the frequency *f* were to be decreased, the particle might be made to continually enter and leave the dee ahead of the zero phase and receive a forward push each time around.

If a particle gets behind the zero phase (*A* to *A″* in the graph), it will not receive a forward pulse and, moving with nearly constant speed, will permit a decreasing frequency of an applied potential difference to catch up and get ahead again. Hence phase stability is assured and forward impulses will occur as the particle spirals outward with increasing energy.

In the 184-in. synchro-cyclotron, the high-frequency potential difference is applied to the dee stem at *P* and the variable capacitor *C*, composed of a fast rotating set of "fan blades" passing between a set of stationary blades, varies the frequency up and down through relatively wide limits. In producing 200 Mev deuterons, the frequency of the dee potential rises and falls 120 times/sec between the limits of 12.5 and 8.5 megacycles/sec.

The positive ions are pulsed into the center of the dee when the frequency is 11.3 megacycles, and they arrive at the outer edge when it has dropped to about 9.6 megacycles. Having made 10,000 turns around the chamber in a period of only one-thousandth of a second, the deuterons have an energy of 300 Mev. The target, located inside the vacuum chamber, is therefore bombarded by pulses coming at the rate of 120 per sec.

A drop from 11.3 to 9.6 megacycles decreases *f* (see Eq. (28d)) by about 15%, thereby increasing a deuteron's energy *E* and mass *m* by 15%, or 0.3 atomic mass unit (abbr. amu). Such an increase, by Eq. (24m), is equivalent to 300 Mev.

28.6. The Synchrotron

This device is an electron accelerator employing the principles of the *cyclotron* and *phase stability*. A cut-away diagram of such an instrument is shown in Fig. 28H. As in the betatron, electrons are injected into a doughnut-shaped vacuum chamber by an electron gun. Operating first as a betatron, the electrons are accelerated in an orbit of fixed radius by a rapidly increasing magnetic field. The rising field is produced by discharging a large capacitor bank through the magnet coils. Part of the field goes through the relatively small *flux bars* near the orbit center, and part through the pole faces and vacuum chamber.

As the electrons quickly approach the speed of light, their mass begins to

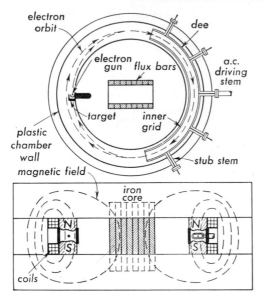

Fig. 28H   *Cross-section diagrams of a 300-Mev synchrotron.*

increase rapidly; from about 2 Mev on, they move at almost constant speed (between 98 % and 100 % the speed of light). Increasing energy is added in this *second phase* by an alternating potential difference applied to the *sector dee* shown in the diagram. Instead of decreasing the frequency, as in the synchrocyclotron, the magnetic induction B is increased (see Eq. (28d)) and as the electron *mass* increases, the stronger field maintains the beam orbit constant.

Upon reaching a maximum energy of 300 Mev, for example, the electrons, with a mass some 600 times their *rest mass*, are caused to spiral inward to strike a tungsten target, where they produce 300-Mev X rays.

28.7. The Linear Accelerator

Although linear accelerators were proposed as early as 1929, and several were constructed, they have not proved satisfactory until recently. Applying the principles of tubular wave guides and resonant cavities, L. Alvarez and his collaborators, immediately after World War II, constructed the first successful linear accelerator. Since a giant machine employing the same principles is soon to be built at Stanford University, the ideas involved are of considerable importance.

A cut-away diagram of part of the Alvarez lineac is given in Fig. 28I. Protons are initially produced and accelerated to 4 Mev by a pressure Van de Graaff generator (see Fig. 28D) and then injected at that energy into one end of a 40-ft tank, as shown at the upper left. Once inside they are further accelerated as they pass through a series of "drift tubes," and arrive at the other end with an energy of about 40 Mev.

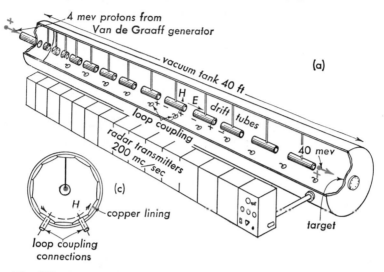

Fig. 28I   *Diagrams showing the first section of a linear accelerator.*

The tank cavity within the copper lining, fed by 30 radar transmitting oscillators, is set resonating at its *dominant mode* at a frequency of 200 megacycles. The "standing-wave" conditions set up are such that the electric field E is parallel to the tube axis and is everywhere rising and falling together. The lengths of the drift tubes gradually increase so that the protons cross each gap when the field E is to the right, and are inside the tubes, in a field free space, when the field is to the left. It now seems possible that additional tank sections can be added end to end to this system to obtain almost any desired energy. Calculations indicate 1 Mev per lineal foot can be expected from such a system.

28.8. Billion Electron Volt Accelerators

The design and construction of an instrument capable of accelerating particles to energies of billions of electron volts (*Bev*) involve many problems. Not the least of these is the economic factor concerned primarily with the initial cost of such an instrument as well as its subsequent maintenance.

A number of accelerators producing particles of 1 Bev or more are now in operation and others of greater and greater energy are being planned.

IN OPERATION (1963)

Pasadena, California	1.1 Bev (electrons)
Ithaca, New York	1.3 Bev (electrons)
Berkeley, California	6.2 Bev (protons)
Dubna, Russia	10 Bev (protons)
Geneva, Switzerland	28 Bev (protons)
Brookhaven, Long Island	32 Bev (protons)
Cambridge, Massachusetts	7.5 Bev (electrons)
Argonne, Illinois	12.5 Bev (protons)

PLANNED OR UNDER CONSTRUCTION
Palo Alto, California   15–45 Bev (electrons)

The basic design of one of these large instruments, the 6.2 Bev *bevatron*, is shown in Fig. 28J. The magnet arrangement consists of four quadrant segments

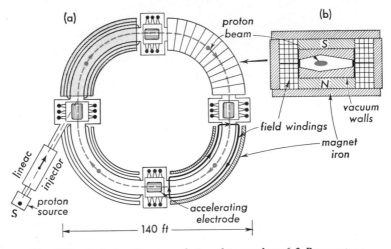

Fig. 28J   *Berkeley Bevatron designed to produce 6.2 Bev protons.*

spaced so that the particle orbits are quarter-circles connected by 20-ft straight sections. The electrical power supplied to the 10,000-ton magnet is provided by a motor generator with a large flywheel. During buildup of the magnetic field, a peak power of 100,000 kilowatts is drawn from the flywheel and stored in the magnet. As the field is reduced between beam pulses, the generator acts as a motor and returns energy to the flywheel.

The protons from a source S are first accelerated to 10 Mev by a linear accelerator, and then injected into the 385-ft race track proper. As they pass through the accelerator electrode in one of the straight sections and are speeded up by the high-frequency potential differences, the magnetic field increases at the proper rate to keep the beam in the same orbit. The output beam consists, as it does in all high-energy accelerators, of a series of pulses. From the time of injection, each pulse of protons takes about 2 sec to acquire its final speed, and in so doing makes about four million revolutions of the orbit and travels about 300,000 miles.

A color photograph of one section of the bevatron is reproduced in Fig. 28K.

QUESTIONS AND PROBLEMS

   1. What is a cyclotron? What is its purpose? What does it accelerate?
   2. What is deuterium? Is deuterium the same as hydrogen?
   3. What is a Van de Graaff generator? What is it frequently used for? Can it be used to accelerate protons and electrons?

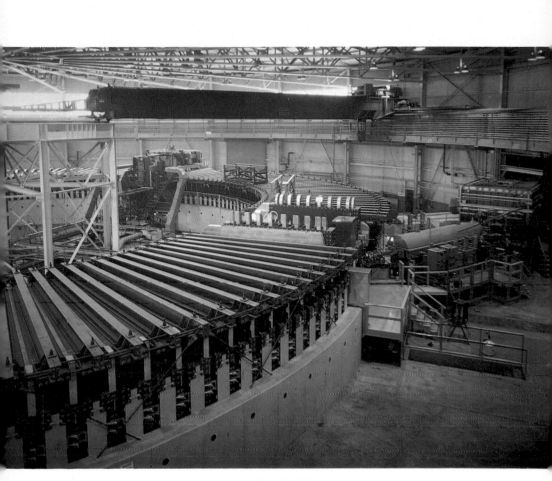

THE BEVATRON

Fig. 28K *A six-billion volt accelerator of atomic particles, located at the University of California, Berkeley.* (Photographed by K. Hildebrand and G. Kagawa. Courtesy of E. Lofgren, D. Cooksey, the Radiation Laboratory at the University of California, and the Atomic Energy Commission.) *Just behind the two men on the platform, at the far right and center, can be seen the rectangular housing of the atomic source and the Cockroft-Walton accelerator. The cylindrical tank section containing the linear accelerator is clearly seen leading into the inflector assembly between quadrants 1 and 4. The main accelerating electrode assembly with the yellow colored ducts leading to it is seen farther back between quadrants 3 and 4. Note the overhead crane used for assembling, repairing, and the handling of massive apparatus and equipment.*

4. What does the abbreviation "Mev" stand for? What does the abbreviation "Bev" stand for?

5. What is a betatron? Where do you think it got its name?

6. Approximately what is the highest energy to which atomic particles have been accelerated in the laboratory?

7. If the frequency of the potential applied to the dees of a cyclotron is 9.2×10^6 cycles/sec, what must be the magnetic induction B to accelerate α particles?

8. Calculate the frequency of the oscillating potential that must be applied to a cyclotron in which deuterons are accelerated. Assume the magnetic induction has a constant value of 25,000 gauss. *Ans.* 19.0 Mc/sec.

9. The frequency applied to the dees of a cyclotron is 8.6 megacycles/sec. What must be the magnetic induction B if protons are to be accelerated?

10. What total accelerator voltage will give protons a velocity of 99% the speed of light? *Ans.* 5.7 Bev.

Transmutation of the Elements

The use of high-speed atomic particles, whether they come from radioactive materials, from cyclotrons, or from cosmic rays has led in recent years to the discovery of many secrets that lie hidden within the atomic nucleus. While the thousands of experiments performed have led to the discovery of many new elementary particles several hundred radioactive isotopes of the known elements have been produced by the process of nuclear bombardment.

29.1. Proton and Deuteron Disintegrations

When high-energy protons or deuterons are used to bombard different known elements, various disintegration products are formed. An experimental arrangement in which the cyclotron acts as the source of high-speed particles is shown in Fig. 29A. To determine the nature of the disintegration taking place within

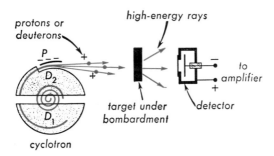

Fig. 29A *Experimental arrangement generally used for bombarding known substances with high-speed deuterons from the cyclotron and for detecting the disintegration products with an ionization chamber as a detector.*

the substance under bombardment, it is common practice to identify the penetrating rays emerging from the other side by the use of suitable detectors.

Numerous experiments have shown that the disintegration products to be looked for may be *protons*, α *particles*, *neutrons*, γ *rays*, *mesons*, *electrons*, *positrons*, etc. For some of these penetrating rays, one kind of detector may be more suitable than another. The *scintillation counter*, for example, is particularly useful in detecting γ rays and electrons, whereas the *Wilson cloud chamber* and *ionization chamber* are useful in detecting protons, α particles, or neutrons.

The detector shown at the right in Fig. 29A represents an ionization chamber. When a Wilson cloud chamber is used to identify disintegration products, charged particles can be identified by the density of their fog tracks, and their energy can be determined by the curvature of the tracks when a magnetic field is applied. This is illustrated in Fig. 29H for positrons. Once the nature of the emerging rays from a bombarded target is known, the recoil product of the disintegration also becomes known by writing down a reaction equation. Six examples of such reaction equations are given by the following:

Q values,
Mev

$$_1H^1 + {}_9F^{19} = {}_8O^{16} + {}_2He^4 \qquad 8.113 \qquad (29a)$$
$$_1H^1 + {}_5B^{11} = {}_6C^{12} + \gamma \text{ ray} \qquad 15.955 \qquad (29b)$$
$$_1H^2 + {}_7N^{14} = {}_6C^{12} + {}_2He^4 \qquad 13.574 \qquad (29c)$$
$$_1H^2 + {}_8O^{16} = {}_7N^{14} + {}_2He^4 \qquad 3.110 \qquad (29d)$$
$$_1H^2 + {}_3Li^6 = {}_3Li^7 + {}_1H^1 \qquad 5.028 \qquad (29e)$$
$$_1H^2 + {}_4Be^9 = {}_5B^{10} + {}_0n^1 \qquad 4.362 \qquad (29f)$$

It is customary to omit the *mass energy* of the bombarding particle from the left-hand side of all reaction equations and to designate the total energy liberated by the disintegration as shown at the right above. The values of Q given above therefore represent the experimentally determined values of the energy over and above that supplied by the incident projectile.

Consider the fifth reaction, which can be taken to represent an experiment in which a beam of 2 Mev deuterons from the cyclotron bombards a target of lithium metal. From the other side of the target a stream of high-energy protons would be detected. If they are sent through a Wilson cloud chamber in a magnetic field, for example, their tracks could be identified as proton tracks and their energy determined by the curvature of the tracks to be 6.0 Mev. The lithium atoms in the target would recoil with the remaining 1 Mev. When the accurate weights of the four nuclei involved are taken into account, there is a total loss of 0.00540 atomic mass units. Multiplying by 931, this is equivalent to 5.0 Mev energy. This value plus the energy of the incident bombarding particle gives 7.0 Mev.

As a second example consider Eq. (29f) in which deuterons, bombarding beryllium metal, produce high-speed neutrons and recoiling boron nuclei. This particular disintegration is important experimentally because it is used as a means of obtaining intense beams of neutrons for use as projectiles in other disintegrations. The nuclear changes are illustrated schematically in Fig. 29B. The available energy from the loss in mass alone is equivalent to 4.4 Mev; so that, if deuterons with an energy of 7 Mev are used to bombard the beryllium target, the available energy becomes 11.4 Mev, 1 Mev going to the recoil boron nucleus and approximately 10.4 Mev to the neutron.

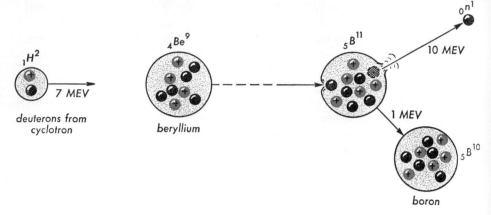

Fig. 29B   *Deuteron disintegration of a beryllium nucleus to produce high-speed neutrons.*

The table of accurately known masses of nuclides given in Appendix IV includes a column 3 of excess mass values in Mev. These values represent the difference between the mass of each nuclide and its *mass number*. These numbers have been multiplied by 931 to give the equivalent energy in Mev in column 4.

As a practical matter column 3 of mass excess energies is very useful. Consider as an illustration the following example.

Example 1. Find the Q value for the reaction shown in Eq. (29e), using the mass energies given in column 3 of Appendix IV.

Solution. The energy equivalents for each of the two nuclides in collision are first added to obtain

$$\begin{array}{ll} {}_1H^2 & 13.135 \text{ Mev} \\ {}_3Li^6 & \underline{14.089 \text{ Mev}} \\ & 27.224 \text{ Mev} \end{array}$$

The energy equivalents for the two product nuclides are next added to obtain

$$\begin{array}{ll} {}_3Li^7 & 14.908 \text{ Mev} \\ {}_1H^1 & \underline{7.288 \text{ Mev}} \\ & 22.196 \text{ Mev} \end{array}$$

The difference between these totals gives directly

$$Q = +5.028 \text{ Mev}$$

29.2. Multiple Disintegrations

A study of certain disintegration experiments shows that some of the unstable nuclei created by the capture of a proton or deuteron by a stable nucleus split up into more than two stable nuclei. Examples of this arise when boron is bombarded by protons and when nitrogen is bombarded by deuterons. In the case of boron (see Fig. 29C), the proton is first captured by a ${}_5B^{11}$ nucleus to form an unstable carbon nucleus, ${}_6C^{12}$. This composite structure, as we will see

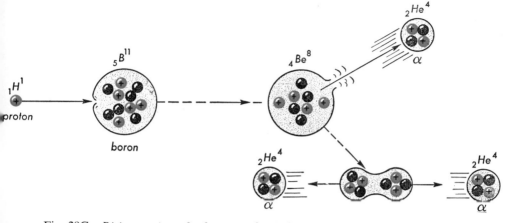

Fig. 29C   *Disintegration of a boron nucleus of mass 11 by a proton to produce three α particles.*

in a later chapter, has too much mass to be a stable $_6C^{12}$ nucleus, and it disintegrates by the expulsion of an α particle with several million volts energy, leaving behind a beryllium nucleus, $_4Be^8$:

$$_1H^1 + {}_5B^{11} = {}_4Be^8 + {}_2He^4 \tag{29g}$$

This nuclear combination is still unstable and splits apart into two more α particles:

$$_4Be^8 = {}_2He^4 + {}_2He^4$$

When the phenomenon was first observed, it was thought that all three α particles came apart simultaneously, but further observations showed that first one and then two were ejected. The total energy liberated has been measured to be about 11 Mev, which checks almost exactly with the value obtained from the loss in mass.

29.3. Branch Disintegrations

It so happens that when certain atoms are bombarded with high-speed particles, two or more disintegration processes may subsequently take place. As an illustration of this phenomenon of *branch disintegration*, consider the proton bombardment of beryllium in which the following two types of disintegration have been identified:

$$_1H^1 + {}_4Be^9 = {}_3Li^6 + {}_2He^4 \tag{29h}$$

$$_1H^1 + {}_4Be^9 = {}_5B^{10} + \gamma \text{ ray} \tag{29i}$$

When a proton is captured by a beryllium nucleus to form an unstable boron nucleus, $_5B^{10}$, there are two ways in which it may split up. The instability of the boron in the first place is due to the presence of too much mass. Such an atom is said to be in an *excited state,* for by the emission of a γ ray it gives up its surplus energy and becomes a stable $_5B^{10}$ nucleus, or by splitting up into two particles it gives up its surplus in the form of kinetic energy to become $_2He^4$ and $_3Li^6$, two stable nuclei.

An abbreviated notation for nuclear reactions is illustrated by the following examples:

$$_1H^1 + {}_9F^{19} = {}_8O^{16} + {}_2He^4 \qquad \text{Abbr.} \quad F^{19} (p, \alpha) O^{16}$$
$$_1H^2 + {}_7N^{14} = {}_6C^{12} + {}_2He^4 \qquad \text{Abbr.} \quad N^{14} (d, \alpha) C^{12}$$
$$_1H^2 + {}_3Li^7 = {}_2He^4 + {}_2He^4 + {}_0n^1 \qquad \text{Abbr.} \quad Li^7 (d, 2\alpha n)$$

29.4. Discovery of Induced Radioactivity

The discovery of induced radioactivity was made in 1934 by F. Joliot and I. Curie Joliot.*

For years the Curie-Joliots, as they are now often called, had been exposing various substances to the α rays from naturally radioactive elements and had been studying the various disintegrations that took place. In the specific instance referred to above, they bombarded aluminum with α *particles from polonium* and measured the energies of the ejected neutrons by the recoiling of protons from paraffin (see Fig. 29D). They observed that, even after the polonium source

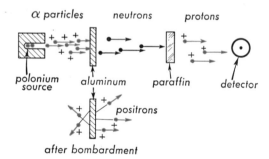

Fig. 29D *Experimental arrangement used by the Curie-Joliots when they discovered induced radioactivity.*

was taken away, the detector continued to respond to some kind of penetrating radiation. Upon investigating the nature of these rays, they found positively charged electrons coming from the aluminum.

Repeating the experiments to make certain of the results, they came to the conclusion that, under the bombardment of α particles, the aluminum had become radioactive in its own right. What was happening has since been verified: α particles striking aluminum nuclei are captured, and the resulting nuclei disintegrate with the violent ejection of neutrons:

$$_2He^4 + {}_{13}Al^{27} = {}_{15}P^{30} + {}_0n^1 \tag{29j}$$

* Irene Curie, daughter of the most famous woman physicist, Marie Curie, is the wife of Frederick Joliot. Because of the now famous name of Curie, the physicists of the world hyphenate the name and call them Mme. Curie-Joliot and M. J. F. Joliot, or for short the Curie-Joliots. For years they worked with radioactive substances in the famous laboratory of the late Mme. Marie Curie at the Radium Institute in Paris.

The newly created recoil particles, with a charge of $+15$ and mass 30 have been identified as phosphorus nuclei which are not stable but radioactive. Spontaneously disintegrating, these radioactive phosphorus nuclei $_{15}P^{30}$ shoot out positrons, leaving behind them stable silicon atoms of charge $+14$ and mass 30:

$$_{15}P^{30} = {}_{14}Si^{30} + {}_1e^0 + \nu \tag{29k}$$

The *half-life* of this activity, which measures the rate of decay of the phosphorus into silicon (for the meaning of half-life see §24.2), is only 2.5 min. The emission of a positively charged electron is accompanied by a neutrino, just as in the case of β-emission. Oftentimes, however, the neutrino is omitted from the reaction since it carries no charge and no rest mass.

Although the mass of the electron is not zero, it is so small compared with unit mass (the mass of one electron, it will be remembered, is $\frac{1}{1836}$th of the mass of the proton) that e is written with a zero superscript. According to this notation a positron is written $_1e^0$ and an electron $_{-1}e^0$.

Because the phosphorus does not all disintegrate immediately, it has been possible to identify the activity as coming from the newly created phosphorus atoms in the following way. A piece of aluminum metal, immediately after being bombarded, is dissolved in hydrochloric acid together with some ordinary inactive phosphorus, and then a standard chemical separation is made. Testing each part separately, the radioactivity is found to be present with the phosphorus residue, and not with the aluminum.

29.5. The Discovery of Radioactive Sodium

Immediately after the discovery of induced radioactivity by the Curie Joliots, Lawrence bombarded sodium with 2-Mev deuterons from the cyclotron and found that it too, like aluminum, became radioactive (see Fig. 29E). Upon

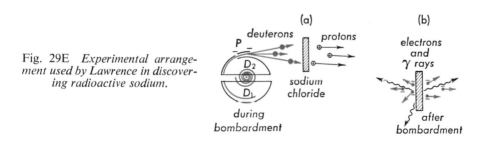

Fig. 29E *Experimental arrangement used by Lawrence in discovering radioactive sodium.*

testing for the nature of the rays given off during bombardment, Lawrence found protons with an energy of about 7 Mev. When the sodium target was removed from the deuteron beam, as shown in diagram (b), and then tested for activity, it was found to be emitting both electrons and γ rays. The bombarding reaction, therefore, is

$$_1H^2 + {}_{11}Na^{23} = {}_{11}Na^{24} + {}_1H^1 \tag{29l}$$

followed by the radioactive decay of the unstable sodium nuclei,

$$_{11}Na^{24} = {}_{12}Mg^{24} + {}_{-1}e^0 + \gamma \text{ ray} \tag{29m}$$

The first stage of the disintegration process is shown at the left in Fig. 29F, and the radioactive decay is shown at the right.

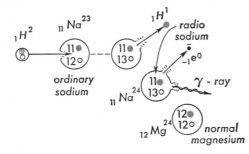

Fig. 29F *The production and disintegration of radiosodium.*

The residual nucleus $_{11}Na^{24}$ of the first disintegration is called *radiosodium.* Having a charge of $+11$ and mass of 24, it must be an isotope of sodium not found in nature. Since measurements of the activity of radiosodium give a *half-life* of only 15 hr, it is clear why such atoms are not found in nature. If they were formed some time in the ages past, they would all have disintegrated by this time.

An energy level diagram for the decay of radioactive sodium is shown in Fig. 29G. The left-hand energy scales are given equally well in *amu* or in *Mev.*

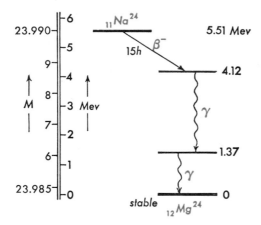

Fig. 29G *Energy-level diagram for radioactive decay of sodium-24.*

After emitting an electron, the nuclear charge increases by unity, and we have a magnesium nucleus in an excited state or energy level. With two successive transitions in which 2.75 Mev and 1.37 Mev γ rays are emitted, the nucleus becomes a stable system, a normal isotope of magnesium $_{12}Mg^{24}$.

Up to the present time more than six hundred different kinds of radioactive atoms have been produced in the laboratory. Two examples, in addition to those already given, are illustrated by the following reactions:

$$_1H^2 + {}_{15}P^{31} = {}_{15}P^{32} + {}_1H^1, \qquad {}_{15}P^{32} = {}_{16}S^{32} + {}_{-1}e^0 \tag{29n}$$

$$_1H^2 + {}_6C^{12} = {}_7N^{13} + {}_0n^1, \qquad {}_7N^{13} = {}_6C^{13} + {}_1e^0 \tag{29o}$$

The first of these reactions forms *radioactive phosphorus* $_{15}P^{32}$ which is *electron active* with a half-life of 15 days.

A cloud-chamber photograph of the positrons emitted by radioactive nitrogen $_7N^{13}$ is reproduced in Fig. 29H. The magnetic field bends all the rays in the

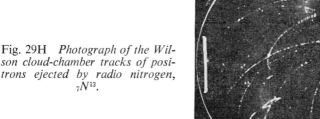

Fig. 29H　*Photograph of the Wilson cloud-chamber tracks of positrons ejected by radio nitrogen,* $_7N^{13}$.

same direction, indicating that all are positive charges. The low-density fogdrops forming the tracks indicate particles with the mass of an electron.

29.6. Electron Capture

Many radioactive nuclei, when created by some collision process, are unstable to the extent of one extra positive charge. Although many such nuclei disintegrate and become stable by the emission of a positron, others draw to them an orbital electron from the K shell of the same atom. Inside the nucleus this negative charge neutralizes a positive charge, whereas outside an L or M electron jumps into the K shell vacancy with the simultaneous emission of a characteristic X ray. See Fig. 29I. Beryllium-7 and gallium-65 are specific examples of unstable nuclei in which K *capture* occurs. The reactions for these are

$$_4Be^7 + {}_{-1}e^0 = {}_3Li^7$$
$$_{31}Ga^{65} + {}_{-1}e^0 = {}_{30}Zn^{65} \tag{29p}$$

An energy-level diagram involving the radioactive decay of actinium-226 is shown in Fig. 29J. This nucleus is β-active; 80% of its atoms disintegrating by emitting β particles with an end-point energy of approximately 1 Mev, each followed by the emission of one or two γ rays to become thorium-226. The other 20% of the atoms capture an electron from the K shell, and immediately emit one or two γ rays to become radium-226.

Electron capture (abbr. E.C.) changes one of the nuclear protons into a

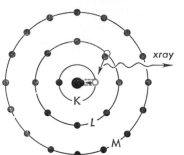

Fig. 29I *Unstable nucleus captures an electron from the K-shell of orbital electrons. As a consequence, an X ray is emitted.*

neutron, and the atomic number drops by unity from 89 to 88. The emission of a β particle changes one of the neutrons into a proton, and the atomic number increases by unity from 89 to 90. In some few isotopes, E.C. is the only means of radioactive decay.

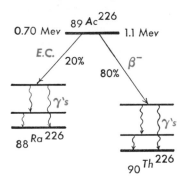

Fig. 29J *Energy-level diagram of the two-way radioactive decay of actinium-226.*

29.7. Nuclear Stability

We now know that if we bombard targets with atomic particles of sufficiently high energy, every known stable element can be converted into radioactive isotopes of that element or of neighboring elements. Furthermore, a number of isotopes can be produced for any one element, some with masses smaller than any of its stable isotopes, and some with greater masses. In the case of copper, for example, the following isotopes have been produced.

Copper $Z = 29$

MASS	ACTIVITY	HALF-LIFE			
58	β^+	$3s$			
59	β^+	$81s$			
60	β^+	$24m$			
61	β^+	$3.3h$			
62	β^+	$10m$			
63	Stable	—	70%		
64	$\beta^-/\beta^+ = 2$	$12.8h$		42%	E.C.
65	Stable	—	30%		
66	β^-	$5.1m$			
67	β^-	$58s$			
68	β^-	$32s$			

The first five isotopes have too many protons to be stable nuclides, and by positron emission convert a proton into a neutron and become an isotope of nickel. The last three isotopes have too few protons, and by electron emission convert a neutron into a proton and become an isotope of zinc.

Isotope $_{29}Cu^{64}$, lying between two stable isotopes, is an interesting combination of nucleons, since these nuclei are found to decay in any one of four different ways. All four processes are illustrated in an energy-level diagram in Fig. 29K.

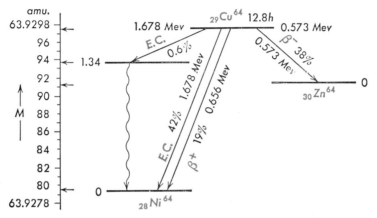

Fig. 29K   *Energy-level diagram of a four-way radioactive decay of copper-64.*

Approximately 38% of the copper-64 nuclei decay by the emission of a β^- and a neutrino, to become zinc-64:

$$_{29}Cu^{64} = {_{30}}Zn^{64} + {_{-1}}e^0 + \nu \qquad +0.573 \text{ Mev}$$

The β^- end-point energy of 0.573 Mev corresponds to a decrease in nuclear mass of 0.000616 amu. *The masses of all nuclides include the masses of all orbital electrons for the neutral atoms.* When $_{29}Cu^{64}$ ejects an electron from the nucleus, the daughter product $_{30}Zn^{64}$ picks up a stray electron to become a neutral atom. The term *nuclide* is applied to any known assembly of nuclear particles called a nucleus.

Approximately 42% of the copper-64 nuclei decay by *electron capture* with the simultaneous emission of a neutrino:

$$_{29}Cu^{64} + {_{-1}}e^0 = {_{28}}Ni^{64} + \nu \qquad +1.678 \text{ Mev}$$

The end-point energy of this process is 1.678 Mev, and the neutrino spin of $\frac{1}{2}\hbar$ compensates for the spin of $\frac{1}{2}\hbar$ given the nucleus by the captured electron.

Some 19% of the copper-64 nuclei emit a β^+ and a neutrino to become stable nuclei, nickel-64:

$$_{29}Cu^{64} = {_{28}}Ni^{64} + {_1}e^0 + \nu \qquad +0.656 \text{ Mev}$$

The end-point energy for these positrons is 0.656 Mev. The difference between the two end-point energies of 1.678 Mev and 0.656 Mev, or 1.022 Mev, is just the mass energy equivalence of two electrons. When $_{29}Cu^{64}$ ejects a positron from the nucleus, the daughter product also drops one of its orbital electrons to become a neutral atom. The mass energy equivalence of one electron ($m = 0.00549$ amu) is 0.511 Mev.

Finally, a few copper-64 nuclei, approximately 0.6%, capture an orbital electron and emit a neutrino, leaving the daughter product, $_{28}Ni^{64}$ in an *excited state*. This process is immediately followed by the emission of a 1.34 Mev gamma ray:

$$_{29}Cu^{64} + _{-1}e^0 = _{28}Ni^{64} + \nu + \gamma \text{ ray} \qquad 1.678 \text{ Mev}$$

The end-point energy for this electron capture process alone is only 0.338 Mev.

A small section of a chart of the known nuclides of all the elements is shown in Fig. 29L.* Each hexagon represents one nuclide, and includes such information as relative abundance, atomic number, mass, half-life, activity, etc. *Isotopes*, nuclides having the same number of protons, lie along a line inclined at 30° with the horizontal. *Isotones*, nuclides having the same number of neutrons, lie along the other 30° line. *Isobars*, nuclides having the same number of nucleons, lie along vertical lines. The small black rectangles represent the stable nuclides; all others are radioactive.

The stability of every nucleus is associated with the relative numbers of neutrons and protons bound together. If N represents the number of neutrons and Z the number of protons in any nucleus, and we plot a graph for all the known stable nuclides, we obtain a chart like the one shown in Fig. 29M. It will be noted that along any isobaric line, constant A ($A = Z + N$), there is but one nuclide for *odd-A*. There are only two exceptions to this rule, and these are found at $A = 113$ and $A = 123$. These pairs are

$$_{48}Cd^{113} \quad \text{and} \quad _{49}I^{113}$$
$$_{51}Sb^{123} \quad \text{and} \quad _{52}Te^{123}$$

For *even-A*, there are usually two, and occasionally three, stable nuclides having the same A.

If, in Fig. 29M, we plotted vertically upward out of the page, as a third dimension, the accurately known atomic masses M for all known stable as well as radioactive nuclides, we would obtain a kind of valley running diagonally up the chart, with the lowest points near the center of the stable nuclides. This is illustrated by a cross-section diagram in Fig. 29N for $A = 87$. These are the isobars given in the second column from the right in Fig. 29L:

$_{35}Br^{87}$	86.927310 amu	β^-
$_{36}Kr^{87}$	86.919311 amu	β^-
$_{37}Rb^{87}$	86.915410 amu	β^-
$_{38}Sr^{87}$	86.915130 amu	Stable
$_{39}Y^{87}$	86.916820 amu	E.C.
$_{40}Zr^{87}$	86.920330 amu	β^+
$_{41}Nb^{87}$	86.926380 amu	β^+

* The complete chart of nuclides, prepared by William H. Sullivan, of the Oak Ridge National Laboratory, can be purchased at small cost from the Superintendent of Documents, U. S. Government Printing Office, Washington 25, D. C.

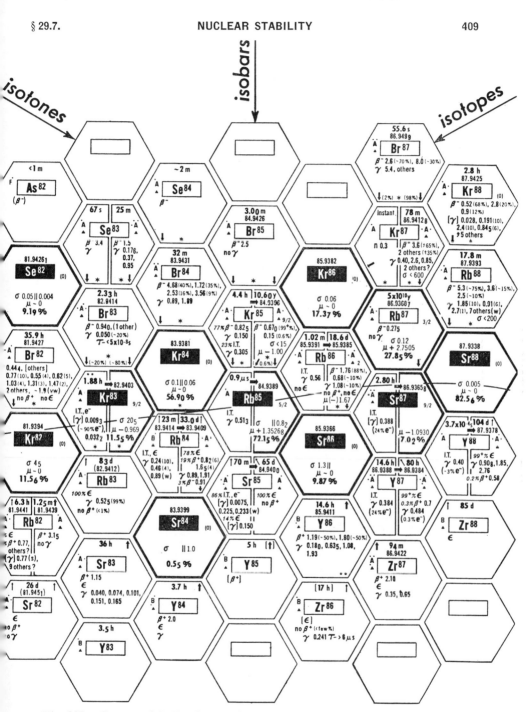

Fig. 29L *Section of the "Trilinear Chart of Nuclides," by W. H. Sullivan. (Complete chart is obtainable from Supt. of Documents, U. S. Gov't. Printing Office, Washington 25, D. C.) Masses given are based upon O¹⁶ as having a mass of 16 even. For masses based upon carbon-12, see Appendix IV.*

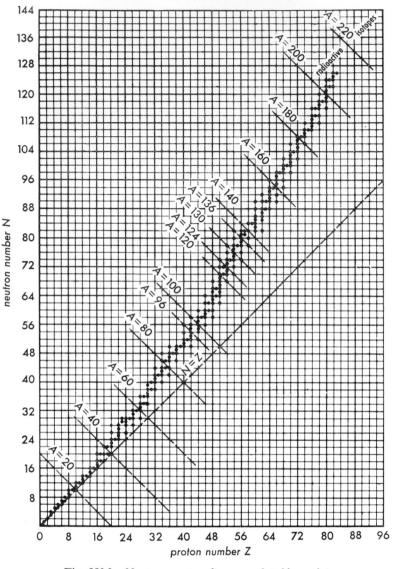

Fig. 29M *Neutron-proton diagram of stable nuclei.*

Note that the only stable nuclide, $_{38}Sr^{87}$, lies deepest in the valley curve.

Figure 29O is a typical atomic mass graph for isobars of *even A*. The isobars of *even Z* and *even N* fall on the lower parabola and therefore represent more tightly bound nuclear structures than the *odd Z odd N* isobars.

Among the stable nuclides there are 54 pairs of *even-even* isobars and only two cases of three. The latter are: $_{50}Sn^{124}$, $_{52}Te^{124}$, $_{54}Xe^{124}$, and $_{54}Xe^{136}$, $_{56}Ba^{136}$, $_{58}Cs^{136}$.

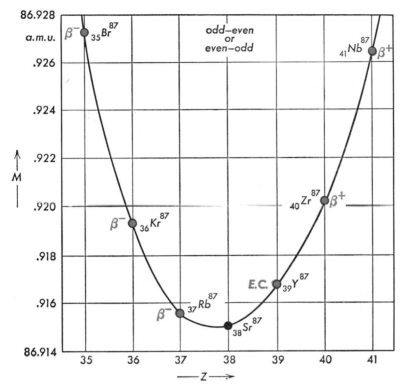

Fig. 29N   *Atomic mass relations for nuclides with A = 87: Odd Z and even N nuclides fall on the same parabola with even Z and odd N.*

The isobar parabolas given in Fig. 29O are drawn from the following known masses:

$_{38}Sr^{92}$	91.916650 amu	β^-
$_{39}Y^{92}$	91.914730 amu	β^-
$_{40}Zr^{92}$	91.911130 amu	Stable
$_{41}Nb^{92}$	91.913210 amu	E.C.
$_{42}Mo^{92}$	91.912720 amu	Stable
$_{43}Tc^{92}$	91.919100 amu	β^+

29.8. The Mechanics of Recoiling Particles

When a nucleus disintegrates by ejecting a particle, classical laws of mechanics are found to apply to the resultant motions. Consider, for example, a radio-active nucleus as shown in Fig. 29P.

If the compound nucleus is initially at rest, and the masses of the two fragments after disintegration are m_1 and m_2, conservation of momentum requires that

$$m_1 v_1 = m_2 v_2 \qquad (29q)$$

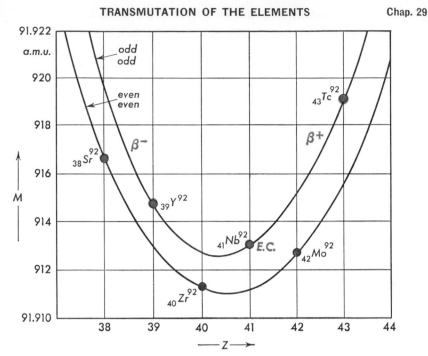

Fig. 29O   *Atomic mass relations for nuclides with A = 92: Odd Z and odd N nuclides form one parabola and even Z and even N nuclides form another.*

If an amount of energy E is liberated by the disintegration process, the law of conservation of energy requires that

$$E = E_1 + E_2 \qquad (29r)$$

where

$$E_1 = \tfrac{1}{2}m_1v_1{}^2$$

and

$$E_2 = \tfrac{1}{2}m_2v_2{}^2$$

The ratio of the two energies is, therefore,

$$\frac{E_1}{E_2} = \frac{\tfrac{1}{2}m_1v_1{}^2}{\tfrac{1}{2}m_2v_2{}^2} \qquad (29s)$$

If we solve Eq. (29q) for v_1, we can substitute m_2v_2/m_1 for v_1 in Eq. (29s), and obtain

$$\frac{E_1}{E_2} = \frac{m_2{}^2}{m_1m_2} \qquad (29t)$$

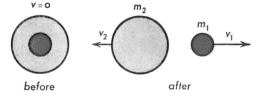

Fig. 29P   *Mechanics involved in the disintegration of a compound nucleus.*

Example 2. If an energy of 20 Mev is liberated in a reaction in which the nuclear recoil mass is 100 times the mass of the ejected particle, find (a) the energy ratio of the two recoiling particles, and (b) the energy given each particle.

Solution. Since $m_2 = 100m_1$, direct substitution in Eq. (29t) gives

(a)
$$\frac{E_1}{E_2} = 100 \quad \text{or} \quad E_1 = 100E_2$$

By substituting $100E_2$ for E_1, and 20 Mev for E, in Eq. (29r), we obtain

$$20 \text{ Mev} = 100E_2 + E_2$$
or
$$E_2 = 0.20 \text{ Mev}$$
and
$$E_1 = 19.8 \text{ Mev}$$

In all such nuclear reactions the momentum imparted to both bodies is always equal, yet approximately 99% of the energy in this example is carried away by the lighter particle. In other words, conservation of energy and momentum divides the energy approximately inversely as the masses.

If the compound nucleus is moving with a velocity v at the time of disintegration, the above equations apply to a *center of mass* point continuing on with the same initial velocity v.

29.9. Medical Applications

Since the time of its discovery, radiosodium has found numerous and important applications in many branches of science. It has, for example, been used as a means of tracing certain organic and inorganic chemicals passing through the human body, through plants, and through chemical experiments of one kind or another. To give a simple example, one can show, by drinking water containing radiosodium in the form of common table salt (NaCl), that within 2 min some of it has entered the blood stream and has been distributed to all parts of the body. The presence of the salt in the finger tips or the toes can be demonstrated by detecting the electrons and γ rays from the sodium with a scintillation counter. Similar experiments can be performed with trees and plants to see how rapidly their roots take up certain plant foods and distribute them to the leaves and branches. At the present time radiophosphorus 32, because of its chemical properties and its comparatively long half-life, is being used in a number of medical researches as a possible cure for certain diseases, as well as for tracing the migration of phosphorus through the body.

One useful medical technique is the use of radioactive elements to obtain *radioautographs*. A plant or animal is given a single dose of liquid containing the "tagged" atoms; at various times thereafter thin sections of tissue from the plant or animal are placed in direct contact against a photographic film. After several hours of exposure to any possible rays from "tagged" atoms in the section, the film is developed. Fig. 29Q shows at the left a thin cross section of the mouse, previously fed radiophosphorus (see Eq. 29n), and at the right the resulting radioautograph of this same section. The dark areas indicate the uptake of phosphorus by the *spleen* and *liver* sections, and nowhere else.

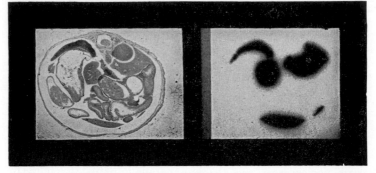

Fig. 29Q   *Radioautograph of the type used in medical studies, showing selective absorption of phosphorus by the liver and spleen.*

Figure 29R shows an *X ray* and *strontium radioautograph* of a section of an amputated leg from a patient with osteogenic sarcoma (bone cancer). The patient received radiostrontium orally two days before removal of the extremity.

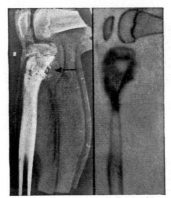

Fig. 29R   *X-ray and radioautograph of a human leg.*

Courtesy, Dr. J. G. Hamilton

The X ray shows the bone tumor at the upper end of the tibia, while the radioautograph shows selective deposits of strontium in the tumor, with small amounts in the surrounding bone and relatively little in the soft tissues. These studies indicate that radiostrontium may prove useful in clinical therapy.

Since the process of life is one of continual change, there is a rapid turnover of elements in living systems. When common molecules are chemically formed with one or more of their atoms radioactive, they are said to be tagged. In living systems, such as the various organs of the body, these tagged molecules can change places with others of like kind. In time, however, these molecules disappear by natural elimination processes or by decay.

If $_{15}P^{32}$ is taken orally, for example, one finds that after four hours 44% is in the bones, 25% in the muscles, and 2.4% in the blood. When $_{53}I^{131}$ is taken orally, turnover studies show that 80 times as much iodine has been taken up by the thyroid as by any other per gram mass of the body.

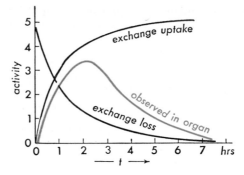

Fig. 29S   *Typical exchange curves for radioactive molecules in living systems.*

A typical graph of the behavior of radioactive elements by living systems is shown in Fig. 29S.

29.10. Radiation Damage

Energy absorbed by the passage of radiation through matter gives rise to structural changes called *radiation damage*. The nature and amount of damage produced is closely related to the number of ion pairs created, and this in turn depends upon the absorbing material and the characteristics of the radiation.

Since biological effects vary widely between body tissues and different types of radiation, it is customary to measure radiation absorption by the number of ion pairs produced in air. The exposure of specimens to any and all kinds of radiation is measured in terms of a unit called the *roentgen*.

Roentgen. The roentgen unit, abbreviated 1 r, is that quantity of X or γ radiation that produces one electrostatic unit of negative, or positive, charge in 0.001293 grams of dry air. Under standard conditions air has a density of 0.001293 gm/cm³, and one coulomb $= 3 \times 10^9$ esu. Since the charge of a single electron, or ion, is 1.60×10^{-19} coulomb, the roentgen can also be defined as the quantity of X or γ radiation which, under standard conditions, produces in air:

$$\frac{1}{3 \times 10^9 \times 1.6 \times 10^{-19}} = 2.08 \times 10^9 \text{ ion pairs/cm}^3$$

Experimental measurements show that on the average it takes 32.5 electron volts to produce one ion pair. Consequently,

$$1 \text{ r} = 2.08 \times 10^9 \times 32.5 = 67.6 \text{ Bev}$$

This energy is equivalent to 0.108×10^{-7} joules, or 0.108 ergs. In one gram of air 1 r would produce

$$\frac{0.108 \text{ ergs}}{0.001293} = 83.5 \text{ ergs}$$

This same radiation will produce 93 ergs per gram of water (the main constituent of tissue), approximately 40 ergs/gm in fat, and 900 ergs/gm in bone.

Roentgen Equivalent, Physical (rep). This is a unit applied to statements of dose of ionizing radiation not covered by the definition of the roentgen. One *rep* has been defined as the dose that produces an energy absorption of 93 ergs per gram of tissue for the particular radiations in question.

Roentgen Equivalent, Man (rem). A dose of any ionizing radiation producing the same biological effect in tissue as that produced by 1 *roentgen* of high-voltage, X radiation is called 1 *rem.* The roentgen is frequently divided into one thousand equal parts called the *milliroentgen:*

$$1 \text{ roentgen} = 1000 \text{ milliroentgens}$$
$$1 \text{ r} = 1000 \text{ mr}$$

Radiation Dose. The total radiation entering a specified area or volume is called the radiation dose and is measured in *roentgens per hour* or *milliroentgens per hour.*

One effect of penetrating radiation is to bring about gene mutation. Such mutations are of greatest concern to the individual man or woman from birth through the reproductive years. In human beings this is assumed to be about 30 years. Experiments show that a total of 50 r per person received during the reproduction period of 30 years doubles the number of mutations that occur naturally in the absence of radiation.

The amount of radiation to which an individual can be subjected with no adverse effects is called a *permissible dose.* This is generally accepted to be 250 mr per week, or 6.25 mr per hr based on a 40 hr working week.*

Curie. The *curie,* as a unit of radioactivity, was originally defined as the number of disintegrations per second emanating from one gram of radium. Because of experimental difficulties the unit was redefined by the International Commission on Radiological Units (July 1953) as follows: The curie is a unit of radioactivity defined as the quantity of any radioactive nuclide in which the number of disintegrations per second is 3.700×10^{10}. The *millicurie,* mc, and *microcurie,* μc, are smaller units in frequent use. Medical doses are usually expressed in millicurie-hours.

PROBLEMS

1. When 5-Mev α particles bombard Na^{23}, protons are observed being ejected. (a) Write down the disintegration equation, and (b) find the energy liberated.

2. If a 4-Mev deuteron on N^{14} produces a proton, what is the energy liberated? Assume that 94% of the energy goes into the proton. What will be (a) its maximum energy, and (b) its range in air? (See Fig. 24G.) *Ans.* (a) 11.8 Mev, (b) 152 cm.

3. When 10-Mev deuterons bombard B^{10}, protons, neutrons, and α particles are observed as disintegration products. Assuming these are branch disintegrations write down the three reactions and give the liberated energies.

4. If 6-Mev deuterons on N^{14} produce C^{12} and α particles, (a) how much en-

---

* See *Pile Neutron Research,* by D. J. Hughes, Addison Wesley Press.

References: *Radioactivity and Nuclear Physics,* by James Cork, D. Van Nostrand Co.; *Introduction to Atomic Physics,* by Henry Semat, Farrar and Rinehart. *The Atomic Nucleus,* by R. D. Evans, McGraw-Hill Book Co.

ergy is liberated? If the available energy is divided between the two particles in the inverse ratio of their respective masses, what is (b) the maximum energy of the α particles, and (c) their range in air? *Ans.* (a) 19.6 Mev, (b) 14.7 Mev, (c) 20 cm.

5. If 5-Mev protons are incident on Be^9, α particles are observed emitted as high-speed disintegration products. (a) Write down the reaction, and (b) find the energy liberated. If the liberated energy is divided between the disintegration products in the inverse ratio of their respective masses, what is (c) the energy of the α particles, and (d) their maximum range in air?

6. If the accurately known mass of copper-64 is 63.929761 amu, find the accurate masses of the stable isobars nickel-64 and zinc-64. (See Fig. 29K.) *Ans.* nickel-64, 63.927959 amu; zinc-64, 63.929145 amu.

7. Cesium-130 is a radioactive isotope with a half-life of 30 m. Some of the atoms decay by positron emission with an end-point energy of 1.97 Mev, while the others decay by electron emission with an end-point energy of 0.44 Mev. Write down the reactions, and draw an energy level diagram. (See Fig. 29J.)

8. There are five isobars of mass number $A = 104$, whose masses are known from measurements. They are as follows:

$_{44}Ru^{104}$	103.912000	Stable
$_{45}Rh^{104}$	103.912610	β^-
$_{46}Pd^{104}$	103.910170	Stable
$_{47}Ag^{104}$	103.914450	β^+
$_{48}Cd^{104}$	103.913800	β^+

(a) Plot an atomic mass graph for these nuclides similar to that shown in Fig. 29O. (b) At what value of Z do you find the minimum mass for the even-even isobars? (c) Extrapolate to find the approximate mass of $_{43}Tc^{104}$. *Ans.* (b) 45.6. (c) 103.925 amu.

9. There are five isobars of mass number $A = 134$, whose masses are known from measurements. They are as follows:

$_{53}I^{134}$	133.916020	β^-
$_{54}Xe^{134}$	133.911884	Stable
$_{55}Ba^{134}$	133.912930	E.C. and β^-
$_{56}La^{134}$	133.910870	Stable
$_{57}Ce^{134}$	133.914570	E.C. and β^+

(a) Plot an atomic mass graph for these nuclides similar to that shown in Fig. 29O. (b) At what value of Z do you find a minimum mass for the even-even isobars? (c) Extrapolate to find the approximate mass of $_{58}Pr^{134}$.

10. There are five isobars of mass number $A = 124$, whose masses are known from measurements. They are as follows:

$_{50}Sn^{124}$	123.911740	Stable
$_{51}Sb^{124}$	123.912340	β^-
$_{52}Te^{124}$	123.909420	Stable
$_{53}I^{124}$	123.912610	E.C. and β^+
$_{54}Xe^{124}$	123.912560	Stable

(a) Plot an atomic mass graph for these nuclides similar to that shown in Fig. 29O. (b) At what value of Z do you find the minimum mass for the even-even isobars? (c) Extrapolate to find the approximate masses of $_{49}In^{124}$ and $_{55}Cs^{124}$. *Ans.* (b) $Z = 51.8$. (c) 123.923 amu, and 123.924 amu, respectively.

11. If a boron-11 nucleus is at rest when it liberates a neutron, and the total energy liberated is 12 Mev, find (a) the energy ratio of the two recoiling particles, and (b) the energy of each particle in Mev.

12. If a neon-20 nucleus is at rest when it liberates a proton, and the total energy liberated is 12 Mev, find (a) the energy ratio of the two recoiling particles, (b) the energy of each particle, and (c) the range of the proton. *Ans.* (a) 19. (b) 0.6 Mev and 11.4 Mev. (c) 145 cm.

Neutron and Gamma Ray Reactions

While the bombardment of all targets by high-energy-charged atomic particles, such as protons, deuterons, and alpha particles, is found to produce disintegrations and transmutations of various kinds, it is also possible to bring about similar reactions with neutrons and gamma rays.

30.1. Neutron Reactions

The first disintegrations produced by high-speed neutrons as atomic projectiles were announced in 1932 by the English physicist Feather. Immediately following Chadwick's discovery of these neutral particles, Feather allowed neutrons from beryllium (see Fig. 30A) to enter a Wilson cloud chamber containing pure nitrogen gas. Numerous expansions of the chamber and the simultaneous clicks of a camera shutter gave many photographs of the ion tracks left by recoiling nitrogen atoms.

Although most of the photographs indicated elastic collisions between nitrogen atoms and neutrons, an occasional photograph showed a forked track, indicating a disintegration of a nitrogen nucleus:

$$_0n^1 + {}_7N^{14} - {}_6C^{14} + {}_1H^1 \qquad (30a)$$

Two photographs of several such disintegrations are reproduced in Fig. 30A. Although hundreds of neutrons enter the cloud chamber every second, they do not ionize atoms as charged particles do, and hence leave no tracks. When a head-on nuclear collision occurs, however, the disintegrated nuclei, possessing as they do high speeds and positive charges, leave a trail of ions behind them. The fork in each photo shows a proton track of considerable length originating at the same point as the more dense, short-ranged track of the recoiling carbon nucleus.

Strong sources of neutrons are produced by inserting a thin plate of beryllium metal in the intense beam of deuterons coming from the cyclotron as shown in

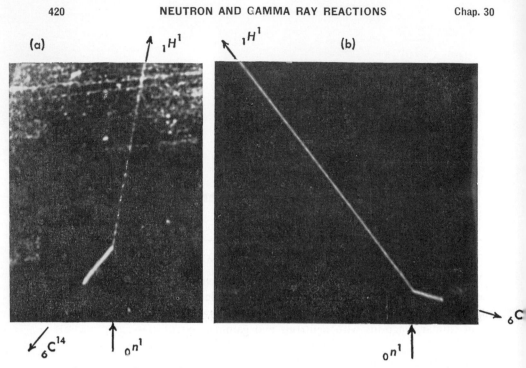

Fig. 30A   *Cloud-track photographs of neutron disintegrations of nitrogen. (After Feather and Rasetti.)*

Fig. 30B. The disintegration process, giving rise to the neutrons, is the reaction Eq. (29f).

Into such a beam of chargeless particles, numerous substances of known chemical constitution have been inserted, and the disintegration products

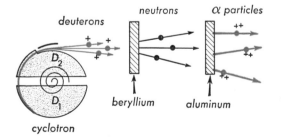

Fig. 30B   *Experimental arrangement for producing intense beams of neutrons by bombarding beryllium with deuterons. The neutrons are then used as projectiles for further disintegrations, as illustrated here for aluminum.*

studied with suitable detectors. To illustrate by an example, suppose that a thin sheet of aluminum is inserted into the beam of neutrons as shown in the figure. In this particular instance, α particles are observed emerging from the aluminum, thus enabling one to write down the following neutron reaction:

$$_0n^1 + {}_{13}Al^{27} = {}_{11}Na^{24} + {}_2He^4 \tag{30b}$$

Thus *radioactive sodium*, produced originally by the deuteron bombardment of ordinary sodium, is here produced by a different reaction. See Eq. (29l). As proof of the result, the bombarded aluminum target is found to be β-ray and γ-ray active with a half-life of 15 hr. There are at least two other known disintegration processes by which radiosodium is produced: one by the neutron bombardment of silicon, and the other by the α-particle bombardment of magnesium.

This is but one example of the many known radioactive elements that can be manufactured in four different ways. As a matter of fact, with sufficiently energetic atomic bullets, it is now possible to produce hundreds of atomic nuclei not found in nature. While most of them have short half-lives and do not last very long, there are many with long half-lives, some extending into thousands of years.

Examples of other neutron disintegrations are illustrated by the following reactions:

$$_0n^1 + {}_9F^{19} = {}_7N^{16} + {}_2He^4 \tag{30c}$$

$$_0n^1 + {}_{20}Ca^{42} = {}_{19}K^{42} + {}_1H^1 \qquad {}_{19}K^{42} = {}_{20}Ca^{42} + {}_{-1}e^0 \tag{30d}$$

Equation (30d) represents a typical case of the capture of a neutron to form a radioactive isotope $_{19}K^{42}$ which, by the ejection of an electron, reverts back to the original stable element, $_{20}Ca^{42}$. Many such reactions are known, particularly among the heavier elements in the first half of the periodic table.

When 100- to 200-Mev deuterons from a large cyclotron strike almost any target, neutrons with energies of 100 Mev or more are produced. Figure 30C is a Wilson cloud chamber photo showing the result of a 100-Mev neutron impact with an oxygen nucleus. The paths of the recoiling fragments, four α particles, are bent in the magnetic field.

Not all disintegration processes liberate more energy than that required to produce them. This is illustrated by the following example:

$$_0n^1 + {}_6C^{12} = {}_4Be^9 + {}_2He^4 \qquad Q = -5.7 \text{ Mev} \tag{30e}$$

To carry out this disintegration, the bombarding neutrons must have an energy of 5.7 Mev or greater. The sum of the masses produced is greater by this amount than those that went to make them. This is an example of what is called a fast-neutron reaction. This reaction is the reverse of the reaction by which neutrons were first produced in large quantity. (See §26.5 and Eq. (26g).)

30.2. Slow Neutron Reactions

The fact that neutrons, slowed down to very low speeds, have the ability to disintegrate certain atoms was first discovered and investigated by the Italian physicist, Enrico Fermi, and his collaborators. A neutron approaching the nucleus of an atom does not experience a repulsive force, as does a proton, deuteron, or α particle, and consequently its chances of penetration into a nucleus and of being captured by it are relatively large. It is for this reason that

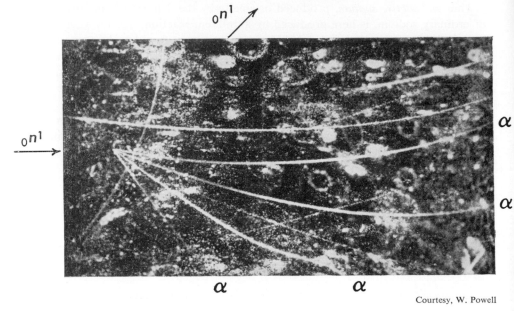

Fig. 30C   *Wilson cloud-chamber photo showing the result of a 100-Mev neutron impact on an oxygen nucleus.*

slowly moving neutrons are able to bring about disintegrations that slowly moving charged particles cannot.

The customary method of producing *slow neutrons* is to surround a source of fast neutrons with paraffin or some material containing large quantities of hydrogen or deuterium. As neutrons pass through this "moderator" material, elastic collisions with hydrogen nuclei continually slow them down, until at a distance of several centimeters from the source most of them have lost all of their original energy. What little energy they do have is picked up by regular thermal collisions with other atoms.

Since their resultant motions become quite the same as the random motions of the atoms and molecules of the moderator, they are called *thermal neutrons.* Thermal neutrons are defined as neutrons in equilibrium with the substance in which they exist, commonly, neutrons of kinetic energy of 0.0253 electron volts. Compared with fast neutrons moving with almost the speed of light, like those from a target of beryllium bombarded by the beam from a cyclotron, thermal neutrons have a velocity of only 2200 m/sec. This is essentially the velocity of hydrogen molecules in a gas at normal temperature and pressure.

One of the many known slow-neutron reactions is one in which boron-10 captures a neutron and ejects an α particle:

$$_0n^1 + {}_5B^{10} \rightarrow {}_3Li^7 + {}_2He^4 \tag{30f}$$

30.3. A Thermal Neutron Source

A very good laboratory source of thermal neutrons can be made with a small quantity of radium and beryllium. The radium sample is surrounded by a thin jacket of beryllium metal and an outer jacket of lead, and then this unit is imbedded at the center of several cubic feet of paraffin. See Fig. 30D. The α particles from radium produce neutrons by the reaction

$$_2\text{He}^4 + {}_4\text{Be}^9 \rightarrow {}_6\text{C}^{12} + {}_0\text{n}^1 \qquad Q = 5.6 \text{ Mev}$$

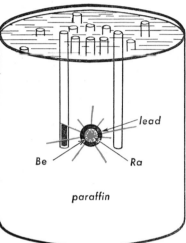

Fig. 30D *Diagram of a Ra-Be source of thermal neutrons for laboratory experiments.*

The fast neutrons given off by the beryllium metal bounce around in the paraffin, quickly slowing down to thermal velocities, while the unwanted β and γ rays from the radium are absorbed by the lead.

Any source to be radiated by slow neutrons is placed in a holder and lowered through any one of several ports shown in the top surface.

An effective demonstration of slow neutron reactions can be made by rolling a thin sheet of silver metal into a hollow cylinder, about 1 in. in diameter and 4 in. long, and subjecting it to the Ra-Be source (Fig. 30D). After several minutes the cylinder is withdrawn and quickly placed around a Geiger-counter tube, or near the fluor of a scintillation counter. The activity detected will be strong at first and then will slowly die out over a period of several minutes.

If a scaler counting system like the one shown in Fig. 25O is employed, and the number of counts registered on the tube faces is recorded at 10 sec intervals, a decay curve like the one shown in Fig. 30E can be plotted. Such a graph with two distinct straight sections is found to be a composite of two half-life curves, one for 24 sec and the other for 2.3 min. These two half-lives are attributed to the two stable isotopes of silver, one of mass 107 amu and one of 109 amu.

The capture of a neutron by each of these nuclides produces the radioactive

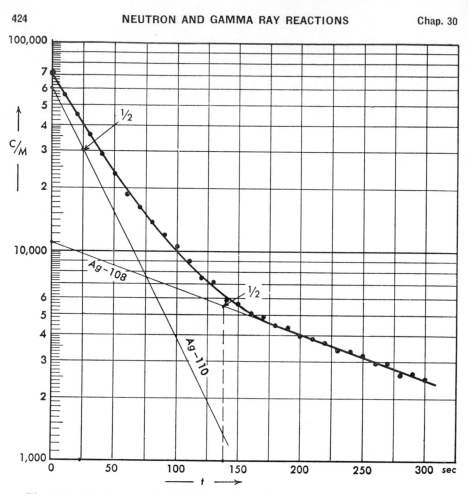

Fig. 30E *Semilog graph of the β rays from the two silver isotopes, made active by thermal neutrons.*

isotopes, silver-108 and silver-110. Both of these are electron active, and produce stable cadmium nuclides as follows:

$$_{47}Ag^{108} \rightarrow {}_{48}Cd^{108} + {}_{-1}e^0 \qquad (2.3m)$$

$$_{47}Ag^{110} \rightarrow {}_{48}Cd^{110} + {}_{-1}e^0 \qquad (24.2s)$$

The procedure for finding the two half-lives from the graph in Fig. 30E is to start with the lower straight section and extrapolate back to zero time. This straight line is then subtracted from the upper curve at two points of time, such as $t = 0$ and $t = 100$ sec, and the other straight line drawn in as shown. The summation of these two straight lines should produce the observed curve. Note that the straight lines drop to half their initial values at $t = 24$ sec and $t = 138$ sec.

Another effective demonstration of slow neutron reactions is to roll a thin sheet of indium metal into a hollow cylinder, subject it to the slow neutron source (Fig. 30D) and then place it over a Geiger-counter tube or scintillation counter. Like silver, it too will show a strong activity at first and then die out over a period of several minutes.

Indium has two stable isotopes, $_{49}In^{113}$ and $_{49}In^{115}$. Upon neutron capture they will become the radioactive isotopes, $_{49}In^{214}$ and $_{49}In^{216}$. Both of these nuclides decay by electron emission into stable tin isotopes. $_{49}In^{116}$ decays in either one of two ways, as follows:

$$_{49}In^{114} \rightarrow {}_{50}Sn^{114} + {}_{-1}e^0 \quad (72 \text{ sec})$$
$$_{49}In^{116} \rightarrow {}_{50}Sn^{116} + {}_{-1}e^0 \quad (54 \text{ m})$$
$$_{49}In^{116} \rightarrow {}_{50}Sn^{116} + {}_{-1}e^0 \quad (13 \text{ sec})$$

The longer-lived In^{116} is an excited state of the nucleus, called an *isomeric state*, and beta decay with an end-point energy of 1 Mev is followed by several gamma rays. The shorter-lived I^{116} is the normal state of the nucleus, and beta decay with an end-point energy of 2.9 Mev takes it directly to the normal state of Sn^{116} without the emission of gamma rays.

30.4. Neutron Diffraction

We have seen in a preceding chapter that wave properties are associated with all moving bodies. The De Broglie wavelength of any mass m is given by Eq. (22i) as

$$\lambda = \frac{h}{mv} \tag{30g}$$

which for thermal neutrons with a velocity of 2200 m/sec gives $\lambda = 1.8 \times 10^{-10}$ meters, or 1.8 A. This value is comparable to the spacings of atoms in solids and suggests the possibilities of diffraction, as in the case of X rays and Laue patterns. (See Figs. 14F and 14G.)

The diffraction of strong beams of neutrons by the atoms of a crystal has been studied by many people. Wollan and Shull, using a sodium chloride crystal and an experimental arrangement similar to that shown for X rays in Fig. 14F, obtained the picture reproduced in Fig. 30F. Since neutrons have little or no effect upon photographic films, the front face of the film was covered with a sheet of indium metal 0.5 mm thick.

Neutrons captured by the indium nuclei produce radioactive isotopes, as shown in the preceding section. These unstable isotopes disintegrate, with the emission of electrons. Electrons do affect a photographic emulsion and upon development produce the spots shown. Neutrons, like X rays, therefore, become useful tools in the study of the atomic structure of matter.

30.5. Photon Interactions

The term "photon interaction" refers to the disintegration of atomic nuclei brought about by photons. In general, the photons to be used must have a very

Fig. 30F   *Neutron diffraction pattern of sodium chloride.*

Courtesy, E. O. Wollan and C. G. Shull

high energy hv, and this means their frequency v must be high and their wavelength λ short: $(c = v\lambda)$.

X rays and γ rays are electromagnetic waves in character, and both are composed of photons. The terms X ray or γ ray only signify their origin and, in reactions with nuclei, are of no consequence.

Instead of specifying a photon by giving its frequency v, it is customary in nuclear reactions to express its energy in electron volts. By the general energy equation, Eq. (24k), we have

$$Ve = hv \quad \text{or} \quad V = hv/e \qquad (30h)$$

For reference purposes, only the following wavelengths and frequencies are calculated from this equation.

V (volts)	λ (m)	v (1/sec)
1 Kev	1.24×10^{-9}	2.4×10^{17}
1 Mev	1.24×10^{-12}	2.4×10^{20}
1 Bev	1.24×10^{-15}	2.4×10^{23}

To obtain sources of high-energy photons, it is common practice to use γ rays from radioactive sources, or to produce X rays with the high-energy electron beam of a betatron or synchrotron. Thorium C'', for example, produces 2.62-Mev γ rays that are useful for many kinds of experiments.

As an example of the production of high-energy photons with the beam from a large electron accelerator, consider the 1.3-Bev electrons of the Cornell synchrotron (see Fig. 30G). Each time the beam in this machine is allowed to strike the metal target inside the vacuum chamber, the 1.3-Bev electrons produce X rays

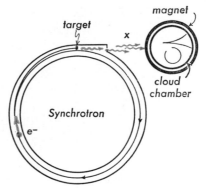

Fig. 30G   *Photons produced by the electron beam from a synchrotron are used to study photon reactions in a cloud chamber.*

of extreme penetrating power. Target composition is of little importance in this process, since the phenomenon called "Bremsstrahlung" is responsible for the high-energy X-ray beam that emerges. (See Fig. 14M.) In such a beam one finds a continuous band of photons ranging in energy from practically 0 up to 1.3 Bev.

As such a beam is made to traverse any form of matter, there are several processes by which the rays are absorbed or scattered from the beam. These, as shown in Fig. 30H, are:

- Photoelectric effect
- Compton effect
- Pair production

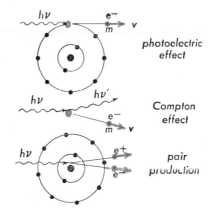

Fig. 30H   *The three major processes responsible for the absorption of high-energy photons in matter.*

For photons of low energy the photoelectric effect is chiefly responsible for absorption. As the energy increases, the Compton effect becomes more and more important until, at energies of 1 Mev, pair production sets in and eventually becomes predominant.

A diffusion cloud-chamber photograph of a Compton recoil electron and an electron pair is shown in Fig. 30I. The hundreds of photons that passed through

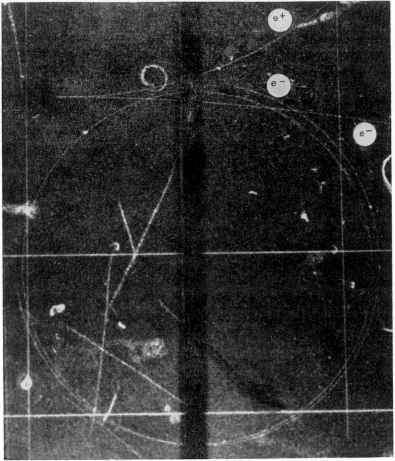

Fig. 30I   *Cloud-chamber photograph showing a Compton recoil electron (circle) and an electron pair (V-track) produced by photons.*

this gas-filled cloud chamber, from left to right, left no tracks since they were electrically neutral. One photon, however, collided with an atomic electron near the upper center of the field, and the forward recoiling electron, in the magnetic field applied to the chamber, made $1\frac{3}{4}$ turns around a circle. Another photon, coming close to a nucleus at the upper left, created an electron pair. From the curvatures of the tracks it is clear that the positron curving upward had more energy than the Compton electron, but less energy than its electron associate.

The production of a pair of electrons by a photon passing close to an atomic nucleus can also occur in the region close to an electron. See Fig. 30J. In both of these processes, the force exerted by the photon on the charged particle during the pair creation causes that particle to recoil forward. Because of its

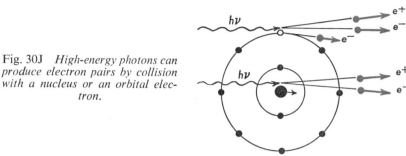

Fig. 30J   *High-energy photons can produce electron pairs by collision with a nucleus or an orbital electron.*

relatively small mass, the electron recoil may be large enough to appear as a third track in a cloud chamber, whereas with the far heavier nucleus the recoil is so small that a track is seldom observed.

A cloud-chamber photograph showing an electron triplet and an electron pair is reproduced in Fig. 30K.* Of the many photons traversing the chamber from

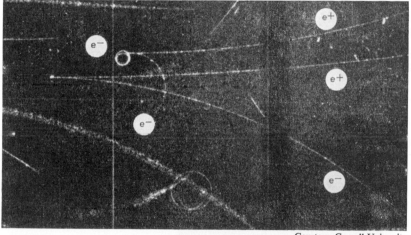

Courtesy, Cornell University

Fig. 30K   *Cloud chamber photograph showing an electron triplet and an electron pair produced by high-energy photons. See Fig. 30J.*

left to right, one came extremely close to an atomic electron, and there in the strong electric field created a pair, the positron bending upward, the slower electron downward, and the still slower recoiling electron spiraling around in a small circle. The second photon coming close to the nucleus of a hydrogen atom produced another pair, the positron bending upward and the electron downward. Note the tiny recoil track of the nucleus.

* The photographs in Figs. 30I, 30K, 30M, 30O, and 30Q were supplied by the diffusion cloud chamber research group, B. Chasan, G. Cocconi, V. Cocconi, E. Hart, R. Schectman, and D. White, at Cornell University.

30.6. Photon-Deuteron Interactions

If a cloud chamber is filled with deuterium, that is, with gas in which all molecules are composed of the heavy hydrogen atoms, $_1H^2$, interesting photon interactions can be observed. One process is shown in Fig. 30L. A deuteron,

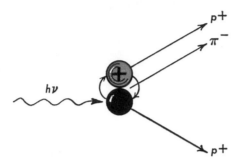

Fig. 30L *One type of reaction produced by photon collisions with deuterons. See Fig. 30M.*

the nucleus of a deuterium atom, is composed of one neutron and one proton.

If a photon collides with the neutron of a deuteron, it may produce a proton p and a pi-minus meson, π^-, as shown. Since the two particles leave so quickly after the collision, the remaining "spectator proton" continues on with whatever momentum it had at the moment of impact. The reaction for this can be written

$$h\nu + d \rightarrow p + \pi^- + \underbrace{p}_{\text{spectator}} \tag{30i}$$

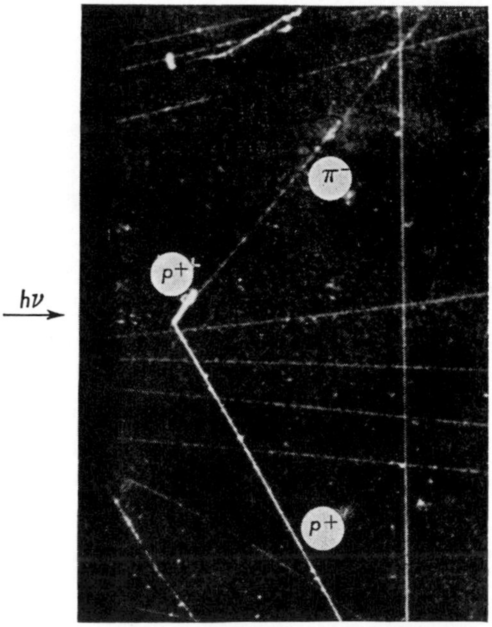

Fig. 30M *Cloud-chamber photograph showing one type of photon reaction. See Fig. 30L.*

Courtesy, Cornell Laboratories

Such an explanation accounts for the cloud chamber event shown in Fig. 30M. A photon entering from the left collides with the neutron of a deuteron. The neutron is split into a proton p and a π^- meson, with the low momentum "spectator proton" stopping in the chamber.

Since everyone knows that a proton or deuteron has unit positive charge, and neutrons and neutrinos have none, their charge exponents are usually omitted in reactions.

30.7. Photo Production of Meson Pairs

If a high-energy photon comes close enough to a proton in a nucleus, a π meson pair may be created in much the same way that an electron pair is produced with a photon of lower energy. If the nucleus is a deuteron, as shown in Fig. 30N, the proton recoils from the impact while the neutron continues on with whatever momentum it had at the moment of impact.

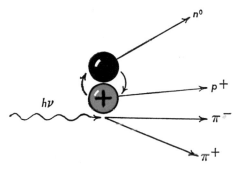

Fig. 30N *Diagram of a photon collision with the proton of a deuteron, producing a pair of π mesons. See Fig. 30O.*

As a reaction, we can write,

$$h\nu + d \rightarrow p + \pi^- + \pi^+ + \underbrace{n}_{\text{spectator}} \qquad (30j)$$

This reaction accounts for the cloud-chamber event shown in Fig. 30O. A photon, entering from the left, comes close to the proton of a deuteron. A pair of π mesons is created, the proton recoils under the impact, and the neutron "spectator," not being able to form a track, goes on in some unknown direction.

In all such reactions the conservation laws of energy, momentum, and charge must hold. Measurements of track curvatures give velocities and momenta of particles, and calculations are made to see that the conservation of energy and momentum hold. We have seen previously that approximately 1 Mev is required to create an electron pair. Since a π meson has a mass 275 times that of an electron, an energy of approximately 275 Mev is required to create a pair of π mesons. This is the threshold energy for meson pair-production.

Another interesting reaction involving the impact of a high-energy photon with a deuteron is one in which the processes shown in Figs. 30L and 30N are combined. As shown schematically in Fig. 30P, a photon collides with the

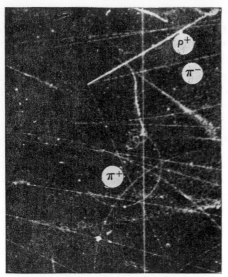

Fig. 30O   *Photograph of a photon collision with the proton of a deuteron, producing a pair of π mesons.*

Courtesy, Cornell Laboratories

neutron, splits it apart into a proton, *p*, and a π⁻meson, and simultaneously creates a π meson pair, while the proton "spectator" continues on with its relatively low energy. The reaction is written

$$h\nu + d \rightarrow p + p + \pi^- + \pi^+ + \pi^- \tag{30k}$$

All five of these particles have a charge and may be identified in the cloud-chamber event reproduced in Fig. 30Q. The laboratory-measured angles, the ionization along the tracks of the five particles, as well as the momentum calculations, readily verify the dynamics of this event.

QUESTIONS AND PROBLEMS

1. Charged particles show their tracks in a Wilson cloud chamber. Neutrons and γ rays do not. Why?

2. How are strong beams of neutrons produced? Can neutrons be accelerated in a cyclotron?

3. When silicon-28 is bombarded by neutrons, protons are found coming from the target. Write down the resultant reaction.

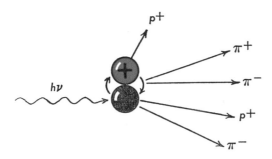

Fig. 30P   *Diagram of a photon collision with a deuteron, splitting the neutron and producing a pair of π mesons. See Fig. 30Q.*

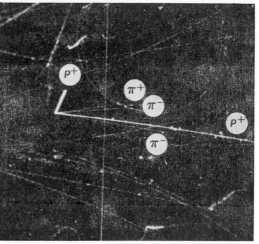

Fig. 30Q *Cloud chamber photograph of a photon collision with a deuteron, splitting the neutron and producing a pair of π mesons.*

Courtesy, Cornell Laboratories

4. When neutrons collide with oxygen-16 nuclei, α particles are observed to be given off. Write down the reaction.

5. (a) If radioactive phosphorus-32 is produced by the neutron bombardment of sulfur-32, what is the reaction equation? (b) What happens to the phosphorus-32?

6. How are photons with an energy of 1 Bev produced in the laboratory? Make a diagram and explain the atomic process involved.

7. What are thermal neutrons? What are slow neutrons? What are their velocities in m/sec?

8. Make a diagram of a radium-beryllium source of thermal neutrons. Explain the function of each substance used.

9. When slow neutrons are captured by lanthanum-139, the resulting nuclei give off electrons with an end-point energy of 1.43 Mev. Write down the two reactions.

10. If the mass of iodine-127 is 126.9109 amu, find the masses of the two other nuclei resulting from slow-neutron capture described in Problem 9. *Ans.* (a) $I^{128} = 127.9199$, (b) $Xe^{128} = 127.9181$.

11. If indium metal is subjected to thermal neutrons for some time and then placed close to a scintillation counter, the initial counting rate is found to be 60,000 counts per min. Plot a semilog graph for the activity from this metal for the succeeding 2 min.

12. If a photon in coming close to a nucleus were to create a pair of protons, what would the threshold energy be in electron volts? (Note: Proton mass is 1.007276 amu.) *Ans.* 1.876 Bev.

Special Atomic and Nuclear Effects

It is well known from detailed observations of spectrum lines that whether radiant energy arises from the external structure of atoms or from the nucleus, the breadths of most lines are independent of the spectrograph used to observe them. In many cases narrow and broad lines are observed in the same spectrum. The sharp and diffuse series of the optical spectra of the alkali metals and alkaline earth elements are good examples of this. (See Fig. 13A.) Although special sources of light have been devised that are capable of sharpening most lines to within the limits of the resolving power of the finest spectrographs, some lines have resisted all efforts to make them narrow.

In this chapter we are concerned with the principal causes of the observed breadth of lines, and with certain atomic and nuclear processes associated with the emission and absorption of photons. The principal causes for the breadth of spectrum lines are:

- Doppler effect
- Natural breadth
- Stark effect

31.1. Doppler Broadening

One of the most classical of all atomic phenomena is the Doppler principle as it applies to the observed frequencies of light. This is well illustrated by certain lines in the solar spectrum shown in Fig. 9J. The light emitted from atoms on the side of the sun approaching the earth are shifted to higher frequencies, while light emitted from atoms on the side receding from the earth are shifted to lower frequencies.

Quite similar effects are well known to exist in the case of an electrical discharge tube where, owing to thermal agitation, most of the atoms emitting light have high velocities. The random motions of the atoms and molecules in a gas, however, produce a net broadening of the line with no apparent shift in its central maximum.

Figure 31A gives a classical representation of two identical atoms emitting

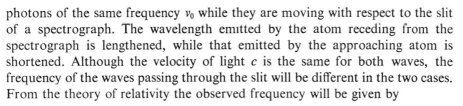

Fig. 31A   *Diagram of the Doppler effect for light emitted by excited atoms, receding from, and approaching the slit of a stationary spectrograph.*

photons of the same frequency v_0 while they are moving with respect to the slit of a spectrograph. The wavelength emitted by the atom receding from the spectrograph is lengthened, while that emitted by the approaching atom is shortened. Although the velocity of light c is the same for both waves, the frequency of the waves passing through the slit will be different in the two cases. From the theory of relativity the observed frequency will be given by

$$v = v_0 \frac{\sqrt{1 - \dfrac{v^2}{c^2}}}{1 - \dfrac{v}{c}}$$

This is the relativistic Doppler equation, where v is the observed frequency, v_0 is the emitted frequency, and v is the relative velocity of the atom with respect to the spectrograph, $+v$ for an approaching atom and $-v$ for a receding atom.

By a mathematical expansion this equation can be written in the more useful form of a series, as follows:

$$v = v_0 \left(1 + \frac{v}{c} + \frac{1}{2}\frac{v^2}{c^2} + \frac{1}{2}\frac{v^3}{c^3} + \cdots \right) \tag{31a}$$

For velocities that are low compared to the velocity of light the third and succeeding terms of this series are extremely small and can be neglected. The equation then becomes

$$v = v_0 \left(1 + \frac{v}{c}\right) \tag{31b}$$

which is just the Doppler formula derived for sound waves. For velocities where v is high and not greatly different from c, the first three or four terms are also required for calculations and the Doppler formula from Newtonian mechanics is not valid.

Since the emitting atoms and molecules of a gas are moving with random velocities, the thousands of photons entering the slit of a spectrograph produce a broadened spectrum line having an intensity contour like that shown in Fig. 31B.

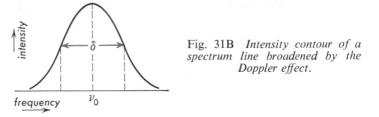

Fig. 31B *Intensity contour of a spectrum line broadened by the Doppler effect.*

The frequency spread of this line is called the *half-intensity breadth*, and is given by the formula

$$\delta = 1.67 \frac{\nu_0}{c} \sqrt{\frac{2RT}{m}} \qquad (31c)$$

The half-intensity breadth is defined as the interval between two points where the intensity drops to half its maximum value. In Eq. (31c) R is the general gas constant, T is the absolute temperature of the gas, and m is the atomic weight. This equation shows that the Doppler broadening is (1) proportional to the frequency ν_0 (2) proportional to the square root of the temperature, and (3) inversely proportional to the square root of the atomic weight. As an illustration of the use of Eq. (31c) consider the following example:

Example. Find the Doppler half-intensity breadth of the sodium yellow line $\lambda = 5890$ A if the light source temperature is 500°K.

Solution. The general gas constant can be found in any book of tables of physical constants and has the value 8.32 joules/mole/K°, and the molecular weight of sodium is 23×10^{-3} Kg/mole.* By direct substitution in Eq. (31c), we obtain

$$\delta = \frac{1.67}{5.89 \times 10^{-7} \text{ m}} \sqrt{\frac{2 \times 8.32 \dfrac{\text{Kg m}^2}{\text{mole K}^\circ \text{ sec}} 500 \text{K}^\circ}{23 \times 10^{-3} \dfrac{\text{Kg}}{\text{mole}}}}$$

$$\delta = \frac{1.67}{5.89 \times 10^{-7} \text{ m}} \sqrt{3.62 \times 10^5 \dfrac{\text{m}^2}{\text{sec}^2}}$$

$$\delta = 1.68 \times 10^9 \frac{\text{vib}}{\text{sec}}$$

To change this to wave numbers, divide by $c = 3 \times 10^{10}$ cm/sec.

$$\delta = 0.056 \text{ cm}^{-1}$$

Since $\lambda = 5890$ A corresponds to a wave-number frequency of 16,980 cm^{-1}, the wave-number breadth of 0.056 cm^{-1} corresponds to a wavelength breadth of only

* A mole is defined as a mass of a substance equal to the molecular weight. Since sodium has an atomic weight of 23, 1 mole of pure sodium weighs 23 grams. The number of molecules in one mole of a substance is given by Avogadro's number, i.e., 6.02×10^{23} molecules. This is the number of sodium atoms in 23 gm of sodium metal.

$$\delta_\lambda = \frac{0.056 \times 5890}{16,980} = 0.0194 \text{ A}$$

Experimental observations are in keeping with Eq. (31c); to produce sharper lines in any given spectrum the temperature must be lowered. Furthermore the lines produced by the lighter elements in the periodic table are in general broader than those produced by the heavy elements.

Figure 31C is a graph of the same spectrum line produced by a high-temper-

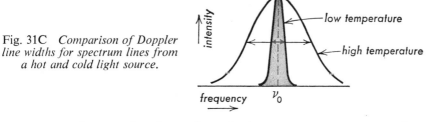

Fig. 31C *Comparison of Doppler line widths for spectrum lines from a hot and cold light source.*

ature source such as an electric spark, and a low-temperature source such as a liquid-air-cooled gaseous discharge.

31.2. Natural Breadths of Spectrum Lines

According to quantum mechanics an energy level diagram of an atom is not to be thought of as a set of sharp discrete levels but a sort of continuous distribution of energy possibilities with pronounced peaks as shown in Fig. 31D. This

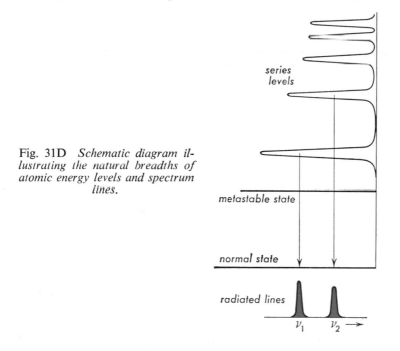

Fig. 31D *Schematic diagram illustrating the natural breadths of atomic energy levels and spectrum lines.*

is a kind of graph in which the energy is plotted vertically and the probability of an electron's being excited to an upper level is plotted horizontally.

When an electron is excited to an upper level, it has a high probability of arriving at an energy value at or near one of the peaks before it returns to the normal state and emits light. Since there is a high probability it may go slightly higher or lower than the center of each peak, the spectrum lines arising from transitions to lower levels will give rise to a so-called *natural breadth* of the observed spectrum lines.

The longer an electron remains in an excited state, the greater is the probability it will arrive at the peak energy value before jumping back down. For this reason *metastable states*, with their relatively long average lifetime t, will be quite narrow, as is the *normal state*. The shorter the average lifetime in an excited state, the wider it will be. The natural half-value widths of energy levels are given by the simple relation

$$\delta_n = \frac{1}{2\pi t} \tag{31d}$$

where for many levels in neutral atoms, t is approximately 10^{-8} sec. For the yellow lines of the sodium spectrum the natural half-intensity breadth is only 0.000116 A, a value a hundred times smaller than the Doppler width from normal gas discharge tubes.

In most spectrum lines the natural width is so narrow as to be completely masked by Doppler broadening.

An explanation of the natural breadth of energy levels is to be found in *Heisenberg's uncertainty principle*. (See §22.8.) Instead of writing Planck's constant h as the product of *momentum times velocity*, it may be written as the product of *energy times time:* $h = 6.6238 \times 10^{-34}$ joule sec:

$$\Delta E \times \Delta t = h \tag{31e}$$

In this equation ΔE represents the uncertainty of the energy when the uncertainty of the time is Δt. If an atom remains in the normal state for a long time, the uncertainty of the energy value is small and the level is sharp. If the electron is excited to an upper level where it remains for an extremely short time, the uncertainty of the energy value is greater and the level is widened. In other words, an electron excited toward an upper level may have little time to settle into the peak position before jumping down again.

31.3. The Stark Effect

Although the splitting up of spectrum lines in a magnetic field was discovered by Zeeman as early as 1897, some sixteen years elapsed before anyone succeeded in showing that a similar effect is produced when a source of light is placed in an electric field.

In 1913 J. Stark demonstrated that every line of the Balmer series of hydrogen, when the source was located in a uniform electric field of 100,000 volts/cm, is split into a number of components. A greatly enlarged photograph of the red hydrogen line λ = 6562 A is shown in Fig. 31E.

The origin of the Stark effect is to be found in the energy levels themselves.

Fig. 31E *Observed Stark effect pattern for the red line of hydrogen, H_α, λ = 6562 A.*

In an electric field many of the energy levels that are otherwise single are split up into a number of component levels. As in the Zeeman effect, where the atoms are in a magnetic field, the levels form a pattern that is symmetrical about the field free level. The stronger the electric field intensity, the wider the splitting is.

Not all levels of the same atoms are split the same amount in any given field. Some levels are widely spaced, while others are close together and some remain single. In general the widest splitting occurs for levels arising from nonpenetrating electron orbits. These are the orbits of valence electrons that are nearly circular on the orbital model and correspond to the D and F levels of the elements in the first two columns of the periodic table.

In an ordinary arc, where no external electric field is applied to the light source, many ions are produced which upon collision with other atoms give rise to strong electric fields. The effect of these intermolecular inhomogeneous fields is to produce varying amounts of splitting of the energy levels of atoms, resulting in a broadening of the observed spectrum lines. This is the origin of the diffuseness of the so-called *diffuse series* of lines observed in many spectra. The sharp series, on the other hand, involve levels that show little or no splitting in strong electric fields.

31.4. Gamma-Ray Spectrum Lines

Gamma rays emitted by radioactive nuclei, like visible light emitted by the external electron structure of atoms, show a finite breadth when photographed with an appropriate γ-ray spectrograph. Gamma-ray spectrographs are identical in principle and similar in construction to X-ray spectrographs, and γ-ray spectrograms are similar in appearance to X-ray spectrograms. (See Figs. 14I and 14J.) In a gas at room temperature radioactive atoms, like stable atoms, move about with different velocities. The emission of γ rays by randomly moving nuclei will, therefore, give rise to a Doppler broadening as shown in Fig. 31B.

Because γ rays have extremely high frequencies, their photon energies $h\nu$ and momenta $h\nu/c$ are thousands of times greater than for visible light, and upon emission by a nucleus give rise to an appreciable recoil energy and momentum. This is illustrated schematically in Fig. 31F.

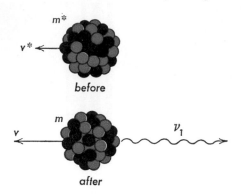

Fig. 31F *Diagram of γ-ray frequency change due to Doppler effect and nuclear recoil.*

Before disintegration the excited nucleus of mass m^* is moving with a random velocity v^*. As a result of γ-ray emission the recoiling nucleus acquires additional momentum and the γ ray loses momentum. Momentum balance therefore requires

$$m^*v^* = \underset{\text{After}}{mv + \frac{h\nu_1}{c}} \tag{31f}$$

$$\underset{\text{Before}}{}$$

Conservation of energy also applies to the process, giving

$$E^*{}_n = E_n + E_\gamma$$

or

$$\underset{\text{Before}}{\tfrac{1}{2}m^*v^{*2}} = \underset{\text{After}}{\tfrac{1}{2}mv^2 + h\nu_1} \tag{31g}$$

It is important to note that the frequency ν_1 is not the frequency ν_0 that would be observed if the nuclear mass were at rest.

Since the nuclei producing any given line all have the same mass, and emit γ rays of the same energy $h\nu_0$, the observed frequency loss r due to recoil and the broadening δ due to the Doppler effect have the result shown in Fig. 31G.

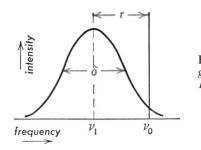

Fig. 31G *Frequency-intensity graph for γ-ray emission showing Doppler width δ and recoil frequency shift r.*

If a nuclear mass could be steadily increased, momentum and energy balance would still hold, but the energy of nuclear recoil would decrease, and the maximum of v_1 would approach v_0.

31.5. Optical and Gamma-Ray Resonance Absorption

When a beam of yellow light from a sodium lamp is made to pass through a glass vessel containing sodium vapor, strong absorption occurs. The atomic processes taking place under these conditions are a kind of resonance phenomenon commonly observed with sound waves.

A good demonstration in sound is shown in Fig. 31H. Two tuning forks with

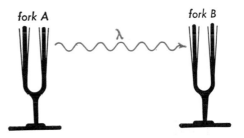

Fig. 31H   *An experiment for demonstrating resonance between two identical tuning forks.*

exactly the same natural frequency, that is, with the same pitch, are mounted on separate sounding boards. Fork A is set vibrating for a moment and then stopped. Fork B will then be found to be vibrating.

Each sound pulse that emerges with each wave from fork A passes by the other fork, pushing with just the right frequency to set fork B vibrating. Such resonance absorption will fail if there is a frequency mismatch between the second fork and the passing waves. If fork A is moved rapidly away from or toward fork B, such a mismatch due to the Doppler effect will prevent resonance.

An analogous demonstration of resonance absorption with visible light is shown in Fig. 31I. Light from a sodium lamp in passing through a sodium flame of a Bunsen burner casts a pronounced dark shadow on a nearby screen. A small piece of asbestos paper soaked in common table salt, NaCl, and placed in an ordinary gas flame can be used to produce an abundance of free sodium atoms.

The atomic process of resonance absorption taking place in this experiment is shown in Fig. 31J. An excited atom in the sodium lamp emits, for example, a

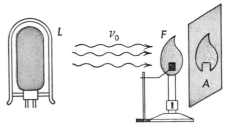

Fig. 31I   *An experiment for demonstrating resonance between sodium atoms.*

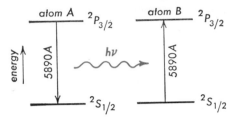

Fig. 31J *Energy-level diagram illustrating light emission and resonance absorption between two sodium atoms.*

wave λ = 5890 A by the downward transition from the $3^2P_{3/2}$ excited level to the $3^2S_{1/2}$ normal state. (See Fig. 12G.)

This wave on coming close to a normal sodium atom in the flame will be absorbed and raise the single valence electron to the $3^2P_{3/2}$ level. This second atom will in turn emit the same frequency again, to be absorbed by another atom in the flame, or to escape from the flame in some random direction. Because re-emission will be in a random direction and seldom in the original direction from the lamp, a shadow is cast.

When an experiment analogous to this is performed with gamma rays, various difficulties are encountered. Because an appreciable amount of the energy of a gamma ray emitted by a radioactive atom is given up to the recoiling nucleus, the frequency of the emitted wave is reduced sufficiently from ν_0 to cause a mismatch between the gamma-ray frequency ν and the natural frequency ν_0 of another nucleus of the same isotope. See Fig. 31G. Furthermore, the recoil of a second nucleus resulting from a gamma-ray impact would give rise to an even greater mismatch.

Intensity-frequency graphs for gamma emission and absorption by identical nuclides are given in Fig. 31K. If the radioactive source, and the target on which

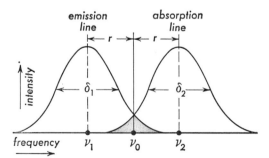

Fig. 31K *Frequency-intensity graphs for γ-ray emission and absorption lines, showing Doppler widths δ, recoil shifts r, and overlap frequencies.*

the γ rays impinge, are at the same temperature, the Doppler broadening will be the same for both.

The area of overlap of the two curves indicates conditions where resonance absorption might be expected to occur. For example, a point at ν_0 corresponds to an event in which a radioactive nucleus is approaching the target at the time of emission and has sufficient kinetic energy to compensate for the energy loss required for recoil. If this emitted γ ray of frequency ν_0 encounters a normal target nucleus moving toward the oncoming wave with an appropriate kinetic energy, resonance absorption can occur. The small overlap area would indicate, however, that resonance absorption under the conditions described should be very small.

31.6. The Mossbauer Effect

The phenomenon of resonance absorption with γ rays was predicted by W. Kuhn in 1921 and first observed by P. B. Moon in 1951. A major breakthrough occurred in 1958, when the phenomenon of recoilless nuclear resonance absorption was discovered by R. L. Mossbauer.

This new absorption effect provided a means for eliminating the energy losses by recoil, both at the source and at the receiver. As an additional feature of the discovery, gamma-ray spectrum lines of extreme narrowness have been produced, which may be useful for other research experiments.

According to the principles described in §31.4 the loss of energy to a recoiling nucleus could be suppressed by rigidly fastening the emitting nuclei, as well as the absorbing nuclei, to an infinitely heavy mass. An experimental method for doing this is the binding of nuclei to a crystal lattice.

To reduce Doppler broadening the γ-ray source and absorber are cooled to extremely low temperatures. This is accomplished by immersing the radioactive crystal source and the crystal absorber in a liquid helium bath.

At temperatures close to absolute zero, atomic vibrations within the crystal lattice are reduced to practically zero, so that during emission or absorption the atoms have little or no motion. Under these conditions Doppler broadening is reduced to practically zero and the observed line reveals its natural line breadth. See Figs. 31C and 31D.

The elimination of atomic vibrations within the crystal lattice has another important effect. Atomic vibrations are quantized, that is, vibration energies have discrete values which may be represented by spaced energy levels. When atoms are in their lowest vibrational energy state, as they are when the crystal is at absolute zero, the recoil energy of a gamma ray is not always sufficient to raise an atom to the next vibrational state. Under these conditions the binding of a nucleus with its neighbors is such that an appreciable part of the crystal lattice takes up any γ-ray recoil momentum. Because the crystal mass is so large, the recoil energy is practically *zero*, and the value of r in Fig. 31K is reduced to zero. Gamma-ray emission or absorption under these conditions is said to be *recoilless*.

With the Doppler effect and nuclear recoil both reduced to zero, only the natural line breadths are involved in nuclear resonance absorption.

A diagram of Mossbauer's apparatus is shown in Fig. 31L. Gamma rays from a source at M pass through a narrow channel C and a thin target A, to where they fall on a scintillation counter S. The lead shielding around the source and

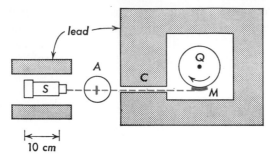

Fig. 31L *Diagram of Moss-bauer's apparatus with which he discovered recoilless emission and absorption of γ rays.*

the detector is for the purpose of reducing all extraneous radiation to a minimum.

To detect resonance absorption in *A*, the source is set into rotation as indicated by an arrow and the detector turned on only when the source is moving in line with the absorber. This motion introduces a Doppler shift *d*, that is, a slight increase or decrease in the frequency of the gamma rays. The effect is illustrated by shifting the red curve in Fig. 31M to the right or to the left. The

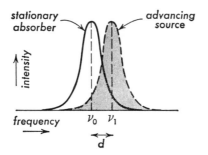

Fig. 31M *Graphs of the natural breadths of γ-ray lines from radio-active crystals at low temperatures near absolute zero.*

mismatch in frequency between the γ rays from the moving source and the nuclei in the stationary absorber reduces the number of photons absorbed. The faster the source moves toward or away from the target, the greater the Doppler shift is and the greater the number of γ rays reaching the counter.

Figure 31N is a graph from Mossbauer's original experiment, performed by using the 0.129-Mev gamma rays from iridium-191. Note that the greatest absorption occurs when the source is at rest. Observe also the extremely low velocities required to produce a Doppler mismatch in frequency. The latter is a clear indication of the narrowness of γ-ray spectrum lines and shows that one is observing natural line breadths rather than Doppler broadening. See Figs. 31B and 31D. The mean lifetime of the Ir[191] transition can be calculated from the half-width of the curve in Fig. 31N and Eq. (31e).

To make this calculation we first find the γ-ray frequency v_0 by means of the equation $Ve = hv$.

$$v_0 = \frac{Ve}{h} = \frac{1.29 \times 10^5 \times 1.6 \times 10^{19}}{6.6 \times 10^{-34}} = 3.13 \times 10^{19} \frac{\text{vib}}{\text{sec}}$$

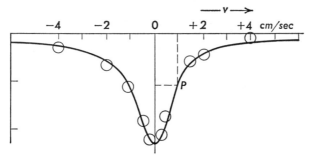

Fig. 31N *Transmission of the 0.129 Mev γ radiation through a resonance absorber of iridium-129, measured at different velocities of the source. (After Mossbauer.)*

The half-width at half-intensity of the resonance curve in Fig. 31N, point P, is approximately 1 cm/sec, or 1×10^{-2} m/sec. If we now use the Doppler formula, Eq. (31b), to find the corresponding change in frequency Δv, where $\Delta v = v - v_0$, we obtain

$$\Delta v = v_0 \frac{v}{c} = \frac{3.13 \times 10^{19} \times 1 \times 10^{-2}}{3 \times 10^8} = 1.04 \times 10^9 \frac{\text{vib}}{\text{sec}}$$

Since $E = hv$, we can write $\Delta E = h\,\Delta v$, and use Eq. (31e) to find Δt:

$$\Delta t = \frac{h}{\Delta E} = \frac{h}{h\,\Delta v} = \frac{1}{\Delta v}$$

or

$$\Delta t = \frac{1}{1.04 \times 10^9} = 9.6 \times 10^{-10} \text{ sec}$$

The mean life of this gamma-ray energy state is, therefore, 9.6×10^{-10} sec.

The nuclear energy levels of Ir^{191} involved in Mossbauer's experiment are shown in Fig. 31O.

Fig. 31O *Energy-level diagram illustrating γ-ray emission and resonance absorption between two iridium-191 nuclei.*

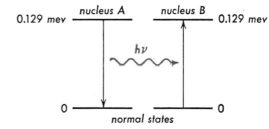

Many experimenters have confirmed Mossbauer's discovery and recoilless γ-ray transitions have been observed in fifteen or more radioactive nuclei. For his discovery Mossbauer was awarded, jointly with Hofstader, the 1961 Nobel Prize in physics. See Appendix X.

PROBLEMS

1. A distant star is found to be approaching the earth at a speed of 1.6×10^6 m/sec. Its spectrum shows that its surface contains a large amount of calcium. In the laboratory the violet line from a calcium discharge tube has a wavelength of 3968 A. Find (a) the frequency of this violet light, (b) the difference between this frequency and the frequency of the light coming from the star, (c) this frequency difference in wave numbers, and (d) the wavelength shift in angstroms.

2. A distant star is found to be approaching the earth at a speed of 4.5×10^6 m/sec. Its spectrum reveals that its surface is composed largely of hydrogen. In the laboratory the red line of the Balmer series, observed with a hydrogen discharge tube, has a wavelength of 6562 A. What is (a) the frequency of this laboratory light, (b) the difference between this frequency and the frequency of the light coming from the star, (c) this same frequency shift in wave numbers, and (d) the wavelength shift in angstroms? *Ans.* (a) 4.57×10^{14} vib/sec. (b) 6.86×10^{12} vib/sec. (c) 229 cm^{-1}. (d) 98.4 A.

3. When the spectrum lines from a distant star are measured they are all found to be shifted to shorter wavelengths than when they are produced in the laboratory. The yellow line of sodium, for example, is not at its normal wavelength of 5890 A, but at 5870 A. Find (a) the frequency of the normal sodium light, (b) the frequency of the light from the star, (c) the frequency difference, and (d) the velocity of the star with respect to the earth.

4. When the spectrum of the light from a distant star is observed and measured the lines are found to be shifted to shorter wavelengths than when they are produced in the laboratory. The green line of iron, for example, is not at its normal wavelength of 5270 A, but at 5240 A. Find (a) the frequency of the normal green line, (b) the frequency of the corresponding line from the star, (c) the frequency difference, and (d) the velocity of the star with respect to the earth. *Ans.* (a) 5.692×10^{14} vib/sec. (b) 5.725×10^{14} vib/sec. (c) 3.3×10^{12} vib/sec. (d) 1.74×10^6 m/sec.

5. Find the Doppler half-intensity breadth of the hydrogen red line $\lambda = 6562$ A, if the light source is at 1000°K: (a) in vib/sec, (b) in wave numbers, and (c) in angstroms.

6. Find the Doppler half-intensity breadth of the mercury green line $\lambda = 5461$ A, if the light source is at 1000°K; (a) in vib/sec, (b) in wave numbers, and (c) in angstroms. *Ans.* (a) 8.81×10^8 vib/sec. (b) 0.0294 cm^{-1}. (c) 0.00877 A.

7. If the Mossbauer experiment is performed with a 5-Mev gamma ray from a radioactive source and the half-width at half-intensity of the resonance curve (see Fig. 31N) is 2.5 cm/sec, find (a) the frequency of the gamma ray, (b) the Doppler shift in vib/sec, and (c) the mean life of the gamma-ray state.

8. If the Mossbauer experiment is performed with a 1.80-Mev gamma ray from a radioactive source and the half-width at half-intensity of the resonance curve (see Fig. 31N) is 1.2 cm/sec, find (a) the frequency of the gamma ray, (b) the Doppler shift in vib/sec, and (c) the mean life of the gamma-ray state. *Ans.* (a) 4.36×10^{20} vib/sec. (b) 3.63×10^{10} vib/sec. (c) 2.75×10^{-11} sec.

The Atomic Nucleus

Although the disintegrations of different nuclei give rise to many different kinds of particles or units of energy, it would appear that we need assume only two kinds of particles existing within the nucleus—*neutrons* and *protons*. If this is correct, our task becomes the difficult one of explaining not only the disintegration mechanism of an unstable nucleus but the binding forces that hold a stable nucleus together. An answer to the latter question will serve as a starting point for the following presentations.

32.1. Nuclear Binding Forces

According to the *neutron-proton theory* of the atomic nucleus (see Fig. 32A), the deuteron nucleus contains but one neutron and one proton. Let us compare,

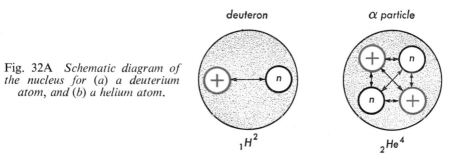

deuteron $\qquad$ α *particle*

Fig. 32A *Schematic diagram of the nucleus for (a) a deuterium atom, and (b) a helium atom.*

$_1H^2$ $\qquad$ $_2He^4$

therefore, the mass of one free proton and one free neutron with their mass when combined as a deuteron (for masses, see Appendix IV):

$$
\begin{array}{lll}
\text{Neutron mass} & _0n^1 = & 1.008665 \\
\text{Hydrogen mass} & _1H^1 = & \underline{1.007825} \\
& \text{Sum} = & 2.016490 \\
\text{Deuteron mass} & _1H^2 = & 2.014103
\end{array}
$$

The difference in mass of 0.002387 atomic mass unit (abbr. amu) is not due to inaccurate measurements of mass but is a real difference to be accounted for as the annihilation energy that binds the two particles together. When a neutron and proton come together to form a deuteron, a small part of their mass—namely, 0.002387 amu (equivalent to 2.22 Mev energy)—is radiated from the newly formed nucleus. At close approach, in other words, the two particles attract each other so strongly that once together, it takes the equivalent of a little more than 2 Mev of energy to pull them apart. This has been confirmed by a nuclear photoelectric effect, an experiment in which γ rays of 2.22 Mev energy or greater are found to break up deuterium nuclei into their constituent parts, while γ rays of lower energy have no effect. How neutrons and protons attract each other when very close together is a question of great importance, for we now realize that the stability of all the universe as we know it depends upon these forces.

Consider as a second example, the attractive forces between the four nucleons of a helium nucleus, i.e., the two neutrons and two protons of an α particle as shown in Fig. 32A. By combining the masses of the four free particles and comparing them with the mass of the helium atom, we obtain

$$2_0n^1 + 2_1H^1 = 4.032980 \text{ amu}$$
$$He^4 = 4.002604 \text{ amu}$$
$$\text{Mass difference} = 0.030376 \text{ amu}$$
$$E = 28.3 \text{ Mev}$$

This value of 28.3 Mev indicates a binding energy of approximately 7 Mev per nucleon, a value considerably higher than 2.2 Mev for the deuteron. This is a measure of the energy that must be expended in breaking the attractive bonds shown in the diagram.

If we make similar calculations for other nuclides near the beginning of the periodic table, and plot a graph of the binding energy per nucleon, E/A, where A is the mass number, we obtain Fig. 32B. If this same procedure is carried out for the entire periodic table, a graph like the one shown in Fig. 32C is obtained. By drawing a horizontal line across the upper part of the graph at 8 Mev, as

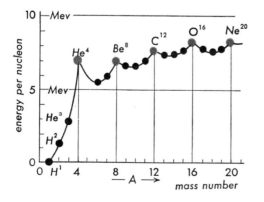

Fig. 32B   *Binding energy per nucleon for the lightest elements in the periodic table.*

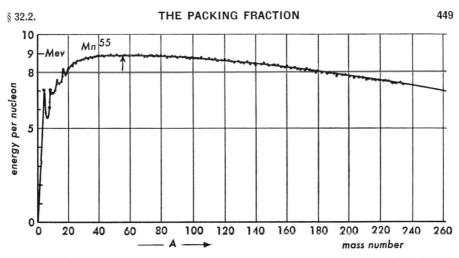

Fig. 32C   *Binding energy per nucleon for the stable isotopes of the periodic table.*

shown by the colored line, we obtain a kind of average binding energy per nucleon for nearly all the elements. This value is approximately the difference between unit atomic mass and the mass of a free neutron or proton.

Collisions between high-energy photons and nuclei, giving rise to several kinds of reactions, are given in the preceding chapter. One process not described there, and called the *nuclear photoelectric effect*, is shown in Fig. 32D. A photon

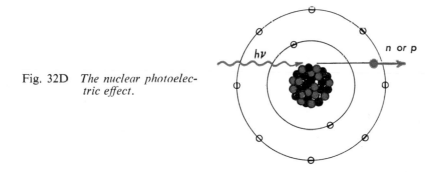

Fig. 32D   *The nuclear photoelectric effect.*

of energy $h\nu$, passing close to a nucleus, is absorbed; part of the energy is used to remove a neutron or proton from the nucleus, and the remainder given to the particle as kinetic energy. See Eq. (10a):

$$h\nu = W + \tfrac{1}{2}mv^2 \tag{32a}$$

The work function W, as we have seen above, is about 8 Mev and is essentially the energy required to create the extra mass needed by the nucleon to set it free.

32.2. The Packing Fraction

An informative method of displaying the mass differences and binding energies of stable nuclei is to plot a graph of all packing fractions. The *packing fraction P* of any nuclide is obtained from the relation

$$P = \frac{M - A}{A} \qquad (32b)$$

where $M - A$ for any nuclide is called its *mass defect:*

$$M = \text{mass of nuclide}$$
$$A = \text{mass number} \qquad (32c)$$

In other words, P is the difference between the average nucleon mass for that nuclide and unit atomic mass. It will be seen in Fig. 32E that the very light and

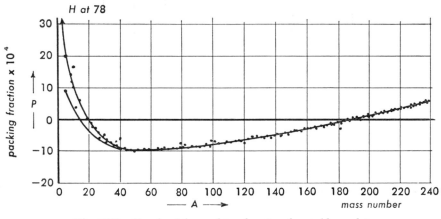

Fig. 32E   *Graph of the packing fraction for stable nuclei.*

the heavy nuclides have an average mass per nucleon greater than unity, whereas those near the middle of the graph have a mass per nucleon less than unity. In the next chapter we shall see that these differences are directly related to the availability of nuclear energy.

32.3. The Nuclear Potential Barrier

Early in the development of ideas concerning nuclear disintegration, Gamow proposed a model by which one might represent the atomic nucleus. This model is based upon the forces acting between two positive charges and is an extension of the nuclear model described in §26.2, and illustrated in Figs. 26D and 26E.

Picture again a proton or an α particle, with its positive charge, approaching a positively charged nucleus. As the two charges come closer and closer together, they repel each other with greater and greater forces as given by Coulomb's law. This repulsion cannot continue to increase all the way to zero separation, however, for as the two charges come very close together we know, from what has been said in the preceding section, there must be an attraction. Gamow proposed, therefore, that at close approach there is another law of force that comes into play and that this force is one of attraction for neutrons as well as protons and is very strong.

Since Coulomb's law is known to hold quite accurately at large distances, this added factor is called a *short-range force*. Graphs of Coulomb's law for repulsion and the short-range force for attraction are shown in Fig. 32F. Note that when

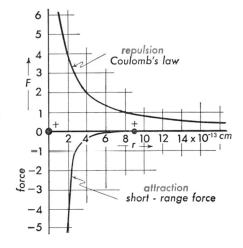

Fig. 32F   *The two force laws for a positive charge close to a nucleus.*

the particle is at the distance shown, 9×10^{-13} cm, Coulomb's law of repulsion is predominant, whereas, at a distance less than 2×10^{-13} cm, the short-range force of attraction predominates. Since both of these forces are effective for all distances, they should be combined into one graph as shown by the F curve in Fig. 32G. If instead of the force F we plot the potential energy stored between

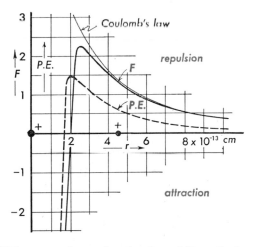

Fig. 32G   *Graphs of the force and potential energy of a positively charged particle close to a nucleus.*

the two particles, we obtain the P.E. curve shown by the dotted line. Such a curve represents what is called the *potential barrier* of the nucleus. The highest point of the barrier is frequently called the edge of the nucleus which, for heavy atoms in the periodic table, occurs at, and gives a nuclear radius of, from 1 to 8×10^{-13} cm. See Table 34A.

A very good model of the nucleus, having the form of a *well* or the *crater of a volcano*, can be made by rotating the dotted line around the vertical axis. When this is done we obtain the cross-section diagram in Fig. 32H. With this

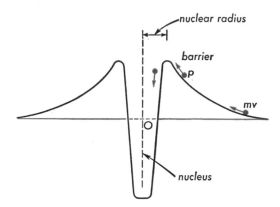

Fig. 32H *A graphical model of the atomic nucleus as proposed by Gamow. The potential barrier of a nucleus to an approaching positive charge is analogous to the crater of a volcano.*

model the electrical potential energy, P.E., between the two positively charged particles, is analogous to the potential energy of a ball at any point on the side of the hill, and the electrostatic force of repulsion is analogous to the force of gravity.

32.4. Bohr's Nuclear Model

In 1937 Niels Bohr, the famous Danish physicist, made another outstanding contribution to modern physics when he improved Gamow's model of the nucleus by extending what is sometimes called the *waterdrop model* of the nucleus. Bohr and his collaborator Kalkar imagine the many particles in a heavy nucleus as moving about within a spherical enclosure with motions analogous to the molecules in a drop of water. The surface of the spherical enclosure, which is the top of the potential barrier as represented in Fig. 32I, is

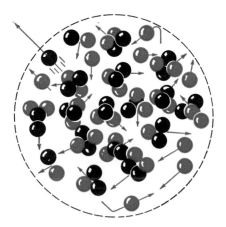

Fig. 32I *Nucleus in the act of ejecting a neutron, based upon the Bohr-Gamow waterdrop model.*

analogous to the surface tension which holds a small waterdrop to its spherical form.

Just as the rapid motion of the molecules in water is a measure of the temperature, so Bohr speaks of the rapid motion of the neutrons and protons within the spherical boundary of the nucleus as a sort of *pseudotemperature*. To explain disintegration, the analogy is drawn that the ejection of a particle from the nucleus is like the evaporation of a water molecule from a drop of water. Just as a rise in temperature brings about a more rapid evaporation of water, so an increase in the motions within the nucleus gives rise to a higher probability of disintegration.

In a stable nucleus, the particles within are moving about with very little kinetic energy and are in the analogous state of a relatively low temperature. When a high-speed particle from outside penetrates the potential barrier, it is accelerated toward the center of the nucleus and acquires a very high kinetic energy before it collides with one or more of the particles inside. Soon the energy becomes divided among the many particles, and the nucleus takes on a higher temperature state. The potential-well model of this same condition is shown in Fig. 32J.

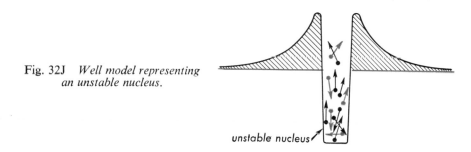

Fig. 32J *Well model representing an unstable nucleus.*

unstable nucleus

Now, as the particles move about inside, there is a certain probability or chance that, within a given interval of time, some one particle will be hit by several particles, giving it a sufficiently high velocity in an outward direction to permit an escape through the potential barrier. The more rapid the internal motions, that is, the higher the temperature, the greater is this chance of escape.

A direct disintegration may be described in this way: If upon entering the nucleus a high-speed particle like a proton adds sufficient energy to give the nucleus a high temperature, another particle like a neutron or an α particle may be ejected immediately. Since such an ejected particle has to be supplied with a certain minimum energy to get free, the remaining particles will be slowed down, and the nucleus will have a lower temperature.

32.5. Nuclear Demonstration Models

A demonstration model illustrating the capture of a high-speed proton or an α particle by a nucleus, prior to disintegration, is shown in Fig. 32K. Marbles

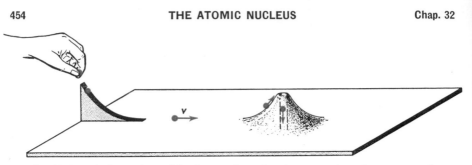

Fig. 32K *Mechanical model of a nucleus for demonstrating the capture of a high-speed proton, deuteron, or α particle, prior to disintegration.*

rolled down the incline represent the speeding-up of atomic projectiles by an accelerator like the cyclotron. Approaching the potential barrier, a marble may roll part way up and then be deflected off to one side, illustrating elastic scattering of the kind observed by Rutherford. If the initial velocity is high enough the marble may go over the top of the barrier and drop into the crater opening at the top, representing a capture prior to disintegration.

A demonstration of what happens inside the nucleus is illustrated by another model as shown in Fig. 32L. In this case the vertical scale of the barrier has of

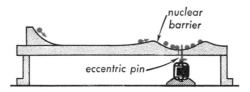

Fig. 32L *Mechanical model of a nucleus for demonstrating (a) the increased kinetic energy of nuclear particles after a capture and (b) the chance probability of radioactive decay or disintegration by the ejection of a particle.*

necessity been reduced, i.e., flattened out. When a marble is rolled down the incline and into the group of marbles at the center of the barrier, there may be several collisions before another particle usually goes bouncing out on the other side. This corresponds to a direct disintegration where one particle like a proton goes in and a neutron comes out.

If a single particle does not emerge, most of the particles inside take on random motions, colliding with each other much the same as do the molecules or atoms in a gas or liquid. To prevent friction from stopping them (there is no friction in an atom), the marbles are continually agitated by a small pin protruding from underneath the barrier. This pin is mounted slightly off center at the end of the shaft of a small electric motor. If the motor is left running for some time, a single marble will eventually be hit by several particles moving in the same direction and will recoil with sufficient speed to carry it over the barrier and out. This corresponds to a disintegration or radioactive decay, which takes place according to the *laws of chance*, and to the resultant drop in "temperature" of the nucleus.

The faster the motor runs, the greater is the internal agitation and chance of ejection, and the shorter is the so-called half-life of the element.

32.6. Nuclear Model for Neutron Disintegrations

When a neutron approaches a nucleus prior to a disintegration, it does not encounter a potential barrier of the type already described for protons and α particles. A neutron has no charge, so that at large distances it is not repelled by the positively charged nucleus. It may, therefore, approach a nucleus with very little speed of its own and be captured when it comes too close. At very close range the short-range attractive force shown in Fig. 32F sets in, and draws the two together.

To an approaching neutron, the nucleus acts as though it were a pit into which the particle will fall. This is illustrated by the flat potential curve in Fig. 32M.

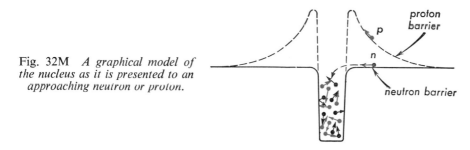

Fig. 32M *A graphical model of the nucleus as it is presented to an approaching neutron or proton.*

The marble rolling along the horizontal plane toward the pit represents the influence of the nucleus upon the neutron's motion, whereas the marble rolling up the hill (dotted line) represents the influence of the same nucleus upon the motion of a proton. A mechanical model patterned after Fig. 32K, and made with the neutron barrier shape, offers an excellent demonstration of neutron scattering and capture.

32.7. Nuclear Spin

Approximately $\frac{1}{3}$ of all known stable nuclei, and many of the radioactive nuclides, are known to have a mechanical moment p_I and a magnetic moment μ_I. The earliest evidence for a nuclear spin, as it is called, was found in atomic spectra. Many spectrum lines show a fine structure which in §12.4 was explained as being due to electron spin. Many of these fine-structure lines show a still finer structure, now known to be due to a nuclear spin. (See §18.5, and Figs. 18J and 18K.)

As an additional example of this *hyperfine structure*, as it is called, several of the spectrum lines of the element praseodymium are shown in Fig. 32N. In a normal spectrogram these look like single lines, but under the high dispersion and magnification shown here, each line reveals six hyperfine components. The six components signify that the nucleus of praseodymium has a spin angular momentum $p_I = \frac{5}{2}\hbar$ and a large magnetic moment.

The angular momenta of nuclei in general are given by

$$p_I = I\hbar \tag{32d}$$

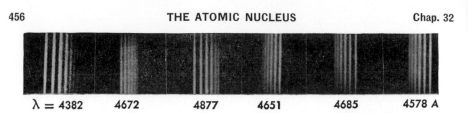

$\lambda = 4382 \qquad 4672 \qquad\qquad 4877 \qquad 4651 \qquad\qquad 4685 \qquad 4578\ A$

Fig. 32N *Photographs of the hyperfine structure of a few of the spectrum lines of the element praseodymium, Z = 59. Wavelengths are in Angstrom units: structure produced by nuclear spin.*

where *I* is the nuclear spin quantum number, and $\hbar$ represents unit angular momentum given by $h/2\pi$. See Eq. (12b). Quantitative experiments show that this *quantum number*, *I*, can take half-integral as well as whole-number values. (See Table 32A.)

The magnetic moments of nuclei in general are extremely small as compared with those associated with electrons in the outer structures of atoms. For all nuclei the magnetic moment is given by the relation

$$\mu_I = g\hbar\,\frac{e}{2M} \tag{32e}$$

where the *g*-factor varies from nucleus to nucleus, $\hbar$ is unit angular momentum, and *M* is the mass of the proton. The last two factors involve fixed atomic constants, and combined are called the *nuclear magneton*. One nuclear magneton is given by

$$\mu_n = \hbar\,\frac{e}{2M} \tag{32f}$$

or

$$\mu_n = 5.050 \times 10^{-27}\ \text{ampere meters}^2$$

The *g*-factor therefore gives the nuclear magnetic moment in nuclear magnetons. *Note that one nuclear magneton is $\frac{1}{1836}$ of one Bohr magneton.* (See Eq. (12f).)

A list of a few nuclides in Table 32A shows typical values of the nuclear spin quantum number *I*, and the measured magnetic moments in nuclear magnetons.

Table 32A. Typical Nuclear Spins and Magnetic Moments

	I	*g*-FACTOR		*I*	*g*-FACTOR
$_1H^1$	$\frac{1}{2}$	$+2.792$	$_{23}V^{51}$	$\frac{7}{2}$	$+5.15$
$_1H^2$	1	$+0.857$	$_{25}Mn^{55}$	$\frac{5}{2}$	$+3.47$
$_3Li^6$	1	$+0.82$	$_{50}Sn^{119}$	$\frac{1}{2}$	-1.05
$_3Li^7$	$\frac{3}{2}$	$+3.26$	$_{55}Cs^{135}$	$\frac{7}{2}$	$+2.73$
$_4Be^9$	$\frac{3}{2}$	-1.18	$_{73}Ta^{181}$	$\frac{7}{2}$	$+2.1$
$_5B^{10}$	3	$+1.80$	$_{80}Hg^{198}$	0	0
$_7N^{14}$	1	$+4.01$	$_{80}Hg^{199}$	$\frac{1}{2}$	$+0.50$
$_8O^{16}$	0	0	$_{80}Hg^{200}$	0	0
$_8O^{17}$	$\frac{5}{2}$	-1.89	$_{80}Hg^{201}$	$\frac{3}{2}$	-0.56
$_8O^{18}$	0	0	$_{83}Bi^{209}$	$\frac{9}{2}$	$+4.08$

32.8. Proton and Neutron Spin

Every proton, whether it is bound to an atom or is free, has a spin angular momentum of $\frac{1}{2}\hbar$. See Fig. 32O:

$$p_{\text{proton}} = \tfrac{1}{2}\hbar \tag{32g}$$

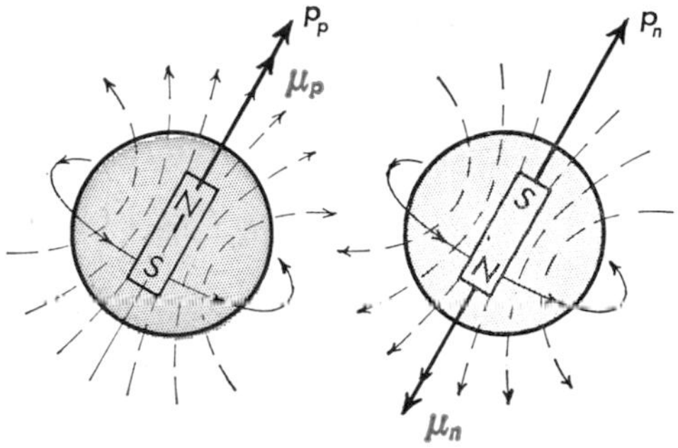

Fig. 32O *Schematic diagrams showing proton and neutron spins and associated magnetic moments.*

Having a positive charge, the magnetic field around a proton is parallel to its mechanical moment, instead of oppositely directed as is the case with an electron. (See Fig. 12F.) Furthermore, since the mass of a proton is 1836 times the mass of an electron, it maintains the same angular momentum by spinning much slower, and this reduces its magnetic moment to a relatively small value. The magnetic moment of a proton is found by precision experimental measurements to be

$$\mu_{\text{proton}} = 2.792 \text{ nuclear magnetons} \tag{32h}$$

The neutron, like the proton and electron, has a spin angular momentum of $\frac{1}{2}\hbar$:

$$p_{\text{neutron}} = \tfrac{1}{2}\hbar \tag{32i}$$

Having no net charge, the neutron might well be expected to have zero magnetic moment. Experimentally, however, its spin does produce a magnetic field, and one that is oppositely directed to angular momentum. See Fig. 32O. The neutron's magnetic moment is

$$\mu_{\text{neutron}} = -1.913 \text{ nuclear magnetons} \tag{32j}$$

A negative magnetic moment is a clear indication that the neutron is a complex particle containing negative and positive charges in equal amounts, and that the negative charge is, on the average, farther from its axis of rotation.

32.9. The Deuteron

The deuteron is a nuclear particle composed of one proton and one neutron. Its known spin of $I = 1$ (see Table 32A) indicates that the spins of the two particles are parallel to each other as shown in Fig. 32P. Since the two magnetic moments are oppositely directed, the resultant magnetic moment of the combi-

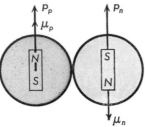

Fig. 32P *Schematic diagram of the deuteron, showing proton and neutron spins parallel and magnetic moments opposing.*

nation should be $2.792 - 1.913$, or 0.879 nuclear magnetons. Precision measurements, however, show that μ_d is slightly smaller than this, and is

$$\mu_{\text{deuteron}} = 0.857 \text{ nuclear magnetons} \tag{32k}$$

32.10. Nuclear Shell Model

A reasonably successful theory* of nuclear spins and magnetic moments is based upon the electron spin-orbit model so successful in explaining the outer structure of atoms. (See §12.5.) Experimental evidence for this idea stems in large measure from a classification of all stable nuclides as to their odd or even numbers of protons and neutrons. (See Table 32B.)

While all nucleons have a spin of $\frac{1}{2}\hbar$, neutrons and protons appear to pair off with opposing spins, thus canceling both the mechanical moments and magnetic moments. While the positive magnetic moments (see Table 32A) are probably

Table 32B. Classification of Known
Stable Nuclides

Z PROTONS	N NEUTRONS	KNOWN STABLE NUCLIDES	NUCLEAR SPIN I
odd	odd	4	$1, 2, 3, 4$
odd	even	50	$\frac{1}{2}, \frac{3}{2}, \frac{5}{2}, \frac{7}{2}, \frac{9}{2}$
even	odd	55	$\frac{1}{2}, \frac{3}{2}, \frac{5}{2}, \frac{7}{2}, \frac{9}{2}$
even	even	165	0

due to *odd protons*, and the negative magnetic moments to *odd neutrons*, the wide range in values and the half-integral values of I in the *odd-even* and *even-odd* nuclides suggests orbital motions.

By analogy with the electron structure of atoms, orbital quantum numbers $l = 0, 1, 2, 3, \ldots$, for $s, p, d, f, g, \ldots$, respectively, are assigned to all nucleons. If the odd orbital nucleon is a proton, its positive charge and spin will give rise to two positive magnetic moments. See Fig. 32Q. If l and s are parallel as in

* For their development of the theory of the nature of the shell structure of the atomic nucleus, Maria Goeppert-Mayer of California, J. Hans D. Jensen of Heidelberg, and Eugene Wigner of Princeton were jointly awarded the 1963 Nobel Prize in physics.

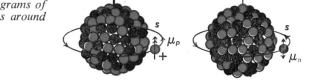

Fig. 32Q *Schematic diagrams of proton and neutron orbits around nuclei.*

the diagram, so that $j = l + s$, the magnetic moments will add; if l and s are oppositely directed, $j = l - s$, the magnetic moments will subtract.

For a proton the orbital angular momentum is given by

$$p_l = l\hbar$$

and the associated orbital magnetic moment by

$$\mu_l = l\hbar \frac{e}{2M}$$

By adding or subtracting the spin magnetic moment μ_{proton} (see Eq. (32h)), we obtain the expected total nuclear magnetic moment μ_I. While experimentally determined values of μ_I do not agree exactly with the values calculated by this simple model, they are in surprisingly good agreement.

If the orbital nucleon is a neutron, its orbital motion cannot give rise to an orbital magnetic moment. See Fig. 32Q. Neutron spin, however, does have a negative magnetic moment, and the magnitude of its contribution to the entire nucleus will depend upon its orientation with respect to the nuclear spin axis.

Some success with the application of the Pauli exclusion principle, applied to the filling of proton and neutron subshells, has been achieved. (See §12.9.) The experimental evidence that such closed subshells exist in nuclei is to be found in a number of nuclear properties. If the binding energy, angular momentum, or magnetic moment is plotted against proton number Z or neutron number N, discontinuities occur when either Z or N has the value 2, 8, 14, 20, 28, 50, 82, and 126. While these so-called "magic numbers" seem to represent closed shells and subshells, much is yet to be learned about the structure of atomic nuclei.

32.11. Mesic Atoms

Because of the coulomb attractive forces between unlike charges, negatively charged mesons have a far greater probability of being captured by atomic nuclei than have their positively charged antiparticles. See Fig. 35A. There is a certain probability that as π^- and μ^- particles pass through matter they may be captured in any one of a number of outside orbits to form a kind of Bohr atom.

All equations derived for hydrogen can be applied to such mesic atoms by replacing the proton charge $+e$ by the nuclear charge $+Ze$, where Z is the atomic number. When this is done, Eq. (11c) for the radius of circular orbits becomes

$$r = \frac{n^2 h^2}{4\pi^2 m e^2 k Z} \tag{32l}$$

where m is the particle mass and n is the principal quantum number. The innermost orbit for hydrogen and its electron, where $n = 1$ and $Z = 1$, has a radius $r_1 = 5.28 \times 10^{-9}$ cm.

Since the radius varies inversely as the mass of the orbiting particle, the radius of the allowed orbits, $n = 1, 2, 3, 4$, etc., for π and μ mesic atoms will be about $\frac{1}{273} r_1 n^2$ and $\frac{1}{207} r_1 n^2$, respectively. Around a proton, the π^- orbit, $n = 1$, would have a radius of 1.93×10^{-11} cm, and for $n = 2$ a radius four times this. In an atom like $_{35}\text{Br}^{79}$, the innermost π^- orbit would be only $1/35 \times 273$, or $\frac{1}{9555}$ as large as the normal hydrogen atom, and the meson would be just skimming the surface of the nucleus with a radius of 5.53×10^{-13} cm. See Fig. 32R.

Fig. 32R *Schematic orbital diagram of a π mesonic atom—phosphorus, $_{15}P^{31}$—in the $n = 1$ circular orbit.*

Since the orbital energy is proportional to the particle mass m (see Eq. (11k)), transitions between orbits will have energies $h\nu$ comparable to X rays. X rays arising from jumps from the $n = 2$ to $n = 1$ orbits of π^- mesic atoms were first observed in 1952 by Cemak, McGuire, Platt, and Schulte. These were produced by a beam of π^- mesons from the University of Rochester cyclotron traversing the beryllium, carbon, and oxygen target.

Similar observations were made by Rainwater and Fitch in 1953 for μ^- particles from a cyclotron. Confirmation of the origin of these X rays is attributed to the agreement between measured X-ray wavelengths and those calculated from the Bohr formula.*

PROBLEMS

1. Using the atomic masses given in Appendix IV, calculate the binding energy per nucleon for (a) sulfur-36 and (b) calcium-42.

2. Using the atomic masses given in Appendix IV, calculate the binding energy per nucleon for (a) oxygen-16, and (b) silicon-29. *Ans.* (a) 7.97 Mev. (b) 8.44 Mev.

3. Calculate the packing fraction for (a) magnesium-24, and (b) silicon-28. (See Appendix IV for atomic masses.)

4. Calculate the packing fraction for (a) boron-10, and (b) chlorine-35. *Ans.* (a) 0.00161 amu. (b) 0.00057 amu.

5. Substitute the values of the atomic constants in Eq. (32f) and calculate the nuclear magneton.

6. Calculate the ratio between the angular velocity of electron spin and

* For more details of nuclear structure, see *Introduction to Atomic and Nuclear Physics* by H. Semat, Rinehart and Company.

proton spin. Assume uniform homogeneous spheres of equal size. *Ans.* 1836.

7. If the odd proton in a nucleus were in a d-orbit, with its spin oppositely directed to l, what would you expect (a) the nuclear spin value I, and (b) the nuclear magnetic moment μ_I to be?

8. If the odd proton in a nucleus were in a g orbit, $l = 4$, with its spin parallel to l, what would you expect (a) the nuclear spin value I, and (b) the nuclear magnetic moment μ_I, to be? *Ans.* (a) $I = \frac{9}{2}$. (b) 6.792 nuclear magnetons.

9. Calculate the radius of the μ^- orbit, $n = 1$, for an oxygen atom.

10. Calculate the radius of the π^- orbit, $n = 1$, for the nuclide $_{20}Ca^{40}$. *Ans.* 9.7 $\times 10^{-13}$ cm.

11. Calculate the radius of the π^- orbit, $n = 1$, for the nucleus shown in Fig. 32R.

Fission and Fusion

In 1937 Fermi, Segré, and their collaborators subjected uranium to the bombardment of neutrons. From the radioactivity produced they believed they had succeeded, for the first time, in producing a series of new elements, 93, 94, 95, etc., beyond uranium 92. The reason for their belief was that the uranium, after bombardment, gave off electrons with a number of different half-lives. If one attributed these different half-lives to the successive disintegrations of the same atoms, a single nucleus should emit several electrons, one after the other. With each emission the nuclear charge would increase by unity, thus producing an atom of higher and higher atomic number. Similar observations were later made by the Curie-Joliots, in France.

33.1. The Discovery of Fission

In 1939, just prior to World War II, Otto Hahn in Germany, and his two associates Lise Meitner and F. Strassmann made a new and important discovery. After bombarding uranium metal with neutrons they carefully performed a series of chemical separations of the uranium sample to determine the element to which the newly produced radioactivity belonged. To their amazement they found the radioactive atoms to be identical chemically with a number of different elements, nearly all of which are near the center of the periodic table. In other words, a uranium nucleus, after the capture of a single neutron, seemed to be splitting apart into two nearly equal fragments, as illustrated in Fig. 33A.

In the few weeks that followed this discovery, many observers in different laboratories the world over not only confirmed the results, but extended the observations by studying in detail the products of the disintegrations. To explain the phenomenon in simple words, consider the details of the process illustrated in Fig. 33A. An original uranium nucleus, $_{92}U^{235}$, with its 92 protons and 143 neutrons is shown in (a) as it captures a slowly moving neutron.

In diagram (b), the newly formed nucleus is unstable and starts to separate

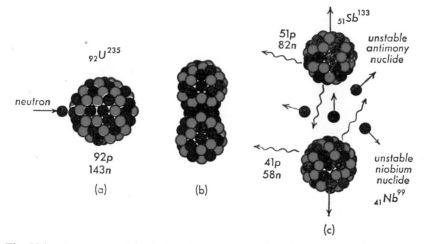

Fig. 33A   *Diagrams of the fission of a uranium nucleus into two unstable nuclides.*

into two nearly equal parts. Because this process resembles cell division in the science of biology, the phenomenon is called *fission*. In coming apart, the uranium nucleus, behaving like the analogous waterdrop, splashes out small drops, that is, neutrons and γ rays. So great is the energy liberated by this explosion of the nucleus that each of the two heavy nuclei fly apart in opposite directions with tremendous speeds. That they do so has been confirmed by many Wilson cloud chamber photographs, one of which is reproduced in Fig. 33B.

To obtain this photograph the cloud chamber contained a thin film of material coated with uranium. The fission of one uranium nucleus reveals two tracks of the same density, showing clearly that the particles traveled outward in opposite directions. The heavy forks near the ends of the tracks are characteristic of highly charged fission fragments that have made several collisions with other nuclei before coming to rest.

Not all of the uranium nuclei divide into antimony and niobium as shown in Fig. 33A, but into any one of many pairs of fragments corresponding to elements near the center of the periodic table. The experimental evidence seems to favor pairs of slightly unequal mass, accompanied by from one to five or more neutrons, as shown in diagram (c).

Numerous measurements of the masses of fission fragments have made it possible to construct the graph shown in Fig. 33C. The fission fragment yield is plotted vertically to a logarithmic scale, and the mass number A is plotted horizontally to a uniform scale. The curve is seen to rise sharply between $A = 75$ and 90 and to drop equally fast between $A = 145$ and 160. The most probable values for the mass numbers of the two fragments are 95 and 139 with a minimum of $\frac{1}{10}$ of 1% at $A = 117$.

In general, fission fragments are not stable nuclei but contain an excess number of neutrons. Typical ensuing events that occur to most fragments are

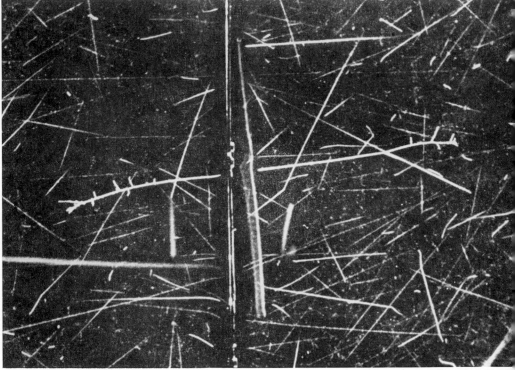

Courtesy, I. K. Bog

Fig. 33B *Wilson cloud-chamber photograph showing a pair of fission fragments recoiling in opposite directions. Note the δ-ray forks near the ends. Fission was produced by a neutron beam from a cyclotron.*

shown in Fig. 33D. After a series of β-emissions, in which neutrons are converted into protons in the nucleus, a stable nuclide results.

Starting at the left in the diagram with the unstable antimony nuclide of charge $+51$ and mass 133, the successive emissions of four electrons raise the nuclear charge by four unit steps, ending with a stable cesium nuclide, $_{55}Cs^{133}$. The other fragment, $_{41}Nb^{99}$, in Fig. 33A(c), carries out a similar series of β-ray emissions, ending up with $_{44}Ru^{99}$.

As proof that the above series is produced by fission, previously bombarded uranium has been chemically analyzed for elements near the center of the periodic table. After each chemical separation is performed, a test of the β-ray activity is made by a measurement of the half-life. A comparison of this measured half-life with the values already known for the same element from other disintegration experiments has made it possible to identify some of the radioactive nuclei produced. Such tests, for example, have been made by Wu* for the series of four elements in Fig. 33D. Note the increasing half-lives she

* C. S. Wu (Mrs. C. L. Yuan), *Physical Review*, Vol. 58, 1940, p. 925.

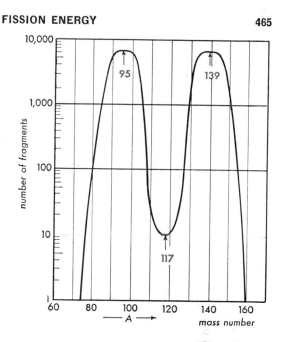

Fig. 33C   *Semilog graph showing the fission fragment yield for slow neutrons on uranium-235.*

identified for this series, indicating increased stability as the stable nucleus cesium is approached.

Approximately 99% of the neutrons ejected as the result of the fission of uranium occur within an extremely short time interval and are called *prompt neutrons* or *secondary neutrons*. About one out of a hundred neutrons are emitted one or more seconds later, and these are called *delayed neutrons*. Delayed neutrons originate from unstable fragments that decay by neutron emission on their way to becoming stable nuclei.

33.2. Fission Energy

The energy liberated in the fission of uranium is due largely to the U-235 isotope. Uranium found in the earth's crust has three principal isotopes with the following relative abundance:

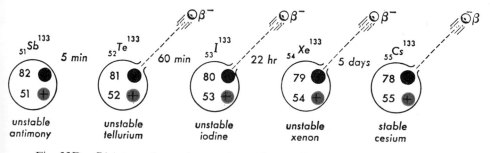

Fig. 33D   *Disintegration series starting with unstable antimony, one of the fragments of the fission of a uranium-235 nucleus.*

U-238	99.280	4.51×10^9 y
U-235	0.714	7.10×10^8 y
U-234	0.006	2.48×10^5 y

All three of these nuclides are radioactive and decay by α-emission. (See Appendix III.)

When a slow or fast neutron is captured by a U-235 nucleus, the two fission fragments as well as the neutrons fly apart with a tremendous amount of kinetic energy. This energy release can best be illustrated by graphs. In Fig. 33E

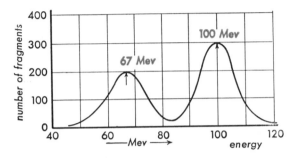

Fig. 33E   *Graph of the energies of the fission fragments from uranium-235.*

the number of fission fragments from a given quantity of U-235 is plotted vertically, and their kinetic energy in Mev is plotted horizontally. The result is a double-peaked curve with maxima at 67 Mev and 100 Mev. While the greatest probability is for a particle of 100 Mev, the areas under the two peaks represent the total numbers of particles produced, and these are approximately equal.

When U-235 nuclei undergo fission as the result of slow neutron capture, the average number of neutrons liberated is found to be 2.5 neutrons per fission. Some may yield as many as five neutrons, but two and three are the most probable numbers. When the initial kinetic energies of prompt neutrons are measured, a graph of the kind shown in Fig. 33F is obtained. Although the maximum probability is for a neutron of 0.7 Mev, the median energy is approximately 2.0 Mev.

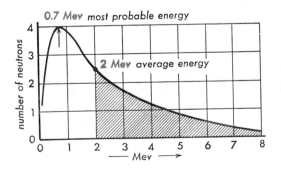

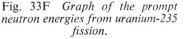

Fig. 33F   *Graph of the prompt neutron energies from uranium-235 fission.*

The average γ-ray energy emitted in the fission of U-235 is approximately 23 Mev.

A rough calculation of the average energy liberated in the fission of U-235 can be made from known atomic masses. For U-235, the atomic mass, $M = 235.043933$ amu, gives a mass defect $M - A$ of 0.043933 amu. The incoming neutron with a mass of 1.008665 amu has a mass defect of 0.008665 amu. The average mass defect $M - A$ for isotopes near the middle of the periodic table is approximately -0.094 amu. If the fission products, for example, have mass numbers 100 and 133, and these are accompanied by three prompt neutrons, the mass balance gives the following:

BEFORE FISSION		AFTER FISSION	
U-235	235.043933	2 frag.	232.812000
n	1.008665	3 n	3.025995
	236.052598		235.837995

The total mass converted is the difference, 0.214603 amu. After multiplying by 931 Mev/amu, we find the energy released to be 200 Mev. This value is consistent with the measured energies given in Figs. 33E and 33F. For the two fragments the average measured energy is 171 Mev. For the three neutrons the average measured energy is 6 Mev. These, added to the γ-ray energy release of 23 Mev, give a measured total of 200 Mev.

33.3. Bohr's Liquid Drop Model

It was Bohr who first proposed that a heavy nucleus behaves like a liquid drop, and that fission may be explained as the result of oscillations brought about by the impinging neutron. See Fig. 33G. Attractive forces between nucleons, like the attractive forces between liquid molecules, give rise to surface tension and the spheroidal state. As raindrops grow in size while falling through the air, their spherical stability decreases; they respond more readily to dis-

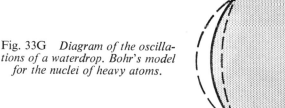

Fig. 33G *Diagram of the oscillations of a waterdrop. Bohr's model for the nuclei of heavy atoms.*

ruptive forces of the air stream and break up into smaller drops. A similar instability could be expected in heavy nuclei where the attractive forces give rise to a similar kind of surface tension.

Consider a nucleus like $_{92}U^{235}$, containing 92 protons and 143 neutrons, closely packed so that attractive forces between neighboring nucleons are all the same. This nucleus should be slightly larger than six nucleons in diameter, as shown in Fig. 33A(a). Each nucleon in the interior of such a system will be in contact with, and therefore will be bound by, 12 others, as shown in Fig. 33H.

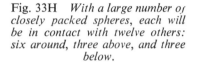

Fig. 33H   *With a large number of closely packed spheres, each will be in contact with twelve others: six around, three above, and three below.*

Since the binding energy U between each pair of nucleons belongs equally to both, the total binding energy of each nucleon should be $6U$:

$$B_v = +6U \tag{33a}$$

This quantity B_v applies only to the interior of the nucleus and is called the *volume binding energy*. Experimentally, B_v is about 14 Mev per nucleon for heavy nuclides.

Nucleons on the surface of the sphere are attracted by only half as many neighbors as those on the interior. For this reason the *surface binding energy* per nucleon should be approximately half as great as the volume binding energy. As more and more particles are added to build up larger and larger nuclides, the number inside and on the surface both increase, but the number inside increases more rapidly. The surface area of a sphere is proportional to the square of the radius, while the volume is proportional to the cube of the radius.

If, therefore, we assign the volume binding energy B_v to all nucleons, we must subtract some surface binding energy B_s to take care of surface nucleons. The relative amount to be subtracted will decrease as A, the total number of nucleons, increases. (See Fig. 33I.)

While the short range forces of attraction between nucleons account for nuclear stability, the weaker Coulomb forces between protons are also present. Since these Coulomb forces are repulsive, the binding energy per nucleon is further reduced. The amount by which it is reduced is called the *Coulomb energy*.

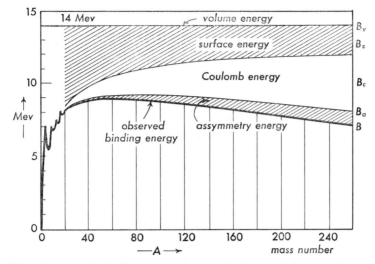

Fig. 33I   *Graph of the binding energy factors for the liquid drop model of atomic nuclei.*

Finally, the total binding energy is also reduced in heavy nuclei by the preponderance of neutrons. If we tried to construct a heavy nucleus out of equal numbers of protons and neutrons, the repulsive Coulomb forces between protons would not be overcome by the attractive short-range forces between all nucleons, and the nucleus would be unstable. The excess number of neutrons provides the additional attractive forces necessary to ensure stability. However, since subshells of neutrons and protons are filled according to the Pauli exclusion principle (see §32.10) the excess neutrons must be placed in higher energy levels. Being in higher energy levels means they are less tightly bound to the nucleus, thus reducing the average binding energy per nucleon.

Hence there are four factors affecting the stability of nuclei:

$$+B_v = \text{volume binding energy}$$
$$-B_s = \text{surface binding energy}$$
$$-B_c = \text{Coulomb energy}$$
$$-B_a = \text{asymmetric energy}$$

The curves and their shaded areas in Fig. 33I show how, starting with the volume binding energy of 14 Mev for each nucleon, each of the other three factors reduces this value in going to heavier and heavier nuclides. This gives but a rough accounting of the experimentally determined curve shown by the heavy line, and reproduced from Fig. 32C.

33.4. The Transuranic Elements

All elements with atomic numbers greater than 92 are called the *transuranic elements*. The first transuranic element, neptunium, atomic number 93, was identified in 1939 by McMillan and Abelson. A beam of neutrons incident on a target of uranium metal gave rise to several known nuclear reactions. A neutron captured by one of the most abundant nuclei, U-238, forms U-239 and a γ ray, followed by β emission, to yield neptunium, element 93:

$$_0n^1 + {}_{92}U^{238} \rightarrow {}_{92}U^{239} + \gamma \text{ ray}$$
$$_{92}U^{239} \rightarrow {}_{93}Np^{239} + {}_{-1}e^0 + \gamma \text{ ray}$$

Ten different radioactive isotopes of neptunium are now known. Ranging in mass number from 231 to 240, isotope $_{93}Np^{237}$ emits α particles and has the longest half-life, 2.2 million years. While slow or fast neutron capture by U-238 is almost always followed by β emission, about one out of a hundred fast neutrons will cause fission.

Plutonium (Pu), element 94, was first identified by Kennedy, McMillan, Seaborg, Segrè, and Wahl, as arising from the spontaneous emission of β particles from $_{92}Np^{239}$:

$$_{92}Np^{239} \rightarrow {}_{94}Pu^{239} + {}_{-1}e^0 + \gamma \text{ ray}$$

Schematic diagrams of the above processes are given in Fig. 33J. Fifteen different radioactive isotopes of plutonium are now known. Ranging in mass numbers from 232 to 246, several have long half-lives:

$_{94}Pu^{239}$	α-active	24,300 y
$_{94}Pu^{240}$	α-active	6000 y
$_{94}Pu^{242}$	α-active	380,000 y
$_{94}Pu^{244}$	α-active	7.6×10^7 y

Americium (Am), element 95, and curium (Cm), element 96, were discovered in 1944 by Seaborg, James, Morgan, and Ghiorso, in collaboration with Hamilton. A sizable quantity of $_{94}Pu^{239}$ was bombarded in the Berkeley cyclotron by 40-Mev helium nuclei and two isotopes, $_{96}Cm^{240}$ and $_{96}Cm^{242}$, formed by (α, n) and $(\alpha, 3n)$ reactions. These nuclei are α emitters with half-lives of one month and five months, respectively. $_{95}Am^{241}$ was first discovered arising from a relatively long-lived β emitter, $_{94}Pu^{241}$, and emits α particles with a half-life of about 500 years.

Ten radioactive isotopes of americium are now known. Ranging in mass

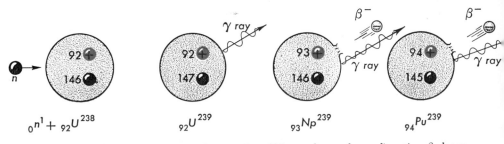

Fig. 33J *Neutron capture by uranium-238 produces, by radioactive β-decay, neptunium and plutonium.*

number from 237 to 246, $_{95}Am^{243}$ has the longest half-life of 8000 years and emits α particles.

Thirteen radioactive isotopes of curium are now known. Ranging in mass number from 239 to 250, several isotopes have long half-lives. These are

$$_{96}Cm^{245} \quad \alpha\text{-active} \quad 8000 \text{ y}$$
$$_{96}Cm^{247} \quad \alpha\text{-active} \quad 4 \times 10^7 \text{ y}$$
$$_{96}Cm^{248} \quad \alpha\text{-active} \quad 5 \times 10^5 \text{ y}$$

Berkelium (Bk), element 97, was discovered in 1949 by Thompson, Ghiorso, and Seaborg. $_{95}Am^{241}$ bombarded by 35-Mev α particles yields $_{97}Bk^{243}$. K capture with a half-life of 4.7 hours turns 99.9% of these nuclei into $_{96}Cm^{243}$, with 0.1% emitting α particles to become $_{95}Am^{239}$.

Eight radioactive isotopes of berkelium are known. Ranging in mass number from 243 to 250, $_{97}Bk^{247}$ has a half-life of 1000 years and is α-active.

Californium (Cf), element 98, was discovered in 1950 by Thompson, Street, Ghiorso, and Seaborg. $_{96}Cm^{242}$ bombarded by 35-Mev α particles yields $_{98}Cf^{244}$. With a half-life of 45 min, $_{98}Cf^{244}$ ejects α particles with an energy of 7.1 Mev to produce $_{96}Cm^{240}$.

Eleven radioactive isotopes ranging in mass number from 244 to 254 are known. Isotope $_{98}Cf^{251}$ has the longest half-life, 800 years, and is α-active.

Einsteinium (E), element 99, and fermium (Fm), element 100, were discovered as a joint project by 16 scientists.

In November 1952 the first full-scale thermonuclear explosion was detonated on an island in the Pacific Ocean. The explosion produced a mile-wide crater in the coral island site, and the radioactive cloud rose ten miles and reached a diameter of approximately 100 miles. Drone planes, radio-controlled, flew through this cloud collecting samples for laboratory study. Among other things it was found that some of the uranium nuclei not undergoing fission had captured as many as 17 neutrons, and formed $_{92}U^{255}$. This very heavy isotope proceeds to give off electrons one after the other, each one raising the nuclear charge by one. Elements 99 and 100 were identified by their predicted chemical properties.

Ten radioactive isotopes of einsteinium are now known, and range in atomic mass number from 246 to 255. Isotope $_{99}E^{254}$ has the longest half-life, 480 days, and is α-active.

Seven radioactive isotopes of fermium are now known. They range in atomic mass number from 250 to 256. Isotope $_{100}Fm^{253}$ has the longest half-life, 7 days, and is α-active.

Mendelevium (Mv), element 101, was discovered by Ghiorso, Harvey, Chappin, Thompson, and Seaborg in 1955. One radioactive isotope, $_{101}Mv^{256}$, is known. Its half-life is approximately one hour. This nuclide was produced by bombarding einsteinium-253 with helium nuclei from a cyclotron beam.

It is now known that all heavy nuclei, starting approximately with Th-232, are fissionable, i.e., under proper excitation conditions they split apart with great violence into almost equal pair fragments. Some of them, like U-233, U-235, and Pu-239, fission by the capture of a slow neutron as well as a fast neutron, whereas others like U-238 and Pu-241 fission only by the capture of fast neutrons. The capture of a slow neutron by U-238 is followed by β decay to produce Np and Pu, whereas fast-neutron capture is followed by fission.

33.5. Photofission of Heavy Nuclei

The fission of uranium and thorium initiated by γ rays was first discovered by Haxley, Schoupp, Stevens, and Wells in 1941. These experimenters bombarded a uranium target with 6.2-Mev γ rays and observed fission fragments. Later experiments by others have shown that photofission can be demonstrated with many of the heavy elements, and that the threshold energy for nuclides Th-230, U-233, U-235, U-238, and Pu-239 is just over 5 Mev.

The discovery that some nuclei undergo fission spontaneously was made by Petrzhak and Flerov in 1940. Many of the heavier isotopes of the transuranic elements show spontaneous fission.

33.6. The Meson Theory of the Nucleus

The possible existence of mesons was first proposed by the Japanese mathematical physicist Yukawa in 1935. He proposed that the short-range forces between protons and neutrons inside the nucleus are to be attributed to relatively smaller particles.

According to Yukawa, nucleons emit and absorb mass-quanta, called *mesons*, just as electrons in the outer structure of the atom emit and absorb photons. The fact that the nuclear forces extend over only a short range can be shown to mean that the meson, unlike the massless photon, would have a finite rest mass. Furthermore, some mesons are charged, and some are neutral. See Fig. 27O.

The present concept of nucleons is that they consist of some sort of common core surrounded by a pulsating cloud of π mesons called *pions*.

Since pions are charged +, 0, or −, the rapid jumping back and forth between nucleons changes the nucleon identity equally fast and at the same time binds the two nucleons together. See Fig. 33K. This diagram might well represent a deuteron, with both spins in the same direction to give $I = 1$, and the negative pion charge at a greater distance from the center of the neutron to produce its negative magnetic moment. See Fig. 32P.

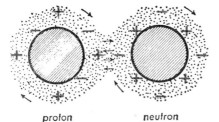

Fig. 33K   *Schematic diagram of the π mesonic clouds around each of two nucleon cores, the charge exchange accounting for the strong attractive force U.*

proton   neutron

To liberate two pions as free particles, sufficient energy must be imparted to the nucleus to create their rest mass of 270 m_e each, or a total of 0.503 Bev. See Fig. 30N.

33.7. Fusion

Measurements of solar radiation reaching the earth each day not only make it possible to calculate the surface temperature of the sun but also to determine its total radiation. The fact that the sun, over a period of many years, shows no signs of cooling off, has long been an unsolved mystery. With the discovery of nuclear disintegration and the development of methods of producing many new types of atoms, this mystery has in a measure been recently solved.

Although there is no direct way known of observing the interior of a star like our sun, mathematical calculations based upon well-established physical laws show that down deep within such a mass the temperature is so extremely high that matter must be a conglomeration of atoms, electrons, and light waves, all moving about at tremendously high speeds.

Near the center of the sun where the temperature is about 20 million degrees, the atoms are stripped of their electrons and the light waves produced are of such high frequencies that they should be classified as γ rays and X rays. Here, where the average particle velocity is so high, nuclear reactions must be taking place on a large scale and the liberated energy must be filtering up through to cooler and cooler layers as light waves of lower and lower frequency. At the surface, most of the radiations escaping are of sufficiently low frequency to be classified as *visible*, *ultraviolet*, and *infrared*.

A careful study of all known nuclear reactions led Bethe in 1938 to propose the following set of chain processes as those most probably responsible for the generation of energy at the sun's central core:

$$(1) \quad {}_1H^1 + {}_6C^{12} = {}_7N^{13} + \gamma \text{ ray}$$
$$(2) \qquad\qquad\quad {}_7N^{13} \rightarrow {}_6C^{13} + {}_1e^0$$
$$(3) \quad {}_1H^1 + {}_6C^{13} = {}_7N^{14} + \gamma \text{ ray}$$
$$(4) \quad {}_1H^1 + {}_7N^{14} = {}_8O^{15} + \gamma \text{ ray} \qquad (33b)$$
$$(5) \qquad\qquad\quad {}_8O^{15} \rightarrow {}_7N^{15} + {}_1e^0$$
$$(6) \quad {}_1H^1 + {}_7N^{15} = {}_6C^{12} + {}_2He^4$$

By summing up the equations it will be seen that four hydrogen atoms are consumed and that two positrons, three γ rays, and one helium nucleus, are created. The other nuclei cancel out, since the original carbon atom in the first reaction is returned unaltered in the last reaction. Hence hydrogen is burned and helium is liberated. The loss in mass for each such cycle of reactions is, therefore, as follows:

$$4_1H^1 = 4.031300 \qquad {}_2He^4 = 4.002604 \qquad 2_1e^0 = 0.001098$$

Subtracting gives $4.031300 - 4.002604 - 0.001098 = 0.027598$ amu. This is equivalent to 25.7 Mev energy.

More recent experiments and calculations* indicate that the proton-proton cycle given below is of even greater importance in the creation of solar and stellar energy than the above carbon cycle:

$$_1H^1 + {}_1H^1 \rightarrow {}_1H^2 + {}_1e^0 + 0.93 - 2 \text{ Mev} \tag{33c}$$

$$_1H^1 + {}_1H^2 \rightarrow {}_2He^3 + \gamma \text{ ray} + 5.5 \text{ Mev} \tag{33d}$$

$$_2He^3 + {}_2He^3 \rightarrow {}_2He^4 + 2{}_1H^1 + 12.8 \text{ Mev} \tag{33e}$$

The net result is the same as before, four hydrogen atoms have been converted into one helium atom. Note that since two $_2He^3$ nuclei are involved in the reaction, Eq. (33e), two proton reactions of the type of Eqs. (33c) and (33d) are required to form one $_2He^4$ nucleus. Six protons are used and two are returned.

The rates at which these reactions should take place are not only consistent with the temperature of 20 million degrees, calculated from other considerations, but hydrogen and helium are known to be the most abundant elements of which stars are made.

In order for the sun to radiate 3.8×10^{26} joules of energy per second, Einstein's equation $E = mc^2$ shows that mass must be annihilated at the rate of 4.2×10^9 Kg/sec (or 4,500,000 tons/sec). While this result indicates that the sun is losing mass at a tremendous rate, the amount is small when compared with the sun's total mass of 1.98×10^{30} Kg. To illustrate, in one million years the sun should lose one ten-millionth of its total mass.†

QUESTIONS AND PROBLEMS

1. Make a diagram like Fig. 33A, showing a plutonium-239 nucleus capturing a slow neutron and undergoing fission. Assume that four prompt neutrons are emitted, and that the fragments have initial charges of 139 and 97, respectively.

2. If one of the fission fragments of U-235 is the radioactive isotope xenon-140, what stable isotope will it become if a series of β particles ensues? Make a diagram like Fig. 33D. *Ans.* $_{57}La^{139}$.

3. If one of the fission fragments of U-235 is the radioactive isotope krypton-97, what stable isotope will it become if a series of β-decays takes place? Make a diagram like Fig. 33D.

4. If one of the fission fragments of U-235 is the radioactive isotope krypton-95, what stable isotope will it become if a series of β particles ensues? Make a diagram like Fig. 33D. *Ans.* $_{42}Mo^{95}$.

5. If a small quantity of americium-244 is bombarded by deuterons and β particles are emitted, what is the reaction?

6. When americium-243 is bombarded by deuterons from a cyclotron, α particles are emitted. What is the reaction? *Ans.* $_1H^2 + {}_{95}Am^{243} \rightarrow {}_{94}Pu^{241} + {}_2He^4$.

7. What are the reactions by which neptunium and plutonium were discovered?

* See "Nuclear Reactions in Stars" by E. E. Salpeter, *Physical Review*, 88, 547 (1952).

† For a more complete treatment of nuclear energy, see *Atomic Energy* by S. Glasstone, D. Van Nostrand, Princeton. For more details of the general subject of nuclear physics, see *The Atomic Nucleus* by R. D. Evans, McGraw-Hill, New York.

8. If nuclei are composed of closely packed spherical shells of nucleons, how many nucleons are required to form a nucleus with three complete shells around one nucleon at the center? Assume nucleon centers to lie on the surface of a sphere. *Ans.* 176.

9. If nucleons have a diameter of 2.4×10^{-13} cm, what is the diameter of the nucleus fermium-254?

10. (a) How many surface nucleons are there in Problem 8? (b) How many interior nucleons are there? *Ans.* (a) 113. (b) 63.

34

Nuclear Energy

Not long after the discovery of fission in 1939, it became evident to many scientific groups in America and in Europe that if a sufficient quantity of pure uranium-235 (U-235) could be isolated from its more abundant isotope uranium-238 (U-238), it might have explosive powers many times greater than anything heretofore known. The reasons for believing this appeared at the time to be somewhat as follows.

34.1. A Chain Reaction

Suppose that a given mass of uranium metal, all composed of U-235 atoms, was brought together into one lump. The first cosmic ray that penetrated this mass and produced a neutron might well set off the chain reaction shown schematically in Fig. 34A. A U-235 nucleus would capture the neutron and in splitting apart with great violence would liberate one or more additional neutrons. These in turn would be quickly absorbed by other nearby atoms, which in turn would split up, at the same time liberating other neutrons. Hence a rapidly growing kind of avalanche might occur, a kind which, if fast enough, would have the characteristics of an explosion.

A graph showing the rate of growth of such a chain process is given in Fig. 34B. Since even the slowest of neutrons in solid matter will have average speeds of hundreds of thousands of centimeters per second, about the same as hydrogen atoms in a gas at ordinary temperatures, and since many neutron collisions will, on the average, occur within several centimeters, the graph shows how quickly the growth reaches gigantic proportions. The *time* scale is of the order of microseconds.

The escape of neutrons from any quantity of uranium is a *surface effect* depending on the area of the surface, whereas fission capture occurs throughout the body and is therefore a *volume effect*. If the assembled mass of uranium is too small, the probability that most neutrons liberated by fission would escape

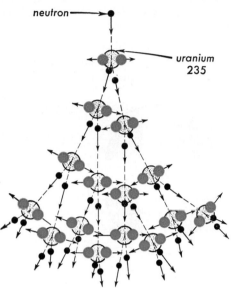

Fig. 34A   *Schematic diagram of a chain reaction in pure uranium-235.*

through the surface before being captured might well be so large that a growing chain reaction cannot occur. Since the volume of a sphere increases with the cube of the radius while the surface area increases with the square of the radius, the *probability of escape* would decrease with increasing size. In other words, if

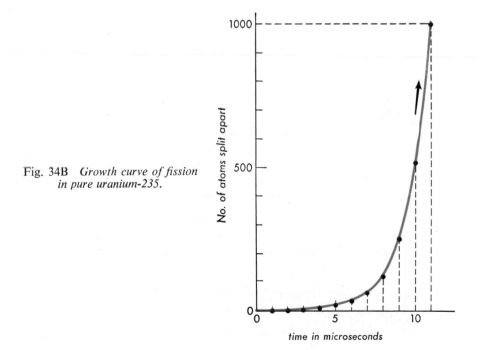

Fig. 34B   *Growth curve of fission in pure uranium-235.*

the uranium mass were too small, the growth process shown in Figs. 34A and 34B would be cut off before it became very large, and only if the mass were greater than some critical value would an explosion take place.

Consider, therefore, a large quantity of U-235 in two or more units, each smaller than the critical size, and separated by a short distance. See Fig. 34C.

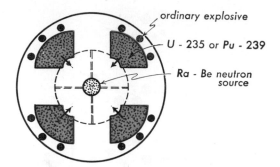

Fig. 34C *A schematic diagram of a hypothetical atomic explosion device, based upon a fast chain reaction in pure uranium-235 or plutonium-239.*

Because of the relatively large surface area of each unit, neutrons readily escape and a chain reaction cannot develop. Suddenly, an explosive like TNT is detonated behind the separated blocks, driving them together as indicated at the center of the diagram. Neutrons entering this greater-than-critical mass from a Ra-Be source will now initiate a rapid chain reaction which results in a violent explosion. The tremendous energy release of such a device can only be fully realized by those who have actually seen and heard one detonated.

The first atomic bomb ever produced was assembled by scientists under the direction of the University of California, and was successfully detonated at Alamogordo, New Mexico, on July 16, 1945. This device was composed of practically pure U-235. Many of the atomic devices exploded since that time have used plutonium-239 as the fissionable material, and peaceful uses of all kinds are now being studied by many.

One such promising development makes use of the tremendous amount of heat energy liberated. Repeated explosions underground can be confined to relatively small volumes of space, and the heat can be tapped off through some heat-transfer system, such as circulating steam pipes.

34.2. Nuclear Radius and Geometrical Cross Section

Many experiments concerned with the collisions between atomic particles indicate that a nucleus may be considered a conglomerate of closely packed spheres of the same size. The average nucleon radius is now believed to be

$$r_0 = 1.2 \times 10^{-13} \text{ cm} \tag{34a}$$

If spheres of this size are packed together into a ball-like structure as shown in Fig. 33A(a), the approximate radius of the combination will be given by

$$R = r_0 \sqrt[3]{A} \tag{34b}$$

where the number of nucleons is given by the mass number A. For a uranium nucleus, U-238, for example, $A = 238$ and $R = 7.4 \times 10^{-13}$ cm, a value only six times that of the proton.

The concept of nuclear cross section is one of considerable importance in nuclear studies. If a nucleus is set up as a target for other atoms to hit, its *geometrical cross section* serves as a reasonably good measure of the target size, and is given by

$$\sigma_g = \pi R^2 \tag{34c}$$

For U-238 this cross section is approximately

$$\sigma_g = 1.73 \times 10^{-24} \, \text{cm}^2$$

The unit area for nuclei has been arbitrarily set at 1×10^{-24} cm², and is called the barn. The geometrical cross section for U-238, for example, would be written $\sigma_g = 1.73$ barns:

$$1 \text{ barn} = 1 \times 10^{-24} \, \text{cm}^2 \tag{34d}$$

Calculated cross sections for a few nuclides are given in Table 34A, along with the heights of the nuclear barrier B for incident α particles. See Fig. 32H.

Table 34A. Nuclear Radii, Geometrical Cross Sections and Potential Barrier Heights

NUCLIDE	R $(10^{-13}$ cm)	σ_g (barns)	B (Mev)
$_2$He^4	1.9	.11	2.4
$_8$O^{16}	3.2	.32	6.0
$_{30}$Zn^{64}	4.8	.72	14.0
$_{48}$Cd^{113}	5.8	1.06	19.0
$_{100}$Fm^{252}	7.6	1.82	29.0

34.3. Cross Sections and Mean Free Path

When neutrons are incident upon some material, each nucleus within the target area does not always behave as though its cross section is a constant. For example, in the capture of a neutron prior to radioactive decay, the nuclear size may appear to be quite different from its calculated geometric cross section. For fast neutrons, a nuclear cross section might be relatively small, while for slow neutrons it may be quite large.

With this explanation we see that cross section is not a target area in the literal sense, but is a figurative concept expressing the *interaction probability*.

The probability of interaction between a neutron and a nucleus is called the *microscopic cross section* σ, and is regarded as the effective target area of a nucleus. If a beam of neutrons is incident on 1 cm³ of material the total effective target area will be $n\sigma$, where n is the total number of atoms per cm³. See Fig. 34D. This product is called the *macroscopic cross section*, and is designated Σ.

$$\Sigma = n\sigma \tag{34e}$$

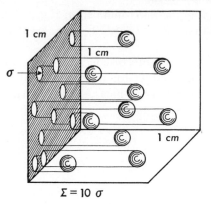

Fig. 34D *Diagram showing the microscopic cross section and the meaning of the macroscopic cross section.*

The total cross section σ_t of a single nuclide is the sum of several cross sections, one for each different kind of process that alters the bombarding particles' energy or momentum:

$$\sigma_t = \sigma_s + \sigma_c \tag{34f}$$

where σ_s is the effective nuclear area that produces measurable scattering, and σ_c is the effective area for capture. The latter is sometimes called the *radiative capture cross section*, since radioactivity usually follows capture. If fission is also one of the effects of capture, as it is with some of the heaviest nuclides like U-235 and Pu-239, we have

$$\sigma_t = \sigma_s + \sigma_c + \sigma_f \tag{34g}$$

As an illustration of these various σ's, the cross sections for U-235 for fast neutrons (2 Mev) and thermal neutrons (0.0253 ev) are given in Table 34B.

Table 34B. Cross Sections for Pure U-235

CROSS SECTIONS IN BARNS	FAST NEUTRONS 2 Mev $\left(\dfrac{2 \times 10^7}{\text{m/sec}}\right)$	SLOW NEUTRONS 0.0253 ev $\left(\dfrac{2200}{\text{m/sec}}\right)$
Scattering $\sigma_s =$	5.0	10
Capture $\sigma_c =$	0.25	107
Fission $\sigma_f =$	1.27	580

U-235 is one of a number of exceptional nuclei, for most cross sections are small and comparable with geometrical cross sections. The whole subject of nuclear energy and the design of nuclear reactors and explosives is very dependent upon the smallness of the cross sections of some elements and the largeness of others.

The average distance a neutron travels between nuclear events is given by the reciprocal of the macroscopic cross section. This average distance is called the *mean free path:*

$$\text{Mean free path } \lambda = \frac{1}{\Sigma} \tag{34h}$$

Example 1. Find (a) the total macroscopic cross section, and (b) the mean free path for 2-Mev neutrons in pure U-235.

Solution. From Table 34B, and by Eq. (34g), we obtain

$$\sigma_t = 6.52 \text{ barns}$$

To find n for Eq. (34e), we take the density of uranium metal, $\rho = 18.7$ gm/cm³, and divide by the mass in grams of one uranium atom. Since unit atomic mass is 1.66×10^{-24} gm (see Appendix V), the mass of one U-235 atom is

$$M - 235 \times 1.66 \times 10^{-24} - 3.90 \times 10^{-22} \text{ gm}$$

Dividing ρ by M, gives

$$n = \frac{18.7}{3.9 \times 10^{-22}} = 4.79 \times 10^{22} \frac{\text{atoms}}{\text{cm}^3}$$

Substituting σ_t and n in Eq. (34e), we obtain

$$\Sigma_t = 4.79 \times 10^{22} \times 6.52 \times 10^{-24}$$

or $(a) \quad \Sigma_t = 0.312 \text{ cm}^{-1}$

To find the mean free path, Eq. (34h) gives

$$(b) \quad \lambda_t = 3.21 \text{ cm} \tag{34i}$$

By using the separate value σ_s, σ_c, or σ_f alone, and the procedure given in Example 1, the mean free paths for scattering, radiative capture, or fission alone can be calculated. For the scattering and fission of U-235, by 2 Mev neutrons, for example,

$$\lambda_s = \quad 4 \text{ cm} \tag{34j}$$

and

$$\lambda_f = 16 \text{ cm} \tag{34k}$$

Since the scattering cross section of U-235 is four times the fission cross section, secondary neutrons will collide several times with U-235 nuclei before producing fission. A schematic diagram of this process is illustrated in Fig. 34E.

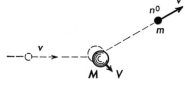

Fig. 34E   *Elastic scattering of a neutron by a nucleus.*

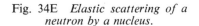

34.4. Neutron Scattering

There are, in general, two ways in which particles may be scattered by impacts with atomic nuclei, one is by *elastic scattering*, and the other is called *inelastic scattering*. In cases of elastic scattering, the total kinetic energy as well as the total momentum before impact are equal, respectively, to the total kinetic energy and the total momentum after impact. In mechanics this means that the coefficient of restitution $r = 1$.

If a fast neutron collides elastically with a light particle like a proton, deuteron, or an α particle, considerable energy may be imparted to the recoiling nuclide. If the neutron collides with a heavy nucleus, on the other hand, conservation laws show that little kinetic energy can be imparted to the recoiling heavy nucleus and that the neutron will rebound in some new direction with most of its original energy. See Fig. 34E.

If, therefore, fast neutrons are to be slowed down to relatively low velocities by elastic scattering, materials composed of large quantities of atoms of low atomic weight, such as hydrogen, will be most effective. Any material used for this purpose is called a *moderator*.

Since the function of a moderator is to reduce the speeds of fast neutrons to low velocities by elastic collisions, the material used to do this will best serve its purpose if the nuclei scatter elastically and have small capture cross sections. Furthermore, the lighter the atoms the greater will be the recoil energy of the moderator nuclei when they do collide, and the fewer will be the impacts necessary to reduce the neutrons to thermal energies. See Table 34C.

Table 34C. Number of Impacts to Reduce 2-Mev Neutrons to Thermal Energies, 0.0253 ev

ELEMENT	NO. OF IMPACTS
Hydrogen	18
Deuterium	25
Beryllium	87
Carbon	115
Uranium	2160

Capture cross sections of a few moderators are given in Table 34D.

Table 34D. Neutron Capture Cross Sections of Moderators and Structural Materials

ELEMENT	σ_c (barns)
Hydrogen	0.33
Deuterium	0.00046
Carbon	0.0032
Beryllium	0.010
Aluminum	0.23
Zirconium	0.18
Molybdenum	2.4
Iron	2.5
Copper	3.6

Note the extremely low cross section and atomic mass of the first four elements.

When fast neutrons collide with uranium nuclei, either U-235 or U-238, some of them are scattered inelastically, while others are captured. The inelastic scattering process is illustrated in Fig. 34F. In passing close to one of these heavy

Fig. 34F   *Inelastic scattering of a neutron by a uranium nucleus.*

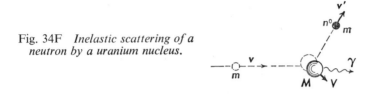

nuclei, a considerable amount of energy is taken from the neutron, and the nucleus is raised to an excited state. The neutron leaves the nucleus in some new direction, with perhaps less than half its initial kinetic energy.

The excited nuclei, in returning to their normal state, emit γ rays. While inelastic impacts always obey the law of conservation of momentum, thus giving rise to the heavy-particle recoil shown, *conservation of kinetic energy does not hold.*

34.5. Explosive Chain Reactions

Uranium metal as it is usually refined is composed of three radioactive isotopes.

URANIUM ISOTOPE	RELATIVE ABUNDANCE
$A = 238$	99.280%
235	0.714%
234	0.006%

Since both U-235 and U-238 undergo fission with fast neutrons, one might think ordinary uranium in a large enough mass might explode. One good reason for believing this is that fission cross sections for both isotopes are about the same for fast neutrons. (σ_f for U-235 is 1.27 barns. See Table 34B.) Another reason is that fast neutrons might be expected to shorten the time between fission events.

The reason that ordinarily refined uranium in a mass of any size will not explode is that the scattering cross sections are larger than fission cross sections, thus making the mean free path between scattering events relatively small. After one or two inelastic impacts with uranium nuclei, the secondary neutrons have lost most of their initial energy of 2 Mev, and the fission capability of U-238 has dropped to an extremely low value. The fission cross section of U-238 for slow neutrons is practically zero.

Since the fission cross section for U-235 increases as the neutron velocity decreases and becomes extremely large at thermal energies (see Table 34B), only relatively pure U-235 can develop an explosive chain reaction. Thermal neutron cross sections for fissionable nuclides are given in Table 34E.

Table 34E. Thermal Neutron Cross Section for Fissionable Materials
(in Barns, at 2200 Meters/Sec)

NUCLIDE	σ_s	σ_c	σ_f
U-235	10	107	580
Pu-239	9.6	315	750
U-233	—	52	533
U-238	8.3	3.50	0

34.6. Critical Mass Factors

For a mass of fissionable material like pure U-235 or Pu-239 to be explosive, the time between fission events must be very small, and relatively few neutrons must escape through the surface, or be lost somewhere by capture. A schematic diagram of possible paths of neutrons between two consecutive fission events is shown in Fig. 34G.

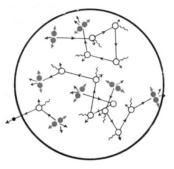

Fig. 34G *Schematic diagram of inelastic scattering between fission events in uranium-235 or plutonium-239.*

Starting out as fast neutrons, the neutrons lose considerable energy with each inelastic impact. As the neutron velocity decreases, the U-235 fission cross section increases and fission capture becomes more and more probable.

To find the time T_f between fission events, we note that the velocity of 2 Mev neutrons is approximately 2×10^9 cm/sec, while the average velocity of thermal neutrons (0.0253 ev) is 2.2×10^5 cm/sec. If we adopt a geometric average velocity of 2×10^7 cm/sec, and an average mean free path for scattering of 4 cm, as given by Eq. (34j), *the average time between scattering events will be*

$$T_s = \frac{4 \text{ cm}}{2 \times 10^7 \text{ cm/sec}} = 2 \times 10^{-7} \text{ sec} \tag{34l}$$

Allowing an average of five scattering events before fission capture, we obtain

$$T_f = 1 \times 10^{-6} \text{ sec} \tag{34m}$$

Owing to the random directions of the inelastic scattering (see Fig. 34G) the average diffusion distance or *straight-line distance* between fission events is from 5 to 8 cm.

The critical mass for a nuclear explosive device should lie somewhere between the size of a marble (2 cm diameter) and the size of a basketball (24 cm diameter). Visualize, therefore, two spheres of pure U-235 or Pu-239, one large and one small, as shown in Fig. 34H. It is clear that if the average straight-line distance

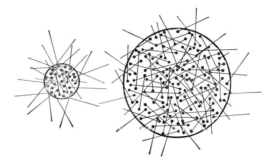

Fig. 34H *Schematic diagram of pure uranium-235, or plutonium-239, showing escape of most of the neutrons for a subcritical size, and capture of most of the neutrons for greater than critical size.*

between fission events is 6 cm (see **Fig. 34G**) few neutrons will be captured in the small sphere, while many will be captured in a mass the size of a basketball.

Whether or not any mass will sustain a chain reaction at all is determined by what is called the *reproduction factor*. The reproduction factor k is given by the ratio,

$$k = \frac{\text{rate of neutron production}}{\text{rate of neutron disappearance}} \tag{34n}$$

If the rate of neutron production equals the rate at which neutrons disappear, the mass is said to be *critical*, and $k = 1$.

Curves showing the growth in an assembly where k is slightly greater, and slightly smaller, than unity are given in Fig. 34I. Since one generation of neutrons requires about 1 microsecond (see Eq. (34m)), the horizontal scale can represent time in microseconds.

In five generations the curve $k = 1.1$ rises to 1.61, since

$$1.1 \times 1.1 \times 1.1 \times 1.1 \times 1.1 = 1.61 \text{ neutrons}$$

In ten generations this same curve rises to $(1.1)^{10}$ or 2.59 neutrons, and in 100 generations to $(1.1)^{100}$ or 1×10^5 neutrons. In 1000 generations, or approximately $\frac{1}{1000}$ of a second, the number rises to approximately 1×10^{41} neutrons. Since this represents more atoms than would be available in any given assembly, 1000 generations would not materialize.

In pure U-235, or Pu-239 the size of a marble, the reproduction factor is approximately 0.1, while for a sphere the size of a basketball, $k = 2.4$.

34.7. Nuclear Reactors

A nuclear reactor, formerly called an atomic pile, is an apparatus in which nuclear fission can be maintained as a self-supporting yet controlled chain re-

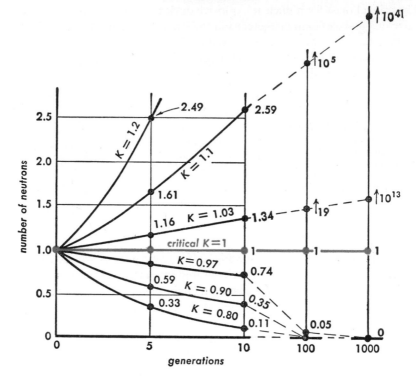

Fig. 34I   *Growth curves for neutrons with reproduction factors barely above and below critical, K = 1.*

action. It is a kind of furnace in which uranium is the fuel burned, and many useful products such as heat, neutrons, and radioactive isotopes are products.

Reactors are of many kinds, sizes, and shapes, the two principal ingredients of them being a quantity of fissionable material, and a moderating substance for slowing down the neutrons to thermal velocities. A reactor is often designated according to the moderator, or coolant, used within it. Because of the immensity of the subject, only the simplest elements of these devices will be described here, and these as illustrations of the basic principles of many others.

A self-sustaining chain reaction cannot be maintained in pure uranium alone, no matter how large the mass. By properly combining or surrounding the metal with a moderator, however, the 2-Mev neutrons produced in the fission of U-235 can be slowed down by elastic scattering to thermal energies of 0.0253 ev. At these relatively low velocities of 2200 m/sec, the fission cross section of U-235 has the enormous value of 580 barns (see Table 34B).

The cross sections of a few nuclear fuels at thermal energy are given in Table 34E.

The first self-sustaining chain reaction ever created by man was put into operation at the University of Chicago on December 2, 1942. This device consisted of a huge "pile" of small carbon blocks, carefully laid together to form one solid mass about the size of a normal school room. During construction, lumps of pure uranium metal were inserted at regular intervals throughout the mass.

A schematic diagram of a pile constructed of large carbon blocks is shown in Fig. 34J. Long cylindrical holes through the blocks provide for the insertion or

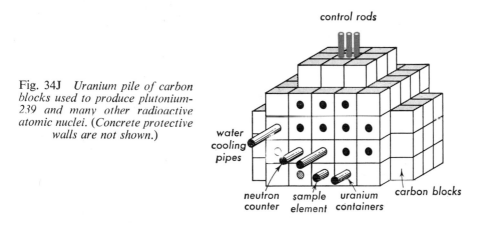

Fig. 34J　*Uranium pile of carbon blocks used to produce plutonium-239 and many other radioactive atomic nuclei. (Concrete protective walls are not shown.)*

removal of fuel elements, control rods, detecting devices, samples to be irradiated, etc. The fuel elements consist of pure uranium metal sealed in thin-walled aluminum cylinders.

The distance between uranium fuel elements in the moderator material is of importance in reactor design. The slowing-down distances for three commonly used moderators are as follows:

Ordinary water,	H_2O	5.7 cm
Heavy water,	D_2O	11.0 cm
Carbon blocks,	C	19.0 cm

When, within the uranium metal, a few U-235 nuclei undergo fission, fast neutrons are liberated. Most of these enter the surrounding carbon (the moderator), where they collide elastically with carbon nuclei and slow down. Eventually, many of them enter the uranium metal as thermal neutrons and are captured by U-235 nuclei to cause fission. A diagram showing this process is given in Fig. 34K.

Not all neutrons produced within a reactor result in capture by U-235. Some are lost by escape through the surface, some by radiative capture by U-235 as well as U-238, and some by capture by structural materials and fission products.

The neutron balance in a natural uranium reactor, operating at the critical rate $k = 1$, is shown in Fig. 34L. Of the millions of fast neutrons produced in the reactor in each microsecond, the diagram starts with 1000 fast neutrons at the center, and shows what might reasonably happen to them in regenerating 1000 more fast neutrons. The average number of neutrons produced per fission is 2.5. This factor is directly involved in the bottom four squares where it is

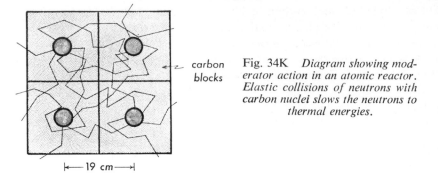

carbon
blocks

Fig. 34K   *Diagram showing moderator action in an atomic reactor. Elastic collisions of neutrons with carbon nuclei slows the neutrons to thermal energies.*

|←— 19 cm —→|

used in accounting for reproduction. It should be noted that 60% of the neutrons disappear by other than fission processes.

If the reproduction factor k of the reactor assembly is greater than unity, see Eq. (34n), the total number of neutrons will rise, and along with it the temper-

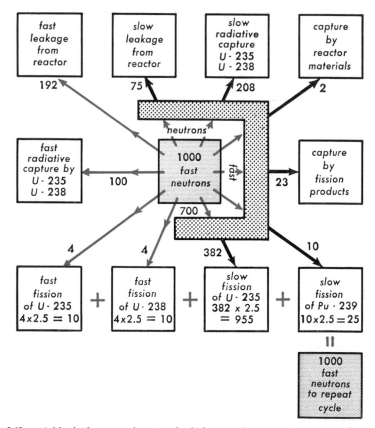

Fig. 34L   *A block diagram showing the balance of neutrons in a natural uranium reactor that is just critical. (Calculated numbers by courtesy of Westinghouse Electric Corp.)*

ature. To prevent the temperature from rising too high, control rods, which are strong neutron absorbers, are lowered into the central core.

The most widely used control rods, or plates, contain boron or cadmium. Both of these elements have enormous capture cross sections for slow neutrons. When *normal cadmium metal* is subjected to thermal neutrons, the *average cross section* is 2500 barns. Experiments with separated isotopes show that the six principal isotopes of cadmium have the individual cross sections shown in Table 34F.

Table 34F. Thermal Neutron Capture Cross Sections for Cadmium Isotopes

ISOTOPE (A)	ABUNDANCE (%)	CROSS SECTION (barns)
110	12.4	.20
111	12.7	.26
112	24.1	.03
113	12.3	20,000
114	28.9	.14
116	7.6	1.5

Direct experimental evidence for the large cross section of Cd-113 is shown in Fig. 34M. The upper photograph is a mass spectrograph record of the five

110 111 112 113 114   116

Fig. 34M   *Normal cadmium mass spectrogram (above) and isotopes altered by neutron absorption (below). (After Dempster.)*

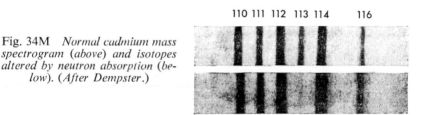

principal stable isotopes of cadmium made with the metal as normally refined. The lower reproduction is a similar record made with cadmium metal after it had been subjected to intense neutron irradiation in a reactor. Note that the isotope 113 is missing and that isotope 114 is enhanced. The capture of a neutron by Cd-113 produces Cd-114.

An excellent demonstration can be made by placing a piece of metallic silver in a thin-walled box made of cadmium metal and subjecting it to a Ra-Be neutron source as shown in Fig. 30D. After some time the silver is removed from the cadmium box and placed near a Geiger or scintillation counter. No counts will be recorded. If the silver metal is subjected to the neutrons without the cadmium shield, it will show considerable β activity, as expected.

34.8. Power Reactors or Nuclear Power Plants

To utilize the natural heat developed in a uranium reactor as a source of great power has long been recognized as a feasible enterprise. The basic princi-

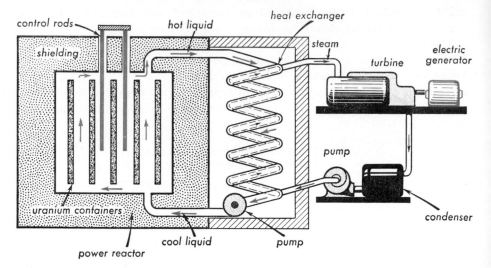

Fig. 34N   *Schematic diagram of one type of nuclear power plant.*

ples of one type of "power reactor" are shown in Fig. 34N. A quantity of enriched uranium, in the form of a pure metal or in the form of a solution of soluble salt in water forms the center of the heat energy source.

The energy released by fission produces great quantities of heat, and the rising temperature is regulated to a predetermined value by cadmium rods. To reduce the fission rate, and thereby lower the temperature, the central rods are pushed in a little farther to absorb more neutrons, while to raise the temperature they are pulled out a little farther.

Because of the harmful effects of the intense neutron radiation upon men and equipment, it is not reasonable to vaporize a liquid directly as in a steam boiler; it is better to circulate a fluid through the shielded reactor and heat-exchanger, as shown in the diagram.

The hot liquid flowing through the heat-exchanger vaporizes a more volatile liquid like water; the resulting hot gas or steam under pressure drives a turbine of special design. The turbine in turn drives an electric generator, developing power that can be used to light our cities and factories or to drive ships and submarines through the water and large planes through the air.

One of the problems connected with such power reactors is the effect of the intense neutron radiation on the metal structures. The neutrons change some atoms and permanently displace others from their normal positions in the crystal lattice of the solids, and as a result weaken certain crucial mechanical parts. Intensive studies of the properties of various materials under conditions likely to be encountered in power reactors are continually carried on in our research laboratories.

Another important problem concerns the nature of the coolant; it must be able to withstand high temperatures and must not absorb neutrons and become radioactive to any appreciable extent; yet it must be efficient in the transfer of heat in both the reactor and the heat-exchanger. Certain metals with low melting points appear to be most promising in these and other respects.

34.9. Swimming-Pool Reactor

The swimming-pool type of reactor derives its name from the fact that a large tank of ordinary water is used as a protective shield for the operating personnel. Figure 34O is a cut-away diagram of a typical reactor of this kind, and one that

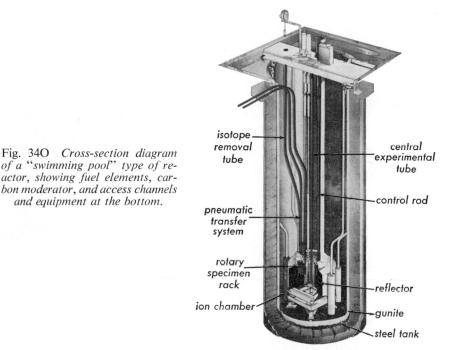

Fig. 34O   *Cross-section diagram of a "swimming pool" type of reactor, showing fuel elements, carbon moderator, and access channels and equipment at the bottom.*

isotope removal tube

pneumatic transfer system

rotary specimen rack

ion chamber

central experimental tube

control rod

reflector

gunite

steel tank

Courtesy, General Atomics

is designed as a multipurpose instrument. The several fuel elements at the bottom of the tank are in the form of small cylindrical rods; each is composed of a solid homogeneous alloy of uranium and zirconium hydride moderator, clad in their aluminum cylinders. The uranium is enriched to 20% of U-235. The three control rods are of boron carbide.

Full physical and visual access to the core is possible at all times from the top, as shown in Fig. 34P. Samples to be irradiated by neutrons can be lowered into the water in the region of the core, and an observer can readily see the blue glow of the water around it, which is caused by the Cerenkov radiation.

A rotary specimen rack ("lazy susan") located just above the large carbon or graphite moderator block, provides a water-tight facility for radioactive isotope production. A pneumatic tube running to the bottom of the tank permits a

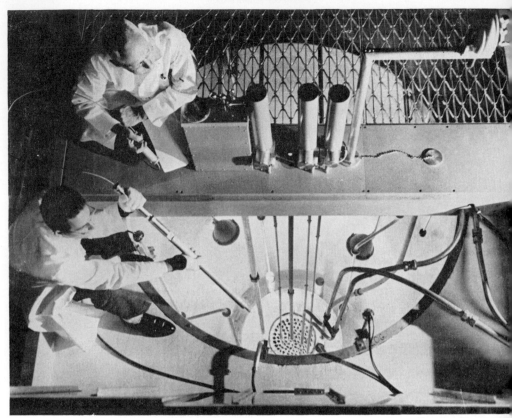

Fig. 34P *Photograph looking down through the water to the principal elements at the bottom of a "swimming pool" type of reactor.*

sample element, in a small container called a *rabbit*, to be subjected to neutrons and quickly removed for the measurement of very short half-lives.

34.10. Uncontrolled Fusion

We have seen in §33.7 how the sun, and other stars, by means of certain nuclear reaction cycles and temperatures of millions of degrees, are able to fuse protons into α particles (hydrogen into helium), with the simultaneous emission of great quantities of energy. While all of the ingredients needed in these reactions are plentiful on the earth's surface, and can be purified and assembled in the research laboratory, the temperature of several million degrees required to cause them to fuse cannot be produced by any of the standard laboratory methods.

Here the atomic bomb employing the process of fission has come to our aid and made such temperatures possible. When an atomic bomb, containing U-235 or Pu-239, explodes, the temperature reached at the central core, although it may last only a small fraction of a second, is comparable to that reached at the center of the sun.

While superatomic bombs containing hydrogen and other light elements have been highly successful, the exact materials used, their proportions, the physical size and shape of the devices, and the mechanical and electrical systems involved are all classified by governments as secret military information. These same government agencies, however, do conduct, through their civilian directed laboratories, scientific research projects on the peaceful uses of nuclear explosions.

In the interest of knowledge itself, it should be said that through the study of explosive reactions we can learn more of nature's fathomless mysteries. We can look, for example, at some of the lightest of nuclides and select several reactions that look promising from an energy standpoint. Five promising reactions involving the hydrogen and lithium isotopes are

$$_1H^2 + {}_1H^2 \rightarrow {}_2He^3 + {}_0n^1 \ + \ 4.0 \ \text{Mev}$$
$$_1H^2 + {}_1H^3 \rightarrow {}_2He^4 + {}_0n^1 \ + 17.6 \ \text{Mev}$$
$$_1H^2 + {}_3Li^6 \rightarrow {}_2He^4 + {}_2He^4 + 22.1 \ \text{Mev}$$
$$_1H^1 + {}_3Li^7 \rightarrow {}_2He^4 + {}_2He^4 + 17.5 \ \text{Mev}$$
$$_0n^1 + {}_3Li^6 \rightarrow {}_2He^4 + {}_1H^3 \ + \ 4.6 \ \text{Mev}$$

Note the particularly large energies released with the fusion of deuterium with tritium, deuterium with lithium-6, and hydrogen with lithium-7.

Lithium is relatively abundant on the earth, and lithium-6 can be separated in reasonable quantities from its more abundant isotope, lithium-7. Tritium, $_1H^1$, on the other hand, is a radioactive isotope of hydrogen not found in nature, because of its relatively short half-life of 12.2 years. It is only produced in quantities, at considerable expense, in atomic reactors.

34.11. Controlled Fusion

The conviction on the part of many physicists that controlled fusion is possible has led, these past few years, to the expenditure of a great deal of scientific manpower and money.

In principle one would visualize a jet of fusible material, such as deuterium, being fed from a nozzle into a cavity where, upon fusion, great quantities of energy in the form of heat would be continuously generated and tapped off.

From the very start of all projects working on this problem, it has been realized that, owing to the temperature requirements of millions of degrees, no material walls can be close to the region in which fusion is to be consummated. This has led many to the use of a gaseous discharge called a *plasma*, held suspended in space by the magnetic lines of force of an electromagnet.

A plasma is an electrically neutral stream or mass of ionized atoms, molecules, and electrons and may be produced in various ways. A high-current arc, such as that used in a searchlight, is a good plasma source. Plasma in many devices is confined by what has been termed a *magnetic bottle*. Visualize, as shown in Fig. 34Q, a stream of ionized atoms injected into the central region of a hollow solenoid. Moving with the high speeds of ions in a hot gas, these particles spiral around in the field with the lines of force acting as guides.

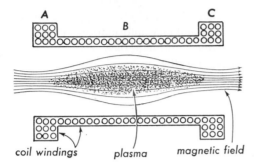

Fig. 34Q   *Schematic diagram of a plasma held suspended in a magnetic field: a "magnetic bottle."*

coil windings   plasma   magnetic field

As particles spiral into the stronger field at either end, the force components drive them back again toward the center. By increasing and decreasing the currents in the field coils *A, B,* and *C,* the field shape can be modified at will and the plasma widened or narrowed. By decreasing the field at the right and progressively increasing the field from left to right the plasma can be transported lengthwise from one tube to another.

If all fields are increased, the plasma volume is compressed and the temperature rises. A rising temperature means higher ion velocities and greater probability that impacts will produce fusion. If the field is decreased at one end and increased at the other, the plasma may be quickly moved along the cylinder axis, thus constituting a jet. Plasma jet studies by various aircraft and other research laboratories offer promising results for the near future. While much has been learned from the various controlled-fusion study projects, there is still much to be discovered before successful power sources are realized. There are many who now believe that greater research effort should be placed on studies of the basic principles of the plasma itself and that eventual rewards will be well worth the effort.

PROBLEMS

1. Since the critical size for a uranium-235 explosion lies between a marble and a basketball, find (a) the ratio of these sphere diameters, (b) the ratio of their surface areas, (c) the ratio of their volumes, and (d) the ratio of volume to surface area. See § 34.6 for sizes.

2. Find (a) the radius and (b) the geometrical cross section for the nucleus of silver-109. *Ans.* (a) 5.73×10^{-13} cm. (b) 1.03 barns.

3. Find (a) the radius and (b) the geometrical cross section for the nucleus of barium-138.

4. If the reproduction factor for fission in a mass of U-235 is 1.20, how many secondary neutrons will be produced in (a) 5 generations, (b) 10 generations, (c) 100 generations, and (d) 1000 generations? (Note: Use logarithms.) *Ans.* (a) 2.49. (b) 6.19. (c) 8.28×10^7. (d) 1.52×10^{79}.

5. If the reproduction factor for fission in a mass of Pu-239 is 1.45, how many secondary neutrons will be produced in (a) 5 generations, (b) 10 generations, and (c) 100 generations? (Note: Use logarithms.)

6. The reproduction factor in a mass of U-235 is 1.05. (a) How many neutron generations will occur in 2 milliseconds, and (b) how many secondary neutrons will be released? *Ans.* (a) 2000. (b) 2.39×10^{42}.

7. The reproduction factor in a mass of Pu-239 is 1.35. (a) How many neutron generations will occur in $\frac{1}{10}$ of a millisecond, and (b) how many secondary neutrons will be released?

8. (a) Find the total macroscopic cross section for thermal neutrons in pure U-235. (b) Find the total mean free path. *Ans.* (a) 33.4 cm^{-1}. (b) 0.299 mm.

9. (a) Find the macroscopic fission cross section for thermal neutrons in pure U-235. (b) Find the corresponding mean free path.

10. (a) Find the macroscopic fission cross section for thermal neutrons in pure Pu-239. (b) Find the corresponding mean free path. (Assume the density of Pu to be 17 gm/cm³.) *Ans.* (a) 46.0 cm^{-1}. (b) 0.21 mm.

11. (a) Find the macroscopic scattering cross section for thermal neutrons in pure Pu-239. (b) Find the corresponding mean free path. (Assume the density of Pu to be 17 gm/cm³.)

12. Calculate the average cross section in barns for thermal neutrons on the six cadmium isotopes given in Table 34F. *Ans.* 2460 barns.

35

Elementary Particles

For many decades now, physicists have been searching for the ultimate particles of which all matter is composed. From the atom and its electron structure, the search has extended into the nucleus, and from the nucleus to the structure of nucleons themselves.

The purpose behind the planning and building of atomic accelerators that will produce higher and higher energies, reaching into the Bev and hundreds of Bev ranges, has been, and continues to be, the hitting of nuclei harder and harder. When a nucleon is hit by a very high-energy particle, a variety of particles are frequently produced.

The questions concerning the origin and nature of these particles present challenging problems to experimentalist and theorist alike. Do these particles exist in some strange form within the nucleus, or are they created from the release of mass and impact energy? If they are created, what is the mechanism involved? How long do they last, and what becomes of their energy?

In this last chapter we will take a brief look at some recent discoveries and observations in "high-energy physics," in the hopes that such glimpses may instill some interest and give you, the reader, some clue as to what new experiments might be performed that would shed new light on these "elementary particles."

35.1. Elementary Particles

With the discovery of the neutron by Chadwick in 1932, the number of elementary particles became four in number: the *electron*, the *proton*, the *neutron*, and the *photon*. The first three are the atomic particles of which atoms are built, while the photon is the quantum unit of radiation emitted or absorbed by the electrons in the outer structure of atoms or by the particles within the nucleus.

The photon can only exist when traveling with the speed of light, and because of its motion possesses energy $h\nu$. By the mass-energy relation, $E = mc^2$, a photon also has mass $h\nu/c^2$. It possesses mass by virtue of its motion, for at rest it would have no energy and no mass.

The electron, proton, and neutron, on the other hand, have a definite rest mass m_0 and a rest energy m_0c^2. When they are set into motion, their mass increases, and their total energy is given by mc^2.

35.2. Antiparticles

The positron, discovered by Anderson in 1932, is a positively charged electron. We have seen in §27.11 how electron pairs can be created and how a free positron in coming together with an electron is annihilated and becomes two γ rays. It is this very property that gives the positron the name *antiparticle*—it destroys itself along with an electron and becomes another form of energy.

We have also seen in Fig. 30N that, when a high-energy photon comes close to the nucleus, mesons may be produced. If a pair of mesons is produced, one positive and one negative, one is the antiparticle of the other. These π-mesons may react with other nuclei or they may decay into μ-mesons as shown in Fig. 27Q. The muons in turn may decay into electrons and positrons, or they may be captured by some other nucleus.

The discovery of the positron, three pi mesons π^+, π^0, and π^-, the two mu mesons μ^+ and μ^-, along with the two neutrinos ν^0 and $\bar{\nu}^0$, raised the number of elementary particles by 1950 to twelve.

35.3. Particle Classification

The energy-level diagram in Fig. 35A represents an orderly array of eighty-nine *strongly interacting* and *elementary particles* known today. Some of these particles were predicted before they were identified in laboratory experiments.

The vertical column of symbols and numbers on the right gives the *particle designations* suggested for general adoption by G. F. Chew, M. Gell-Mann, and A. H. Rosenfeld, along with the *rest mass energy* in millions of electron volts, Mev, and the particles *spin angular momentum J* in units of $\hbar$. The numbers in the third row across the top give the *baryon number*, or *atomic mass, A*, and the numbers in the fourth row give the *particle charge Q* in units of the electronic charge e. All mesons (center) and leptons (bottom) have mass number $A = 0$.

One scale on the left gives the rest mass in multiples of the electronic mass m_e, and the other scale the rest mass in Mev. The specifically labeled dots, with the exception of ω^0, represent the thirty so-called elementary particles identified prior to 1957. Note that the left half of the table is a mirror image of the right half.

$$m_e c^2 = 0.511 \text{ Mev}$$

The term *baryon* applies to all elementary particles having a mass equal to or greater than a nucleon, while the term *hyperon* refers to particles whose masses are greater than those of nucleons.

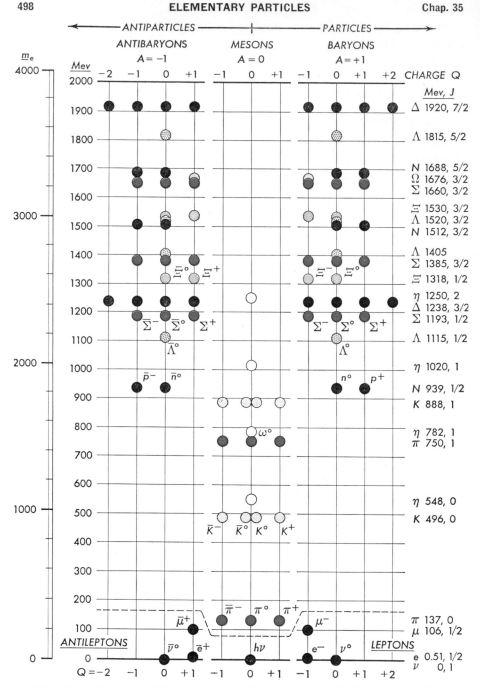

Fig. 35A　*Energy-level diagram of strongly interacting and elementary particles as compiled by Geoffrey F. Chew, Murray Gell-Mann, and Arthur H. Rosenfeld.* (Scientific American, *February 1964*)

The photon $h\nu$, the π^0 mesons, and the η^0 mesons are placed on the center line because they differ from all the rest in one respect: each of these particles acts as its own antiparticle.

The very interesting idea put forward that some distant galaxies in the heavens might be made up of antimatter might well be true. In such a galaxy, all hydrogen atoms would be made of antiprotons and positrons, and the protons and electrons would be their antiparticles. Thus Fig. 35A carries a double meaning, and *either half represents the antiparticles for the other half.*

Working up from the bottom of the table, it will be seen that we have already studied the very lightest-weight particles, called *leptons,* as well as π mesons and nucleons. The others, briefly, are as follows.

35.4. K Particles

These elementary particles belong to the meson class and have no spin angular momentum. Being heavier than the π mesons they have more possible modes for decay into other particles. K^+ particles have a mean life of 0.85×10^{-8} sec and decay with the following modes. For a definition of mean life, see end of §24.2.

$$
\begin{aligned}
K^+ &\rightarrow \pi^+ + \pi^+ + \pi^- \;\; (\tau \text{ mode}) \\
K^+ &\rightarrow \pi^+ + \pi^0 \quad\quad (\theta \text{ mode}) \\
K^+ &\rightarrow \pi^+ + \pi^0 + \pi^0 \\
K^+ &\rightarrow \mu^+ + \nu \\
K^+ &\rightarrow \mu^+ + \nu + \pi^0 \\
K^+ &\rightarrow e^+ + \nu + \pi^0
\end{aligned}
\tag{35a}
$$

The neutral K has similar decay modes and a very much shorter mean life.

35.5. Baryons and Antibaryons

Λ PARTICLES. These neutral particles belong to the baryon class and, being heavier than nucleons, are also called *hyperons.* Their mass is about 80 electron masses greater than a proton plus a pion, and they have a spin of $\frac{1}{2}\hbar$. Their principal decay modes are

$$
\begin{aligned}
\Lambda^0 &\rightarrow p + \pi^- \\
\Lambda^0 &\rightarrow n + \pi^0 \\
\Lambda^0 &\rightarrow p^- + \pi^+
\end{aligned}
\tag{35b}
$$

Σ PARTICLES. These hyperons are of three kinds, Σ^+, Σ^0, and Σ^-, and belong to the baryon class. They have a spin of $\frac{1}{2}\hbar$ and decay principally by the following modes:

$$
\begin{aligned}
\Sigma^+ &\rightarrow p + \pi^0 \\
\Sigma^+ &\rightarrow n + \pi^+ \\
\Sigma^- &\rightarrow n + \pi^- \\
\Sigma^0 &\rightarrow \Lambda^0 + h\nu
\end{aligned}
\tag{35c}
$$

Ξ PARTICLES. These hyperons are about 130 electron masses heavier than $\Lambda + \pi$ and belong to the baryon class. They also have a spin of $\frac{1}{2}\hbar$ and decay principally by the following modes:

$$\Xi^- \rightarrow \Lambda^0 + \pi^-$$
$$\Xi^0 \rightarrow \Lambda^0 + \pi^0 \tag{35d}$$

Most collision events requiring particles of high energy are called *strong interactions*, and are to be contrasted with natural decay processes, such as $\pi^+ \rightarrow \mu^+ + \nu \rightarrow e^+ + \nu + \bar{\nu}$, which are called *weak interactions*.

35.6. Strong and Weak Interactions

It is not difficult to make a clear distinction between strong interactions such as those that occur between protons, neutrons, and hyperons and weak interactions that occur between muons, electrons, and neutrinos. We know that neutrons and protons are subject to strong attractive forces because of the strong binding of these particles in nuclei. See §32.1.

When two high-energy nucleons approach each other in collision, for example, the very strong *action* and *reaction* forces they exert on each other occur at such short distances that the collision time is of extremely short duration. Photons, electrons and positrons, muons and neutrinos, do not involve such strong short-range forces, but exert weaker forces at greater distances. Collisions or disintegrations involving these particles are, therefore, of longer duration.

Collision times and *decay times* for strong interactions are of the order of 10^{-23} second, while weak interactions require time intervals billions of times longer and of the order of 10^{-13} second. The lightest-weight strongly interacting particles are pi mesons. They with their rest mass of 137 Mev are in direct contrast with electrons and their rest mass of 0.5 Mev, and photons and neutrinos with zero rest mass. One exception to this involves the weakly interacting muons with a rest mass of 106 Mev.

A useful quantum number called the *strangeness number* was introduced in elementary particle theory by Gell-Mann* to predict which new particles could be produced by impact, that is, by strong interactions. See § 35.8. Strangeness numbers are conserved in strong interactions, but not in weak interactions. All interactions obey the ordinary conservation laws of *energy, momentum, angular momentum*, and *charge*.

35.7. Nuclear Spin and Isotopic Spin

The concept of nuclear spin *I* was introduced in §32.7 to explain the hyperfine structure of spectrum lines as well as the measured magnetic moments of neutrons and protons. The neutrino was then introduced into β decay in order

---

* See *Scientific American*, pp. 72–88, July 1957.

to maintain the conservation laws of energy, momentum, and angular momentum.

In writing down nuclear reactions as equations, the *law of conservation of electrical charge* is always imposed as part of the balancing process. Since elementary particles carry a charge of $+1$, 0, or -1, Heisenberg was the first to apply quantum numbers to charge and refer to the concept as *isotopic spin.*

While both words in this term are misnomers, they arose from the idea that pairs of particles like nucleons, and triplets like the π mesons, may be thought of as isotopes and that their charges, differing from each other by unity, suggest space quantization like electron spin and orbit in a magnetic field. (See Fig. 12P.)

Chew, Gell-Mann, and Rosenfeld now use the symbol J for the spin angular momentum quantum number, and the symbol I to represent the isotopic spin quantum number proposed by Heisenberg.

Within the nucleus, short-range forces between neutrons and protons are exactly alike, but the ever-present coulomb forces between protons removes this symmetry and we are able to differentiate between the two different kinds of nucleons. In an effort to formulate a theory of these differences, particle charge has been likened to a vector, here called isotopic spin, and differences in charge have been compared with the space quantization of a spin vector in a field. (See Fig. 35B.)

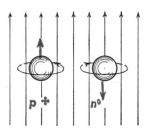

Fig. 35B *Schematic diagram of isotopic spin, a kind of charge quantization for elementary doublets like the neutron and proton.*

By this analogy a nucleon may be thought of as a particle with an isotopic spin of $I = \frac{1}{2}$ and a *spin average* or center of $\overline{Q} = +\frac{1}{2}$. If the particle spin lines up with the spin average, we have a proton with isotopic spin $I = +1$; if it lines up in the opposite direction we have a neutron with isotopic spin $I = 0$.

The isotopic spin of antinucleons is $I = -\frac{1}{2}$, and with a spin average or center of $\overline{Q} = -\frac{1}{2}$, this gives $Q = 0$ for the antineutron and $Q = -1$ for the antiproton.

The π mesons form a triplet, with charges $+1, 0, -1$. This particle is therefore assigned an isotopic spin of $I = 1$. Applying the concept of space quantization to this vector magnitude, we obtain three possible orientations as shown in Fig. 35C. These three positions, centered at charge $\overline{Q} = 0$, give $Q = +1, 0$, and -1, corresponding to π^+, π^0, and π^-.

It would seem that this fiction of comparing electrical charge with spin orientation in a field might have some basic foundation and that additional experimental data will lead to a more acceptable theory of the basic concept we call electric charge.

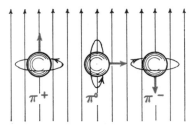

Fig. 35C *Schematic diagram of isotopic spin, a kind of charge quantization for elementary triplets like π mesons.*

35.8. Particle Symbols and Quantum Numbers

It will be observed in Fig. 35A that all particles occur in families, or charge multiplets: *singlets, doublets, triplets,* and *quadruplets.* Within each multiplet, particles differ only in electrical charge Q.

Only ten different patterns of baryons and mesons are known or predicted and each is represented by a different set of three quantum numbers, A, Y, and I.

Quantum number A, called the *baryon number,* has the value $+1$, 0, or -1 for all particles, and represents the familiar atomic mass number long used for nuclei. For uranium 238, $A = 238$, etc.

Quantum number Y, called the *hypercharge,* is assigned the whole number value given by twice the average charge of a multiplet.

$$Y = 2\overline{Q}$$

For the nucleon doublet p^+ and n^0, for example, the average charge $\overline{Q} = (1 + 0)/2 = +1/2$. This gives $Y = 2\overline{Q} = +1$.

Quantum number I, called *isotopic spin,* gives the number of particles in a multiplet. The multiplicity is given by

$$M = 2I + 1$$

For Δ particles, for example, $I = \frac{3}{2}$ and the multiplicity $M = 2 \times \frac{3}{2} + 1 = 4$.

The quantum numbers assigned each of the ten known family patterns are:

BARYONS	A	Y	I
Lambda Λ	1	0	0
Omega Ω	1	-2	0
Nucleon N	1	1	$\frac{1}{2}$
Xi Ξ	1	-1	$\frac{1}{2}$
Sigma Σ	1	0	1
Delta Δ	1	1	$\frac{3}{2}$

MESONS	A	Y	I
Eta H	0	0	0
Kappa K	0	1	$\frac{1}{2}$
Kappa bar $\overline{K}$	0	-1	$\frac{1}{2}$
Pi π	0	0	1

As an example of the proposed symbolism a particle designated N^0, 939, $\frac{1}{2}$, is a neutron with quantum numbers $A = 1$, $Y = 1$, $I = \frac{1}{2}$, $J = \frac{1}{2}$, and a rest mass energy of 939 Mev. Similarly a particle designated Σ^-, 1193, $\frac{1}{2}$, is an antisigma baryon with quantum numbers $A = 1$, $Y = 0$, $I = 1$, $J = \frac{1}{2}$, and a rest mass of 1193 Mev. A particle designated $\overline{K}^0$, 496, 0, is an antiparticle *kappa bar* with quantum numbers $A = 0$, $Y = -1$, $I = \frac{1}{2}$, $J = 0$, and rest mass energy of 496 Mev.

All baryons and mesons, along with their antiparticles, are frequently referred to as *strange particles*. The quantum number called the *strangeness number S* was introduced in 1957 to predict which new particles might be produced by impact, that is, by strong interactions. It is now known that the strangeness number assigned to each particle family is related to the hypercharge quantum number Y and baryon number A as follows:

$$S = Y - A$$

The strangeness number for the predicted and recently found particle Ω^-, 1676, $\frac{3}{2}$, for example, is $S = -2 - 1 = -3$.

35.9. Photographic Emulsions

When a photon or ionizing particle traverses the sensitive emulsion of a photographic film, the clear silver bromide crystal grains that are penetrated are turned into black silver upon development by regular film-developing processes.

Since the sensitive emulsion on most film is extremely thin, satisfactory emulsions up to 1 mm in thickness, containing about 80% silver bromide, were first developed by C. F. Powell in England. These can be stacked in layers to build up larger volumes, exposed to high-energy nuclear beams, or cosmic rays, and then developed. See Fig. 35D. By studying consecutively numbered films sepa-

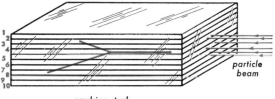

emulsion stack

Fig. 35D	*Numbered stack of extra-thick photographic emulsions used in photographing high-energy nuclear events.*

rately, under a suitable measuring microscope, one can observe nuclear collision events and make measurements of the different particles, their directions, ranges, track densities, etc.

Track densities vary widely with particle *charge* and *velocity*, as shown in Fig. 35E. The energies required of various particles to travel 1 mm in an average nuclear emulsion are

e	0.7 Mev	p	14.0 Mev
μ	5.5 Mev	d	20.0 Mev
π	6.1 Mev	α	55.0 Mev

Because emulsion densities are far greater than the gas in a cloud chamber, the track ranges are extremely small. This high-density and short-range feature is particularly useful, therefore, to the study of nuclear events involving very high-speed particles.

The procedure used for displaying emulsion tracks is to make enlargements of microscope photographs of neighboring sections, and to piece the photographic prints together as shown in Fig. 35F. These print assemblies were made from a stack of 46 emulsions, 15 cm × 15 cm, exposed at high altitude over England in 1952.

The first assembly (a) shows a "star" of tracks in emulsion 17, one track being that of a K^+ meson as indicated. Six cm along this track in the emulsion (this corresponds to 60 ft on the enlarged scale shown here) we come to assembly (b) from emulsion 6. There the K^+ meson disintegrated by the τ mode into π mesons. (See Eq. (35a).) Two cm along the π^- track, in assembly (c), the meson was captured by a heavy nucleus, resulting in a nuclear explosion. Among the secondary products were four singly charged particles and a $_3Li^8$ nucleus. The latter, on reaching the end of its range, emitted a β particle forming $_4Be^8$ which

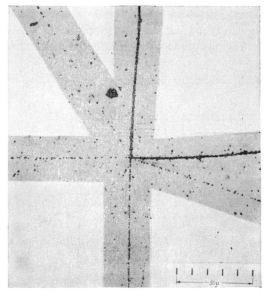

Fig. 35E   *Charged particle tracks in a photographic emulsion showing differences in track density.*

Courtesy, C. F. Powell, P. H. Fowler, and D. H. Perkins, from *The Study of Elementary Particles*, Pergamon Press Ltd.

(a)　　　　　　(b)　　　　　　(c)

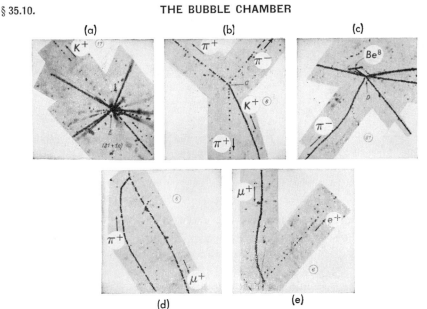

(d)　　　　(e)

Courtesy of C. F. Powell, P. H. Fowler, and D. H. Perkins, from *The Study of Elementary Particles*, Pergamon Press Ltd.

Fig. 35F　*Five directly related events traced through a photographic emulsion stack.*

spontaneously splits up into two α particles to produce the "hammer track" shown.

About 3 cm along the π^+ track, in assembly (d), the meson decays into a μ^+ and a neutrino. A short distance along this track, in assembly (e), the μ^+ decays into a positron e^+, a neutrino ν, and an antineutrino $\bar{\nu}$.

The nuclear emulsion photograph in Fig. 35G shows a "star" of 22 tracks. A primary cosmic ray, a proton, enters from the top, and collides with a silver or bromine nucleus in the emulsion, causing an explosion. Most of the tracks were made by π mesons, and the others probably by K mesons and protons.

While the number of stars of this general nature is small relative to other kinds of nuclear events observed in emulsions, a great many have been observed and studied.

35.10. The Bubble Chamber

The bubble chamber, invented in 1952 by D. H. Glaser, has become one of the most valuable instruments for studying the minute details of high-energy nuclear events. In a cloud chamber, fogdrops form on the ions produced by charged atomic particles that have just previously traversed the gas-filled chamber. In the bubble chamber, the ions, formed by charged particles traversing the liquid, form local heat centers in which tiny gas bubbles develop and grow.

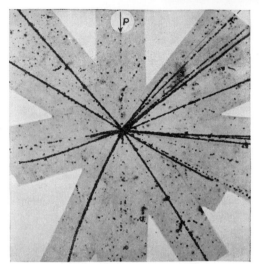

Fig. 35G   *Photographic emulsion star produced by a cosmic-ray proton.*

Courtesy of C. F. Powell, P. H. Fowler, and D. H. Perkins, from *The Study of Elementary Particles,* Pergamon Press Ltd.

The basic principles of the bubble chamber involve the superheating of a liquid and the bubbles that form in the process of boiling. Water, for example, boils at 100°C at standard atmospheric pressure. If the pressure is increased as in a pressure cooker boiling will not begin until a higher temperature is reached. If the pressure is then suddenly reduced, boiling begins with the sudden formation of tiny bubbles that grow quickly in size.

A simplified diagram of a bubble chamber is shown in Fig. 35H. A box with thick glass walls, filled with a liquid, is connected to a pressure system and then heated to some predetermined temperature. High-energy particles enter the liquid through a thin window. A sudden release of a valve in the pressure system

Fig. 35H   *Cross-section diagram of a bubble chamber showing track illumination, camera, and incident particle beam.*

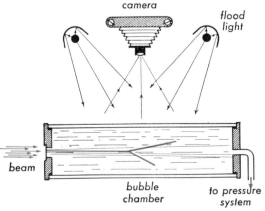

is quickly followed by the flash of floodlights and the snap of a camera shutter. If the chamber is operating properly, and events are correctly timed, sharply defined trails of bubbles that have formed on the paths of ions made by the traversing particles will be photographed. See Figs. 35K, 35L, 35N, and 35P.

The extensive use of liquid hydrogen in bubble chambers has been particularly effective in the study of elementary and strange particle events. The difficulties of handling liquid hydrogen at −253°C in large bubble chambers have been overcome by L. Alvarez and his colleagues at the University of California over a period of several years. Liquid hydrogen is particularly useful in that it provides a high concentration of "target protons," the simplest of atomic nuclei, and at the same time greatly shortens the distance between events that would be required in the gas-filled space of a cloud chamber. See Figs. 35I and 35J.

It is customary to locate a bubble chamber in the strong field of a large electro-

Fig. 35I   *Cutaway model of the 72-inch liquid hydrogen bubble chamber at the Lawrence Radiation Laboratory.*

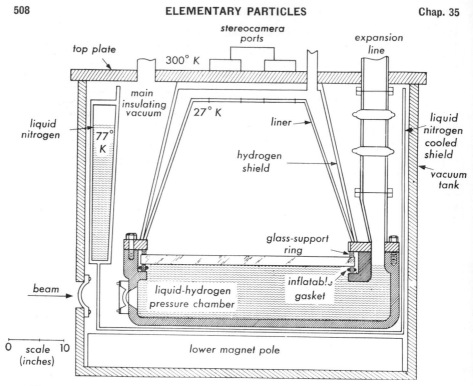

Fig. 35J *Cross-section diagram of the 72-inch liquid hydrogen bubble chamber.*

magnet so that particle charge and momentum relations can be obtained from track curvature. (See Figs. 35K and 35L.)

35.11. Antiprotons

The existence of antiprotons was discovered in 1955 by Chamberlain, Segrè, Wiegand, and Ypsilantis at the University of California. This discovery came as the result of long-range plans laid for the purpose of answering the question, "Is there in nature, or can there be created by a strong interaction, a negatively charged particle with the mass of a proton?" One of the principal objectives in building the 6.2-Bev proton accelerator in 1953 was to find an answer to this question. Protons accelerated to a high enough energy should, in colliding with heavier nuclei, impart sufficient energy to create, if such were possible, a pair of protons, one plus, the other minus.

The first antiprotons were discovered as high-energy, negatively charged particles emerging from a copper target in the proton beam of the bevatron and having all of the anticipated properties. By means of a strong magnet the antiprotons were bent away from the protons and into a scintillation counter. Today antiproton events are commonly observed and studied by means of a bubble chamber.

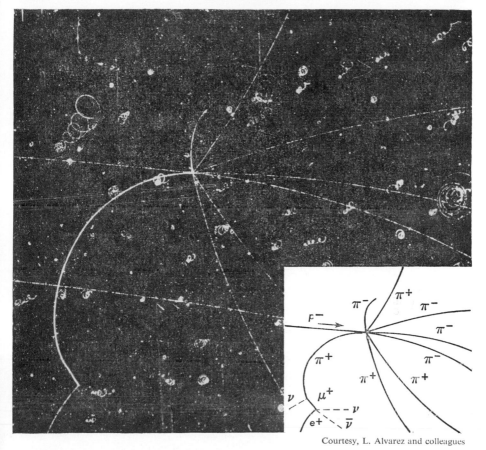

Fig. 35K　*Pi-meson star produced by the proton capture of an antiproton in a liquid hydrogen bubble chamber.*

Fig. 35K is a photograph in which an antiproton entered the liquid hydrogen bubble chamber from the left. Near the center of the picture it interacted with a proton, and both particles were annihilated in the production of four pairs of π mesons. Since the bubble chamber operated in the uniform field of a strong electromagnet, the π^- tracks curve clockwise, and the π^+ tracks curve counterclockwise. The π^+ that curved back to the left and down decayed into a μ^+ and an e^+ as shown in the accompanying legend diagram.

The reactions for this event are written as follows:

$$p^- + p \rightarrow 4\pi^- + 4\pi^+$$
$$\pi^+ \rightarrow \mu^+ + \nu \tag{35e}$$
$$\mu^+ \rightarrow e^+ + \nu + \bar{\nu}$$

35.12. Baryon Reactions

An antiproton event involving the production of a baryon-antibaryon pair is reproduced in Fig. 35L. As shown in the legend diagram, an antiproton from the bevatron target enters the bubble chamber at the left and interacts with a proton.

The track disappears at this point because the interaction with a proton in the liquid hydrogen produces a neutral lambda-antilambda pair. With but a short lifetime of about 10^{-10} seconds, each of these Λ^0 particles disintegrates, one into π^- and p^+, and the other into p^- and π^+. The antiproton is captured by another proton at the upper right and is immediately annihilated in the production of two pion pairs. The reactions for these events are

$$p^- + p \rightarrow \bar{\Lambda} + \Lambda$$
$$\Lambda \rightarrow p + \pi^-$$
$$\bar{\Lambda} \rightarrow p^- + \pi^+ \tag{35f}$$
$$p^- + p \rightarrow 2\pi^+ + 2\pi^-$$

As high-energy protons from a large accelerator traverse any target material whatever, a considerable number of π mesons are produced along with other particles. These charged mesons, π^+ or π^-, can be channeled, by bending in a

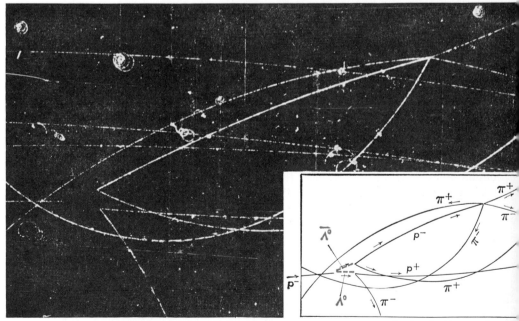

Courtesy, L. Alvarez and colleag

Fig. 35L   *Bubble-chamber photograph and legend diagram of correlated events involving two antiprotons, two Λ° particles, and seven π mesons.*

magnetic field, to any other area for experimentation. Directed into a bubble chamber, for example, their reactions with protons can be studied.

Fig. 35M is a photograph in which a π^- meson enters a liquid hydrogen bubble chamber from the left. As shown in the legend diagram, interaction with a proton cancels the charges and ends the track by creating two neutral particles.

These two recoiling particles K^0 and Λ^0 do not produce tracks in the chamber but, having short lifetimes, quickly disintegrate. The neutral K-particle decays into a pair of pions, π^+ and π^-, while the Λ-particle decays into a proton and a π^-. As reactions

$$\pi^- + p \rightarrow K^0 + \Lambda^0$$
$$K^0 \rightarrow \pi^+ + \pi^-$$
$$\Lambda \rightarrow p + \pi$$

(35g)

The bubble chamber photograph in Fig. 35N(a) shows a series of related events involving 14 particles: π^+, $2\pi^0$, $2\pi^-$, p, n, K^+, Σ^+, μ^+, e^+, $\bar{\nu}$ and 2ν. As shown by the labeled diagram in Fig. 35N(b), one of the high energy π^--particles enters from the left and interacts with a proton. This strong interaction produces Σ^-

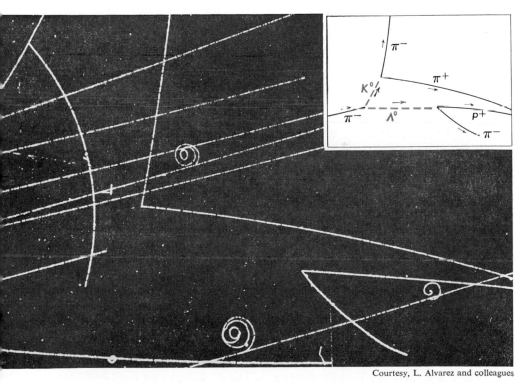

Courtesy, L. Alvarez and colleagues

Fig. 35M *Bubble-chamber photograph with π^-, p, K°, Λ°, π^+, and π^- tracks, all in one related event. Legend diagram at upper right.*

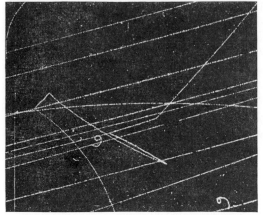

Courtesy, L. Alvarez and colleagues

Fig. 35N(a) *Bubble-chamber photograph (in liquid hydrogen) of correlated events involving Σ^- and K^+ particles. See legend diagram below.*

Fig. 35N(b) *Labeled track pattern for bubble-chamber photograph in Fig. 35N(a).*

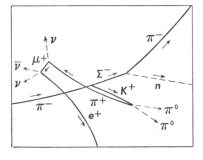

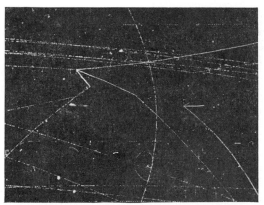

Courtesy, Wilson Powell and colleagues

Fig. 35O *Propane bubble-chamber photograph of an event involving K-mesons and a Ξ particle.*

and K^+ particles. The Σ^- decays into a π^- and a neutron, while the K^+ decays into a π^+ and two (not showing) π^0's. The π^+ decays into a μ^+ and then e^+ in the customary way.

The reactions involved may be written

$$\pi^- + p \rightarrow \Sigma^- + K^+$$
$$\Sigma^- \rightarrow \pi^- + n$$
$$K^+ \rightarrow \pi^+ + \pi^0 + \pi^0 \qquad (35h)$$
$$\pi^+ \rightarrow \mu^+ + \nu$$
$$\mu^+ \rightarrow e^+ + \nu + \bar{\nu}$$

The photograph reproduced in Fig. 35O shows K^- meson tracks entering from the left. In the upper left, one of these particles collides with a carbon nucleus and produces three charged particles, as shown by the three-pronged fork. Energy and momentum calculations from the measured tracks show that the reaction is

$$K^- + {}_6C^{12} \rightarrow K^+ + \Xi^- + p + {}_4Be^{10}$$

The lower prong of the fork is the K^+ particle, which decays into a neutrino, and a μ^+ recoiling down to the left:

$$K^+ \rightarrow \mu^+ + \nu$$

The middle prong of the fork is the Ξ^- particle which at the bend in the middle of the picture decays into a π^- and a Λ^0:

$$\Xi^- \rightarrow \pi^- + \Lambda^0$$

The neutral Λ-particle, recoiling to the right, creates no track but quickly decays into a π^- and a p to form the V-shaped track at the right:

$$\Lambda^0 \rightarrow \pi^- + p \qquad (35i)$$

35.13. Parity

Parity is a mathematical treatment of what is best described as a mirror symmetry of many natural phenomena. Up until recently it was believed that the mirror image of any physical phenomenon or laboratory experiment is just as true to nature as the direct image itself. This is consistent with the principle called *conservation of parity*.

According to the conservation of parity, it was believed that if one observed an experiment by looking in a mirror and was not told he was looking in a mirror, there would be no way in which he would know it. As an example of parity, many crystalline structures show mirror symmetry. With many cubic crystals there is no question, since the image and mirror image have identical structures.

Diagrams of two kinds of quartz crystals found in nature are shown in Fig. 35P. One, called right-handed quartz, has its silicon and oxygen atoms lined up in clockwise spirals around the axis, while the left-handed quartz is a mirror image of the other and has its molecules lined up in a counterclockwise direction. One will rotate plane-polarized light clockwise as it traverses the crystal along the optic axis, while the other will rotate it counterclockwise. The existence of both of these crystals is consistent with the conservation of parity.

The first serious question regarding the conservation of parity arose in the minds of two young theoretical physicists, Yang and Lee, when they pointed

quartz crystals

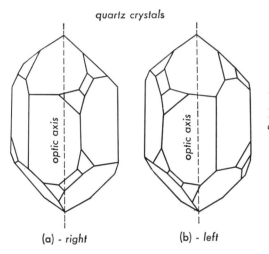

(a) - right (b) - left

Fig. 35P Right- and left-handed quartz crystals. Each is a mirror image of the other.

out in 1956 that K^+ mesons can decay into two pions (θ-mode) or three pions (τ-mode). (See Eq. (35a).) An experiment suggested by Yang and Lee to test for parity was carried out by C. S. Wu, W. E. Ambler, R. W. Hayward, D. D. Hoppes, and R. P. Hudson. This experiment showed that, when radioactive cobalt-60 nuclei were lined up with their spin axes parallel to each other, more β particles were emitted along this axis in one direction than in the other.

To realize the significance of this result, consider the common decay of a π^+ meson into a μ^+ and a ν. The π^+ has zero spin, while the μ^+ and ν have spins of $\frac{1}{2}$. To conserve angular momentum, the two particles must fly apart spinning in opposite directions as shown in Fig. 35Q. Moving in the direction of the

real image - correct

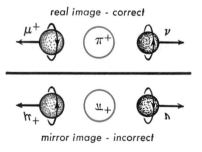

Fig. 35Q Schematic diagram showing the left-handed, or negative, helicity of μ^+ and ν particles in the decay of π^+ mesons.

mirror image - incorrect

arrows, both are advancing, as would left-handed screws. The mirror image in (b), which is apparently forbidden by nature, shows both particles advancing as right-handed screws. *A spinning particle advancing as a right-handed screw is said to have positive helicity; advancing as a left-handed screw it is said to have negative helicity.*

From the observed spin directions of β particles, it is now known that all *neutrinos have negative helicity* while *antineutrinos have positive helicity.* With this rule and spin conservation, we see from Fig. 35R that in π^- decay both μ^- and $\bar{\nu}^-$ has positive helicity.

It should be pointed out that helicity has meaning for neutrinos only, since they move with the speed of light. To an observer moving faster than a muon, for example, the spin direction would appear to be the reverse of that seen by an observer moving slower than the muon.

If all the particles in one or the other of the two mirror images of Figs. 35Q and 35R are changed to their corresponding antiparticles, the resulting inter-

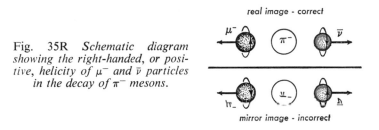

real image - correct

Fig. 35R *Schematic diagram showing the right-handed, or positive, helicity of μ^- and $\bar{\nu}$ particles in the decay of π^- mesons.*

mirror image - incorrect

action becomes the real interaction of the other. Thus there is an over-all symmetry between the two decay reactions.

When a μ^+ decays in the customary way into a positron, a neutrino, and an antineutrino, helicity for the neutrinos as well as the μ^+ are already fixed. What helicity is imparted to the electron and the μ^+ will depend on the relative directions of all recoiling particles. If the neutrinos go off together as shown in Fig. 35S, the positron must recoil with negative helicity. If the antiprotons

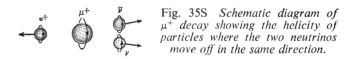

Fig. 35S *Schematic diagram of μ^+ decay showing the helicity of particles where the two neutrinos move off in the same direction.*

recoil in opposite directions, the positron, to conserve spin, can have either positive or negative helicity. Drawings for these cases will be left as exercises for the student.

If a distant world is made of antimatter, there is no known way whereby we on the earth can determine this. Furthermore, though μ^- and μ^+ have positive and negative helicity, respectively, there is no information we could impart to a man on a distant world that would tell him which is which.

35.14. Conservation Laws

We have seen in earlier chapters on nuclear processes that certain conservation laws apply to all reactions. These laws relate measurable quantities existing just before impact with those existing immediately after impact. For convenience we can state them as follows:

LAW 1. *The total energy, including rest mass, remains constant.*

LAW 2. *The total linear momentum remains constant.*

LAW 3. *The total angular momentum remains constant.*

LAW 4. *The total charge remains constant.*

To these four laws we add four more to apply to elementary particle reactions.

LAW 5. *The total baryon number remains constant.*

This fifth law has greater implications when it is stated another way. "The total number of baryons in the world remains constant." Here we count an antibaryon as minus one baryon. This law is not pertinent for reactions involving only mesons and leptons. All mesons and leptons have baryon number zero.

LAW 6. *The total lepton number remains constant.*

The leptons ν, e^-, and μ^- have lepton number $+1$ and their antiparticles have lepton number -1. These particles are involved in the decay of the muons. The decay of a negative muon, for example, will illustrate:

$$\mu^- \rightarrow e^- + \nu + \bar{\nu}$$

Lepton number $+1 = +1 +1 -1$

LAW 7. *The total strangeness number remains constant.*

This last law does not apply to weak reactions as in the decay of unstable particles, but does apply to strong interactions.

LAW 8. *The over-all symmetry of antiparticle parity is preserved.*

This is represented by the reactions in Fig. 35Q.

Example. Apply the rules of conservation of baryons and conservation of strangeness to the strong reaction involved in Fig. 35N.

Solution. From the first reaction in Eq. (35h), the baryon numbers in Fig. 35A, and strangeness numbers given in §35.8, we can write

$$\pi^- + p \rightarrow \Sigma^- + K^+$$

Baryon number $0 + 1 = 1 + 0$

Strangeness number $0 + 0 = -1 + 1$

35.15. Omega Meson

The ω^0 meson, $(\eta, 782, 1)$ in Fig. 35A, is an atomic particle composed of three π mesons, and was first detected by L. W. Alvarez, B. C. Maglic, A. H. Rosenfeld, and M. L. Stevenson in 1961. Antiproton reactions in a hydrogen bubble chamber frequently give rise to four-pronged tracks of the kind shown in Fig. 35T. Momentum and energy measurements of many such events clearly show that two pairs of oppositely charged π mesons are responsible for the four emergent tracks. Frequently, however, one pair of π mesons is produced and the remaining mass quickly decays (in about 10^{-22} sec) into three π mesons, $\pi^+ + \pi^0 + \pi^-$. Before decaying, this uncharged particle, with a mass of 1540 m_e, is called an ω^0 *meson.* The alternate reactions for this event are

$$p^- + p \rightarrow 2\pi^+ + 2\pi^-$$

or

$$p^- + p \rightarrow \pi^+ + \pi^- + \omega$$

followed by

$$\omega \rightarrow \pi^+ + \pi^0 + \pi^-$$

An excellent illustration of the decay of a π^+ meson into a μ^+ and e^+ is shown at the lower right in Fig. 35T. See Fig. 27P.

In the upper right in Fig. 35T, a π^- meson is captured by a proton to produce a neutron and a π^0. Normally a π^0 decays into γ rays, but about one out of

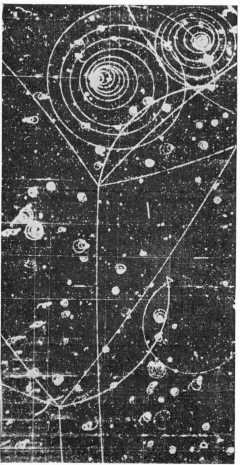

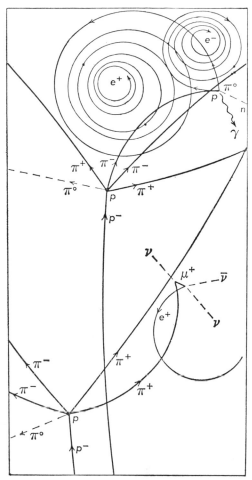

Courtesy of the Lawrence Radiation Laboratory, and L. W. Alvarez and associates

Fig. 35T(a) *Photograph of antiproton reactions in a hydrogen bubble chamber showing meson pairs in the kind of reaction that first led to the identification of the ω meson.*

Fig. 35T(b) *Legend diagram for Fig. 35T(a) showing the uncharged trackless particles as dotted lines.*

eighty such neutral particles converts one of the γ rays into an electron pair. Such a *Dalitz pair*, as it is called, is shown by the oppositely directed spirals. The reactions for this event are

$$\pi^- + p \rightarrow \pi^0 + n$$
$$\pi^0 \rightarrow e^+ + e^- \ \gamma \ \text{ray}$$

35.16. Spark Chamber Detectors

One of the most recent instruments developed for use in the study of high-energy nuclear reactions is known as the *spark chamber*. This device is composed of a stack of equally spaced conducting plates as shown by the cross-section diagram in Fig. 35U.

A typical chamber will have 25 to 100 plates, each about 1 mm thick and 1 m square, and accurately spaced about 6 mm apart. Alternate plates are connected together and the two sets connected to a 10,000 to 15,000 volt d.c. source. The airtight vessel in which the plates are mounted contains pure helium, or a mixture of approximately 90 percent neon and 10 percent helium, at a pressure of 1 atmosphere.

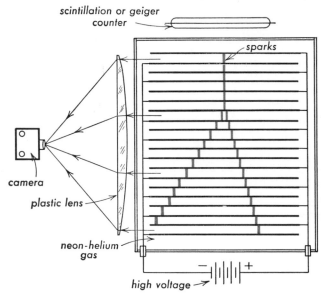

Fig. 35U   *Cross-section diagram of a spark chamber detector used for observing high energy nuclear reactions involving elementary particles.*

When a high-energy charged particle enters such a chamber, a detector outside the chamber, or inside, triggers the high voltage to the plates, and opens the camera shutter. Traversing the chamber a charged particle produces many ion pairs along its path, and sparks jump between the pairs of oppositely charged plates as indicated. A large lens, cut from clear plastic, makes it possible for a camera to photograph the light from sparks between all plates. A second large lens and camera "looking into" an adjacent side of the box permits the simultaneous taking of two pictures, thereby recording stereoscopic photos of each event.

A photograph of a proton scattered by a carbon nucleus is shown in Fig. 35V. This picture shows but one of several events diagramed above. A beam of π^+ mesons, produced by the 6.3-Bev proton beam of the bevatron, passes through a target containing hydrogen. Upon collision with another proton, a K^+ meson and a Σ^+ baryon are produced. The K^+ decays into a μ^+ and a neutrino. See Eq. (35a), fourth equation. The Σ^+ decays into a proton and a π^0. See Eq. (35c), first equation. The proton, detected as it enters the carbon plate spark chamber,

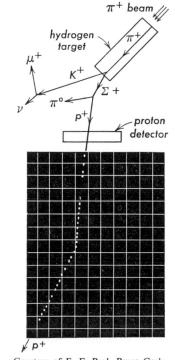

Fig. 35V   *Spark chamber photo-graph of the scattering of a proton by a carbon nucleus.*

Courtesy of E. F. Beal, Bruce Cork, D. Keefe, P. G. Murphy, and W. A. Wenzel, and the Lawrence Radiation Laboratory

is photographed as shown. The horizontal and vertical white lines in the photo are the image of a grid used for making measurements.

35.17. Mean Lives of Elementary Particles

The discovery of so many elementary particles in recent years has greatly complicated all previous theories of the atomic nucleus. The fundamental questions arising in the minds of the physicists include not only the different properties of the individual particles but of how they fit into the structure of the atomic nucleus. At the present stage of development, in trying to unravel the many mysteries concerning elementary particles, some important questions stand out.

Are all of the particles listed in Fig. 35A really elementary, or are some of them just composites of others? Could it be that some of the stars in the universe, or entire galaxies, are made of antimatter, with hydrogen atoms composed of antiprotons and positrons, etc.? Why are elementary charges limited to values of $+1$, 0, and -1?

Every physicist believes that some day we will know the answers to these and many other questions about the nature of elementary particles, the building blocks of the universe

One interesting aspect of the nature of the particles themselves concerns their *mean life*. See Table 35A.

Table 35A. Mean Lives of the Unstable Elementary Particles

PARTICLE	MEAN LIFE (sec)	PARTICLE	MEAN LIFE (sec)
Ξ^+	1.3×10^{-10}	K^+	1.2×10^{-8}
Ξ^0	1.5×10^{-10}	K^0	1.0×10^{-11}
Ξ^-	1.3×10^{-10}	K^-	1.2×10^{-8}
Σ^+	0.8×10^{-10}	π^+	2.5×10^{-8}
Σ^0	1.0×10^{-9}	π^0	2.2×10^{-16}
Σ^-	1.6×10^{-10}	π^-	2.5×10^{-8}
Λ^0	2.5×10^{-10}	μ^+	2.2×10^{-6}
n	1.0×10^3	μ^-	2.2×10^{-6}

The best and most direct means of determining the lifetime of any identified particle is to measure the length of the track it produces in a cloud chamber, bubble chamber, or photographic emulsion, and determine its velocity from conservation laws or the curvature of its path in a magnetic field. The lifetime τ is then given by the distance traveled divided by the velocity.

The lifetime determined in this way is the time the particle existed in the laboratory frame of reference. To find τ_0, the lifetime of the particle in its own moving frame of reference, the special theory of relativity must be used. Starting with the transformation equation, Eq. (19c), we write τ_0 for t' and τ for t,

$$\tau_0 = \gamma \left(\tau - \frac{vx}{c^2} \right)$$

In this equation x represents the distance traveled, and τ the time of travel in the laboratory frame of reference. Placing

$$x = v\tau$$

and

$$\gamma = 1 \bigg/ \sqrt{1 - \frac{v^2}{c^2}}$$

we obtain

$$\tau_0 = \tau \sqrt{1 - \frac{v^2}{c^2}} \tag{35j}$$

This formula has been verified with high accuracy up to velocities of 99.5 percent the speed of light, and with approximately 10 percent accuracy at velocities up to 99.95 percent the speed of light.

Consider for example the events shown in Fig. 35M in which a Λ^0 particle existed long enough to travel 8 cm before decaying into a proton, p^+, and a negative pion, π^-. Traveling this distance with 70 percent the speed of light (energy = 0.45 Bev) its lifetime is given by $8 \times 10^{-2}\text{m}/2.1 \times 10^8\text{m/sec}$, or $\tau = 3.8 \times 10^{-10}$ sec. This is the lifetime in the laboratory frame of reference.

In the frame of the moving Λ^0 particle the real lifetime τ_0 is found by direct substitution in Eq. (35j).

$$\tau_0 = 3.8 \times 10^{-10} \sqrt{1 - (0.70)^2}$$
$$\tau_0 = 2.7 \times 10^{-10} \text{ sec.}$$

There is an appropriate limit, however, to the determination of lifetimes by this direct method. For high-energy particles traveling with 70 percent the speed of light an observed track of only 3 microns in a photographic emulsion corresponds to a time $\tau = (3 \times 10^{-6}\text{m})/(2.1 \times 10^8 \text{ m/sec})$, or 1.4×10^{-14} sec. In the reference frame of the moving particle the real lifetime $\tau_0 = 1 \times 10^{-14}$ sec.

One of the most interesting questions that has arisen out of modern physics concerns the real existence of the very short-lived elementary particles. Since particles such as π, K, Λ, Σ, Ξ and ω are known to have mean lives of one-thousandth of a millionth of a second or less, do we have a right to say they exist long enough to call them particles at all?

One approach to finding an answer to this question is to graphically represent the lifetime of a few commonly recognized natural phenomena or bodies of matter, and to compare the time scales of their existence. The generally accepted lifetime of the earth, from the time of its creation to the present, is 4.5×10^9 yr, or 1.4×10^{17} sec. This interval of time is represented by the colored line at the top in Fig. 35W.

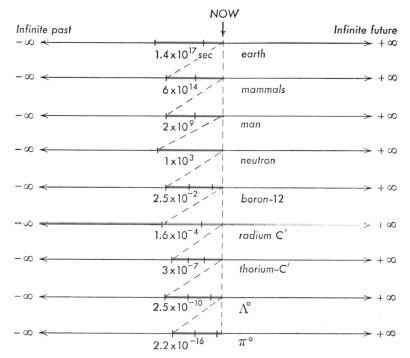

Fig. 35W   *Comparison of time scale graphs for natural bodies from the earth to elementary particles.*

Since we admit the earth was created sometime in the past, we must also admit that time existed before this event, and that time must be extended to infinity in the past. While the earth still exists at the present, and will sometime come to an end in the future, time must continue on with succeeding astronomical events and by logical reasoning we find it necessary to extend the time scale to plus infinity. The line is therefore infinitely long in both directions.

Paleontologists agree that mammals have inhabited this earth for approximately two million years, or about 6×10^{14} sec. On the earth's time scale this is but a tiny dot, but by a 200 fold expanded scale as in the second line, we can show the finite span of 6×10^{14} sec as the colored line. This scale like the one above it, extends to infinity in the past as well as to infinity in the future.

The average life span of individual man is approximately 70 yr, or 2×10^9 sec. On the mammalian scale this is but a dot, but by expanding the scale ten thousand fold or more, we obtain the scale shown in the third line of the diagram. It is on this colored line that all the events of our entire life may be represented timewise.

The mean life of a free neutron is approximately 16 minutes, or 1000 sec, yet on the time scale of man this is so short an interval that it becomes a mere dot. Since intense beams of neutrons can be produced with atomic accelerators, and atomic reactors, in the laboratory, and many different kinds of events involving them at great distances from their origin are observed, their existence is just as real as the existence of man.

Let us now carry these time scales down several more steps, through the short-lived radioactive isotopes like boron-12, radium C' and thorium C', until we reach the time scale of a π^0 meson, 2.2×10^{-16} sec. Here on a sufficiently expanded time scale the interval looks like all the others. It has a finite length representing creation at the left and radioactive decay on the right. The scale extends infinitely far into the past like all the others, and it extends infinitely far into the future.

If we now step down to the ω particle with a mean life of 10^{-22} sec the time of its existence had to be determined from reasonably well developed theory. Moving with three quarters the speed of light, after its creation, this particle exists long enough to travel five to ten nuclear diameters before decaying into three π mesons. It will therefore be left to you the reader to decide whether particles with such short lifetimes really exist, and whether or not there is a limit to the shortness of any time scale.

QUESTIONS AND PROBLEMS

1. Calculate the threshold energy for the production of a proton pair.

2. Calculate the threshold energy for the production of four pairs of π mesons. See Fig. 35K. *Ans.* 110 Mev.

3. Make a list of all hyperons, and another list of all strange-particles.

4. In Fig. 35M(a), a π^- meson and proton come together and create a K^0 and Λ^0. (a) Show that strangeness-numbers are conserved by writing numbers under the reaction symbols. (b) Could the same two original particles have created a Σ^+ and K^-? *Ans.* (a) $0 + 0 = 1 - 1$. (b) No.

5. Write down the first reaction in Eq. (35g), and then replace each particle by its antiparticle. Do the conservation laws (a) obey the baryon law #5, and (b) the strangeness law #7?

6. Write down the first reaction in Eq. (35h), and then replace each particle by its antiparticle. Do the conservation laws (a) obey the baryon law #5, and (b) the strangeness law #7? *Ans.* (a) yes, (b) yes.

7. Apply the law of conservation of baryons and conservation of strangeness number to show why the strongly reacting particles in Fig. 35M(a) could not have resulted in the reaction $\pi^- + p \rightarrow \Sigma^0 + \Lambda^0$.

8. Apply the laws of conservation of baryon number and strangeness number, and show why the strong reaction in Fig. 35M(a) could not have been $p^- + p \rightarrow \Lambda^0 + K^0$. *Ans.* Baryons not conserved.

9. If the measured distance the K^0 particle travels in Fig. 35M is 2.5 cm, and the speed of the particle is 50 percent the speed of light, find (a) the lifetime of the particle in the laboratory frame of reference, and (b) the real lifetime.

10. If the measured distance a Λ^0 particle travels in a bubble chamber is 9.5 cm, between its creation and its decay into a proton and a π^- meson, and it has a velocity of 65 percent the speed of light, find (a) its laboratory lifetime, and (b) its real lifetime. *Ans.* (a) 4.9×10^{-10} sec. (b) 3.7×10^{-10} sec.

11. If the measured distance the Λ^0 particles traveled in Fig. 35L is 2.5 cm, and the speed of the particles is 45 percent the speed of light, what is their lifetime in (a) the laboratory frame of reference, and (b) in their own frame of reference?

12. If the measured distance a Σ^0 particle travels in a bubble chamber is 120 cm, between its creation and its decay into a Λ^0 and a γ ray, and the particle speed is 75 percent the speed of light, find its lifetime in the laboratory frame of reference, and (b) its real lifetime. *Ans.* (a) 5.3×10^{-9} cm. (b) 3.5×10^{-9} sec.

13. A Σ^+ track in a bubble chamber was measured to be 8 cm long, and from the curvature of its track in a magnetic field, its velocity was 80 percent the speed of light. Find (a) its lifetime in the laboratory, and (b) its real lifetime.

14. The Ξ^- particle in Fig. 35P has a speed of 80 percent the speed of light and exists long enough to travel 5 cm before decaying into a Λ^0 and a π^-. Find its lifetime (a) in the laboratory frame of reference, and (b) its real lifetime. *Ans.* (a) 2.1×10^{-10} sec. (b) 1.3×10^{-10} sec.

15. A Λ^0 particle in Fig. 35P has a speed of 60 percent the speed of light and travels a distance of 3.5 cm before decaying into a proton and a π^- meson. Find its lifetime (a) in the laboratory frame of reference, and (b) its real lifetime.

16. Make a legend diagram of the 3-pronged event shown in Fig. 35R. Label each track with the appropriate particle symbols.

17. An atomic particle is designated Σ^-, 1385, $\frac{3}{2}$. Write down the values of the quantum numbers Y, A, I, J, S, and M, and specify the mass in units of the electronic mass m_e.

18. An atomic particle is designated Ω^-, 1676, $\frac{3}{2}$. Write down the values of the quantum numbers Y, A, I, J, S, and M, and give the particles mass in units of the electronic mass m_e.

19. What are the values of the quantum numbers Y, A, I, J, S, and M for an atomic particle designated $\overline{N}^+$, 1512, $\frac{3}{2}$? Write down its mass in units of the electronic mass m_e.

20. What are the quantum numbers Y, A, I, J, S, and M, for an atomic particle designated $\overline{\Delta}^{-2}$, 1920, $\frac{7}{2}$? Write down the mass in units of the electronic mass m_e.

21. Write down the reaction for the decay of a positive muon, and show that Laws 3, 4, and 6 are not violated.

Appendix I—Relative Atomic Weights

Based on the atomic mass of $C^{12} = 12$ even.
The elements are listed in alphabetical order. See § 5.7.

NAME	SYMBOL	ATOMIC NUMBER	ATOMIC WEIGHT	NAME	SYMBOL	ATOMIC NUMBER	ATOMIC WEIGHT
Actinium	Ac	89	. . .	Mercury	Hg	80	200.59
Aluminium	Al	13	26.9815	Molybdenum	Mo	42	95.94
Americium	Am	95	. . .	Neodymium	Nd	60	144.24
Antimony	Sb	51	121.75	Neon	Ne	10	20.183
Argon	Ar	18	39.948	Neptunium	Np	93	. . .
Arsenic	As	33	74.9216	Nickel	Ni	28	58.71
Astatine	At	85	. . .	Niobium	Nb	41	92.906
Barium	Ba	56	137.34	Nitrogen	N	7	14.0067
Berkelium	Bk	97	. . .	Nobelium	No	102	. . .
Beryllium	Be	4	9.0122	Osmium	Os	76	190.2
Bismuth	Bi	83	208.980	Oxygen	O	8	15.9994^a
Boron	B	5	10.811^a	Palladium	Pd	46	106.4
Bromine	Br	35	79.909^b	Phosphorus	P	15	30.9738
Cadmium	Cd	48	112.40	Platinum	Pt	78	195.09
Cesium	Cs	55	132.905	Plutonium	Pu	94	. . .
Calcium	Ca	20	40.08	Polonium	Po	84	. . .
Californium	Cf	98	. . .	Potassium	K	19	39.102
Carbon	C	6	12.01115^a	Praseodymium	Pr	59	140.907
Cerium	Ce	58	140.12	Promethium	Pm	61	. . .
Chlorine	Cl	17	35.453^b	Protactinium	Pa	91	. . .
Chromium	Cr	24	51.996^b	Radium	Ra	88	. . .
Cobalt	Co	27	58.9332	Radon	Rn	86	. . .
Copper	Cu	29	63.54	Rhenium	Re	75	186.2
Curium	Cm	96	. . .	Rhodium	Rh	45	102.905
Dysprosium	Dy	66	162.50	Rubidium	Rb	37	85.47
Einsteinium	Es	99	. . .	Ruthenium	Ru	44	101.07
Erbium	Er	68	167.26	Samarium	Sm	62	150.35
Europium	Eu	63	151.96	Scandium	Sc	21	44.956
Fermium	Fm	100	. . .	Selenium	Se	34	78.96
Fluorine	F	9	18.9984	Silicon	Si	14	28.086^a
Francium	Fr	87	. . .	Silver	Ag	47	107.870^b
Gadolinium	Gd	64	157.25	Sodium	Na	11	22.9898
Gallium	Ga	31	69.72	Strontium	Sr	38	87.62
Germanium	Ge	32	72.59	Sulfur	S	16	32.064^a
Gold	Au	79	196.967	Tantalum	Ta	73	180.948
Hafnium	Hf	72	178.49	Technetium	Tc	43	. . .
Helium	He	2	4.0026	Tellurium	Te	52	127.60
Holmium	Ho	67	164.930	Terbium	Tb	65	158.924
Hydrogen	H	1	1.00797^a	Thallium	Tl	81	204.37
Indium	In	49	114.82	Thorium	Th	90	232.038
Iodine	I	53	126.9044	Thulium	Tm	69	168.934
Iridium	Ir	77	192.2	Tin	Sn	50	118.69
Iron	Fe	26	55.847^b	Titanium	Ti	22	47.90
Krypton	Kr	36	83.80	Tungsten	W	74	183.85
Lanthanum	La	57	138.91	Uranium	U	92	238.03
Lead	Pb	82	207.19	Vanadium	V	23	50.942
Lithium	Li	3	6.939	Xenon	Xe	54	131.30
Lutetium	Lu	71	174.97	Ytterbium	Yb	70	173.04
Magnesium	Mg	12	24.312	Yttrium	Y	39	88.905
Manganese	Mn	25	54.9380	Zinc	Zn	30	65.37
Mendelevium	Md	101	. . .	Zirconium	Zr	40	91.22

^{a} Atomic weights so designated are known to be variable because of natural variations in isotopic composition. The observed ranges are: Hydrogen ±0.00001; Boron ±0.003; Carbon ±0.00005; Oxygen ±0.0001; Silicon ±0.001; Sulfur ±0.003.

^{b} Atomic weights so designated are believed to have the following experimental uncertainties: Chlorine ±0.001; Chromium ±0.001; Iron ±0.003; Bromine ±0.002; Silver ±0.003.

Appendix II—Relative Masses of One Stable Isotope for Each of the Elements

Based on $O^{16} = 16$ and $C^{12} = 12$. See § 5.7.
These values are taken from Everling, König, Mattauch, and Wapstra, *Nucl. Phys.* 18, 529 (1960); and König, Mattauch, and Wapstra, *Nucl. Phys.* 31, 18 (1962).
Space is left between third and fourth decimal place to facilitate reading.

SYMBOL	$O^{16} = 16$ EVEN	$C^{12} = 12$ EVEN	SYMBOL	$O^{16} = 16$ EVEN	$C^{12} = 12$ EVEN
$_1H^1$	1.008 1456	1.007 8252	$_{51}Sb^{121}$	120.942 190	120.903 750
$_2He^4$	4.003 8761	4.002 6036	$_{52}Te^{128}$	127.945 370	127.904 710
$_3Li^7$	7.018 236	7.016 005	$_{53}I^{127}$	126.944 697	126.904 352
$_4Be^9$	9.015 051	9.012 186	$_{54}Xe^{129}$	128.945 765	128.904 784
$_5B^{11}$	11.012 805	11.009 3051	$_{55}Cs^{133}$	132.947 340	132.905 090
$_6C^{12}$	12.003 8150	12 even	$_{56}Ba^{138}$	137.948 860	137.905 010
$_7N^{14}$	14.007 5262	14.003 0744	$_{57}La^{139}$	138.950 220	138.906 060
$_8O^{16}$	16 even	15.995 9149	$_{58}Ce^{140}$	139.949 760	139.905 280
$_9F^{19}$	19.004 444	18.998 4046	$_{59}Pr^{141}$	140.952 186	140.907 390
$_{10}Ne^{20}$	19.998 7964	19.992 4404	$_{60}Nd^{142}$	141.952 493	141.907 478
$_{11}Na^{23}$	22.997 081	22.989 773	$_{61}Pm^{143}$	142.956 240	143.910 800
$_{12}Mg^{24}$	23.992 670	23.985 045	$_{62}Sm^{144}$	143.957 400	143.911 650
$_{13}Al^{27}$	26.990 113	26.981 535	$_{63}Eu^{153}$	152.969 230	152.920 720
$_{14}Si^{28}$	27.985 822	27.976 927	$_{64}Gd^{156}$	155.971 810	155.922 240
$_{15}P^{31}$	30.983 611	30.973 763	$_{65}Tb^{159}$	158.974 800	158.924 300
$_{16}S^{32}$	31.982 232	31.972 074	$_{66}Dy^{164}$	163.980 200	163.928 100
$_{17}Cl^{35}$	34.979 972	34.968 855	$_{67}Ho^{165}$	164.982 000	164.929 600
$_{18}Ar^{40}$	39.975 089	39.962 384	$_{68}Er^{166}$	165.984 700	165.931 900
$_{19}K^{39}$	38.976 101	38.963 714	$_{69}Tm^{169}$	165.985 500	165.932 700
$_{20}Ca^{40}$	39.975 294	39.962 589	$_{70}Yb^{177}$	176.999 710	176.943 940
$_{21}Sc^{45}$	44.970 811	44.955 916	$_{71}Lu^{177}$	176.998 710	176.942 450
$_{22}Ti^{48}$	47.963 491	47.947 948	$_{72}Hf^{180}$	180.002 330	179.945 120
$_{23}V^{51}$	50.960 174	50.943 978	$_{73}Ta^{181}$	181.003 700	180.646 180
$_{24}Cr^{52}$	51.957 027	51.940 514	$_{74}W^{184}$	184.007 630	183.949 150
$_{25}Mn^{55}$	54.958 123	54.941 293	$_{75}Re^{187}$	187.014 410	186.954 980
$_{26}Fe^{56}$	55.952 714	55.934 932	$_{76}Os^{189}$	189.017 290	188.957 220
$_{27}Co^{59}$	58.951 925	58.933 189	$_{77}Ir^{193}$	193.023 680	192.962 340
$_{28}Ni^{58}$	57.953 761	57.935 342	$_{78}Pt^{195}$	195.026 450	194.964 460
$_{29}Cu^{63}$	62.949 600	62.929 594	$_{79}Au^{197}$	197.029 171	196.966 552
$_{30}Zn^{64}$	63.949 469	63.929 145	$_{80}Hg^{202}$	202.034 840	201.970 630
$_{31}Ga^{69}$	68.947 594	68.925 682	$_{81}Tl^{205}$	205.039 627	204.974 462
$_{32}Ge^{74}$	73.944 660	73.921 150	$_{82}Pb^{206}$	206.039 940	205.974 460
$_{33}As^{75}$	74.945 400	74.921 580	$_{83}Bi^{209}$	209.046 856	208.980 417
$_{34}Se^{80}$	79.941 919	79.916 512	$_{84}Po^{214}$	214.063 224	213.995 192
$_{35}Br^{81}$	80.942 069	80.916 344	$_{85}At^{217}$	217.073 639	217.002 405
$_{36}Kr^{84}$	83.938 181	83.911 504	$_{86}Rn^{222}$	222.085 948	222.015 365
$_{37}Rb^{85}$	84.938 700	84.911 710	$_{87}Fr^{221}$	221.084 440	221.014 176
$_{38}Sr^{88}$	87.933 560	87.905 610	$_{88}Ra^{226}$	226.097 210	226.025 360
$_{39}Y^{89}$	88.933 690	88.905 430	$_{89}Ac^{227}$	227.099 990	227.027 814
$_{40}Zr^{90}$	89.932 900	89.904 320	$_{90}Th^{232}$	232.111 980	232.038 211
$_{41}Nb^{93}$	92.935 550	92.906 020	$_{91}Pa^{234}$	234.117 770	234.043 370
$_{42}Mo^{98}$	97.936 640	97.905 510	$_{92}U^{238}$	238.126 440	238.050 760
$_{43}Tc^{99}$	98.938 400	98.907 300	$_{93}Np^{239}$	239.128 937	239.052 938
$_{44}Ru^{102}$	101.936 110	101.903 720	$_{94}Pu^{240}$	240.130 292	240.053 974
$_{45}Rh^{103}$	102.937 520	102.903 800	$_{96}Am^{241}$	241.133 325	241.056 689
$_{46}Pd^{106}$	105.936 870	105.903 200	$_{96}Cm^{242}$	242.135 750	242.058 800
$_{47}Ag^{107}$	106.938 960	106.904 970	$_{97}Bk^{245}$	245.144 150	245.066 240
$_{48}Cd^{112}$	111.928 410	111.902 840	$_{98}Cf^{246}$	246.147 000	246.068 780
$_{49}In^{115}$	114.940 600	114.904 070	$_{99}Es^{249}$	249.155 400	249.076 220
$_{50}Sn^{120}$	119.940 250	119.902 130	$_{100}Fm^{252}$	252.162 790	252.082 650

Appendix III—Complete List of the Stable Isotopes of the Chemical Elements

Atomic weights are based on the atomic mass of $C^{12} = 12$ even. See § 5.7.

AT. NO.	ELEMENT	SYM.	ISOTOPES, MASS. NO.	AT. WT.
1	hydrogen	H	1, (2)	1.00797[a]
2	helium	He	**4**, (3)	4.0026
3	lithium	Li	6, **7**	6.939
4	beryllium	Be	**9**	9.0122
5	boron	B	10, **11**	10.811[a]
6	carbon	C	**12**, (13)	12.01115[a]
7	nitrogen	N	**14**, (15)	14.0067
8	oxygen	O	**16**, (18), (17)	15.9994[a]
9	fluorine	F	**19**	18.9984
10	neon	Ne	**20**, (21), **22**	20.183
11	sodium	Na	**23**	22.9898
12	magnesium	Mg	**24**, 25, 26	24.312
13	aluminum	Al	**27**	26.9815
14	silicon	Si	**28**, 29, 30	28.086[a]
15	phosphorus	P	**31**	30.9738
16	sulfur	S	**32**, 33, 34	32.064[a]
17	chlorine	Cl	**35**, 37	35.453[b]
18	argon	Ar	(36), (38), **40**	39.948
19	potassium	K	**39**, (40), 41	39.102
20	calcium	Ca	**40**, (42), (43), 44	40.08
21	scandium	Sc	**45**	44.956
22	titanium	Ti	46, 47, **48**, 49, 50	47.90
23	vanadium	V	**51**	50.942
24	chromium	Cr	50, **52**, 53, 54	51.996[b]
25	manganese	Mn	**55**	54.9380
26	iron	Fe	54, **56**, 57, (58)	55.847[b]
27	cobalt	Co	**59**	58.9332
28	nickel	Ni	**58**, 60, 61, 62 (**64**)	58.71
29	copper	Cu	**63**, 65	63.54
30	zinc	Zn	**64**, 66, 67, 68, (70)	65.37
31	gallium	Ga	**69**, 71	69.72
32	germanium	Ge	70, 72, 73, **74**, 76	72.59
33	arsenic	As	**75**	74.9216

[a] Atomic weights so designated are known to be variable because of natural variations in isotopic composition. The observed ranges are: Hydrogen ±0.00001; Boron ±0.003; Carbon ±0.00005; Oxygen ±0.0001; Silicon ±0.001; Sulfur ±0.003.

[b] Atomic weights so designated are believed to have the following experimental uncertainties: Chlorine ±0.001; Chromium ±0.001; Iron ±0.003; Bromine ±0.002; Silver ±0.003.

AT. NO.	ELEMENT	SYM.	ISOTOPES, MASS, NO.	AT. WT.
34	selenium........	Se	(74), 76, 77, 78, **80**, 82	78.96
35	bromine........	Br	**79, 81**	79.909[b]
36	krypton........	Kr	(78), 80, 82, 83, **84**, 86	83.80
37	rubidium......	Rb	**85**, 87	85.47
38	strontium......	Sr	(84), 86, 87, **88**	87.62
39	yttrium........	Yt	**89**	88.905
40	zirconium......	Zr	**90**, 91, 92, 94, 96	91.22
41	niobium.......	Nb	**93**	92.906
42	molybdenum...	Mo	92, 94, 95, 96, 97, **98**, 100, 102	95.94
43	technetium.....	Tc	*99*	97.2
44	ruthenium.....	Ru	96, 98, 99, 100, 101, **102**, 104	101.07
45	rhodium.......	Rh	**103**	102.905
46	palladium......	Pd	(102), 104, 105, **106**, 108, 110	106.4
47	silver..........	Ag	107, **109**	107.870[b]
48	cadmium......	Cd	106, (108), 110, 111, **112**, 113, 114, 116	112.40
49	indium........	In	113, **115**	114.82
50	tin............	Sn	112, (114), (115), 116, 117, 118, 119, **120**, 122, 124	118.69
51	antimony......	Sb	**121, 123**	121.75
52	tellurium......	Te	(120), 122, 123, 124, 125, 126, **128**, 130	127.60
53	iodine.........	I	**127**	126.9044
54	xenon.........	Xe	(124), (126), 128, **129**, 130, 131, 132, 134, 136	131.30
55	cesium........	Cs	**133**	132.905
56	barium........	Ba	(130), (132), 134, 135, 136, 137, **138**	137.34
57	lanthanum.....	La	**139**	138.91
58	cerium........	Ce	(136), (138), **140**, 142	140.12
59	praseodynium..	Pr	**141**	140.906
60	neodymium....	Nd	**142**, 143, **144**, 145, 146, (148), (150)	144.24
61	prometeum....	Pm		146.0
62	samarium......	Sa	144, 147, 148, 149, 150, **152**, 154	150.35
63	europium......	Eu	**151, 153**	151.96
64	gadolinium....	Gd	155, **156**, 157, **158**, 160	157.25
65	terbium.......	Tn	**159**	158.924
66	dysprosium....	Dy	161, 162, 163, **164**	162.50
67	holmium......	Ho	**165**	164.930
68	erbium........	Er	**166**, 167, 168, 170	167.26
69	thulium.......	Tm	**169**	168.934
70	ytterbium......	Yb	171, 172, 173, **174**, 176	173.04
71	lutecium.......	Lu	**175**	174.97
72	hafnium.......	Hf	176, 177, 178, 179, **180**	178.49
73	tantalum......	Ta	**181**	180.948
74	tungsten.......	W	182, 183, **184, 186**	183.85
75	rhenium.......	Re	185, **187**	186.2
76	osmium.......	Os	186, (187), 188, 189, 190, 192	190.2
77	iridium........	Ir	191, **193**	192.2
78	platinum......	Pt	(192), 194, **195**, 196, 198	195.09
79	gold..........	Au	**197**	196.967
80	mercury.......	Hg	(196), 198, 199, 200, 201, **202**, 204	200.59

81	thallium	Tl	210 RaC″	203	205	207 AcC″	208 ThC″
82	lead	Pb	214 RaB	210 RaD	206	211 AcB	207	212 ThB	208	209
83	bismuth	Bi	214 RaC	210 RaE		211 AcC		212 ThC	213	209
84	polonium	Po	218 RaA	214 RaC′	210 Po	215 AcA	211 AcC′	216 ThA	212 ThC′	213
85	astatine	At	α			α	α	217
86	radon	Rn	222 Rn			219 An	220 Tn	α
87	francium	Fa	α			α	α	221
88	radium	Ra	226 Ra			223 AcX	228 MsI	224 ThX	225 α
89	actinium	Ac	α		227 Ac	α 227	α 228 MsII	α	225
90	thorium	Th	230 Io	234 UX₁	231 UY	227 RaAc	232 Th	228 RaTh	229
91	proto-actinium	Pa	α 234 UX₂	α	231 Pa		233	α
92	uranium	U	234 UII	238 U	235 AcU			233
93	neptunium	Np	α	α			237 Np
94	plutonium	Pu	238	236	239	K	241 α
95	americium	Am	α	α	α	239	241
96	curium	Cm	242	240	243

97	berkelium	Bk	247	7000 y half-life
98	californium	Cf	249	470 y
99	einsteinium	Es	253	20 d
100	fermium	Fm	252	36 h
101	mendelevium	Mv	256	1 h
102	nobelium	No	258	—
103	lawrencium	Lw	—	—

Appendix IV—Table of Isotopes and Their Properties

Based upon carbon-12 as 12 even.

SYMBOL	ATOMIC MASS (amu)	MASS EXCESS (Mev)	ABUND. OR ACTIVITY	HALF-LIFE
$_{-1}e^0$	0.000 549	0.511	—	stable
$_1p^1$	1.007 276	6.774	—	stable
$_0n^1$	1.008 665	8.071	β^-	13 m
$_1H^1$	1.007 825	7.288	**99.985**	stable
H^2	2.014 103	13.135	**0.015**	stable
H^3	3.016 049	14.949	β^-	12.2 y
$_2He^3$	3.016 030	14.931	**0.00013**	stable
He^4	4.002 604	2.425	**99.9999**	stable
He^5	5.012 296	11.453	$\alpha + n$	inst.
He^6	6.018 900	17.604	β^-	0.82 s
$_3Li^5$	5.012 541	11.681	$\alpha + p$	inst.
Li^6	6.015 126	14.089	**7.52**	stable
Li^7	7.016 005	14.908	**92.48**	stable
Li^8	8.022 488	20.947	β^-	0.86 s
Li^9	9.027 300	25.400	β^-	0.17 s
$_4Be^6$	6.019 780	18.430	β^+	0.4 s
Be^7	7.016 931	15.770	E.C.	53.6 d
Be^8	8.005 308	4.944	2α	inst.
Be^9	9.012 186	11.350	**100**	stable
Be^{10}	10.013 535	12.607	β^-	2.5×10^6 y
$_5B^8$	8.024 612	22.924	β^+	0.6 s
B^9	9.013 335	12.420	$2\alpha + p$	inst.
B^{10}	10.012 939	12.052	**18.6**	stable
B^{11}	11.009 305	8.667	**81.4**	stable
B^{12}	12.014 353	13.369	β^-	0.022 s
$_6C^{10}$	10.016 830	15.670	β^+	19.1 s
C^{11}	11.011 433	10.649	β^+	20.5 m
C^{12}	12.000 000	0	**98.892**	stable
C^{13}	13.003 354	3.124	**1.108**	stable
C^{14}	14.003 242	3.020	β^-	5.6×10^3 y
C^{15}	15.010 600	9.873	β^-	2.3 s

SYMBOL	ATOMIC MASS (amu)	MASS EXCESS (Mev)	ABUND. OR ACTIVITY	HALF-LIFE
$_7N^{12}$	12.018 709	17.426	β^+	0.012 s
N^{13}	13.005 739	5.345	β^+	10.1 m
N^{14}	14.003 074	2.864	**99.635**	stable
N^{15}	15.000 108	0.101	**0.365**	stable
N^{16}	16.006 089	5.672	β^-	7.36 s
N^{17}	17.008 449	7.869	β^-	4.14 s
$_8O^{14}$	14.008 597	8.008	β^+	74 s
O^{15}	15.003 072	2.861	β^+	2.0 m
O^{16}	15.994 915	−4.736	**99.76**	stable
O^{17}	16.999 133	−0.807	**0.04**	stable
O^{18}	17.999 160	−0.782	**0.20**	stable
O^{19}	19.003 577	3.332	β^-	29.4 s
O^{20}	20.004 071	3.792	n	13.6 s
$_9F^{17}$	17.002 098	1.954	β^+	66 s
F^{18}	18.000 950	0.884	β^+	1.87 h
F^{19}	18.998 405	−1.486	**100**	stable
F^{20}	19.999 986	−0.013	β^-	11.2 s
F^{21}	20.999 972	−0.026	β^-	5 s
$_{10}Ne^{18}$	18.005 715	5.323	β^+	1.6 s
Ne^{19}	19.001 892	1.762	β^+	19 s
Ne^{20}	19.992 440	−7.041	**90.92**	stable
Ne^{21}	20.993 849	−5.729	**0.26**	stable
Ne^{22}	21.991 385	−8.025	**8.82**	stable
Ne^{23}	22.994 475	−5.146	β^-	40.2 s
Ne^{24}	23.993 597	−5.964	β^-	3.4 m
$_{11}Na^{20}$	20.008 890	8.280	β^1	0.3 s
Na^{21}	20.997 638	−2.200	β^+	23 s
Na^{22}	21.994 435	−5.183	β^+, E.C.	2.6 y
Na^{23}	22.989 773	−9.526	**100**	stable
Na^{24}	23.990 967	−8.414	β^-	15.0 h
Na^{25}	24.989 920	−9.390	β^-	60 s
$_{12}Mg^{23}$	22.994 135	−5.463	β^+, E.C.	11 s
Mg^{24}	23.985 045	−13.930	**78.60**	stable
Mg^{25}	24.985 840	−13.189	**10.11**	stable
Mg^{26}	25.982 591	−16.215	**11.29**	stable
Mg^{27}	26.998 346	−14.581	β^-	9.45 m
Mg^{28}	27.983 880	−15.015	β^-	21.4 h
$_{13}Al^{24}$	24.000 090	0.090	β^+	2.1 s
Al^{25}	24.990 414	−8.928	β^+	7.6 s
Al^{26}	25.986 900	−12.201	β^+, E.C.	10^5 y
Al^{27}	26.981 535	−17.199	**100**	stable
Al^{28}	27.981 908	−16.851	β^-	2.3 m
Al^{29}	28.980 442	−18.217	β^-	6.6 m
$_{14}Si^{27}$	26.986 701	−12.387	β^+, E.C.	4.9 s
Si^{28}	27.976 927	−21.491	**92.27**	stable
Si^{29}	28.976 491	−21.897	**4.68**	stable
Si^{30}	29.973 761	−24.440	**3.05**	stable
Si^{31}	30.975 349	−22.961	β^-	2.6 h
Si^{32}	31.974 020	−24.200	β^-	7×10^2 y

SYMBOL	ATOMIC MASS (amu)	MASS EXCESS (Mev)	ABUND. OR ACTIVITY	HALF-LIFE
$_{15}P^{28}$	27.991 740	−7.690	β^+	0.28 s
P^{29}	28.981 816	−16.937	β^+	4.4 s
P^{30}	29.978 320	−20.193	β^+	2.5 m
P^{31}	30.973 763	−24.438	**100**	stable
P^{32}	31.973 908	−24.303	β^-	14.3 d
P^{33}	32.971 728	−26.334	β^-	24.4 d
P^{34}	33.973 340	−24.830	β^-	12.4 s
$_{16}S^{31}$	30.979 599	−19.002	β^+	2.6 s
S^{32}	31.972 074	−26.012	**95.054**	stable
S^{33}	32.971 461	−26.583	**0.740**	stable
S^{34}	33.967 865	−29.932	**4.190**	stable
S^{35}	34.969 034	−28.843	β^-	87.1 d
S^{36}	35.967 091	−30.653	**0.016**	stable
S^{37}	36.971 040	−26.980	β^-	5.0 m
S^{38}	37.971 220	−26.800	β^-	6 m
$_{17}Cl^{32}$	31.986 030	−13.010	β^+	0.31 s
Cl^{33}	32.977 446	−21.008	β^+	2.4 s
Cl^{34}	33.973 764	−24.437	β^+	32.4 m
Cl^{35}	34.968 855	−29.010	**75.4**	stable
Cl^{36}	35.968 312	−25.516	β^-, E.C.	3.2 y
Cl^{37}	36.965 898	−31.766	**24.6**	stable
Cl^{38}	37.968 002	−29.804	β^-	37.5 m
Cl^{39}	38.968 003	−29.803	β^-	55.5 m
Cl^{40}	39.970 400	−27.500	β^-	1.4 m
$_{18}Ar^{35}$	34.975 275	−23.030	β^+	1.83 s
Ar^{36}	35.963 548	−30.227	**0.337**	stable
Ar^{37}	36.966 772	−30.950	E.C.	34.1 d
Ar^{38}	37.962 725	−34.720	**0.063**	stable
Ar^{39}	38.964 321	−33.233	β^-	2.6×10^2 y
Ar^{40}	39.962 384	−35.037	**99.600**	stable
Ar^{41}	40.964 508	−33.058	β^-	1.83 h
Ar^{42}	41.963 043	−34.423	β^-	3.5 y
$_{19}K^{38}$	37.969 090	−28.791	β^+	7.7 m
K^{39}	38.963 714	−33.798	**93.08**	stable
K^{40}	39.964 008	−33.524	**0.0119**	10^9 y
K^{41}	40.961 835	−35.548	**6.91**	stable
K^{42}	41.962 417	−35.006	β^-	12.5 h
K^{43}	42.960 731	−36.577	β^-	22.4 h
K^{44}	43.962 040	−35.360	β^-	22.0 m
$_{20}Ca^{39}$	38.970 706	−27.286	β^+	1.0 s
Ca^{40}	39.962 589	−34.846	**96.97**	stable
Ca^{41}	40.962 279	−35.135	E.C.	2×10^5 y
Ca^{42}	41.958 628	−38.536	**0.64**	stable
Ca^{43}	42.958 780	−38.394	**0.145**	stable
Ca^{44}	43.955 490	−41.458	**2.06**	stable
Ca^{45}	44.956 189	−40.807	β^-	164 d
Ca^{46}	45.953 689	−43.136	**0.0033**	stable
Ca^{47}	46.954 512	−42.370	β^-	4.9 d
Ca^{48}	47.952 363	−44.371	**0.185**	10^{16} y
Ca^{49}	48.955 662	−41.298	β^-	8.8 m

For atomic masses in *amu* and mass excess energies in *Mev*, for isotopes in the remainder of the periodic table, see *Nuclear Physics*, March-April, pp. 28–42, 1962, by L. A. König, J. H. E. Mattauch, and A. H. Wapstra.

Appendix V—Values of the General Physical Constants (after Du Mond)

Planck's constant of action. . . . . $h = 6.6238 \times 10^{-34}$ joule sec
Electronic charge. . . . . $e = 1.6019 \times 10^{-19}$ coulombs
Electronic charge. . . . . $e = 4.8022 \times 10^{-10}$ e.s.u.
Specific electronic charge. . . . . $e/m = 1.7598 \times 10^{11}$ coulombs/Kg
Specific proton charge. . . . . $e/M_p = 9.5795 \times 10^{7}$ coulombs/Kg
Electronic mass. . . . . $m = 9.1072 \times 10^{-31}$ Kg
Mass of atom of unit atomic weight. . . . . $M = 1.6600 \times 10^{-27}$ Kg
Mass of proton. . . . . $M_p = 1.6722 \times 10^{-27}$ Kg
Ratio mass proton to mass electron. . . . . $M_p/m = 1836.1$
Wien's displacement-law constant. . . . . $C = 0.28976$ cm deg
Velocity of light. . . . . $c = 299,790$ Km/sec
Square of velocity of light. . . . . $c^2 = 8.9874 \times 10^{10}$ Km²/sec²
Unit atomic angular momentum. . . . . $\hbar = 1.0544 \times 10^{-34}$ joule sec

Appendix VI—Electron Subshells

The table shows the order in which the electron subshells are filled in, in the building up of the elements of the periodic table. Atomic weights are given with respect to carbon-12 as 12 even.

n+1	Sub-Shells	1	2	3	4	5	6	7	8	9	10	11	12	13	14
1	1s	1.0080 H 1	4.003 He 2												
2	2s	6.939 Li 3	9.012 Be 4												
3	2p	10.81 B 5	12.000 C 6	14.007 N 7	15.999 O 8	18.998 F 9	20.183 Ne 10								
3	3s	22.990 Na 11	24.312 Mg 12												
4	3p	26.98 Al 13	28.08 Si 14	30.974 P 15	32.064 S 16	35.453 Cl 17	39.948 Ar 18								
4	4s	39.102 K 19	40.08 Ca 20												
5	3d	44.96 Sc 21	47.90 Ti 22	50.94 V 23	51.996 Cr 24	54.94 Mn 25	55.85 Fe 26	58.93 Co 27	58.71 Ni 28	63.54 Cu 29	65.37 Zn 30				
5	4p	69.72 Ga 31	72.59 Ge 32	74.92 As 33	78.96 Se 34	79.91 Br 35	83.80 Kr 36								
5	5s	85.47 Rb 37	87.62 Sr 38												
6	4d	88.90 Y 39	91.22 Zr 40	92.91 Nb 41	95.94 Mo 42	(99) Tc 43	101.1 Ru 44	102.91 Rh 45	106.4 Pd 46	107.870 Ag 47	112.40 Cd 48				
6	5p	114.82 In 49	118.69 Sn 50	121.75 Sb 51	127.60 Te 52	126.90 I 53	131.3 Xe 54								
6	6s	132.90 Cs 55	137.34 Ba 56												
7	4f	138.91 La 57	140.12 Ce 58	140.91 Pr 59	144.24 Nd 60	(145) Pm 61	150.35 Sm 62	151.96 Eu 63	157.25 Gd 64	158.92 Tb 65	162.50 Dy 66	164.93 Ho 67	167.2 Er 68	168.93 Tm 69	173.04 Yb 70
7	5d	174.97 Lu 71	178.5 Hf 72	180.95 Ta 73	183.92 W 74	186.2 Re 75	190.2 Os 76	192.2 Ir 77	195.09 Pt 78	196.97 Au 79	200.59 Hg 80				
7	6p	204.37 Tl 81	207.19 Pb 82	208.98 Bi 83	210 Po 84	(210) At 85	222 Rn 86								
7	7s	(223) Fr 87	226.05 Ra 88												
8	5f	227 Ac 89	232.04 Th 90	231 Pa 91	238.03 U 92	(237) Np 93	(242) Pu 94	(243) Am 95	(245) Cm 96	(245) Bk 97	(248) Cf 98	(253) Es 99	(254) Fm 100	(256) Mv 101	No 102
8	6d	Lw 103	104	105	106	107	108	109	110	111	112				

Appendix VII—The Periodic Table of Chemical Elements

I	II	III	IV	V	VI	VII	VIII		
1 H 1.0080							2 He 4.0026		
3 Li 6.939	4 Be 9.0122	5 B 10.811	6 C 12.011	7 N 14.007	8 O 15.999	9 F 18.998	10 Ne 20.183		
11 Na 22.990	12 Mg 24.312	13 Al 26.981	14 Si 28.086	15 P 30.974	16 S 32.064	17 Cl 35.453	18 Ar 39.948		
19 K 39.102	20 Ca 40.08	21 Sc 44.956	22 Ti 47.90	23 V 50.942	24 Cr 51.996	25 Mn 54.938	26 Fe 55.847	27 Co 58.933	28 Ni 58.71
29 Cu 63.54	30 Zn 65.37	31 Ga 69.72	32 Ge 72.59	33 As 74.922	34 Se 78.96	35 Br 79.909	36 Kr 83.80		
37 Rb 85.47	38 Sr 87.62	39 Yt 88.905	40 Zr 91.22	41 Nb 92.906	42 Mo 95.94	43 Tc ...	44 Ru 101.07	45 Rh 102.905	46 Pd 106.4
47 Ag 107.870	48 Cd 112.40	49 In 114.82	50 Sn 118.69	51 Sb 121.75	52 Te 127.60	53 I 126.904	54 Xe 131.30		
55 Cs 132.905	56 Ba 137.34	57 La 138.91	72 Hf 178.49	73 Ta 180.95	74 W 183.85	75 Re 186.2	76 Os 190.2	77 Ir 192.2	78 Pt 195.09
79 Au 196.97	80 Hg 200.59	81 Tl 204.37	82 Pb 207.19	83 Bi 208.98	84 Po ...	85 At ...	86 Rn ...		
87 Fr ...	88 Ra ...	89 Ac ...	90 Th 232.04	91 Pa ...	92 U 238.03	93 Np ...	94 Pu ...	95 Am ...	96 Cm ...

58 Ce 140.12	59 Pr 140.91	60 Nd 144.24	61 Pm ...	62 Sm 150.35	63 Eu 151.96	64 Gd 157.25	65 Tn 139.2	66 Ds 162.50	67 Ho 164.93	68 Er 167.26	69 Tm 168.93	70 Yb 173.04	71 Lu 174.97
90 Th 232.04	91 Pa ...	92 U 238.03	93 Np ...	94 Pu ...	95 Am ...	96 Cm ...	97 Bk ...	98 Cf ...	99 E ...	100 Fm ...	101 Md ...		

Atomic weights are given on the carbon-12 equal 12 even.

Appendix VIII—Four-Place Tables of Common Logarithms and Natural Trigonometric Functions

APPENDICES
COMMON LOGARITHMS

(To obtain Naperian logarithm of a number multiply these logarithms by 2.3026.)

N	0	1	2	3	4	5	6	7	8	9
0	....	0000	3010	4771	6021	6990	7782	8451	9031	9542
1	0000	0414	0792	1139	1461	1761	2041	2304	2553	2788
2	3010	3222	3424	3617	3802	3979	4150	4314	4472	4624
3	4771	4914	5051	5185	5315	5441	5563	5682	5798	5911
4	6021	6128	6232	6335	6435	6532	6628	6721	6812	6902
5	6990	7076	7160	7243	7324	7404	7482	7559	7634	7709
6	7782	7853	7924	7993	8062	8129	8195	8261	8325	8388
7	8451	8513	8573	8633	8692	8751	8808	8865	8921	8976
8	9031	9085	9138	9191	9243	9294	9345	9395	9445	9494
9	9542	9590	9638	9685	9731	9777	9823	9868	9912	9956
10	0000	0043	0086	0128	0170	0212	0253	0294	0334	0374
11	0414	0453	0492	0531	0569	0607	0645	0682	0719	0755
12	0792	0828	0864	0899	0934	0969	1004	1038	1072	1106
13	1139	1173	1206	1239	1271	1303	1335	1367	1399	1430
14	1461	1492	1523	1553	1584	1614	1644	1673	1703	1732
15	1761	1790	1818	1847	1875	1903	1931	1959	1987	2014
16	2041	2068	2095	2122	2148	2175	2201	2227	2253	2279
17	2304	2330	2355	2380	2405	2430	2455	2480	2504	2529
18	2553	2577	2601	2625	2648	2672	2695	2718	2742	2765
19	2788	2810	2833	2856	2878	2900	2923	2945	2967	2989
20	3010	3032	3054	3075	3096	3118	3139	3160	3181	3201
21	3222	3243	3263	3284	3304	3324	3345	3365	3385	3404
22	3424	3444	3464	3483	3502	3522	3541	3560	3579	3598
23	3617	3636	3655	3674	3692	3711	3729	3747	3766	3784
24	3802	3820	3838	3856	3874	3892	3909	3927	3945	3962
25	3979	3997	4014	4031	4048	4065	4082	4099	4116	4133
26	4150	4166	4183	4200	4216	4232	4249	4265	4281	4298
27	4314	4330	4346	4362	4378	4393	4409	4425	4440	4456
28	4472	4487	4502	4518	4533	4548	4564	4579	4594	4609
29	4624	4639	4654	4669	4683	4698	4713	4728	4742	4757
30	4771	4786	4800	4814	4829	4843	4857	4871	4886	4900
31	4914	4928	4942	4955	4969	4983	4997	5011	5024	5038
32	5051	5065	5079	5092	5105	5119	5132	5145	5159	5172
33	5185	5198	5211	5224	5237	5250	5263	5276	5289	5302
34	5315	5328	5340	5353	5366	5378	5391	5403	5416	5428
35	5441	5453	5465	5478	5490	5502	5514	5527	5539	5551
36	5563	5575	5587	5599	5611	5623	5635	5647	5658	5670
37	5682	5694	5705	5717	5729	5740	5752	5763	5775	5786
38	5798	5809	5821	5832	5843	5855	5866	5877	5888	5899
39	5911	5922	5933	5944	5955	5966	5977	5988	5999	6010
40	6021	6031	6042	6053	6064	6075	6085	6096	6107	6117
41	6128	6138	6149	6160	6170	6180	6191	6201	6212	6222
42	6232	6243	6253	6263	6274	6284	6294	6304	6314	6325
43	6335	6345	6355	6365	6375	6385	6395	6405	6415	6425
44	6435	6444	6454	6464	6474	6484	6493	6503	6513	6522
45	6532	6542	6551	6561	6571	6580	6590	6599	6609	6618
46	6628	6637	6646	6656	6665	6675	6684	6693	6702	6712
47	6721	6730	6739	6749	6758	6767	6776	6785	6794	6803
48	6812	6821	6830	6839	6848	6857	6866	6875	6884	6893
49	6902	6911	6920	6928	6937	6946	6955	6964	6972	6981
50	6990	6998	7007	7016	7024	7033	7042	7050	7059	7067
N	0	1	2	3	4	5	6	7	8	9

COMMON LOGARITHMS (Cont.)

N	0	1	2	3	4	5	6	7	8	9
50	6990	6998	7007	7016	7024	7033	7042	7050	7059	7067
51	7076	7084	7093	7101	7110	7118	7126	7135	7143	7152
52	7160	7168	7177	7185	7193	7202	7210	7218	7226	7235
53	7243	7251	7259	7267	7275	7284	7292	7300	7308	7316
54	7324	7332	7340	7348	7356	7364	7372	7380	7388	7396
55	7404	7412	7419	7427	7435	7443	7451	7459	7466	7474
56	7482	7490	7497	7505	7513	7520	7528	7536	7543	7551
57	7559	7566	7574	7582	7589	7597	7604	7612	7619	7627
58	7634	7642	7649	7657	7664	7672	7679	7686	7694	7701
59	7709	7716	7723	7731	7738	7745	7752	7760	7767	7774
60	7782	7789	7796	7803	7810	7818	7825	7832	7839	7846
61	7853	7860	7868	7875	7882	7889	7896	7903	7910	7917
62	7924	7931	7938	7945	7952	7959	7966	7973	7980	7987
63	7993	8000	8007	8014	8021	8028	8035	8041	8048	8055
64	8062	8069	8075	8082	8089	8096	8102	8109	8116	8122
65	8129	8136	8142	8149	8156	8162	8169	8176	8182	8189
66	8195	8202	8209	8215	8222	8228	8235	8241	8248	8254
67	8261	8267	8274	8280	8287	8293	8299	8306	8312	8319
68	8325	8331	8338	8344	8351	8357	8363	8370	8376	8382
69	8388	8395	8401	8407	8414	8420	8426	8432	8439	8445
70	8451	8457	8463	8470	8476	8482	8488	8494	8500	8506
71	8513	8519	8525	8531	8537	8543	8549	8555	8561	8567
72	8573	8579	8585	8591	8597	8603	8609	8615	8621	8627
73	8633	8639	8645	8651	8657	8663	8669	8675	8681	8686
74	8692	8698	8704	8710	8716	8722	8727	8733	8739	8745
75	8751	8756	8762	8768	8774	8779	8785	8791	8797	8802
76	8808	8814	8820	8825	8831	8837	8842	8848	8854	8859
77	8865	8871	8876	8882	8887	8893	8899	8904	8910	8915
78	8921	8927	8932	8938	8943	8949	8954	8960	8965	8971
79	8976	8982	8987	8993	8998	9004	9009	9015	9020	9025
80	9031	9036	9042	9047	9053	9058	9063	9069	9074	9079
81	9085	9090	9096	9101	9106	9112	9117	9122	9128	9133
82	9138	9143	9149	9154	9159	9165	9170	9175	9180	9186
83	9191	9196	9201	9206	9212	9217	9222	9227	9232	9238
84	9243	9248	9253	9258	9263	9269	9274	9279	9284	9289
85	9294	9299	9304	9309	9315	9320	9325	9330	9335	9340
86	9345	9350	9355	9360	9365	9370	9375	9380	9385	9390
87	9395	9400	9405	9410	9415	9420	9425	9430	9435	9440
88	9445	9450	9455	9460	9465	9469	9474	9479	9484	9489
89	9494	9499	9504	9509	9513	9518	9523	9528	9533	9538
90	9542	9547	9552	9557	9562	9566	9571	9576	9581	9586
91	9590	9595	9600	9605	9609	9614	9619	9624	9628	9633
92	9638	9643	9647	9652	9657	9661	9666	9671	9675	9680
93	9685	9689	9694	9699	9703	9708	9713	9717	9722	9727
94	9731	9736	9741	9745	9750	9754	9759	9763	9768	9773
95	9777	9782	9786	9791	9795	9800	9805	9809	9814	9818
96	9823	9827	9832	9836	9841	9845	9850	9854	9859	9863
97	9868	9872	9877	9881	9886	9890	9894	9899	9903	9908
98	9912	9917	9921	9926	9930	9934	9939	9943	9948	9952
99	9956	9961	9965	9969	9974	9978	9983	9987	9991	9996
100	0000	0004	0009	0013	0017	0022	0026	0030	0035	0039
N	0	1	2	3	4	5	6	7	8	9

NATURAL SINES

Angle	0′	6′	12′	18′	24′	30′	36′	42′	48′	54′	1′	2′	3′	4′	5′
											Proportional Parts				
0°	.0000	.0017	.0035	.0052	.0070	.0087	.0105	.0122	.0140	.0157	3	6	9	12	15
1	.0175	.0192	.0209	.0227	.0244	.0262	.0279	.0297	.0314	.0332	3	6	9	12	15
2	.0349	.0366	.0384	.0401	.0419	.0436	.0454	.0471	.0488	.0506	3	6	9	12	15
3	.0523	.0541	.0558	.0576	.0593	.0610	.0628	.0645	.0663	.0680	3	6	9	12	15
4	.0698	.0715	.0732	.0750	.0767	.0785	.0802	.0819	.0837	.0854	3	6	9	12	14
5	.0872	.0889	.0906	.0924	.0941	.0958	.0976	.0993	.1011	.1028	3	6	9	12	14
6	.1045	.1063	.1080	.1097	.1115	.1132	.1149	.1167	.1184	.1201	3	6	9	12	14
7	.1219	.1236	.1253	.1271	.1288	.1305	.1323	.1340	.1357	.1374	3	6	9	12	14
8	.1392	.1409	.1426	.1444	.1461	.1478	.1495	.1513	.1530	.1547	3	6	9	12	14
9	.1564	.1582	.1599	.1616	.1633	.1650	.1668	.1685	.1702	.1719	3	6	9	11	14
10°	.1736	.1754	.1771	.1788	.1805	.1822	.1840	.1857	.1874	.1891	3	6	9	11	14
11	.1908	.1925	.1942	.1959	.1977	.1994	.2011	.2028	.2045	.2062	3	6	9	11	14
12	.2079	.2096	.2113	.2130	.2147	.2164	.2181	.2198	.2215	.2233	3	6	9	11	14
13	.2250	.2267	.2284	.2300	.2317	.2334	.2351	.2368	.2385	.2402	3	6	8	11	14
14	.2419	.2436	.2453	.2470	.2487	.2504	.2521	.2538	.2554	.2571	3	6	8	11	14
15	.2588	.2605	.2622	.2639	.2656	.2672	.2689	.2706	.2723	.2740	3	6	8	11	14
16	.2756	.2773	.2790	.2807	.2823	.2840	.2857	.2874	.2890	.2907	3	6	8	11	14
17	.2924	.2940	.2957	.2974	.2990	.3007	.3024	.3040	.3057	.3074	3	6	8	11	14
18	.3090	.3107	.3123	.3140	.3156	.3173	.3190	.3206	.3223	.3239	3	6	8	11	14
19	.3256	.3272	.3289	.3305	.3322	.3338	.3355	.3371	.3387	.3404	3	5	8	11	14
20°	.3420	.3437	.3453	.3469	.3486	.3502	.3518	.3535	.3551	.3567	3	5	8	11	14
21	.3584	.3600	.3616	.3633	.3649	.3665	.3681	.3697	.3714	.3730	3	5	8	11	14
22	.3746	.3762	.3778	.3795	.3811	.3827	.3843	.3859	.3875	.3891	3	5	8	11	13
23	.3907	.3923	.3939	.3955	.3971	.3987	.4003	.4019	.4035	.4051	3	5	8	11	13
24	.4067	.4083	.4099	.4115	.4131	.4147	.4163	.4179	.4195	.4210	3	5	8	11	13
25	.4226	.4242	.4258	.4274	.4289	.4305	.4321	.4337	.4352	.4368	3	5	8	11	13
26	.4384	.4399	.4415	.4431	.4446	.4462	.4478	.4493	.4509	.4524	3	5	8	10	13
27	.4540	.4555	.4571	.4586	.4602	.4617	.4633	.4648	.4664	.4679	3	5	8	10	13
28	.4695	.4710	.4726	.4741	.4756	.4772	.4787	.4802	.4818	.4833	3	5	8	10	13
29	.4848	.4863	.4879	.4894	.4909	.4924	.4939	.4955	.4970	.4985	3	5	8	10	13
30°	.5000	.5015	.5030	.5045	.5060	.5075	.5090	.5105	.5120	.5135	3	5	8	10	13
31	.5150	.5165	.5180	.5195	.5210	.5225	.5240	.5255	.5270	.5284	2	5	7	10	12
32	.5299	.5314	.5329	.5344	.5358	.5373	.5388	.5402	.5417	.5432	2	5	7	10	12
33	.5446	.5461	.5476	.5490	.5505	.5519	.5534	.5548	.5563	.5577	2	5	7	10	12
34	.5592	.5606	.5621	.5635	.5650	.5664	.5678	.5693	.5707	.5721	2	5	7	10	12
35	.5736	.5750	.5764	.5779	.5793	.5807	.5821	.5835	.5850	.5864	2	5	7	9	12
36	.5878	.5892	.5906	.5920	.5934	.5948	.5962	.5976	.5990	.6004	2	5	7	9	12
37	.6018	.6032	.6046	.6060	.6074	.6088	.6101	.6115	.6129	.6143	2	5	7	9	12
38	.6157	.6170	.6184	.6198	.6211	.6225	.6239	.6252	.6266	.6280	2	5	7	9	11
39	.6293	.6307	.6320	.6334	.6347	.6361	.6374	.6388	.6401	.6414	2	4	7	9	11
40°	.6428	.6441	.6455	.6468	.6481	.6494	.6508	.6521	.6534	.6547	2	4	7	9	11
41	.6561	.6574	.6587	.6600	.6613	.6626	.6639	.6652	.6665	.6678	2	4	7	9	11
42	.6691	.6704	.6717	.6730	.6743	.6756	.6769	.6782	.6794	.6807	2	4	6	9	11
43	.6820	.6833	.6845	.6858	.6871	.6884	.6896	.6909	.6921	.6934	2	4	6	8	11
44	.6947	.6959	.6972	.6984	.6997	.7009	.7022	.7034	.7046	.7059	2	4	6	8	10
	0′	6′	12′	18′	24′	30′	36′	42′	48′	54′	1′	2′	3′	4′	5′

NATURAL SINES (Cont.)

Angle	0′	6′	12′	18′	24′	30′	36′	42′	48′	54′	1′	2′	3′	4′	5′
											Proportional Parts				
45°	.7071	.7083	.7096	.7108	.7120	.7133	.7145	.7157	.7169	.7181	2	4	6	8	10
46	.7193	.7206	.7218	.7230	.7242	.7254	.7266	.7278	.7290	.7302	2	4	6	8	10
47	.7314	.7325	.7337	.7349	.7361	7373	.7385	.7396	.7408	.7420	2	4	6	8	10
48	.7431	.7443	.7455	.7466	.7478	.7490	.7501	.7513	.7524	.7536	2	4	6	8	10
49	.7547	.7559	.7570	.7581	.7593	.7604	.7615	.7627	.7638	.7649	2	4	6	8	9
50°	.7660	.7672	.7683	.7694	.7705	.7716	.7727	.7738	.7749	.7760	2	4	6	7	9
51	.7771	.7782	.7793	.7804	.7815	.7826	.7837	.7848	.7859	.7869	2	4	5	7	9
52	.7880	.7891	.7902	.7912	.7923	.7934	.7944	.7955	.7965	.7976	2	4	5	7	9
53	.7986	.7997	.8007	.8018	.8028	.8039	.8049	.8059	.8070	.8080	2	3	5	7	9
54	.8090	.8100	.8111	.8121	.8131	.8141	.8151	.8161	.8171	.8181	2	3	5	7	8
55	.8192	.8202	.8211	.8221	.8231	.8241	.8251	.8261	.8271	.8281	2	3	5	7	8
56	.8290	.8300	.8310	.8320	.8329	.8339	.8348	.8358	.8368	.8377	2	3	5	6	8
57	.8387	.8396	.8406	.8415	.8425	.8434	.8443	.8453	.8462	.8471	2	3	5	6	8
58	.8480	.8490	.8499	.8508	.8517	.8526	.8536	.8545	.8554	.8563	2	3	5	6	8
59	.8572	.8581	.8590	.8599	.8607	.8616	.8625	.8634	.8643	.8652	1	3	4	6	7
60°	.8660	.8669	.8678	.8686	.8695	.8704	.8712	.8721	.8729	.8738	1	3	4	6	7
61	.8746	.8755	.8763	.8771	.8780	.8788	.8796	.8805	.8813	.8821	1	3	4	6	7
62	.8829	.8838	.8846	.8854	.8862	.8870	.8878	.8886	.8894	.8902	1	3	4	5	7
63	.8910	.8918	.8926	.8934	.8942	.8949	.8957	.8965	.8973	.8980	1	3	4	5	6
64	.8988	.8996	.9003	.9011	.9018	.9026	.9033	.9041	.9048	.9056	1	3	4	5	6
65	.9063	.9070	.9078	.9085	.9092	.9100	.9107	.9114	.9121	.9128	1	2	4	5	6
66	.9135	.9143	.9150	.9157	.9164	.9171	.9178	.9184	.9191	.9198	1	2	3	5	6
67	.9205	.9212	.9219	.9225	.9232	.9239	.9245	.9252	.9259	.9265	1	2	3	4	6
68	.9272	.9278	.9285	.9291	.9298	.9304	.9311	.9317	.9323	.9330	1	2	3	4	5
69	.9336	.9342	.9348	.9354	.9361	.9367	.9373	.9379	.9385	.9391	1	2	3	4	5
70°	.9397	.9403	.9409	.9415	.9421	.9426	.9432	.9438	.9444	.9449	1	2	3	4	5
71	.9455	.9461	.9466	.9472	.9478	.9483	.9489	.9494	.9500	.9505	1	2	3	4	5
72	.9511	.9516	.9521	.9527	.9532	.9537	.9542	.9548	.9553	.9558	1	2	3	4	4
73	.9563	.9568	.9573	.9578	.9583	.9588	.9593	.9598	.9603	.9608	1	2	2	3	4
74	.9613	.9617	.9622	.9627	.9632	.9636	.9641	.9646	.9650	.9655	1	2	2	3	4
75	.9659	.9664	.9668	.9673	.9677	.9681	.9686	.9690	.9694	.9699	1	1	2	3	4
76	.9703	.9707	.9711	.9715	.9720	9724	.9728	.9732	.9736	.9740	1	1	2	3	3
77	.9744	.9748	.9751	.9755	.9759	.9763	.9767	.9770	.9774	.9778	1	1	2	3	3
78	.9781	.9785	.9789	.9792	.9796	.9799	.9803	.9806	.9810	.9813	1	1	2	2	3
79	.9816	.9820	.9823	.9826	.9829	.9833	.9836	.9839	.9842	.9845	1	1	2	2	3
80°	.9848	.9851	.9854	.9857	.9860	.9863	.9866	.9869	.9871	.9874	0	1	1	2	2
81	.9877	.9880	.9882	.9885	.9888	.9890	.9893	.9895	.9898	.9900	0	1	1	2	2
82	.9903	.9905	.9907	.9910	.9912	.9914	.9917	.9919	.9921	.9923	0	1	1	2	2
83	.9925	.9928	.9930	.9932	.9934	.9936	.9938	.9940	.9942	.9943	0	1	1	1	2
84	.9945	.9947	.9949	.9951	.9952	.9954	.9956	.9957	.9959	.9960	0	1	1	1	1
85	.9962	.9963	.9965	.9966	.9968	.9969	.9971	.9972	.9973	.9974	0	0	1	1	1
86	.9976	.9977	.9978	.9979	.9980	.9981	.9982	.9983	.9984	.9985	0	0	1	1	1
87	.9986	.9987	.9988	.9989	.9990	.9990	.9991	.9992	.9993	.9993					
88	.9994	.9995	.9995	.9996	.9996	.9997	.9997	.9997	.9998	.9998					
89	.9998	.9999	.9999	.9999	.9999	1.000	1.000	1.000	1.000	1.000					
	0′	6′	12′	18′	24′	30′	36′	42′	48′	54′	1′	2′	3′	4′	5′

NATURAL COSINES

SUBTRACT

Angle	0'	6'	12'	18'	24'	30'	36'	42'	48'	54'	Proportional Parts				
											1'	2'	3'	4'	5'
0°	1.0000	1.000	1.000	1.000	1.000	1.000	**.9999**	**.9999**	**.9999**	**.9999**					
1	.9998	.9998	.9998	.9997	.9997	.9997	.9996	.9996	.9995	.9995					
2	.9994	.9993	.9993	.9992	.9991	.9990	.9990	.9989	.9988	.9987					
3	.9986	.9985	.9984	.9983	.9982	.9981	.9980	.9979	.9978	.9977	0	0	1	1	1
4	.9976	.9974	.9973	.9972	.9971	.9969	.9968	.9966	.9965	.9963	0	0	1	1	1
5	.9962	.9960	.9959	.9957	.9956	.9954	.9952	.9951	.9949	.9947	0	1	1	1	1
6	.9945	.9943	.9942	.9940	.9938	.9936	.9934	.9932	.9930	.9928	0	1	1	1	2
7	.9925	.9923	.9921	.9919	.9917	.9914	.9912	.9910	.9907	.9905	0	1	1	2	2
8	.9903	.9900	.9898	.9895	.9893	.9890	.9888	.9885	.9882	.9880	0	1	1	2	2
9	.9877	.9874	.9871	.9869	.9866	.9863	.9860	.9857	.9854	.9851	0	1	1	2	2
10°	.9848	.9845	.9842	.9839	.9836	.9833	.9829	.9826	.9823	.9820	1	1	2	2	3
11	.9816	.9813	.9810	.9806	.9803	.9799	.9796	.9792	.9789	.9785	1	1	2	2	3
12	.9781	.9778	.9774	.9770	.9767	.9763	.9759	.9755	.9751	.9748	1	1	2	3	3
13	.9744	.9740	.9736	.9732	.9728	.9724	.9720	.9715	.9711	.9707	1	1	2	3	3
14	.9703	.9699	.9694	.9690	.9686	.9681	,9677	.9673	.9668	.9664	1	1	2	3	4
15	.9659	.9655	.9650	.9646	.9641	.9636	.9632	.9627	.9622	.9617	1	2	2	3	4
16	.9613	.9608	.9603	.9598	.9593	.9588	.9583	.9578	.9573	.9568	1	2	2	3	4
17	.9563	.9558	.9553	.9548	.9542	.9537	.9532	.9527	.9521	.9516	1	2	3	4	4
18	.9511	.9505	.9500	.9494	.9489	.9483	.9478	.9472	.9466	.9461	1	2	3	4	5
19	.9455	.9449	.9444	.9438	.9432	.9426	.9421	.9415	.9409	.9403	1	2	3	4	5
20°	.9397	.9391	.9385	.9379	.9373	.9367	.9361	.9354	.9348	.9342	1	2	3	4	5
21	.9336	.9330	.9323	.9317	.9311	.9304	.9298	.9291	.9285	.9278	1	2	3	4	5
22	.9272	.9265	.9259	.9252	.9245	.9239	.9232	.9225	.9219	.9212	1	2	3	4	6
23	.9205	.9198	.9191	.9184	.9178	.9171	.9164	.9157	.9150	.9143	1	2	3	5	6
24	.9135	.9128	.9121	.9114	.9107	.9100	.9092	.9085	.9078	.9070	1	2	4	5	6
25	.9063	.9056	.9048	.9041	.9033	.9026	.9018	.9011	.9003	.8996	1	3	4	5	6
26	.8988	.8980	.8973	.8965	.8957	.8949	.8942	.8934	.8926	.8918	1	3	4	5	6
27	.8910	.8902	.8894	.8886	.8878	.8870	.8862	.8854	.8846	.8838	1	3	4	5	7
28	.8829	.8821	.8813	.8805	.8796	.8788	.8780	.8771	.8763	.8755	1	3	4	6	7
29	.8746	.8738	.8729	.8721	.8712	.8704	.8695	.8686	.8678	.8669	1	3	4	6	7
30°	.8660	.8652	.8643	.8634	.8625	.8616	.8607	.8599	.8590	.8581	1	3	4	6	7
31	.8572	.8563	.8554	.8545	.8536	.8526	.8517	.8508	.8499	.8490	2	3	5	6	8
32	.8480	.8471	.8462	.8453	.8443	.8434	.8425	.8415	.8406	.8396	2	3	5	6	8
33	.8387	.8377	.8368	.8358	.8348	.8339	.8329	.8320	.8310	.8300	2	3	5	6	8
34	8290	.8281	.8271	.8261	.8251	.8241	.8231	.8221	.8211	.8202	2	3	5	7	8
35	.8192	.8181	.8171	.8161	.8151	.8141	.8131	.8121	.8111	.8100	2	3	5	7	8
36	.8090	.8080	.8070	.8059	.8049	.8039	.8028	.8018	.8007	.7997	2	3	5	7	9
37	.7986	.7976	.7965	.7955	.7944	.7934	.7923	.7912	.7902	.7891	2	4	5	7	9
38	.7880	.7869	.7859	.7848	.7837	.7826	.7815	.7804	.7793	.7782	2	4	5	7.	9
39	.7771	.7760	.7749	.7738	.7727	.7716	.7705	.7694	.7683	.7672	2	4	6	7	9
40°	.7660	.7649	.7638	.7627	.7615	.7604	.7593	.7581	.7570	.7559	2	4	6	8	9
41	.7547	.7536	.7524	.7513	.7501	.7490	.7478	.7466	.7455	.7443	2	4	6	8	10
42	.7431	.7420	.7408	.7396	.7385	.7373	.7361	.7349	.7337	.7325	2	4	6	8	10
43	.7314	.7302	.7290	.7278	.7266	.7254	.7242	.7230	.7218	.7206	2	4	6	8	10
44	.7193	.7181	.7169	.7157	.7145	.7133	.7120	.7108	.7096	.7083	2	4	6	8	10
	0'	**6'**	**12'**	**18'**	**24'**	**30'**	**36'**	**42'**	**48'**	**54'**	**1'**	**2'**	**3'**	**4'**	**5'**

The heavy type indicates that the integer changes.

NATURAL COSINES (Cont.)

SUBTRACT

Angle	0'	6'	12'	18'	24'	30'	36'	42'	48'	54'	Proportional Parts				
											1'	2'	3'	4'	5'
45°	.7071	.7059	.7046	.7034	.7022	.7009	.6997	.6984	.6972	.6959	2	4	6	8	10
46	.6947	.6934	.6921	.6909	.6896	.6884	.6871	.6858	.6845	.6833	2	4	6	8	11
47	.6820	.6807	.6794	.6782	.6769	.6756	.6743	.6730	.6717	.6704	2	4	6	9	11
48	.6691	.6678	.6665	.6652	.6639	.6626	.6613	.6600	.6587	.6574	2	4	7	9	11
49	.6561	.6547	.6534	.6521	.6508	.6494	.6481	.6468	.6455	.6441	2	4	7	9	11
50°	.6428	.6414	.6401	.6388	.6374	.6361	.6347	.6334	.6320	.6307	2	4	7	9	11
51	.6293	.6280	.6266	.6252	.6239	.6225	.6211	.6198	.6184	.6170	2	5	7	9	11
52	.6157	.6143	.6129	.6115	.6101	.6088	.6074	.6060	.6046	.6032	2	5	7	9	12
53	.6018	.6004	.5990	.5976	.5962	.5948	.5934	.5920	.5906	.5892	2	5	7	9	12
54	.5878	.5864	.5850	.5835	.5821	.5807	.5793	.5779	.5764	.5750	2	5	7	9	12
55	.5736	.5721	.5707	.5693	.5678	.5664	.5650	.5635	.5621	.5606	2	5	7	10	12
56	.5592	.5577	.5563	.5548	.5534	.5519	.5505	.5490	.5476	.5461	2	5	7	10	12
57	.5446	.5432	.5417	.5402	.5388	.5373	.5358	.5344	.5329	.5314	2	5	7	10	12
58	.5299	.5284	.5270	.5255	.5240	.5225	.5210	.5195	.5180	.5165	2	5	7	10	12
59	.5150	.5135	.5120	.5105	.5090	.5075	.5060	.5045	.5030	.5015	3	5	8	10	13
60°	.5000	.4985	.4970	.4955	.4939	.4924	.4909	.4894	.4879	.4863	3	5	8	10	13
61	.4848	.4833	.4818	.4802	.4787	.4772	.4756	.4741	.4726	.4710	3	5	8	10	13
62	.4695	.4679	.4664	.4648	.4633	.4617	.4602	.4586	.4571	.4555	3	5	8	10	13
63	.4540	.4524	.4509	.4493	.4478	.4462	.4446	.4431	.4415	.4399	3	5	8	10	13
64	.4384	.4368	.4352	.4337	.4321	.4305	.4289	.4274	.4258	.4242	3	5	8	11	13
65	.4226	.4210	.4195	.4179	.4163	.4147	.4131	.4115	.4099	.4083	3	5	8	11	13
66	.4067	.4051	.4035	.4019	.4003	.3987	.3971	.3955	.3939	.3923	3	5	8	11	13
67	.3907	.3891	.3875	.3859	.3843	.3827	.3811	.3795	.3778	.3762	3	5	8	11	13
68	.3746	.3730	.3714	.3697	.3681	.3665	.3649	.3633	.3616	.3600	3	5	8	11	14
69	.3584	.3567	.3551	.3535	.3518	.3502	.3486	.3469	.3453	.3437	3	5	8	11	14
70°	.3420	.3404	.3387	.3371	.3355	.3338	.3322	.3305	.3289	.3272	3	5	8	11	14
71	.3256	.3239	.3223	.3206	.3190	.3173	.3156	.3140	.3123	.3107	3	6	8	11	14
72	.3090	.3074	.3057	.3040	.3024	.3007	.2990	.2974	.2957	.2940	3	6	8	11	14
73	.2924	.2907	.2890	.2874	.2857	.2840	.2823	.2807	.2790	.2773	3	6	8	11	14
74	.2756	.2740	.2723	.2706	.2689	.2672	.2656	.2639	.2622	.2605	3	6	8	11	14
75	.2588	.2571	.2554	.2538	.2521	.2504	.2487	.2470	.2453	.2436	3	6	8	11	14
76	.2419	.2402	.2385	.2368	.2351	.2334	.2317	.2300	.2284	.2267	3	6	8	11	14
77	.2250	.2233	.2215	.2198	.2181	.2164	.2147	.2130	.2113	.2090	3	6	9	11	14
78	.2079	.2062	.2045	.2028	.2011	.1994	.1977	.1959	.1942	.1925	3	6	9	11	14
79	.1908	.1891	.1874	.1857	.1840	.1822	.1805	.1788	.1771	.1754	3	6	9	11	14
80°	.1736	.1719	.1702	.1685	.1668	.1650	.1633	.1616	.1599	.1582	3	6	9	11	14
81	.1564	.1547	.1530	.1513	.1495	.1478	.1461	.1444	.1426	.1409	3	6	9	12	14
82	.1392	.1374	.1357	.1340	.1323	.1305	.1288	.1271	.1253	.1236	3	6	9	12	14
83	.1219	.1201	.1184	.1167	.1149	.1132	.1115	.1097	.1080	.1063	3	6	9	12	14
84	.1045	.1028	.1011	.0993	.0976	.0958	.0941	.0924	.0906	.0889	3	6	9	12	14
85	.0872	.0854	.0837	.0819	.0802	.0785	.0767	.0750	.0732	.0715	3	6	9	12	14
86	.0698	.0680	.0663	.0645	.0628	.0610	.0593	.0576	.0558	.0541	3	6	9	12	15
87	.0523	.0506	.0488	.0471	.0454	.0436	.0419	.0401	.0384	.0366	3	6	9	12	15
88	.0349	.0332	.0314	.0297	.0279	.0262	.0244	.0227	.0209	.0192	3	6	9	12	15
89	.0175	.0157	.0140	.0122	.0105	.0087	.0070	.0052	.0035	.0017	3	6	9	12	15
	0'	6'	12'	18'	24'	30'	36'	42'	48'	54'	1'	2'	3'	4'	5'

NATURAL TANGENTS

Angle	0'	6'	12'	18'	24'	30'	36'	42'	48'	54'	1'	2'	3'	4'	5'
0°	0.0000	.0017	.0035	.0052	.0070	.0087	.0105	.0122	.0140	.0157	3	6	9	12	15
1	0.0175	.0192	.0209	.0227	.0244	.0262	.0279	.0297	.0314	.0332	3	6	9	12	15
2	0.0349	.0367	.0384	.0402	.0419	.0437	.0454	.0472	.0489	.0507	3	6	9	12	15
3	0.0524	.0542	.0559	.0577	.0594	.0612	.0629	.0647	.0664	.0682	3	6	9	12	15
4	0.0699	.0717	.0734	.0752	.0769	.0787	.0805	.0822	.0840	.0857	3	6	9	12	15
5	0.0875	.0892	.0910	.0928	.0945	.0963	.0981	.0998	.1016	.1033	3	6	9	12	15
6	0.1051	.1069	.1086	.1104	.1122	.1139	.1157	.1175	.1192	.1210	3	6	9	12	15
7	0.1228	.1246	.1263	.1281	.1299	.1317	.1334	.1352	.1370	.1388	3	6	9	12	15
8	0.1405	.1423	.1441	.1459	.1477	.1495	.1512	.1530	.1548	.1566	3	6	9	12	15
9	0.1584	.1602	.1620	.1638	.1655	.1673	.1691	.1709	.1727	.1745	3	6	9	12	15
10°	0.1763	.1781	.1799	.1817	.1835	.1853	.1871	.1890	.1908	.1926	3	6	9	12	15
11	0.1944	.1962	.1980	.1998	.2016	.2035	.2053	.2071	.2089	.2107	3	6	9	12	15
12	0.2126	.2144	.2162	.2180	.2199	.2217	.2235	.2254	.2272	.2290	3	6	9	12	15
13	0.2309	.2327	.2345	.2364	.2382	.2401	.2419	.2438	.2456	.2475	3	6	9	12	15
14	0.2493	.2512	.2530	.2549	.2568	.2586	.2605	.2623	.2642	.2661	3	6	9	12	16
15	0.2679	.2698	.2717	.2736	.2754	.2773	.2792	.2811	.2830	.2849	3	6	9	13	16
16	0.2867	.2886	.2905	.2924	.2943	.2962	.2981	.3000	.3019	.3038	3	6	9	13	16
17	0.3057	.3076	.3096	.3115	.3134	.3153	.3172	.3191	.3211	.3230	3	6	10	13	16
18	0.3249	.3269	.3288	.3307	.3327	.3346	.3365	.3385	.3404	.3424	3	6	10	13	16
19	0.3443	.3463	.3482	.3502	.3522	.3541	.3561	.3581	.3600	.3620	3	7	10	13	16
20°	0.3640	.3659	.3679	.3699	.3719	.3739	.3759	.3779	.3799	.3819	3	7	10	13	17
21	0.3839	.3859	.3879	.3899	.3919	.3939	.3959	.3979	.4000	.4020	3	7	10	13	17
22	0.4040	.4061	.4081	.4101	.4122	.4142	.4163	.4183	.4204	.4224	3	7	10	14	17
23	0.4245	.4265	.4286	.4307	.4327	.4348	.4369	.4390	.4411	.4431	3	7	10	14	17
24	0.4452	.4473	.4494	.4515	.4536	.4557	.4578	.4599	.4621	.4642	4	7	11	14	18
25	0.4663	.4684	.4706	.4727	.4748	.4770	.4791	.4813	.4834	.4856	4	7	11	14	18
26	0.4877	.4899	.4921	.4942	.4964	.4986	.5008	.5029	.5051	.5073	4	7	11	15	18
27	0.5095	.5117	.5139	.5161	.5184	.5206	.5228	.5250	.5272	.5295	4	7	11	15	18
28	0.5317	.5340	.5362	.5384	.5407	.5430	.5452	.5475	.5498	.5520	4	8	11	15	19
29	0.5543	.5566	.5589	.5612	.5635	.5658	.5681	.5704	.5727	.5750	4	8	12	15	19
30°	0.5774	.5797	.5820	.5844	.5867	.5890	.5914	.5938	.5961	.5985	4	8	12	16	20
31	0.6009	.6032	.6056	.6080	.6104	.6128	.6152	.6176	.6200	.6224	4	8	12	16	20
32	0.6249	.6273	.6297	.6322	.6346	.6371	.6395	.6420	.6445	.6469	4	8	12	16	20
33	0.6494	.6519	.6544	.6569	.6594	.6619	.6644	.6669	.6694	.6720	4	8	13	17	21
34	0.6745	.6771	.6796	.6822	.6847	.6873	.6899	.6924	.6950	.6976	4	9	13	17	21
35	0.7002	.7028	.7054	.7080	.7107	.7133	.7159	.7186	.7212	.7239	4	9	13	18	22
36	0.7265	.7292	.7319	.7346	.7373	.7400	.7427	.7454	.7481	.7508	5	9	14	18	23
37	0.7536	.7563	.7590	.7618	.7646	.7673	.7701	.7729	.7757	.7785	5	9	14	18	23
38	0.7813	.7841	.7869	.7898	.7926	.7954	.7983	.8012	.8040	.8069	5	9	14	19	24
39	0.8098	.8127	.8156	.8185	.8214	.8243	.8273	.8302	.8332	.8361	5	10	15	20	24
40°	0.8391	.8421	.8451	.8481	.8511	.8541	.8571	.8601	.8632	.8662	5	10	15	20	25
41	0.8693	.8724	.8754	.8785	.8816	.8847	.8878	.8910	.8941	.8972	5	10	16	21	26
42	0.9004	.9036	.9067	.9099	.9131	.9163	.9195	.9228	.9260	.9293	5	11	16	21	27
43	0.9325	.9358	.9391	.9424	.9457	.9490	.9523	.9556	.9590	.9623	6	11	17	22	28
44	0.9657	.9691	.9725	.9759	.9793	.9827	.9861	.9896	.9930	.9965	6	11	17	23	29
	0'	6'	12'	18'	24'	30'	36'	42'	48'	54'	1'	2'	3'	4'	5'

NATURAL TANGENTS (Cont.)

Angle	0'	6'	12'	18'	24'	30'	36'	42'	48'	54'	1'	2'	3'	4'	5'
											Proportional Parts				
45°	1.0000	.0035	.0070	.0105	.0141	.0176	.0212	.0247	.0283	.0319	6	12	18	24	30
46	1.0355	.0392	.0428	.0464	.0501	.0538	.0575	.0612	.0649	.0686	6	12	18	25	31
47	1.0724	.0761	.0799	.0837	.0875	.0913	.0951	.0990	.1028	.1067	6	13	19	25	32
48	1.1106	.1145	.1184	.1224	.1263	.1303	.1343	.1383	.1423	.1463	7	13	20	26	33
49	1.1504	.1544	.1585	.1626	.1667	.1708	.1750	.1792	.1833	.1875	7	14	21	28	34
50°	1.1918	.1960	.2002	.2045	.2088	.2131	.2174	.2218	.2261	.2305	7	14	22	29	36
51	1.2349	.2393	.2437	.2482	.2527	.2572	.2617	.2662	.2708	.2753	8	15	23	30	38
52	1.2799	.2846	.2892	.2938	.2985	.3032	.3079	.3127	.3175	.3222	8	16	24	31	39
53	1.3270	.3319	.3367	.3416	.3465	.3514	.3564	.3613	.3663	.3713	8	16	25	33	41
54	1.3764	.3814	.3865	.3916	.3968	.4019	.4071	.4124	.4176	.4229	9	17	26	34	43
55	1.4281	.4335	.4388	.4442	.4496	.4550	.4605	.4659	.4715	.4770	9	18	27	36	45
56	1.4826	.4882	.4938	.4994	.5051	.5108	.5166	.5224	.5282	.5340	10	19	29	38	48
57	1.5399	.5458	.5517	.5577	.5637	.5697	.5757	.5818	.5880	.5941	10	20	30	40	50
58	1.6003	.6066	.6128	.6191	.6255	.6319	.6383	.6447	.6512	.6577	11	21	32	43	53
59	1.6643	.6709	.6775	.6842	.6909	.6977	.7045	.7113	.7182	.7251	11	23	34	45	56
60°	1.7321	.7391	.7461	.7532	.7603	.7675	.7747	.7820	.7893	.7966	12	24	36	48	60
61	1.8040	.8115	.8190	.8265	.8341	.8418	.8495	.8572	.8650	.8728	13	26	38	51	64
62	1.8807	.8887	.8967	.9047	.9128	.9210	.9292	.9375	.9458	.9542	14	27	41	55	68
63	1.9626	.9711	.9797	.9883	.9970	.0057	.0145	.0233	.0323	.0413	15	29	44	58	73
64	2.0503	.0594	.0686	.0778	.0872	.0965	.1060	.1155	.1251	.1348	16	31	47	63	78
65	2.1445	.1543	.1642	.1742	.1842	.1943	.2045	.2148	.2251	.2355	17	34	51	68	85
66	2.2460	.2566	.2673	.2781	.2889	.2998	.3109	.3220	.3332	.3445					
67	2.3559	.3673	.3789	.3906	.4023	.4142	.4262	.4383	.4504	.4627					
68	2.4751	.4876	.5002	.5129	.5257	.5386	.5517	.5649	.5782	.5916					
69	2.6051	.6187	.6325	.6464	.6605	.6746	.6889	.7034	.7179	.7326					
70°	2.7475	.7625	.7776	.7929	.8083	.8239	.8397	.8556	.8716	.8878					
71	2.9042	.9208	.9375	.9544	.9714	.9887	.0061	.0237	.0415	.0595					
72	3.0777	.0961	.1146	.1334	.1524	.1716	.1910	.2106	.2305	.2506					
73	3.2709	.2914	.3122	.3332	.3544	.3759	.3977	.4197	.4420	.4646					
74	3.4874	.5105	.5339	.5576	.5816	.6059	.6305	.6554	.6806	.7062					
75	3.7321	.7583	.7848	.8118	.8391	.8667	.8947	.9232	.9520	.9812		Use interpolation			
76	4.0108	.0408	.0713	.1022	.1335	.1653	.1976	.2303	.2635	.2972					
77	4.3315	.3662	.4015	.4373	.4737	.5107	.5483	.5864	.6252	.6646					
78	4.7046	.7453	.7867	.8288	.8716	.9152	.9594	.0045	.0504	.0970					
79	5.1446	.1929	.2422	.2924	.3435	.3955	.4486	.5026	.5578	.6140					
80°	5.671	5.730	5.789	5.850	5.912	5.976	6.041	6.107	6.174	6.243					
81	6.314	6.386	6.460	6.535	6.612	6.691	6.772	6.855	6.940	7.026					
82	7.115	7.207	7.300	7.396	7.495	7.596	7.700	7.806	7.916	8.028					
83	8.144	8.264	8.386	8.513	8.643	8.777	8.915	9.058	9.205	9.357					
84	9.514	9.677	9.845	10.02	10.20	10.39	10.58	10.78	10.99	11.20					
85	11.43	11.66	11.91	12.16	12.43	12.71	13.00	13.30	13.62	13.95					
86	14.30	14.67	15.06	15.46	15.89	16.35	16.83	17.34	17.89	18.46					
87	19.08	19.74	20.45	21.20	22.02	22.90	23.86	24.90	26.03	27.27		Interpolation not accurate			
88	28.64	30.14	31.82	33.69	35.80	38.19	40.92	44.07	47.74	52.08					
89	57.29	63.66	71.62	81.85	95.49	114.6	143.2	191.0	286.5	573.0					
	0'	6'	12'	18'	24'	30'	36'	42'	48'	54'	1'	2'	3'	4'	5'

The heavy type indicates that the integer changes.

Appendix IX—Derivation of the Gravitational Potential of a Point in Space

---

The gravitational potential of a point in space, at a distance r from a mass M, may be defined as the work done per unit mass in bringing any mass m from infinity up to that point. See Fig. 1K.

By Newton's universal law of gravitation, the force between two masses M and m is given by

$$F = G\frac{Mm}{x^2} \qquad (1)$$

where x is the distance between their *centers of mass*. *Work done* in mechanics is given by the product of two vector quantities, *force* and *distance*:

$$\text{Work done} = F \times x \qquad (2)$$

Since gravitational forces change with distance, we must integrate over the distance through which the force acts, i.e., from infinity, ∞, up to any selected point at distance r from M. The work done is thereby stored as potential energy:

$$E_p = \int_\infty^r F\,dx \qquad (3)$$

Using Eq. (1), we obtain

$$E_p = \int_\infty^r \frac{GMm}{x^2}\,dx$$

or

$$E_p = GMm \int_\infty^r \frac{1}{x^2}\,dx$$

Integrating

$$E_p = GMm \left[-\frac{1}{x}\right]_\infty^r$$

$$E_p = -\frac{GMm}{r}$$

By Eq. (1q) the gravitational potential V_G at a point is given by E_p/m. We obtain, therefore,

$$V_G = -\frac{GM}{r} \qquad (4)$$

Appendix X—The Nobel Prize Winners in Physics

1901 Röntgen, Wilhelm Conrad (1845–1923), German. *Discovery of X-rays.*
1902 Lorentz, Hendrik Antoon (1853–1928), Dutch, and
Zeeman, Pieter (1865–1943), Dutch. *Zeeman Effect.*
1903 Becquerel, Henri Antoine (1852–1908), French. *Discovery of Radioactivity.*
Curie, Pierre (1859–1906), French, and
Curie, Marie Sklodowska (1867–1934), Polish chemist in France. *Studies of Radioactivity.*
1904 Lord Rayleigh (John William Strutt) (1842–1919), English. *Studies of Gases.*
1905 Lenard, Philipp (1862–1947), German. *Studies of Cathode Rays.*
1906 Thomson, Sir Joseph John (1856–1940), English. *Discharges through Gases.*
1907 Michelson, Albert A. (1852–1931), American. *Precision Optical Instruments.*
1908 Lippmann, Gabriel (1845–1921), French. *Interference Color Photography.*
1909 Marconi, Guglielmo (1874–1937), Italian, and
Braun, Ferdinand (1850–1918), German. *Development of Wireless.*
1910 Van der Waals, Johannes D. (1837–1923), Dutch. *Gas Laws.*
1911 Wien, Wilhelm (1864–1928), German. *Heat Radiation Laws.*
1912 Dalen, Gustaf (1869–1937), Swedish. *Automatic Lighting of Lighthouses.*
1913 Kamerlingh-Onnes, Heike (1853–1926), Dutch. *Liquid Helium and Low Temperatures.*
1914 Von Laue, Max Theodor Felix (1879–1960), German. *Diffraction of X-rays.*
1915 Bragg, Sir W. H. (1862–1942), English, and his son
Bragg, W. L. (1890–   ), English. *Crystal Structure.*
1917 Barkla, Charles G. (1877–1944), English. *Characteristic X-rays of the Elements.*
1918 Planck, Max (1858–1947), German. *Quantum Theory of Radiation.*
1919 Stark, Johannes (1874–1957), German. *The Stark Effect of Spectrum Lines.*
1920 Guillaume, Charles E. (1861–1938), Spanish. *Study of Nickel-Steel Alloys.*
1921 Einstein, Albert (1879–1955), German. *Theory of Relativity and the Photoelectric Effect.*
1922 Bohr, Niels (1885–1962), Danish. *Theory of Atomic Structure.*
1923 Millikan, Robert A. (1868–1953), American. *Charge on Electron and Photoelectric Effect.*
1924 Siegbahn, Karl M. (1886–   ), Swedish. *X-ray Spectroscopy.*
1925 Franck, James (1882–   ), German, and
Hertz, Gustav (1887–   ), German. *Electron Impact on Atoms.*
1926 Perrin, Jean B. (1870–1942), French. *Discovery of the Equilibrium of Sedimentation.*
1927 Compton, Arthur H. (1892–1962), American. *Compton Effect.*
Wilson, Charles T. R. (1869–1959), English. *Wilson Cloud Chamber.*
1928 Richardson, Owen Willans (1879–1959), English. *Studies of Thermal Ions.*

1929   De Broglie, Louis V. (1892–   ), French. *Wave Character of Electrons.*
1930   Raman, Sir Chandrasekhara V. (1888–   ), Hindu. *Raman Effect.*
1932   Heisenberg, Werner (1901–   ), German. *Creation of Quantum Mechanics.*
1933   Schrödinger, Edwin (1887–   ), German, and
         Dirac, P. A. M. (1902–   ), English. *Atomic Theory.*
1935   Chadwick, James (1891–   ), English. *Discovery of the Neutron.*
1936   Hess, Victor F. (1883–   ), Austrian. *Discovery of Cosmic Rays.*
         Anderson, Carl D. (1905–   ), American. *Discovery of Positron.*
1937   Davisson, Clinton J. (1881–   ), American and
         Thompson, George P. (1892–   ), English. *Electron Diffraction by Crystals.*
1938   Fermi, Enrico (1901–1954), Italian. *Slow Neutron Reactions.*
1939   Lawrence, Ernest O. (1901–1958), American. *Development of Cyclotron.*
1940, 1941, 1942, not awarded
1943   Stern, O. (1888–   ), German. *Magnetic Moment of Proton.*
1944   Rabi, I. I. (1898–   ), American. *Magnetic Moments of Nuclei.*
1945   Pauli, W. (1900–1958), German. *Pauli Exclusion Principle.*
1946   Bridgman, P. W. (1882–1961), American. *Physical Effects of High Pressures.*
1947   Appleton, Sir Ed. V. (1892–   ), English. *Exploration of the Ionosphere.*
1948   Blackett, P. M. S. (1897–   ), English. *Discoveries in Cosmic Radiation.*
1949   Yukawa, H. (1907–   ), Japanese. *Theoretical Prediction of Mesons.*
1950   Powell, C. F. (1903–   ), English. *Photographic Cosmic Ray Studies.*
1951   Cockcroft, Sir J. D. (1897–   ), English, and
         Walton, E. T. S. (1903–   ), English. *First Transmutation of Atomic Nuclei.*
1952   Bloch, Felix (1905–   ), American. *Nuclear Magnetic Moments.*
         Purcell, Ed. M. (1912–   ), American. *Radio Astronomy.*
1953   Zernike, Frits (1888–   ), Dutch. *Phase Contrast Microscope.*
1954   Born, Max (1882–   ), German. *Quantum Mechanics, and Wave Functions.*
         Bothe, Walther (1891–1957), German. *Quantum Mechanics, and Wave Functions.*
1955   Kusch, P. (1911–   ), American, and
         Lamb, W. E. (1913–   ), American. *Microwave Spectroscopy and Atomic Structure.*
1956   Shockley, W. (1910–   ), American,
         Brattain, W. H. (1902–   ), American, and
         Bardeen, J. (1908–   ), American. *Semiconductors, and Their Application to Transistors.*
1957   Yang, C. N. (1922–   ), Chinese, and
         Lee, T. D. (1926–   ), Chinese. *Studies of the Concept of Parity in Atomic Physics.*
1958   Cerenkov, P. A. (1904–   ), Russian,
         Tamm, I. E. (1895–   ), Russian, and
         Frank, I. M. (1908–   ), Russian. *Discovery and Study of Cerenkov Radiation.*
1959   Segrè, E. (1905–   ), American, and
         Chamberlain, O. (1920–   ), American. *Discovery of the Antiproton.*
1960   Glaser, D. A. (1926–   ), American. *The Bubble Chamber.*
1961   Hofstadter, Robt. (1916–   ), American, and
         Mossbauer, R. L. (1930–   ), German. *Nuclear Radiation and Absorption.*
1962   Landau, L. D. (1908), Russian. *Mathematical explanation of very low temperature phenomena.*
1963   Maria Goeppert-Mayer (1906–   ), American, and
         J. Hans D. Jensen (1906–   ), German. *Nuclear Shell Model.*
         Eugene Wigner (1902–   ), American. *Theoretical Work on Symmetry and Parity.*

THE GREEK ALPHABET

A	α	Alpha	H	η	Eta	N	ν	Nu	T	τ	Tau
B	β	Beta	Θ	θ	Theta	Ξ	ξ	Xi	Υ	υ	Upsilon
Γ	γ	Gamma	I	ι	Iota	O	o	Omicron	Φ	ϕ	Phi
Δ	δ	Delta	K	κ	Kappa	Π	π	Pi	X	χ	Chi
E	ϵ	Epsilon	Λ	λ	Lambda	P	ρ	Rho	Ψ	ψ	Psi
Z	ζ	Zeta	M	μ	Mu	Σ	σ	Sigma	Ω	ω	Omega

INDEX